ROUTLEDGE HANDBOOK OF ENVIRONMENTAL IMPACT ASSESSMENT

Globally, environmental impact assessment (EIA) is one of the most enduring and influential environmental management tools. This handbook provides readers with a strong foundation for understanding the practice of EIA, by outlining the different types of assessment while also providing a guide to best practice.

This collection deploys a research and practice-based approach to the subject, delivering an overview of EIA as an essential and practical tool of environmental protection, planning, and policy. To best understand the most pertinent issues and challenges surrounding EIA today, this volume draws together prominent researchers, practitioners, and young scholars who share their work and knowledge to cover two key parts. The first part introduces EIA processes and best practices through analytical and critical chapters on the stages/elements of the EIA process and different components and forms of assessment. These provide examples that cover a wide range of assessment methods and cross-cutting issues, including cumulative effects assessment, social impact assessment, Indigenous-led assessment, risk assessment, climate change, and gender-based assessment. The second part provides jurisdictional reviews of the European Union, the US National Environmental Policy Act, recent assessment reforms in Canada, EIA in developing economies, and the EIA context in England.

By providing a concise outline of the process followed by in-depth illustrations of approaches, methods and tools, and case studies, this book will be essential for students, scholars, and practitioners of environmental impact assessment.

Kevin Hanna is Director of the Centre for Environmental Assessment Research at the University of British Columbia, Canada.

ROUTLEDGE HANDBOOK OF ENVIRONMENTAL IMPACT ASSESSMENT

Edited by Kevin Hanna

Routledge
Taylor & Francis Group

LONDON AND NEW YORK

earthscan
from Routledge

Cover image: Sunshine Seeds @ iStock

First published 2022
by Routledge
4 Park Square, Milton Park, Abingdon, Oxon OX14 4RN

and by Routledge
605 Third Avenue, New York, NY 10158

Routledge is an imprint of the Taylor & Francis Group, an informa business

British Library Cataloguing-in-Publication Data
A catalogue record for this book is available from the British Library

Library of Congress Cataloging-in-Publication Data
Names: Hanna, Kevin - editor.
Title: Routledge handbook of environmental impact assessment/ edited by Kevin Hanna.
Other titles: Handbook of environmental impact assessment
Description: Abingdon, Oxon; New York, NY: Routledge, 2022. |
Series: Routledge environment and sustainability handbooks |
Includes bibliographical references and index.
Subjects: LCSH: Environmental impact analysis–Methodology.
Classification: LCC TD194.6 .R68 2022 (print) | LCC TD194.6 (ebook)
| DDC 333.71/4–dc23/eng/20211227
LC record available at https://lccn.loc.gov/2021052262
LC ebook record available at https://lccn.loc.gov/2021052263

ISBN: 978-0-367-24447-7 (hbk)
ISBN: 978-1-032-13001-9 (pbk)
ISBN: 978-0-429-28249-2 (ebk)

DOI: 10.4324/9780429282492

Typeset in Bembo
by Deanta Global Publishing Services, Chennai, India

CONTENTS

FIGURES

TABLES

BOXES

NOTES ON CONTRIBUTORS

Reece Alberts is Senior Lecturer within the Research Unit for Environmental Science and Management at the North-West University, Potchefstroom campus. He holds a Bcom, LLb, LLm, MEnv Man, and a PhD in environmental management. His research focus is centered on the effectiveness of environmental policy implementation instruments.

Lauren Arnold is a PhD candidate at the University of British Columbia's Centre for Environmental Assessment Research, focusing on how to include social, economic, health, and cultural cumulative impacts in environmental assessment policy and practice. Her academic and government research experiences include work on regional and strategic environmental assessment, cumulative effects assessment, impact assessment decision-making, Indigenous-led assessment, and climate change resilience.

Priya Bala-Miller is a senior sustainability specialist with 20 years of experience working in collaboration with UN agencies, the private sector, academia, NGOs, and trade unions. She is the CEO of Palmyra Partners, an initiative that advances justice, equity, diversity, and inclusion in natural resource governance and sustainable development.

Rob Friberg is a professional forester and consultant focused on community-based conservation, rural development, and climate change initiatives in Canada, Central Africa, Latin America, and the Caribbean. Rob is a PhD candidate at the University of British Columbia. His research is on rural community resilience to environmental change.

Alan Bond has been researching and teaching on all forms of impact assessment over the past 30 years. He has published widely in the field, often with a focus on the contribution of impact assessment to sustainability, and has delivered a variety of short training courses for practitioners in many countries.

Chris Buse is Research Scientist with the University of British Columbia's Centre for Environmental Assessment Research. Dr. Buse researches the cumulative environmental, community, and health impacts of natural resource development, and the health impacts of environmental change, including climate change.

Mathieu Bourbonnais is Assistant Professor in the Department of Earth, Environmental and Geographic Sciences at the University of British Columbia, Okanagan campus. His research focuses on the use of GIS, spatial modeling, and remote sensing for environmental monitoring and for understanding ecological processes.

Dirk Cilliers is Associate Professor in Geography and Environmental Management at the Potchefstroom campus of the North-West University in South Africa. His research interests focus on environmental management-related topics, particularly environmental assessment performance evaluation.

Ayla De Grandpré is a graduate student studying in the Interdisciplinary Studies-Sustainability program at the University of British Columbia, Okanagan campus. Her research centers on climate change adaptation, resilience studies, risk and hazard assessment, agriculture, and human–environmental interactions.

Alan Diduck is Professor and Chair of the Department of Environmental Studies and Sciences at the University of Winnipeg. Prior to joining the university, he was a lawyer and executive director of a social profit organization providing public legal education and information services. His research focuses on community involvement in impact assessment and its implications for environmental justice and social learning.

Thomas B. Fischer has over 30 years of international practice, research, and training. He is Professor and Head of the Environmental Assessment and Management Research Centre at the University of Liverpool, UK, and Director of the World Health Organization Collaborating Centre on 'Health in Impact Assessments'. Thomas has published extensively on environmental assessment. He is also an Extraordinary Professor at the Research Unit for Environmental Science and Management at North-West University, Potchefstroom campus, South Africa.

Josh Fothergill is an internationally recognized expert in impact assessment (IA), including recently leading global reviews into the state of digital IA practice and links between IA and the circular economy. He is the UK's leading EIA trainer, training over 500 delegates since 2017, and has co-authored much of the UK's practical guidance on IA over the last decade, including the 3rd edition of the UK's *EIA Handbook*. Josh is the founder and director of Fothergill Training & Consulting Ltd and an honorary associate of the University of Liverpool's Environmental Assessment and Management Research Centre.

Gesa Geißler is Senior Researcher at the University of Natural Resources and Life Sciences, Vienna, and Visiting Fellow at the Environmental Assessment and Management Research Centre at the University of Liverpool. She is interested in the effectiveness of impact assessment, environmental mitigation, and planning tools in the context of current trends such as digitalization, energy system transformation, and climate change.

Marie Grimm is Research Associate and Lecturer at the Environmental Assessment and Planning Research Group at the Berlin Institute of Technology (Technische Universität Berlin). Marie holds a master's in environmental policy and planning. She is interested in the effectiveness of impact assessment and impact mitigation. Her PhD research is focused on the design and implementation of biodiversity offsetting programs.

Kevin Hanna is Director of the Centre for Environmental Assessment Research and Associate Professor in the Department of Earth, Environmental and Geographic Sciences at the University of British Columbia. Dr. Hanna's research focuses on environmental impact assessment, forest resources management, and energy systems and energy development. He works in western and northern Canada and the Arctic.

Alexandra Jiricka-Pürrer is an environmental planner specializing in environmental impact assessment. She is Senior Scientist at the University of Natural Resources and Life Sciences, Vienna. Her teaching and research focuses on the role of impact assessment in climate change adaptation and mitigation for spatial and infrastructure planning, and nature conservation and tourism planning. She has been the principal investigator on international research projects and has published over 30 scientific articles.

Johann Köppel is Professor at the Berlin Institute of Technology (Technische Universität Berlin) and Head of the Environmental Assessment and Planning Research Group at the Berlin Institute of Technology. His research is focused on the evaluation and development of environmental assessment and planning tools. Over the course of the last 20 years, the sustainable development of renewable energy sources has also been a particular research focus. Dr. Köppel is a member of the German EIA Association Advisory Board.

Matt Lindstrom is Edward L. Henry Professor of Political Science at the College of St. Benedict and St. John's University, Minnesota, USA. He is the author of several books, chapters, and articles addressing environmental policy and urban planning.

Alistair MacDonald is Director and Environmental Assessment Technical Lead for The Firelight Group, Edmonton, Alberta. He has over 15 years of experience working primarily with Indigenous groups in Canada on the social, economic, culture and rights side of environmental impact assessment.

Troy McMillan is a student researcher and PhD candidate at the University of British Columbia, Okanagan campus. He holds a master's in disaster and emergency management from Royal Roads University, British Columbia. His research interests primarily revolve around organizational and urban resilience. Troy has worked for the Canadian federal government in various capacities for over 20 years, across disciplines including information security, business continuity, and emergency management policy.

Anne Merrild Hansen is Professor of Planning and Impact Assessment in the Arctic with Aalborg University and Ilisimatusarfik University of Greenland. She has more than 20 years of experience in working with social impact assessment in theory and practice. She has worked for Maersk Oil and Gas and served as a consultant for the European Investment Bank. Her work is focused on enhancing benefits and mitigating undesired impacts in communities when large-scale projects are planned and implemented. Dr. Hansen has published two books and numerous scientific articles on the topic of collaborative engagement and public participation in impact assessment.

Angus Morrison-Saunders is Professor, Environmental Management in the Centre for People Place and Planet at Edith Cowan University, Australia; and Extraordinary Professor in the Research Unit for Environmental Science and Management at North-West University, South Africa. He has been teaching undergraduate and postgraduate level courses on environmental impact assessment at several universities over the past 30 years. He is an active researcher in the field with a special interest in understanding EIA outcomes and its contribution to sustainability.

Jeffrey Nishima-Miller is a PhD student at the University of British Columbia, Okanagan campus. His work explores emergent opportunities and challenges in the field of integrated natural resources and environmental management, with a particular focus on environmental impact assessment and wildlife management in British Columbia, Canada.

Bram Noble is Professor in the Department of Geography and Planning at the University of Saskatchewan, Canada. His work is focused on environmental assessment, specifically regional and strategic solutions for better-practice cumulative effects assessment.

Ciaran O'Faircheallaigh is Professor of Politics and Public Policy at Griffith University, Brisbane. Over the last three decades he has written extensively on participation in EIA and has managed Indigenous impact assessments.

John Parkins is Professor of Environmental Sociology in the Department of Resource Economics and Environmental Sociology at the University of Alberta. John's work on impact assessment encompasses research on public engagement, social impact research methods and cumulative effects assessment.

Maria Partidario is an international consultant in the public and private sectors and full Professor of the Universidade de Lisboa, Portugal. Her work and research centers on strategic thinking for sustainability in integrated, environmental and social assessment approaches.

Nicole Peletz holds a Bachelor of Engineering in Bioresource Engineering from McGill University, and a master's degree from the University of British Columbia. Her research focused on impact assessment in Nunavut, Canada. Nicole is currently Policy Analyst for the British Columbia Public Service.

Jenny Pope is a consultant, researcher, and trainer of environmental and sustainability assessment. She is a current member of the Western Australian Environmental Protection Authority, which is the statutory body responsible for EIA in Western Australia.

Karaline Reimer is a graduate student with the Center for Environmental Assessment Research at the University of British Columbia. She has a bachelor's degree in physical geography, and her background is in marine and lacustrine oil spill response and environmental emergency planning. Karaline currently works for the Environmental Assessment Office of British Columbia.

Francois P. Retief is Professor in Environmental Management within the Research Unit for Environmental Sciences and Management at the North-West University, South Africa. He completed his PhD at the University of Manchester and served as the Director for the School for Geo and Spatial Sciences at North-West University, and is co-editor of *Impact Assessment and Project Appraisal*. His main research interest is in the performance of environmental assessment and management policy instruments.

Claudine Roos is Senior Lecturer at the North-West University, Potchefstroom campus, South Africa. Her research focuses on environmental management and governance approaches, as well as waste management.

John Sinclair is Professor and Director of the Natural Resources Institute at the University of Manitoba. His main research interest focuses on governance and learning as they relate to resource and environmental decision-making. He has written extensively on impact assessment law and policy, with a current focus on next-generation impact assessment. With Meinhard Doelle, John has recently edited a book that critically reviews the Canadian Impact Assessment Act, published by Irwin Law.

Ben West earned a BA from St. John's University, Minnesota, with a computer science major. He is a professional fly-fishing guide and is planning to attend graduate school.

PREFACE

It is now just over 50 years since the United States National Environmental Policy Act (NEPA) was enacted. At the time its potential influence in the US, let alone globally, was not really understood. NEPA is a remarkably short piece of legislation and it was signed into law by a US President thought to be quite conservative – at least at the time. But since NEPA was passed, most nations have developed a form of environmental impact assessment (EIA), and many have passed laws mandating it for a range of projects and activities and even for policy and planning. EIA is now established practice around the world. It has become an indispensable and lasting tool for environmental protection and planning. The practice of EIA has also evolved to meet new challenges, embrace new technologies, and provide important opportunities for community participation and engagement.

Developing the *Handbook* has taken just over two years – longer than we anticipated. The global pandemic reshaped aspects of the global economy, led to new approaches to work, and greatly affected the lives of many. We were affected too. Several authors were unable to finish their contributions and others were delayed because of the impact of the pandemic on their families and work lives. EIA researchers and practitioners also began to question how COVID-19 would affect the work of environmental agencies, shape political support for environmental management, and impact the ability of proponents and regulators to conduct EIAs. It is fair to say we do not yet know how widely the pandemic has affected EIA practice, but we may see technological and methodological advances in communication, participation, and even data collection and monitoring. But we may also find evidence of lagging enforcement of EIA mandates as governments seek to streamline development for economic recovery.

Recent years have also witnessed the reality of climate change directly impact lives around the world. Wildfires, drought, flooding, and severe heat events are showing us what we can expect if we do not seriously and effectively reduce our reliance on fossil fuels (especially eliminating coal from our energy mix), and plan for the impacts that will come our way regardless of attenuation efforts. Writing this preface comes at a time when British Columbia, where I live, has been devastated by a record wildfire season, and now by catastrophic flooding which has broken the province's transportation infrastructure, damaged power distribution and municipal infrastructure, forced evacuations, and led to food and goods shortages. These are clearly harbingers of climate events that we must plan for, and EIA can play a part in doing that.

Such challenges highlight the lasting importance of EIA. As we grapple with the need to respond quickly and effectively to environmental issues, EIA provides the setting, methodologies, and operational experience that can help implement actions to combat global climate change and build climate and weather resilience into projects; address biodiversity loss, fragmentation, and pollution; reduce the physical footprints of projects; support energy transitions; and even help manage the lifecycle impacts of renewable energy technologies – from wind turbines to lithium production. There is a role for EIA in all of these global challenges. The cautionary note is that EIA cannot be the default mechanism for addressing complex policy and planning needs. EIA is not a substitute for the higher-level policies and planning needed to meet climate change goals, realize sustainability, or ensure equity and wellbeing. Development will continue, the vital objective is to do it better – and this is where EIA comes in.

EIA is focused, functional, and operational. At its most basic, EIA is a process for assessing proposed projects, or plans and polices, understand their impacts on the environment, respond accordingly, and support decision-making. EIA is a tool that helps meet such needs, it is anticipatory and prescient, and it can help in implementing policies that advance meaningful and effective changes to the ways we in which we develop our world.

The chapters in this book reflect the tasks we face now. They are a guide to practice and current thinking about EIA, but they are not a 'how to do an EIA' template. What the *Handbook* provides is an illustration of current issues and tools in EIA. It provides a strong foundation for understanding what the practice of EIA is all about and the types of issues that EIA routinely and uniquely addresses. The innovations that distinguish the Handbook include discussions of disaster management, risk assessment, gender-based assessment, health assessment, and Indigenous-led impact assessment. But the book also provides a complete foundation in all the established components of EIA. And there are selective profiles of national and regional practices that help readers understand how EIA is implemented in specific settings – to see how it works on the ground.

Before you move on to the chapters, I want to acknowledge the hard work of the contributors. A range of experiences in practice, scholarship, and life has produced what you will read. There are well-established authors and early career writers. Each brings a unique voice to EIA, and each has helped to produce a handbook that will be of value to readers who will certainly represent the breadth of disciplines that support EIA, make it function, and ensure that it has the capacity to support good development decisions. Jeff Nishima-Miller has helped shepherd and assemble the final chapters – I greatly appreciate his hard work and advice. Finally, this is a peer-reviewed book. On behalf of the authors I thank the peer reviewers for their time and suggestions, and for helping to ensure that our contributions reflect the challenges faced by EIA practitioners.

Kevin Hanna
British Columbia, Canada
November 2021

PART I

Types of assessment, issues, and practices

1

AN INTRODUCTION TO ENVIRONMENTAL IMPACT ASSESSMENT

Kevin Hanna and Lauren Arnold

Environmental impact assessment (EIA) is a process used to assess projected development proposals, understand their impacts, mitigate negative effects, and support decisions about whether or not to allow a project or activity to go ahead. It is a planning and decision support tool, and it can help advance sustainability objectives. Globally, EIA is arguably one of the most influential and consistent tools for environmental management and protection. EIA now appears in some form or the other in the project reviews and permitting processes of most nations.

Beginning with the United States National Environmental Policy Act (NEPA) in the 1970s, EIA has grown into a complex policy area, and the definition of the environment has become more comprehensive to include social and economic environments as well as the biophysical environment (see Chapter 17 for a review of NEPA). EIA now includes the evaluation of cumulative, social, health, economic and cultural impacts, and climate change. It is fundamentally interdisciplinary. In a sense, EIA is the "original version" or "brand". Processes – such as social impact assessment, health impact assessment, sustainability assessment, and gender-based assessment, cumulative effects assessment, and regional and strategic assessments (which are all covered in this book) – have developed to address our growing understanding of the impacts of development.

This chapter provides an overview of the basic stages or actions involved in EIA and consideration of best practices.[1] While the stages outlined in this chapter are common in many EIA systems, the specific requirements and emphasis on each will differ considerably across different jurisdictions. The emphasis in this chapter is on project-based assessment, but other chapters in this book provide discussions of the participatory, social, cultural, and technical qualities that also shape EIA practice and support its interdisciplinary nature and application at regional scales and to more strategic level issues and challenges.

DOI: 10.4324/9780429282492-2

Box 1.1 Environmental assessment or environmental impact assessment?

Environmental impact assessment and environmental assessment (EA) are terms that may respectively denote a specific assessment (an EIA of a pipeline proposal, for example), and the process of assessment (the overall EA system or regulatory review process), but the terms are now often used interchangeably. Here we use EIA to reflect a broader use of the term "EIA" in the assessment field, and to help emphasize a definition or practice that often emphasizes impacts on an environment that should be defined broadly as the biophysical and human realms (social, cultural, and economic). Some jurisdictions and some researchers use the term *impact assessment* to capture the range of assessment processes and scope of impacts considered; for example, environmental, strategic, social, and economic.

A few principles

Building on the fundamental idea that EIA is a process for identifying and considering the impacts of an action, we can articulate some basic norms to explain what EIA does (these are not to be confused with principles for effective EIA, those are discussed below):

1. EIA describes a proposed activity and the baseline conditions in the place where it will happen;
2. It identifies possible or likely environmental effects of the activity;
3. It proposes measures to mitigate or eliminate adverse effects while providing benefits;
4. It provides some sense of the remaining impacts and their significance;
5. It provides for project follow-up and monitoring; and
6. It engages the public and other interests in debate and conversation about development and the nature of growth.

(Hanna, 2016, 2)

What this describes is a process or system for gathering information that helps proponents, communities, and decision-makers design and implement an action with the best available knowledge of its likely impacts, outcomes, and performance (Hanna, 2016). The ability of EIA to shape and influence decisions is contingent upon the principles and values that shape it as a system and the linkages it has to policy processes.

Researchers and practitioners have outlined principles of an effective EIA – one which fulfills its objectives and contributes meaningfully to planning and decision-making. Sadler's (1996) early work on evaluating practice and performance in EIA provides a seminal discussion of the principles and core values of impact assessment, or at least what they should be. Sadler (1996) wrote of five EIA guiding principles (also noted in Hanna, 2009, 2016):

1. *A strong legislative foundation.* EIA should be based on legislation that provides clarity with respect to objectives, purpose, and responsibilities. Application of EIA should be codified, based in law rather than in discretionary guidelines.
2. *Suitable procedures.* The quality, consistency, and outcomes of EIA should reflect the environmental, political, and social context within which EIA operates, and should demonstrate the ability to respond to divergent issues.

3. *Public involvement (participation or engagement)*. Meaningful and effective public involvement must be present. Not only must those affected and interested be consulted, but also their concerns should be able to affect the decision.
4. *Orientation towards problem solving and decision-making*. The context of EIA is inherently practical and applied. Thus, the EIA system should have relevance to issues of importance, it should generate needed information, and it must influence, and be connected to, the settings where conditions of approval are set and decisions are made.
5. *Monitoring and feedback capability*. The consideration of impacts should not end with approval and implementation; rather, the process must have some capacity for insuring compliance, accuracy of impact prediction, and evaluation of project performance. Not only does such a role strengthen EIA, it provides information that can fine-tune the EIA process, provide knowledge of what impacts actually do occur, and measure project performance.

Sadler's principles have been adapted, modified, quoted, and expanded over the years and have been used as the foundation for further principles of effective EIA (see also Gibson, Doelle, & Sinclair, 2015; Joseph, Gunton, & Rutherford, 2015, Senécal, 1999).

Hanna and Noble (2015) developed a set of criteria for effective EIA based on a Delphi study. Their approach provides a good understanding of effective EIA practices, processes, and systems based on the experiences of the public and private sectors, and academic experts. The outcome was a set of nine principles for an *effective* EIA system:

1. There is *stakeholder[2] confidence* in the objectivity, accessibility, clarity, objectives, and unbiased application of the EIA process.
2. The process is *integrative and linked to approval decision-making*, has the capacity to incorporate multiple forms of knowledge, and is connected to other approval processes that must respect the information, or decision, provided by the EIA process.
3. EIA should *promote betterment and longer-term and substantive gains to environmental management and protection*; and it should be preventative, require monitoring and follow-up, and have provisions for reporting on such activities.
4. *Comprehensiveness* is a key quality in the definition of environment (biophysical, social, cultural, and economic). The process should also have the capacity to focus on significant issues and actions, require the consideration of alternatives, and it must account for cumulative effects and impacts.
5. The *evidence-based* decisions that follow the impact assessment process clearly and directly reflect the knowledge and data presented in the assessment and/or review proceedings, and that the process is open to hearing and considering all relevant, supporting, and opposing evidence.
6. The EIA process must be *accountable* to stakeholders and the public. Documentation and information disclosure requirements are binding on the process and its administrators, proponents, and other stakeholders. There is open and easy access to timely, accurate, and full and complete information. And, the process is independent.
7. There is a requirement and opportunities for stakeholder *participation[3]* throughout the process. Proceedings are open to the public and there are no unjustified limitations to open deliberation and the presentation of evidence; and stakeholders can clearly see how participation was accounted for in the decision. Where applicable, the rights and distinct requirements of Indigenous communities are accounted for in the EIA process and its outcomes.

8. A *legal foundation* for impact assessment provides clarity for stakeholders with respect to applicability, assessment requirements, disclosure requirements, process, reporting, and decision-making. The process contains a legal basis for participation and accountability requirements. It provides procedural fairness.

9. The EIA system possesses *capacity and innovation* features, and is administered by competent and impartial authorities with sufficient resources to ensure the integrity and effectiveness of the process. The process and supporting institutional framework should be flexible, adaptive, and open to new and innovative tools and approaches to assessment.

(Hanna & Noble, 2015)

Some of these echo the qualities we see in Sadler's (1996) outline, which suggests that the field has come to accept common qualities for best practice.

In this book, effective EIA is seen as a process that conveys the above principles; essentially that the process is legislated, that it provides complete information about development impacts, includes effective stakeholder engagement and participation, and is connected to and shapes the decision.

Best practices may be applied to the institutional and governance qualities of EIA; specifically, the qualities of the process, system, and frameworks used for assessing, reviewing, and making decisions. Best practices can also refer to technical and other supporting tools for EIA. These include the scientific, analytical, and predictive tools and approaches used to identify baseline conditions, identify and assess impacts, choose mitigation strategies, determine significance, predict impacts and outcomes, and monitor the performance of facilities. This may encompass a broad range of disciplines, and many forms of knowledge (based on science, culture, history and other ways of knowing) and practice across applied-scientific and scientific fields. Professions will have their own concepts of best practices for methods of mitigating, assessing, and predicting.

The purpose and goals of an EIA system will be context-specific, and understandings and interpretations of effectiveness are collectively shaped by the legal and social setting within which EIA functions and may vary from the proponent to the regulator and to public stakeholders (Hanna, 2016). Some jurisdictions, such as Canada's federal government and British Columbia (see Chapter 19), have added Indigenous knowledge requirements to their EIA processes and are looking at how to address the conceptual and operational challenges in weaving such knowledge into EIA (Chapters 19 and 13). Sustainability objectives are also being increasingly merged into assessment processes. While the principles (and characteristics, metrics, and indicators) that define sustainability will vary by jurisdiction, EIA provides an important opportunity to embed these principles into how projects and development are conceived, and how we define what is in the interest of the public, and the environment (Chapter 6). Ultimately, the effective implementation of EIA principles depends greatly on the social-cultural, political, and institutional context within which the EIA system operates.

The EIA process

The process of conducting an EIA involves a series of actions or stages. An EIA process will vary depending on the jurisdiction, and sometimes the actions are combined, specific qualities of a stage (such as early participation opportunities) may be missing in some processes, or different terms are used. Generally, there are seven stages, each composed of various methods and steps.

The need, or problem, and what to do about it

The process begins with a need, or a problem. At this early stage, the focus is on defining and describing the project or actions needed. This is a basic description of the proposed activity – what we want to do and what purposes it serves, or how it addresses the need or problem. This basic understanding of the nature of the proposal description can inform whether a project will require an EIA (the screening stage), or if an EIA will focus on specific parts of the project. In addition to an outline of the project's physical components, location, and operational activities, the description may provide a preliminary description of anticipated effects.

The nature of the project, the way the proposal has been developed, and early engagement with communities and stakeholders will have a great bearing on the way the EIA evolves, if indeed an EIA is required (Hanna, 2016, 9). Ideally, EIA should really begin as early as possible in the project life – at the point that the problem or need is identified and the potential solution, the project, is imagined.

At this early stage, community and public consultation and engagement should also begin. This can help in identifying technology options, project sites, initial social and environmental baseline conditions, valued ecosystem identification, or early conflict mitigation, and building early awareness and understanding that can lead to efficiencies later in the EIA process. Early engagement and communication between proponents, agencies, and communities, and other stakeholders, will help identify and address concerns in an efficient manner and contribute to a timelier EIA process.

Proposals need to have a clear rationale for why the project is needed and are often required to justify the project in reference to alternative options (Pope, Bond, Morrison-Saunders, & Retief, 2013; Steinemann, 2001). In some jurisdictions, considering alternatives may occur at later stages or at multiple stages, or it may not be required. The consideration of alternatives might require identifying options for achieving the overall objective of the project. For example, if alleviating traffic congestion is the core objective, then the alternatives could include expanding existing roads, building light rail transit, expanding bus services, or instituting congestion charges or a combination of these. If electricity demand is projected to increase in a region and a utility needs to respond (demand response planning), then the alternatives might include a range of generation methods, or buying power from other providers; or policy tools to encourage conservation or when people use power throughout the day (demand management), or indeed some combination of these.

In some jurisdictions, a requirement to consider alternatives might not include alternatives to the project itself but alternative ways of doing it. Certain projects might have flexibility in making changes to project design, technologies, routing, operating conditions, or ways of constructing the project. But other types of project (e.g., mines, forestry, energy) are restricted to the location of the resource. For these projects, the consideration of alternatives can examine alternative production approaches, technologies for extraction/harvest or processing, location of supporting infrastructure (e.g., road, rail, or power lines), or location for activities with siting flexibility – such as ore processing, milling, or smelting.

Screening

Screening answers the basic question, is an EIA required? With screening, we decide if a project is subject to an EIA REQUIREMENT, and if so the level of detail required. We might also determine if public hearings (by a review panel, for example) or an internal agency-based review is required; or it could be determined that the project does not require an assessment and can be referred to other permitting processes (Hanna, 2016; Zhang, Kørnøv, & Christensen, 2013; Senécal et al., 1999). This will depend on the jurisdiction and the screening opportunities and objectives it provides.

EIA legislation can apply to a wide range of activities, many of which might be routine and include minor predictable and well-known environmental impacts (Hanna, 2016). In some jurisdictions, projects might be quickly studied at the screening stage to ensure that no larger impact issues are likely. If that is the case, then a project may proceed to other permitting processes and operational conditions assigned – all without further EIA scrutiny. This is a practical need – to determine if a project will require a substantial review, or a condensed one.

Well structured screening ensures that proposals are subject to appropriate assessment rigor, without subjecting small projects, or projects without significant impacts, to unnecessary delays and costs (Gibson et al., 2015; Hanna, 2016; Snell & Cowell, 2006; Wood & Becker, 2005; Zhang et al., 2013). This discretionary quality needs to be used carefully. Assessment not only provides the opportunity to identify and understand impacts, but also for identifying mitigation measures and judging their relative acceptance (see Figure 1.1). Exempting projects from assessment, especially if the potential for significant impacts is present, compromises public confidence in review and permitting processes generally.

Effective screening should include clearly defined criteria and consistent procedures (Zhang et al., 2013). Screening criteria may include legal requirements (is the undertaking subject to EIA legislation?), scale (does its size, cost, or location mean it will require an EIA), the nature of the proponent of the project (is it public or private sector?), the nature or class of the project (a specific technology, e.g., nuclear power), or a combination of these (Hanna, 2016, 9).

There are several approaches to screening criteria. Some jurisdictions pre-determine what types of projects should require an EIA, or conduct a preliminary analysis of the proposed project to understand the potential impacts of development and use this to decide if an EIA is needed (Morrison-Saunders, 2011; Pinho, McCallum, & Cruz, 2010).

In Canada, the federal Impact Assessment Act (2018) and its regulations use qualities such as federal jurisdiction, thresholds for production, activity type (e.g., type/size of mine, pipeline, power line, road, airport), location (federal lands), and impacts on Indigenous interests as triggers for an EIA. The thresholds and activities are outlined in a regulation (see Reviewable Projects Regulation, 2019). British Columbia's (BC) Environmental Assessment Act (2018) uses such thresholds and/or spatial expansion extent to determine EIA application. Other justifications may apply an assessment requirement to government actions. For example, the US National Environmental Policy Act applies when a US federal agency develops a proposal to undertake a major federal action. This covers a broad range of activities across US federal agencies and federal lands.

A challenge that can emerge is what to do about projects that are just below the threshold? It is fair to say that even if a project is just small enough to "escape" EIA, it will still encounter other permitting processes. But these processes may be perfunctory or devoid of any participation requirements, may have few options built into them for identifying impacts and requiring mitigation, or provide little power for regulators to require changes to the project design. One solution seen in BC's new assessment legislation (Environmental Assessment Act, 2018) is to have a notification requirement for projects that fall below the threshold (there are ranges defined for this) so that the agency responsible for the assessment process can look at the project, consider its potential impacts and public concerns, and then decide if, even though it is just below the threshold, it would still need an EIA. But this depends on the legislation to provide such a notification requirement, and then also provide the authority for the assessment agency to require an EIA.

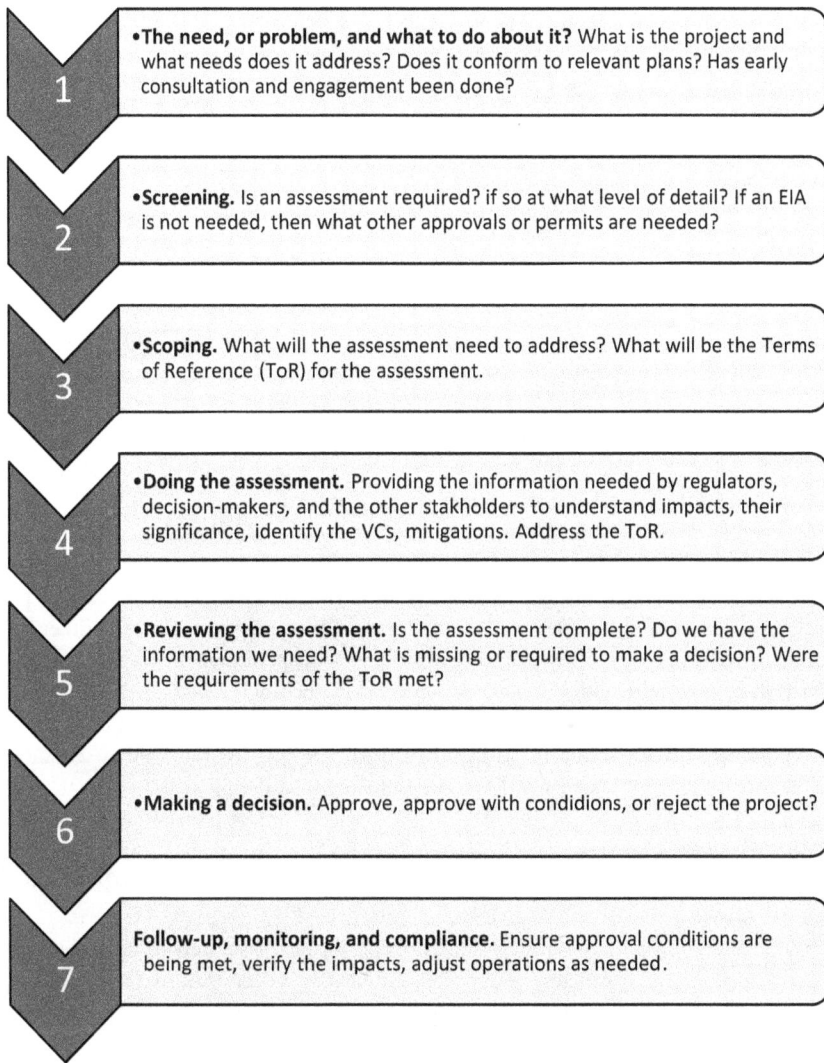

Figure 1.1 The environmental impact assessment process (adapted from Hanna, 2009 and 2016).

Best practice would suggest that regardless of the production capacity or area size, any project with the potential for significant adverse effects should be assessed, or considered for assessment before operations begin. Even if it is determined that effects from small operations are minor, assessment could be an important part of ensuring that cumulative effects are considered (see Chapter 3). Once we know an EIA will be required, we need to know what it will cover – this takes us to scoping.

Scoping

Scoping typically focuses the assessment on the key issues and significant impacts. At this stage, we might establish terms of reference for the assessment (Gibson et al., 2015; Hanna, 2016). This step is important since there may be significant time and resource limitations for conducting

EIA, and we need to decide which potential impacts and environmental qualities we will focus our efforts on.

Some jurisdictions have rules that clearly define what the scope of an EIA must be, while others may provide flexible advice, allowing the EIA to be adapted to the relevant concerns identified through stakeholder participation or negotiation with regulators, which for some projects could be mostly biophysical and for others mostly social issues (Hanna, 2016). The resulting terms of reference would provide a detailed description of the range of what is to be considered for the assessment.

Some EIA processes provide options that allow proponents to consult with the regulator to create *terms of reference* (ToR) that outlines what has to be covered in the EIA documentation, and what the review agencies require and expect to see in the submitted EIA. This contributes to efficiency. The ToR can be binding on proponents and regulators; it reduces uncertainty and provides clarity for all involved in the EIA.

In general, scoping should address the type of project and possible alternatives, the spatial and temporal scales of potential impacts, the availability of baseline data, the consequences of potential impacts for key ecological indicators, and mitigation options (Joseph et al., 2015; Wood, Glasson, & Becker, 2006). The scoping stage can also highlight what additional baseline data is needed and what information will be needed to support decision-making. Baseline data is used to describe the conditions of the area that would be affected by the project and provides an information-foundation foundation for assessment and impact prediction (Duinker & Greig, 2007). Baseline data can be biophysical, economic, and social/cultural.

The activities associated with the development and the ecological and cultural setting dictate the components that would need to be studied. In general, there should be a consideration of social, biophysical, and economic issues in terms of the project itself as well as secondary and regional impacts (Fonseca, McAllister, & Fitzpatrick, 2014; Mulvihill & Baker, 2001). EIA is in part a technical tool for evaluating environmental impacts and as such requires a strong basis in scientific principles; however, there is also a role in EIA for the application of principles and values that define what impacts are important to stakeholders (Gibson et al., 2015; Hanna, 2016, 10; Greig & Duinker, 2011 and 2014; Morrison-Saunders, 2011).

The spatial and temporal scales for assessing impacts should be set beyond the project site and lifespan, and account for reasonably foreseeable cumulative effects. Ultimately, the framework and indicators for the assessment should be based on a strong conceptual framework, rely on recent and reliable data, and utilize standardized methods (where possible) that are transparent and can be evaluated and tested (Fonseca et al., 2014; Joseph et al., 2015). In addition, the process should include tools to assess trends and future scenarios and evaluate impacts with reference to targets, thresholds, and benchmarks (Fonseca et al., 2014).

Public participation is also a critical tool for defining the scope of the EIA (Andre, Enserink, Connor, & Croal, 2006; Snell & Cowell, 2006). Stakeholder participation supports the identification of what is important to those who may be affected by development, and helps to define significant impacts (Hanna, 2016; Mulvihill & Baker, 2001; Zhang et al., 2013). Early participation by the public and stakeholder groups is also vital for building relationships and trust for the assessment and allows key concerns to be addressed before considerable time and resources have been spent on the project (Hartley & Wood, 2005; Sinclair & Diduck, 2016).

Doing the assessment

At this point, more data collection, impact prediction, evaluation of impacts, and mitigation measures, occurs (Hanna, 2016). The mitigation actions identified may be evaluated based on

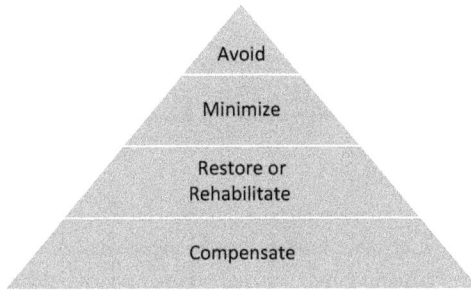

Figure 1.2 The hierarchy of mitigation.

Avoid *(most desirable)*: Can we avoid the impact altogether? This can include the timing or of location temporary activities (e.g. construction activities or staging operations timed to eliminate impacts on migrating wildlife), or choosing a location for infrastructure (e.g. roads, buildings, powerlines, pipelines) so that they avoid impacting a valued component (see Box 2).

Minimize *(somewhat desirable)*: If an impact cannot be avoided, then what measures can be implemented to minimize (decrease, reduce, lessen) the period, intensity or scale (space or size) of impacts? For example, wildlife overpasses or underpasses will allow wildlife to cross a highway to access their range.

Restore of Rehabilitate *(less desirable)*: For impacts that cannot be avoided or minimized, or for significant adverse residual effects, can we rehabilitate or restore or protect places at the project site to achieve no net loss or a net gain in environmental quality? For example, after pipeline construction, we rehabilitate the right of way with new plantings and landscaping.

Compensate *(least desirable)*: This can include measures to compensate for loss of valued components, biodiversity, or social cultural qualities due to the project. For example, biophysical enhancements at another location (such as no net loss or net gain projects for wetland or fisheries habitat), replacement infrastructure, or financial compensation. There may be instances where financial compensation is not acceptable.

their apparent effectiveness and acceptability by stakeholders in the process (see Figure 1.2). This step provides details. Typically the impacts we are most concerned with in EIA are the adverse ones, but there are also beneficial impacts from development. Some jurisdictions may require, or allow for, the inclusion of beneficial impacts in the assessment reports.

During this stage, predicted impacts may also be assessed for their significance – before and after mitigation. Significance can be a subjective notion, determined by the importance that the stakeholders – the proponent, the regulators, the public, and decision-makers – attach to specific value components and impacts (Hanna, 2009, 7). Even if the determination of significance is value-driven, it should be technically supported, transparent, and focused on the key issues and objectives that are important within the jurisdiction and to the stakeholders involved (Lawrence, 2007).

The explicit definition of key valued ecological components (see Box 1.2) and acceptable thresholds for environmental impacts are also helpful for the assessment of impact significance (Lawrence, 2007). Information about natural systems and human–environment systems or the seasonal transitions of places can be lacking in baseline studies, which means they might provide an inventory approach or a very static image of the place where a project will occur. Understanding and accounting for the systemic qualities of the environment can help better define impacts (adverse and beneficial) and ultimately help us understand the cumulative effects of development, even for relatively small "footprint" projects. Seasonality is also important. Key species (migration of birds or animals, estuarine habitat for young fish, or spawning streams) or events (e.g., snowpack, spring runoff, rainy seasons, or reliable dry periods) might be absent from baseline information if data reflects a narrow time span within a year or across years.

Box 1.2 Valued component

"A valued ecosystem component is something that tells you something about something"
(anonymous EIA practitioner).

A valued component (VC) is an environmental component that has ecological, scientific, social, cultural, economic, historical, or other human-defined importance. A valued component may be determined on the basis of scientific study, cultural values, or economic value. Other terms used, sometimes synonymously, are valued ecosystem component (VEC) (which would have a biophysical focus); or valued environmental component (which might have a comprehensive view environment as including the biophysical, social/cultural, and economic qualities of places). A VC/VEC might also be defined in policy, legislation, or plans as requiring special attention or protection. An effective EIA process would ensure that VCs that might be impacted by an action or project are included in the assessments.

During the assessment stage, a monitoring or follow-up program and plan for decommissioning and closure would also be provided. Closure or decommissioning of the project might account for a substantial portion of the assessment as proponents plan for long-term site stewardship and management of impacts that continue or occur well after the operational life of the project. This will vary depending on the project, and some projects have no projected end date. A railway might be seen as part of a long-term transportation strategy, with no set end date, but instead there would be upgrade requirements, possibly even the adding of more lines, or new lines that can carry greater capacities. Large-scale hydropower projects may also be envisioned as "permanent" but can require structural upgrades, equipment replacement, or changes to increase capacity.

Reviewing the assessment

Once an assessment is complete, and information that has been collected and analyzed is brought together and placed in the EIA report, which is presented to the EIA agency for review, recommendation, and then a decision (Hanna, 2016). The contents of the EIA report are often defined by a regulating EIA agency. In some jurisdictions, the layout and content requirements, or expectations, may be given through agency publications, pre-consultation with the proponent, or through the formal provision of terms of reference (Hanna, 2016, 11).

The information provided in the report is assessed by reviewers from government or independent bodies – depending on the approach used in the jurisdiction. Ideally, the review process is independent and transparent (Hanna & Noble, 2015; Joseph et al., 2015).

In addition to being rigorous, review processes must be open and provide settings where stakeholders can bring forward information and express values, which are then clearly considered in the project review (Gibson et al., 2015; Palerm, 2000; Zhang et al., 2013). The review can include assessing the report for completeness, accuracy, adherence to the terms of reference, likely effectiveness of mitigation or risk management (Chapter 5), the quality of public participation (Chapter 14), and compliance with regulations, planning, or other required criteria, or even public acceptance.

The process and requirements for the review will vary depending on the jurisdiction and the nature of the proposed activity. For activities of low public concern, a review and preparation of a recommendation may be completed within an internal administrative setting, all conducted by the EIA agency. But in instances where the project could entail substantial impacts and concerns, the setting for the review may be a public hearing or other formal or quasi-judicial setting (Hanna, 2016). The outcome of the review is usually to provide a recommendation, or in some jurisdictions it may simply be a determination that the EIA is complete or the needed information is present. Regardless of what the outcome is called, the review stage informs, or even directly determines, decision-making.

Making a decision

Decision-making can seem simple, but in practice the factors and the criteria that formally, or informally, shape the decisions can be complex.

The product of an EIA process is often a recommendation, while decision-making power may reside at the political level (Morrison-Saunders, 2011), or perhaps with an administrative body that is independent of the political level. If the process produces a recommendation, this could be to approve a proposal as it is or with conditions, or to reject the proposal outright. But some processes do not provide a recommendation; they only provide information – an outline of impacts and possibly mitigations and a review of what the project entails. In those settings, the information may be used by other permitting processes, or agencies, to inform their decision-making.

The factors that decision-makers need to consider may be provided by the policy or even defined in regulations or legislation. In instances of multi-jurisdictional EIA application, efforts to harmonize reviews or otherwise collaborate in the application of EIA requirements constitute best practice, both in terms of fairness to the proponent (regardless of the final decision) but also to ensure public confidence in the reliability and comprehensiveness of the EIA, and the consistency of EIA practice. We propose seven principles for EIA decision-making:[4]

1. *Transparency*: Are the reasons for the decision and the criteria used clearly explained and justifiable?
2. *Proper use of discretion*: Does the decision reflect the appropriate judgment and provide a justification with reference to evidence and assessment criteria?
3. *Avoidance of fettering*:[5] What is the nature of the requirements imposed on the decision-maker, and does the decision-maker actually have the discretion to decide?
4. *Significance and consequences*: Does the decision acknowledge issues or impacts of significance, if they can be mitigated or not, and the consequences (positive and negative) of the project and the decision?
5. *Procedural fairness*: There are no undue delays, the decision-maker has heard the proponent and other stakeholders, and the decision-maker is impartial.
6. *Reasonableness*: The decision-maker can reflect on context and circumstances.
7. *Sufficient information*: Does the decision-maker have the information needed to make the decision, and are gaps or uncertainties acknowledged?

For EIA to contribute to environmental management and planning, it must support decision-makers and provide the information that decision-makers require (Heinma & Põder, 2010; Sadler, 1996). Management objectives, environmental thresholds, assessment criteria, and the data gathered during the EIA should be clearly framed for use by decision-makers (Joseph et al., 2015). In EIA, the decision to approve a project typically involves setting conditions and seeking commitments from developers, all of which should be supported by legislation to ensure implementation and compliance (Gibson et al., 2015; Morrison-Saunders, 2011). But rejecting a project also requires justification, and transparency. These are all key qualities of EIA practice.

7. Follow-up, monitoring, and compliance

In some settings, it can appear that EIA ends with the decision, but ensuring monitoring and compliance is also key to best practice. This stage helps ensure compliance with any approval conditions, it provides information about how mitigation or other provisions are working, and information that can be used in future assessments, and it helps build the body of baseline information (see Box 1.3) (Gibson et al., 2015; Hanna, 2016; Joseph et al., 2015; Zhang et al., 2013).

The International Association for Impact Assessment (Marshall et al., 2005) presents monitoring as having four qualities:

1. Monitoring includes the collection of activity and environmental data both before (baseline monitoring) and after activity implementation (compliance and impact monitoring);

Box 1.3 Best practices for EIA follow-up

Best practices for EIA follow-up activities are always determined by the nature of the project. However, there are basic principles (see Arts, et al, 2001; Marshall et al., 2005; Morrison-Saunders et al., 2016). The key concepts are well summarized by the set of principles for follow-up best practice developed by Morrison-Saunders (et al 2007) and colleagues for the IAIA (2007), and are adapted here:

1. The proponent is responsible and accountable for implementing follow-up;
2. Regulators are responsible for ensuring that EIA follow-up occurs;
3. Communities have opportunities to be directly involved in the follow-up;
4. All parties, including the proponents, regulators, Indigenous organizations (where present), and civil society, should cooperate;
5. Follow-up should include learning in order to improve future practice;
6. There should be a clear division of responsibilities and roles between the parties for follow-up activities;
7. Follow-up should be "objective led and goal oriented". Follow-up should seek to achieve defined objectives or goals, and defining these goals should be part of follow-up scoping;
8. Follow-up should be "fit for purpose" and tailored to the proposed project or activity;
9. Follow-up should include the "setting of clear performance criteria" to evaluate follow-up practice;
10. Follow-up should be sustained over the entire life of activity, including closure or decommissioning;
11. Adequate resources should be provided for follow-up.

---

2. Evaluation of compliance with standards, predictions or expectations as well as the environmental performance of the project;
3. Management that makes decisions and takes appropriate action in response to issues arising from monitoring and evaluation activities; and
4. Communication that informs stakeholders about the results of EIA follow-up in order to provide feedback on project/plan implementation as well as feedback on EIA processes.

Carrying out monitoring and adhering to approval conditions must be mandatory and enforceable (Arts et al., 2001; Gibson et al., 2015; Marshall et al., 2005; Morrison-Saunders, 2011; Zhang et al., 2013). Regular and sustained dialogue with all stakeholders and utilizing community engaged monitoring will help improve follow-up both during a project's operations and after a project is decommissioned (Appiah-Opoku & Bryan, 2013).

There is another dimension to monitoring – a more audit-oriented approach. Monitoring can also help us understand how effective or appropriate EIA is as a planning and/or environmental management tool. For this, Marshall et al. (2005) note three levels of post-EIA evaluation:

1. Monitoring and evaluation of EIA activities. This is conducted on a project-by-project scale and whether the specific components of EIA were managed in acceptable ways.
2. Evaluation of the effectiveness of the entire EIA system. For instance, this might include the influence of EIA on decision-making or the efficiency of the process.
3. Evaluation of the utility of EIA. This addresses a broad question of whether EIA as a process is beneficial. The key question here is "does EIA work?"

Planning for closure and decommissioning also includes accounting for the monitoring and follow-up after operations cease. For many natural resource projects, especially non-renewables, the activity is temporary. For example, mines and oil and gas extraction sites all have lifespans.

It has been estimated that there are over 10,000 abandoned mines in Canada (Mining Watch, 2000) and over 60,000 in Australia (Campbell et al., 2017). The Canadian province of Alberta has over 73,000 abandoned oil/gas wells, and it has identified 2,124 orphan wells[6] for abandonment and 5,094 sites for reclamation (Alberta, 2021). The effects of such projects, even after they close, can pose long-term environmental and human health impacts without adequate mitigation, plans for long-term monitoring, and knowing who will be responsible for taking care of the project site and its legacies. These are all factors that proponents, regulators, and communities should be thinking about from the first stage of the EIA process.

Knowledge and participation in EIA

A key aspect of an effective EIA system is meaningful public participation and consultation with stakeholders. Ideally, participation in EIA should occur early and be sustained across the stages outlined above. Public participation procedures should be responsive to context, inclusive, cooperative, accountable, and provide accessible information (see Chapter 14 for a discussion of best practices in participation).

In some jurisdictions, distinct engagement with Indigenous peoples may also be legally required. Engagement must recognize the value of local, experiential, and traditional knowledge for a better understanding of complex environmental–societal interactions. This also means engagement with Indigenous groups must not be merely symbolic, but a genuine process, which respects the rights and knowledge of Indigenous people (see Chapter 13). In many settings, there is little opportunity for weaving community and Indigenous knowledge into EIA, and when it is required there may be a lack of transparency from proponents and regulators about how it influences the project or decision, if at all (Appiah-Opoku, 2001; Appiah-Opoku & Bryan, 2013; Nwapi & Ingelson, 2015).

Early and sustained engagement is beneficial for all parties in the EIA process. It provides the opportunity for knowledge to be shared and understood by all stakeholders in the EIA. For the proponent, building relationships with communities can be essential for project approval (Evans, 2015; Nwapi & Ingelson, 2015; Ospina, 2014). Without community support, proponents and governments can face increasing challenges to future developments or expansion of existing operations, and risk losing legitimacy.

Identifying key issues early in the project provides the opportunity for mutual problem solving and adapting the project to better align with community values and livelihoods (Evans, 2015; Franks, 2012). This means providing opportunities, and support, both organizationally and financially, for public input allows issues to be addressed early and more efficiently and can minimize time and financial strains to the proponent through proactive issue identification.

Andre et al. (2006) provide principles that help guide practice:

- Public involvement in EIA should be "*initiated early and sustained*" throughout the process. This helps to build relationships and trust, improves the analysis of the assessment, increases the opportunities for participation and modification of the proposal, improves the image of the proponent and the confidence of regulators in their decision, and improves public knowledge and education on the project and issue.
- The participation process should be "*well planned and focused on negotiable issues*". The credibility of the process is improved if all stakeholders are made aware of the procedures, aims, and expected outcomes. In addition, since consensus may not always be possible, understanding all views should be emphasized.
- The process should be "*supportive to participants*" in that proper assistance for facilitation, resources, and information dissemination should be provided.
- The process should be "*tiered and optimized*". A participation program should be employed at an appropriate level of decision-making for a proposal. Since participation is resource and time consuming, it should be optimized and efficient within the decision-making process.
- The process, information, and decisions must be "*open and transparent*". Information should be communicated in such a way that laypersons and all cultural and ethnic backgrounds can participate.
- "*Context-oriented*". Communities may have specific informal or formal procedures for resource management or public participation.
- "*Credible and rigorous*". Facilitation by a neutral party improves legitimacy and increases public confidence and willingness to participate.

Participation methods also need to be tailored to the demographic and cultural context of the project. Projects may be in remote areas or proposed for places where an industry is unknown , or where an industry does not have a good track record and this can pose significant challenges for a proponent when approaching public participation..

The potential for conflict about development can be intensified without good participation strategies (Evans, 2015; Nwapi & Ingelson, 2015). Participation approaches cannot ignore the social setting, and proponents can find that the tools they use in one setting do not work in another. The ability to participate, and indeed the outcomes of development, will also be different across different communities and demographics, including gender (see Chapter 11), and as a result people will view projects differently (Andre et al., 2006). Cultural or other social limitations to participation may greatly influence the knowledge shared with and voices heard by a proponent. These limitations might present a skewed image of the benefits expected, the impacts of concern, what the valued components are, and the perception of significance.

With all this said, and admittedly it can be easier to flag these issues than deal with them at an operational level, such challenges are seen across global regions regardless of economic, political, or social contexts. But as Healey (1997) notes, the power of public involvement is ultimately in whether or not it has the capacity to affect the decision.

Concluding comments

As EIA research and practice continues to evolve, new areas are emerging that will require attention. Key areas for future knowledge building include continuing to advance data and data management practices and further advancing integration of cumulative effects and strategic assessments as part of the EIA system. Issues such as climate change or sustainability are also increasingly called on to be included in EIA. How best to effectively incorporate these broad issues is an important area for EIA practice (Burdge, 2008).

In an applied process such as EIA, it may be easier to account for objectives such as climate change attenuation or enhancing resiliency and adaptation characteristics of projects, which are measurable and linked to technical and other targets, rather than more conceptual and equivocal objectives such as sustainability, which can sometimes be difficult to define and apply to operationally. Those may be best addressed at a different policy level, or as clear and identifiable aims in strategic planning processes.

A lack of capacity in many settings continues to limit the ability of EIA to contribute to informed environmental management decisions (Appiah-Opoku & Bryan, 2013; Heinma & Põder, 2010; Toro et al., 2010). Capacity issues include:

- Limited legal and administrative support for EIA procedures;
- A lack of practitioner capacity and the need for more EIA training;
- Limited baseline information and other data (e.g., monitoring, follow-up, impact outcomes);
- Weak public and Indigenous participation processes; and
- Limited operational support for monitoring and follow-up requirements.

Another emergent challenge is aligning public and political expectations with the limitations, purpose, and regulatory requirements of EIA. Regardless of such issues, EIA is an essential tool for environmental management. It affords a chance to define practical strategies, helps ensure our growth decisions have positive net environmental and social-economic outcomes, helps reduce negative impacts to the environment, and can provide transparency and community engagement in making decisions about development – if it is applied effectively and well-aligned with decision-making and planning processes.

Effective EIA processes are open to accepting and considering multiple types of information, including traditional knowledge, community-based knowledge, and public opinions and values. And they have the capacity to consider a range of impacts, including biophysical, social, economic, health, and cumulative impacts – a comprehensive and holistic definition of the environment.

In the chapters that follow, the authors examine different facets of EIA practice and illustrate approaches, challenges, and diverse ways of thinking about EIA practice. The principles, practices, and case studies we provide in this book offer a guide to EIA. But the practice of EIA will necessarily vary depending on the social, political, cultural, and environmental circumstances of individual nations and their regions. It can provide an important process for understanding and accounting for the impacts of development, and for enhancing benefits. EIA is not about stopping development or growth; it is not about limiting opportunities for communities – it is about making projects better and supporting informed decisions.

Notes

1 Acknowledgment: This chapter draws on parts of Arnold, L. & Hanna K. (2017). *Best Practices in Environmental Assessment: Cases Studies and Application to Mining.* Vancouver: Canadian International Resources and Development Institute Report 2017-003. These parts are adapted with permission.
2 *Stakeholders* can be defined as those who have direct interest in or may be affected by the project. The *public* can be defined more broadly and may include those who have an interest in the project but may not be directly or indirectly affected by it. We can also use use term *parties to the process* to describe a more formla involvement and role in EIA.
3 We use the terms *participation* and *involvement* to denote the capacity of the public and stakeholders to meaningfully inform the process and shape the decision. Effective participation has the capacity to influence the outcome of the EIA process.
4 We based the principles on our experience drawing on information and knowledge that is observational (the EIA literature, law and regulations, and policy) and conversational (ideas and opinions of colleagues).
5 Fettering is when there is policy or legal constraints on the decision-maker, which means no actual decision-making can occur. If a law or regulation designates a decision-maker(s), that person must have some degree of discretion to decide.
6 A well site without a legally responsible and/or financially viable owner to address its legal decommissioning and reclamation requirements.

References

Alberta (government of). (2021). Oil and gas liabilities management. Retrieved from https://www.alberta .ca/oil-and-gas-liabilities-management.aspx.
Andre, P., Enserink, B., Connor, D., & Croal, P. (2006). *Public participation: International best practice principles: Special Publication Series No. 4.* International Association for Impact Assessment.
Appiah-Opoku, S. (2001). Environmental impact assessment in developing countries: The case of Ghana. *Environmental Impact Assessment Review, 21*(1), 59–71. https://doi.org/10.1016/S0195-9255(00)00063-9
Appiah-Opoku, S., & Bryan, H. C. (2013). EIA follow-up in the Ghanaian mining sector: Challenges and opportunities. *Environmental Impact Assessment Review, 41*, 38–44. https://doi.org/10.1016/j.eiar.2013 .02.003
Arnold, L. & Hanna, K. (2017). Best Practices in Environmental Assessment: Cases Studies and Application to Mining. Vancouver: Canadian International Resources and Development Institute Report 2017-003
Arts, J., Caldwell, P., & Morrison-Saunders, A. (2001). Environmental impact assessment follow-up: good practice and future directions – findings from a workshop at the IAIA 2000 conference. *Impact Assessment and Project Appraisal, 19*(3), 175–185. https://doi.org/10.3152/147154601781767014
Burdge, R. J. (2008). The focus of impact assessment (and IAIA) must now shift to global climate change!! *Environmental Impact Assessment Review, 28*(8), 618–622. https://doi.org/10.1016/j.eiar.2008.03.001
Campbell, R., Linqvist, J., Browne, B., Swann, T., & Grudnoff, M. (2017). *Dark side of the boom: What we do and don't know about mines, closures and rehabilitation.* Canberra: The Australia Institute.
Duinker, P. N., & Greig, L. A. (2007). Scenario analysis in environmental impact assessment: Improving explorations of the future. *Environmental Impact Assessment Review, 27*(3), 206–219. https://doi.org/10 .1016/j.eiar.2006.11.001

Environmental Assessment Act (BC) (2018). https://www.bclaws.gov.bc.ca/civix/document/id/complete/statreg/18051

Evans, M. D. (2015). Escalating social risk around mining: Why does it matter and what can be done? *CIM Journal, 6*(1), 35–41.

Fonseca, A., McAllister, M. L., & Fitzpatrick, P. (2014). Sustainability reporting among mining corporations: A constructive critique of the GRI approach. *Journal of Cleaner Production, 84*, 70–83. https://doi.org/10.1016/j.jclepro.2012.11.050

Franks, D. (2012). Social impact assessment of resource projects. International Mining for Development Centre Mining for Development: Guide to Australian Practice. Retrieved from http://www.csrm.uq.edu.au/Portals/0/Publications/Social-impact-assessment-of-resource-projects1.pdf

Franks, D. M., Brereton, D., & Moran, C. J. (2010). Managing the cumulative impacts of coal mining on regional communities and environments in Australia. *Impact Assessment and Project Appraisal, 28*(4), 299–312. https://doi.org/10.3152/146155110X12838715793129

Gibson, R. B., Doelle, M., & Sinclair, A. J. (2015). Fulfilling the promise: Basic components of next generation environmental assessment. *Journal of Environmental Law and Practice, 29*, 251–276.

Greig, L. A., & Duinker, P. N. (2011). A proposal for further strengthening science in environmental impact assessment in Canada. *Impact Assessment and Project Appraisal, 29*(2), 159–165. https://doi.org/10.3152/146155111X12913679730557

Greig, L., & Duinker, P. (2014). Strengthening impact assessment: What problems do integration and focus fix? *Impact Assessment and Project Appraisal, 32*(1), 23–24.

Gunn, J., & Noble, B. F. (2011). Conceptual and methodological challenges to integrating SEIA and cumulative effects assessment. *Environmental Impact Assessment Review, 31*(2), 154–160. https://doi.org/10.1016/j.eiar.2009.12.003

Hanna, K. S. (2009). Environmental impact assessment: Process, setting and efficacy. In *Environmental impact assessment practice and participation* (2nd ed.). Don Mills: Oxford University Press.

Hanna, K. S. (2016). Environmental impact assessment: Process, Setting and Efficacy. In *Environmental impact assessment practice and participation* (3rd ed.). Don Mills: Oxford University Press.

Hanna, K., & Noble, B. (2011). *A brief to House of Commons Committee on the environment and sustainable development on the Canadian Environmental Assessment Act (CEIA Act).*

Hanna, K. S., & Noble, B. F. (2015). Using a Delphi study to identify effectiveness criteria for environmental assessment. *Impact Assessment and Project Appraisal, 33*(2), 116–125. http://dx.doi.org/10.1080/14615517.2014.992672

Harriman, J. A., & Noble, B. F. (2008). Characterizing project and strategic approaches to regional cumulative effects assessment in Canada. *Journal of Environmental Assessment Policy and Management, 10*(1), 25–50.

Harriman, J., & Noble, B. (2011). Conceptual and methodological challenges to integrating SEIA and cumulative effects assessment. *Environmental Impact Assessment Review, 31*, 154–160.

Hartley, N., & Wood, C. (2005). Public participation in environmental impact assessment – implementing the Aarhus Convention. *Environmental Impact Assessment Review, 25*(4), 319–340. https://doi.org/10.1016/j.eiar.2004.12.002

Healey, P. (1997). *Collaborative planning: Shaping places in fragmented societies.* Hong Kong: Macmillan Press.

Heinma, K., & Põder, T. (2010). Effectiveness of environmental impact assessment system in Estonia. *Environmental Impact Assessment Review, 30*(4), 272–277. https://doi.org/10.1016/j.eiar.2009.10.001

International Association for Impact Assessment and Institute for Environmental Assessment UK. (1999). *Principles of environmental impact assessment best practice.* Fargo, USA, and Lincoln, UK: IAIA and IEIA. Retrieved from http://www.iaia.org/uploads/pdf/principlesEIA_1.pdf (accessed October 26, 2016).

Joseph, C., Gunton, T., & Rutherford, M. (2015). Good practices for environmental assessment. *Impact Assessment and Project Appraisal, 33*(4), 238–254. https://doi.org/10.1080/14615517.2015.1063811

Lawrence, D. P. (2007). Impact significance determination – Back to basics. *Environmental Impact Assessment Review, 27*(8), 755–769. https://doi.org/10.1016/j.eiar.2007.02.011

Marshall, R., Arts, J., & Morrison-Saunders, A. (2005). International principles for best practice EIA follow-up. *Impact Assessment and Project Appraisal, 23*(3), 175–181. https://doi.org/10.3152/147154605781765490

Morrison-Saunders, A. (2011). Principles for effective impact assessment: Examples from Western Australia. In *IAIA11 impact assessment and responsible development for infrastructure, business and industry,* 31st Annual Conference of the International Association for Impact Assessment. Puebla, Mexico. Retrieved from http://researchrepository.murdoch.edu.au/6631/

Morrison-Saunders A., Marshall, R., & Arts, J. (2007). *EIA follow-up international best practice principles*. Special Publication Series No. 6. Fargo, USA: International Association for Impact Assessment.

Morrison-Saunders, A., McHenry, M. P., Rita Sequeira, A., Gorey, P., Mtegha, H., & Doepel, D. (2016). Integrating mine closure planning with environmental impact assessment: challenges and opportunities drawn from African and Australian practice. *Impact Assessment and Project Appraisal*, *34*(2), 117–128. https://doi.org/10.1080/14615517.2016.1176407

Mining Watch. (2000). *Abandoned mines in Canada, prepared by W.O.* Mackasey Associates, Ottawa: Mining Watch and W.O. Mackasey Associates.

Mulvihill, P. R., & Baker, D. C. (2001). Ambitious and restrictive scoping: Case studies from Northern Canada. *Environmental Impact Assessment Review*, *21*(4), 363–384.

Natural Resources Canada. (2007). Aboriginal partnership in mining, information bulletin, partnerships agreements, Raglan Mine. Retrieved from www.nrcan.gc.ca/mms

Noble, B. F. (2009). Promise and dismay: The state of strategic environmental assessment systems and practices in Canada. *Environmental Impact Assessment Review*, *29*(1), 66–75. https://doi.org/10.1016/j.eiar.2008.05.004

Noble, B. F. (2015). Cumulative effects research: Achievements, status, directions and challenges in the Canadian context. *Journal of Environmental Assessment Policy and Management*, *17*(1) 1–7.

Nwapi, C., & Ingelson, A. (2015). Promoting transparency in Central African mineral development. *CIM Journal*, *6*(4), 233–239.

O'Faircheallaigh, C. (2010). Public participation and environmental impact assessment: Purposes, implications, and lessons for public policy making. *Environmental Impact Assessment Review*, *30*(1), 19–27. https://doi.org/10.1016/j.eiar.2009.05.001

Ospina, M. (2014, October). Responsible investment: The reward of managing social risk. *CIM Magazine*, *9*(7), 32–33.

Palerm, J. R. (2000). An empirical-theoretical analysis framework for public participation in environmental impact assessment. *Journal of Environmental Planning and Management*, *43*(5), 581–600. https://doi.org/10.1080/713676582

Pinho, P., McCallum, S., & Cruz, S. S. (2010). A critical appraisal of EIA screening practice in EU Member States. *Impact Assessment and Project Appraisal*, *28*(2), 91–107. https://doi.org/10.3152/146155110X498799

Pope, J., Bond, A., Morrison-Saunders, A., & Retief, F. (2013). Advancing the theory and practice of impact assessment: Setting the research agenda. *Environmental Impact Assessment Review*, *41*, 1–9. https://doi.org/10.1016/j.eiar.2013.01.008

Reviewable Projects Regulation (BC) (2019). Retrieved from https://www.bclaws.gov.bc.ca/civix/document/id/complete/statreg/243_2019

Ross, W. A. (1994). Assessing cumulative environmental effects: Both impossible and essential. In *Cumulative effects assessment in Canada: From concept to practice* (pp. 1–9). Calgary, Alberta: Alberta Society of Professional Biologists.

Sadler, B. (1996). *International study of the effectiveness of environmental assessment. Environmental assessment in a changing world: Evaluating practice to improve performance*. Final report. Ottawa: Ministry of Supply and Services Canada.

Senécal, P., Sadler, B., Goldsmith, B., Brown, K., & Conover, S. (1999). *Principles of environmental impact assessment, best practice*. Fargo, ND: International Association for Impact Assessment; Lincoln, UK: Institute of Environmental Assessment.

Shared Values Solutions. (2012). *Aboriginal participation in environmental monitoring and management in the Canadian mining sector-a scan for current best practices*. Retrieved from http://sharedvaluesolutions.com/wp-content/uploads/2012/09/SVS-Aboriginal-Participation-in-EM-Report-Final-Abridged.pdf

Sinclair, A. J., & Diduck, A. P. (2016). Reconceptualizing public participation in environmental assessment as EIA civics. *Environmental Impact Assessment Review*, *62*: 174–182. https://doi.org/10.1016/j.eiar.2016.03.009

Snell, T., & Cowell, R. (2006). Scoping in environmental impact assessment: Balancing precaution and efficiency? *Environmental Impact Assessment Review*, *26*(4), 359–376. https://doi.org/10.1016/j.eiar.2005.06.003

Status of Women Canada. (2016). Gender-based Analysis Plus. Retrieved from http://www.swc-cfc.gc.ca/gba-acs/index-en.html.

Steinemann, A. (2001). Improving alternatives for environmental impact assessment. *Environmental Impact Assessment Review*, *21*(1), 3–21. https://doi.org/10.1016/S0195-9255(00)00075-5

Toro, J., Requena, I., & Zamorano, M. (2010). Environmental impact assessment in Colombia: Critical analysis and proposals for improvement. *Environmental Impact Assessment Review*, *30*(4), 247–261. https://doi.org/10.1016/j.eiar.2009.09.001

Wood, G., & Becker, J. (2005). Discretionary judgement in local planning authority decision making: Screening development proposals for environmental impact assessment. *Journal of Environmental Planning and Management*, *48*(3), 349–371. https://doi.org/10.1080/09640560500067467

Wood, G., Glasson, J., & Becker, J. (2006). EIA scoping in England and Wales: Practitioner approaches, perspectives and constraints. *Environmental Impact Assessment Review*, *26*(3), 221–241. https://doi.org/10.1016/j.eiar.2005.02.001

Zhang, J., Kørnøv, L., & Christensen, P. (2013). Critical factors for EIA implementation: Literature review and research options. *Journal of Environmental Management*, *114*, 148–157. https://doi.org/10.1016/j.jenvman.2012.10.030

2

STRATEGIC ENVIRONMENTAL ASSESSMENT

A spectrum of understandings

Maria Rosario Partidario

Introduction

While raised as a technical exercise, and still strongly associated with the need to inform decision-making through the delivery of written reports, strategic environmental assessment (SEA) is increasingly acknowledged as a sociopolitical, knowledge-brokerage, and governance exercise (Sheate and Partidário 2010; Partidario and Sheate 2013; Lobos and Partidário 2014; Monteiro et al. 2018; Rodriguez and Partidário 2021). This chapter elaborates on the evolving nature of SEA as a public policy instrument that facilitates the integration of environmental and sustainability issues, in a broader sense, into strategic decision-making, with the purpose of creating better contexts for sustainable development.

Over the years, academics and professionals of SEA have come to accept that there is not only one form of doing and thinking in SEA. The variety of types of SEA systems, applications, and experiences that have contributed to the advancement of SEA has grown significantly over time. This is recognized by the editors of the *Handbook of SEA* (Fischer and González 2021) where forms of SEA rooted in environmental impact assessment (EIA) are still visible, but also where a range of other understandings of SEA, elaborated by several of the handbook authors, are also presented.

SEA is complex. Organizational cultures and governance models across the world determine different policy and planning requirements and practices, inevitably placing different demands on SEA. Processes of learning, knowledge creation, and experience will shape the concepts and the applications of SEA accordingly. Consequently, existing forms of SEA can be observed along a large spectrum of types and experiences. Many times SEA have been very operational, technical, and locally based, concerned with the assessment of impacts of concrete proposals just before projects are designed. But SEA has also been used strategically to help build futures, driven by long-term and broad visioning, assessing policy and planning options that may enable more sustainable development, long before concrete proposals are formulated. This creates a large span of possibilities with SEA.

This chapter aims to contribute to clarifying how SEA reached this considerable versatility, how it is being defined and applied, and what are the key elements and ingredients of practice depending on its positioning in the spectrum of understandings. The chapter

DOI: 10.4324/9780429282492-3

recognizes SEA more as a sociopolitical than a technical exercise, discusses challenges and looks ahead into innovative forms of SEA, and into the role that SEA should play in face of global challenges.

On the history and evolution of SEA

The early days

The term SEA was proposed in the late 1980s by Wood and Djeddour (1989) (see Box 2.1), inspired by the late 1970s strategic planning system in Britain. But the idea of initiating environmental assessment before the project level came earlier. It was built on the potential of policy and planning systems triggered by the new opportunities brought up by the emerging EIA system. The 1969 US National Environmental Policy Act (NEPA) clearly mentioned its application to "proposals for legislation and other major Federal actions significantly affecting the quality of the human environment". EIA, as the first born, was evidently not limited to the assessment of development projects.

A tiered system of EIA, applied to a sequence of levels of decision-making starting with policies and finishing in projects, was first suggested by Lee and Wood (1978). Since 1979 the US Council of Environmental Quality (CEQ) regulated programmatic environmental impact studies to evaluate the effects of broad proposals or planning-level decisions. Soon after, in 1981, the US Housing and Urban Development Department adopted the area-wide impact assessment to assess urban development proposals ahead of project proposals. After 1976 in the UK, and also in Denmark and Sweden, environmental issues were integrated into spatial planning (e.g., environmental zoning in Denmark). The need for an environmental assessment above project level was underlined by O'Riordan and Sewell (1981). These are a few examples of predecessors of SEA under the flagships of EIA, policy review and spatial planning, shaping the subsequent pathways SEA would follow.

One year before advancing the term SEA, Wood (1988) was still thinking about the advantages of an EIA system not confined to projects. In line with NEPA and writing about EIA in plan-making, Wood addressed the EIA system potential to apply to all actions likely to have significant environmental impacts irrespective of their type and level of action. The author endorsed the 1978 tiering concept of an EIA system (Figure 2.1), claiming that "higher levels

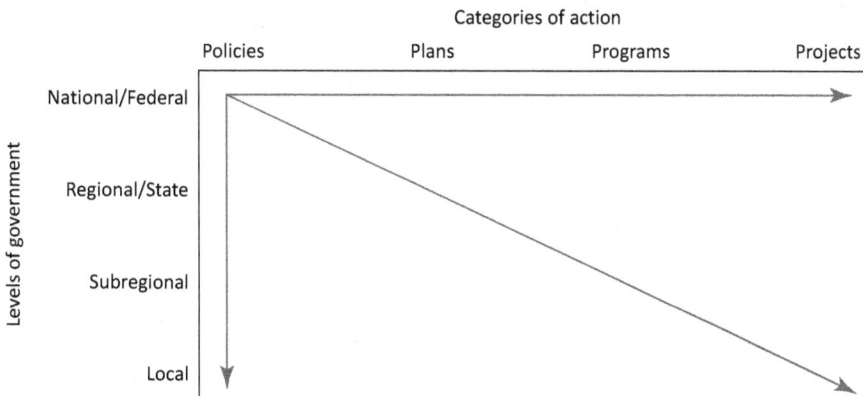

Figure 2.1 The tiering EIA system (based on Wood, 1988).

of action [...] may generate projects [...] which lack sufficient alternatives" (Wood 1988, 98) and that "there may be savings if environmental data are collected when a higher-order action is proposed rather than after an urgent project is suggested or if the assessment of a plan or program obviates the necessity to undertake numerous project EIAs" (Wood 1988, 98–99). The improvement of projects' EIA was clearly advocated by the logic of an EIA in plan-making.

SEA as an innovation

But soon after, Wood, together with his student Djeddour, carved a new term and referred to a more strategic instrument in nature (Box 2.1).

Box 2.1 First time the "SEA" term is mentioned in the literature

The environmental assessments appropriate to policies, plans and programmes are of a more strategic nature than those applicable to individual projects and are likely to differ from them in several important respects [...] We have adopted the term "strategic environmental assessment" (SEA) to describe this type of assessment.

(Wood and Djeddour 1989)

Many scholars recognize in this move the need to shift to a new assessment instrument with a more strategic nature, away from the logic of impact assessment applied to individual actions (projects, plans, or programs). Twenty years of application of EIA had shown that while being a very useful and powerful instrument to assess the environmental impacts of concrete, individual, project proposals, the process and methods, as well as the data needed to accomplish EIA were not compatible with the timing and nature of decisions at more strategic levels of action (Lee and Walsh 1992; Thérivel et al. 1992; Partidário 1996, 2000; Sadler and Verheem 1996).

These were among many of the difficulties then found of applying EIA to higher levels of decision-making. Basically, EIA (including EIA in plan-making) did not adapt to the forward-looking, strategic nature of policy and planning decision-making. That is, policy decisions made incrementally, with vaguely defined intentions of action, less concrete decisions taken with imperfect information, carrying enormous levels of uncertainty, and hence requiring more flexibility. This was not compatible with the procedural rigidity and standard reactive approaches that characterize EIA.

Wood and Djeddour (1989) suggestion of a new instrument, called SEA, to apply to decisions of a more strategic nature was an innovation. Professionals and organizations urged to use SEA, eager to adopt and apply the new idea. SEA rocketed as the grand environmental assessment alternative. But while the term was quickly adopted, only small changes happened to the toolbox and to the mentalities of those using it. The concepts known, or at hand, at the time, especially of EIA and management science, but to some extent also of policy and planning, became the toolbox of SEA. As often is the case, new catchy terms are easily and quickly adopted, but innovation in practice takes time. As well stated in Thérivel (1993, 145–146): "In most countries SEA has evolved upward from...EIA of projects, rather than as a means of tickling down the objectives of [...] environmental policy".

The contrasting directions taken by SEA

Some scholars immediately embraced SEA as an innovative concept, a proactive instrument to all development strategic conceptualizations, along the lines of O'Riordan and Sewell (1981). Earlier advocates (Boothroyd 1995; Partidário 1996, 1999; Sadler 1999; Clark 2000) argued that SEA should address policy and institutional frameworks, be a driver toward sustainability, and integrate societal values in decision processes.

Strong arguments in that direction were provided by Clark (2000). For this author, SEA had different features to other types of impact assessment:

> many decisions are really made incrementally and, more often than not, decisions are made with imperfect information […] environmental assessment prepared at a truly strategic level will look different from traditional EIA […] it will be a short, concise analysis […] [that] will focus on paths, not places.
>
> *(Clark 2000, 16)*

According to the same author, while high-quality assessment of cumulative effects makes EIA richer and assessment of social impacts makes EIA deeper, SEA is a different kind of analysis: "with SEA […] there are different principles involved than the manner in which EIA has evolved" (Clark 2000, 16).

This line of thought represented a new understanding and practice of SEA, away from EIA and aligned with policy and more strategic planning levels of decision-making (Nitz and Brown 2001; Wallington 2002; Bina 2003; Sheate et al. 2003). This was an opportunity for disruption in established environmental assessment practices, a paradigm shift toward more positive and constructive approaches, and to embrace a range of other areas of assessment concerned with social, health, cultural, and overall sustainability that were evolving under the broader impact assessment (IA) concept.

Progressively a growing number of scholars raised their voices questioning the strategic meaning in SEA (Noble 2000; Partidário 2000; Cherp et al. 2007; Hacking and Guthrie 2008; Stoeglehner 2019) and the need for more proactive and strategic approaches in its practice (Nilsson and Dalkmann 2001; Nooteboom 2006; Bina 2007; Wallington, Bina and Thissen 2007; Ahmed and Sanchez-Triana 2008; Tetlow and Hanush 2012; Azcárate 2015; Noble and Nwanekezie 2017; Hayes 2019). Tetlow and Hanush (2012) recognized the need to adapt SEA to environmental, social, economic, cultural, and political contextual factors. For Noble and Nwanekezie (2017), a transition in SEA away from its EIA roots is needed, while recognizing the difference between EIA and SEA is a crucial condition for understanding SEA and allowing process and practice improvement.

Most scholars and practitioners, however, continued endorsing SEA as an expansion of EIA (Fischer 2003; Thérivel 2004; Polido and Ramos 2011; several chapters in Fischer and González 2021), which is currently the dominant SEA practice. The theoretical tiering logic of continuity between policies and projects shows a clear anchor in projects, albeit connected to sustainability, as in the words of Thérivel et al. (1992, 130): "In an ideal world SEA should be based on sustainability and in turn it would cascade down to project planning". Multiple citations encouraging SEA for project planning can be drawn from the literature. Such a type of SEA was created to improve the conditions for EIA performance.

With such a SEA approach prevails a greater emphasis on the instrument per se (broad EIA, now called SEA), but less on the nature of the decision (Lyhne et al. 2020) and the need to make it "fit for purpose" (Sadler and Verheem 1996). This would become the EIA-based model of SEA, also known as traditional SEA, that follows a standard impact assessment logic learned with EIA, applying to levels of decision (such as policies, plans and programs, (PPP)) prior to projects.

The European Directive 2001/42/EC is a stated evidence of this effects or EIA-based model as it refers explicitly to the assessment of the environmental effects of plans and programs that set the context for projects' development: "an environmental assessment shall be carried out for all plans and programs […] which set the framework for future development consent of projects listed in […] (a) in para 2, art 3)". The new instrument did not involve a full methodological and procedural change, and the well-established EIA standard procedure and activities – screening, scoping, baseline data, alternatives, mitigation, reporting, decision, monitoring – were carved in stone for all followers of the European EIA-based SEA model.

Challenges, opportunities, and paradoxes

The rise of sustainable development leading priorities, after UNCED 1992, created an opportunity for SEA as a way of implementing the sustainability agenda, as well argued by Thérivel et al. (1992). It led to the emergence of sustainability assessment (SA) (Sadler 1999) first in Britain (DETR 1999), then quickly replicated in Canada (Gibson et al. 2005; Gibson 2006; White and Noble 2013) and Australia (Pope et al. 2004). Many authors acknowledged the role of SEA to promote sustainable development (Sadler 1999; Partidário 1996; Partidário and Moura 2000), others preferred to avoid mixing SA with SEA (Pope et al. 2004, Gibson 2006) to avoid misleading sustainability with the typical biophysical scope of SEA, but also to emphasize the relevance of sustainability assessment at project level (see Chapter 6 in this book).

SEA has come a long way since its early days. Fischer and González (2021) talk about an evolution of SEA away from the EIA-based concept, arguing on the lack of applicability of EIA-based approaches to higher levels of strategic action, which had been recognized for quite some time (Partidário 1993; Sadler 1994; Boothroyd 1995; Nilsson and Dalkmann 2001). For example, Boothroyd (1995) stated the need to move away from approaches that evolved from EIA-based methods, or to break away from the tendency to simply extend from either impact assessment or policy analysis, while Nilsson and Dalkmann (2001) argued that application of EIA-based approaches in SEA had generally proven to be inadequate.

The fact is that the current literature still recognizes that SEA is practiced largely as an "EIA-based" tool (Verheem and Dusik 2011), a fact confirmed by the empirical results shared by McCluskey and João (2011), Lobos and Partidário (2014), and Bidstrup and Hansen (2014). Noble and Nwanekezie (2017, 165) stated that "the practice of SEA remains deeply rooted in the EIA tradition and scholars and practitioners often appear divided on the nature and purpose of SEA". Other authors found the SEA and planning processes to be disjointed (Hayes 2019). Scholars addressed the apparent paradoxes with SEA, given contradictions between what was expected with SEA in its early days and what is its actual delivery (Noble 2009; McCluskey and João 2011; Bidstrup and Hansen 2014; Sadler and Dusik 2016). Bidstrup and Hansen (2014, 29) elaborated on the paradox of SEA "as the methodological ambiguity of non-strategic SEA" while Sadler and Dusik (2016) spoke about "the paradox of progress and performance" to refer to a lack of connection between methodologies and implementation.

While multiple variations of SEA have been advanced (such as regional assessments, integrated assessments, and environmental appraisals), it seems evident that SEA evolved in two main directions: one that percolates EIA essential principles, elements, and methods into SEA conceptualization; and another, rather more disruptive yet slower, promoting a more strategic conceptualization of SEA. The next section addresses this duality of SEA conceptualization in a spectrum of understandings, and the resulting diversity of SEA definitions and practices.

Spectrum of understandings of SEA and how SEA is defined and applied

The evolution of SEA, as described above, has led to the expansion of multiple SEA interpretations and definitions, evident in the 106 definitions of SEA captured by Silva et al. (2014). This versatility of SEA, and the nature of its various interpretations, led to the suggestion of a spectrum of SEA approaches (Partidário 2005, 2015), ranging between two main schools of thought, influencing the theory and the practice of SEA: the project's evaluation school at one end of the spectrum, and the policy science/strategic planning school at the other end of the spectrum (Partidário 2000). These schools lay behind the two dominant models of SEA recognized in the literature (Partidário 2007; Tetlow and Hanush 2012; Noble and Nwanekezie 2017): an EIA-based or -type model, aligned with the logic of traditional impact assessment, and the strategic thinking model, aligned with a logic of forward-looking, strategic planning and policy sciences.

The concepts and practices of SEA across the spectrum share the mix of influences received from each model. The extreme representing the traditional impact assessment model reflects EIA influence and is led by formal and technical procedures, with the aim of providing information on the assessment and mitigation of effects. The other extreme representing the strategic thinking model reflects the influence of strategic planning and policy science and is led by governance, through increased capacities and dialogues, searching for optional pathways for sustainability (Figure 2.2).

This duality in SEA is acknowledged by several authors. For example, Tetlow and Hanush (2012) stated that: "SEA has evolved from a largely EIA-based and responsive mechanism, to a far more proactive process of developing sustainable solutions as an integral part of strategic planning activities" (Tetlow and Hanusch 2012, 17). Tonk and Verheem (1998) referred to SEA as one concept with multiple forms. More recently, Noble and Nwanekezie (2017), capturing recent thinking about SEA as a process for driving institutional change (Noble and Gunn 2015; Partidário 2012), elaborated on the notion of a SEA spectrum, with impact assessment-based SEA on one end (distinguishing compliance-based SEA and the EIA-like SEA), and strategy-based SEA on the other end (distinguishing strategic futures SEA and strategic transitions SEA).

The purpose of SEA, how it functions, and what it is expected to deliver, are core aspects that distinguish SEA approaches across this spectrum, at each point mixing different features of each of the two extremes. What is expected with SEA includes a whole range of possibilities between fulfilling a legal obligation to getting strategic orientations for long-term development. Generally, all types of SEA share concerns with *participation*, with a fundamental role in *promoting sustainable development*, and with the need to ensure earlier *integration of environmental aspects in decision-making* processes.

But the two SEA models have fundamental differences. It is argued that the impact assessment model of SEA applies mostly to operational plans and programs, often designed as multiple projects, while the strategic thinking model of SEA applies to the formation and formulation

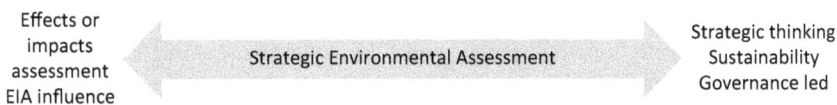

Effects or impacts assessment EIA influence	Strategic Environmental Assessment	Strategic thinking Sustainability Governance led

Figure 2.2 Spectrum of SEA perspectives (after Partidário 2005).

of strategies, irrespective of the decision level. Consequently, its specific ingredients can also be quite different. Each, in turn, will be addressed below (with core ingredients underlined).

SEA as a traditional impact assessment instrument

The concept of SEA as an environmental assessment instrument whose purpose is to integrate environmental considerations into policies, plans, and programs (PPP), and to inform decision-making on the environmental impacts/effects of PPP is vastly acknowledged. This concept is reflected in the first two formal definitions of SEA published in the literature (Lee and Walsh 1992; Thérivel et al. 1992). Subsequent definitions, such as that provided by the OECD-DAC (2006), perhaps the top-cited definition of SEA, make sustainability and participatory dimensions more explicit within the same traditional impact assessment logic (Box 2.2).

Box 2.2 Earlier and most popular definitions of SEA

Strategic environmental assessment (SEA) is the term used to describe the environmental assessment process for policies, plans, and programs which are approved earlier than the authorisation of individual projects.

(Lee and Walsh 1992, 126)

SEA can be defined as the formalised, systematic and comprehensive process of evaluating the environmental impacts of a policy, plan or programme and its alternatives, including the preparation of a written report on the findings of that evaluation, and using the findings in publicly accountable decision-making.

(Thérivel et al. 1992, 19–20)

This Guidance uses the term SEA to describe analytical and participatory approaches that aim to integrate environmental considerations into policies, plans and programs and evaluate the inter linkages with economic and social considerations.

(OECD-DAC 2006, 30)

Acknowledged as the "big brother" of EIA (an expression cast by Fischer 2003, 156), SEA perceived as the application of a project's EIA follows what Lynton Caldwell called "the anatomy of rational policy-making: analysis-assessment-decision" (Caldwell 1991), expressing the influence of the rational-technocratic paradigm in the theory and practice of SEA. "EIA for policies, plans, and programs – also known as strategic environmental assessment (SEA)" (Thérivel 1998, 39), among other similar quotes, would strengthen SEA as an EIA similar assessment.

Both legal frameworks and applications with SEA reveal the standard EIA pattern in the process, procedure and methods. The core process activities and steps of *screening, scoping, assessment, review, decision-making, and follow-up* are an outstanding symbol of this standard pattern. European Directive 2001/42/EC, as well as the United National Economic Commission for Europe (UNECE) Protocol on SEA, portray these activities within the SEA procedure. As of

May 9, 2021, on the Netherlands Commission for Environmental Assessment (NCEA) website, it was stated that the main objectives and key stages of EIA and SEA are similar; it is the scope and relevant stakeholders that can be quite different (https://www.eia.nl/en/our-work/why-esiasea/sea). In China, for example, Plan EIA (Zhu and Lam 2009), or the EIA applied to plans, perform exactly the same activities (2009 Plan EIA Regulations, https://www.eia.nl/en/countries/china/sea-profile).

Together with these standard activities, other key ingredients Feature the traditional impact assessment SEA model. It aims at a comprehensive collection of *baseline data*, consideration of *alternatives*, *predictions*, and *forecasts* of potential *environmental impacts* and the *mitigation* of identified and expected environmental effects. It thoroughly documents all the information through the formal delivery of a *report*, subsequently verified, or controlled by systematic *quality review* processes, before the decision to go ahead is taken. It requires public participation at certain stages of the process, desirably in the scoping stage and indispensable during the quality review. Its implementation is then followed by *monitoring* and *post-evaluation*.

Traditional impact assessment SEA is therefore carried out to inform decision-making by giving a detailed account of the current situation and its expected evolution and in the absence of the proposed plan or program, and assess the effect of program and planning proposals and their alternative solutions, and mitigate environmental effects. It is a *systematic* process, reactive to the formulation of planning and program intention but proactive to project decisions, therefore *anticipating the environmental assessment of development projects*.

SEA driven by strategic thinking

The strategic thinking model of SEA is a framework of activities that do not need detailed set procedures; instead, it is designed to be fully molded to the nature, timing, and dynamics of each decision-making process. Such an idea of SEA was also supported by other authors, for example, Thissen (2001) (Box 2.3). SEA in a strategic sense was first defined in Partidário (2000), eventually consolidating into a definition of strategic thinking SEA in Partidário (2012) (Box 2.3). This definition is intended to clearly mark a drift away from the EIA-based SEA, leaving behind techno-rationality and embracing more constructive and collaborative strategic philosophies (Partidário 2012, 2015, 2021).

Box 2.3 Definition of SEA with a more strategic thinking line of thought

Is SEA an instrument to safeguard environmental concerns in decision-making, or is it intended to foster sustainability, or to support balanced decision-making with respect to all normative views and interests concerned?

(Thissen 2001, 40)

SEA is defined as a strategic framework instrument that helps to create a development context towards sustainability, by integrating environment and sustainability issues in decision-making, assessing strategic development options and issuing guidelines to assist implementation.

(Partidário 2012, 11)

This strategic line of thought suggests that SEA must *act directly upon the process of formation and formulation of strategies* (whether in policies, plans, programs, or even major projects). Strategic thinking SEA aims to increase the capacity to *influence decision priorities* and facilitate environmental and sustainability integration in decision-making (Kørnøv and Thissen 2000; Sheate et al. 2003; Feldman and Khademian 2008). Such SEA is a facilitator of more integrated decisions, not a verifier that decisions are integrated. With the purpose of adding-value to decision-making, strategic thinking SEA reinforces its structural and strategic nature as a process framework, *concerned with paths and not with places*, bringing *strategic focus, flexibility*, and *adaptation* to development and assessment processes.

With strategic thinking, SEA strong motivation is to help build future contexts for development. For that, it needs to act as a governance exercise, engaging multiple relevant perspectives in frequent dialogues, using good communication, managing conflicts, and creating partnerships, acting within macro policy frameworks as drivers of strategic direction. Through multi-stakeholder dialogues and with decision-makers, SEA sets to understand the context and structure the problem, to ensure a strategic focus upon priorities that are collectively identified, to explore environmental and sustainable strategic options as possible pathways to address the problem and meet intended strategic sustainability objectives. The assessment of risks and opportunities of strategic options will help create environmental and sustainability contexts within which development proposals are sought. Planning, management and monitoring guidelines will assist the ongoing (cyclical) facilitation of the planning and assessment joint processes, as continuous dialogues throughout strategies implementation. All these are key ingredients in strategic thinking SEA.

Strategic thinking SEA is expected to perform a fundamentally new attitude in strategic development processes, establishing a relationship with the decision-making process, with a fresh and constructive look, centered in the strategic dimensions of the decisions to be taken. The key is to understand how decision processes function (Kørnøv and Thissen 2000; Nitz and Brown 2001) and what are core decision moments, what is it that decision-makers need to know, how to address their issues in their own timing (Lyhne et al. 2020). Such strategic decision moments (Partidário 1996) are decision windows for SEA, the right time to facilitate dialogues with core stakeholders, to bring in the information that is relevant and useful. SEA in a strategic sense is not only triggered by the formulation and approval of plans, policies or programs; it is triggered by a strategic problem that needs to be understood, structured, and addressed.

Under this model, SEA has an important transformative role (Partidário 2020); it must act as a change agent (Kørnøv 2020), less to inform about impacts and mitigate effects, and more to avoid impacts by searching for more integrated and sustainable development contexts and increased governance capacities (Monteiro et al. 2018; Partidário 2021).

What is SEA for?

The above paragraphs highlighted core differences between the two ends of the SEA spectrum, in synthesis suggesting that:

- Impact assessment SEA, in line with traditional approaches, aims to integrate environmental and social issues by informing decision-making about the effects of proposals – therefore being more appropriate to assess and mitigate the effects of proposed actions in more operational plans or programs that miss a strategic dimension;
- Strategic thinking SEA aims to integrate environmental and sustainability issues by facilitating decision-making in building future development directions that are strategic to the achievement of long-term sustainability vision and objectives – therefore being more appropriate to guide and support strategy formation and formulation, and assist in setting sustainability directions.

Reflecting on "what is SEA for?" requires exemplification on how impact assessment SEA and strategic thinking SEA unfold. This section briefly elaborates on the differences between doing a SEA of any initiative X following a traditional impact assessment model or following a strategic thinking model.

If we use a traditional impact assessment SEA: this is what we will get

- **What is SEA for (motivation of SEA)**: to provide environmental information regarding the policy planning system related to initiative X to support the identification of environmental impacts and propose specific mitigation measures of major interventions and implementation actions proposed by the initiative.
- **What will be assessed (object of assessment)**: alternative proposals of major interventions, implementation measures, and projects identified related to initiative X (policy and legislation, coordination with relevant sectors, cooperation and investment, and other initiative-specific related aspects).
- **How to assess (method, tools)**: through (a) characterization of measures or actions proposed by initiative X; (b) identification of expected significant environmental impacts/effects; (c) baseline information concerning major environmental issues; (d) definition of alternative proposals and analysis of environmental impacts; (e) identification of mitigation and management measures to reduce the magnitude of the negative environmental impacts; (f) consultation of key stakeholders; and (g) monitoring program to follow the operationalization of the strategy proposals (interventions and implementation measures and projects).
- **What are the outcomes of SEA (results of SEA):** To inform decision-makers, public authorities and the public in general documenting technical aspects concerning the environmental baseline, assessment and mitigation of effects of initiative X on the environment, delivering a final report.

But if we use strategic thinking SEA: this is what we will get

- **What is SEA for (motivation of SEA)**: to support the formulation of the initiative X strategy in order to meet its vision of a sustainable future, based on the added-value of the natural and social capital; engage in dialogues with core stakeholders throughout the process to explore broader matters and implications of initiative X within a sustainability framework.
- **What will be assessed (object of assessment)**: a range of strategic development options related to initiative X led by the future aspiration established in the Vision, namely in terms of capacity building, policy development, governance system, and other initiatives specific related aspects; assessment is driven by a strategic focus, with the collective views and inputs of core stakeholders.
- **How to assess (method, tools)**: within the framework of a problem-based approach, and in dialogue with stakeholders, (a) define a strategic focus based on long-term priorities of the initiative X related system, within a macro-policies and strategies framework; (b) develop an institutional analysis of the initiative X governance system; (c) identify strategic options and assess their risks and opportunities considering analysis of trends (strengths and weaknesses, conflicts and development potentials); and (d) suggest recommendations and guidelines for strategy implementation.
- **What are outcomes of SEA (results of SEA)**: an environmental and sustainability integrated decision process resulting from ongoing advice to decision-makers on technical aspects: (a) strategic priorities, (b) opportunities and risks, c) follow-up strategy; governance aspects: (a) institutional context, (b) capacity building, (c) multi-level integrated governmental and

governance systems; communication aspects: (a) communication and engagement strategies, (b) networking, (c) active participation and engagement; process aspects: earlier integration of environmental and sustainability priorities into the formulation of the initiative strategy.

SEA in decision-making

Reflecting on "what is SEA for?" also requires considerations on the relationship of SEA with decision-making. In order to be relevant to decision-making, SEA needs to target decision concerns and priorities and ensure it will bring an added-value to decision-makers. It is important that decision-makers recognize SEA as an ally, an approach that can bring benefits. But it does not happen that way all the time.

It is relatively easy for more strategic minds to recognize a role for SEA in decision processes, and the value it can bring. A relatively limited number of such cases have turned into successful SEA cases (REN/IST 2008; Partidário 2016; Ministério de Energía 2015; Rodriguez and Partidário 2021). But what added-value SEA can represent to actual decision-making is less evident for the project's driven mindsets, when compliance is usually the trigger.

This section looks at the role SEA is expected to play in decision-making and also at different models of relating SEA to decision processes. It will finally elaborate on how we can make a "business case" for SEA.

The advocacy role of SEA in decision-making

The advocacy role of SEA in decision-making can be categorized into three approaches (Box 2.4): a marginal approach, a compliance approach, and a constructive approach, as a function of the role played and outcome expected with SEA.

Box 2.4. SEA advocacy role in decision-making (adapted from Partidário 2009)

- A *marginal approach* – SEA is an end in itself (focus on the instrument); there is limited effort to understand its need and how it can support decision-making; SEA collects environmental and social baseline information, assesses the effects of proposals and delivers a formal report. The outcome of such SEA is marginal and likely to be irrelevant to decision-making.
- A *compliance approach* – SEA is a support instrument to decision-making; it assures legal and policy compliance; environmental and social baseline studies support effects assessment and formal procedures. The outcome of such a SEA will likely fulfill decision needs by providing evidence of legal compliance; the added-value of SEA as environmental and sustainability benefits result from fewer negative impacts and the enhancement of positive impacts.
- A *constructive approach* – SEA is designed to help set direction to strategies, and create development contexts; SEA priority is to understand the complexity of decision-making, its needs and priorities, and to assist a mutual, collective learning process about how environmental and sustainability issues can be constructively built into decision-making. SEA outcomes are embedded in the decision-making cycle, inputs are made at key moments (decision windows), and SEA is an integral part of the decision process.

Drawing on observations in practice, these categories are still valid a decade later. A mix of compliance and constructive approaches can be observed when SEA is legally mandated; going beyond what is required, for example, when used to assist policy and planning in exploring future options. But the marginal and compliance approach still appears to be the most common approach, which may be responsible for much of the frustrations with the application of SEA.

As stated by Bidstrup and Hansen (2014, 34): "SEA is performing as a non-strategic tool, failing on its inherent promise". This is when SEA is reactive to decision intentions, is dominated by extensive baseline descriptions, provides data and information but very little analysis and even less advice to decision-making, when it offers a short-term view of effects, is report-driven, and a necessary standard process to obtain permits. This practice of SEA carries many of the burdens and limitations that it was supposed to resolve.

SEA relationship with decision-making

A taxonomy of four models on the SEA relationship with decision-making were theoretically conceptualized almost two decades ago (Partidário 2004, 2007). These models aimed to represent theoretical ways of designing SEA to fit decision-making (Figure 2.3). Models 1 and 2 related more to the traditional impact assessment SEA and models 3 and 4 to strategic thinking SEA.

Designing SEA to fit decision-making as a single opportunity model (model 1) would most probably deliver a marginal approach to SEA and clearly illustrates bad practice. Regrettably, it is still observable in practice. On the other hand, a full integration (model 3) could be considered the ideal form of designing SEA to fit decision-making in a constructive way, arguably perhaps the best model to aim for. But as it stands, given that SEA is yet far from a full acknowledgment of its added-value by decision-makers, most likely model 3 SEA contribution might be anecdotal in the near future.

The two models that seem to better represent current SEA theory and practice are models 2 and 4. In the parallel model (model 2) SEA inherits the basic elements of traditional impact assessment practices, activities, and processes to ensure that environmental issues are considered in decision-making, by assessing and validating the environmental quality of proposals in policy, planning, and program development. Model 2 is in line with the compliance approach, and may, in best practice cases, somehow reveal constructive approaches.

In the decision-centered model (model 4), SEA has no formal procedure or standard process; it is used strategically to fit decisions of a strategic nature. SEA is shaped as a framework of key functions and activities that have to be structured and designed as a process, but tailor-made to each case that it applies to. It cannot therefore be defined as a standard procedure or as a universal streamlined process. Each case has its particularities and, once objectives are set, the decision-making process leads the way, and SEA becomes fully adjusted to the dynamics of the

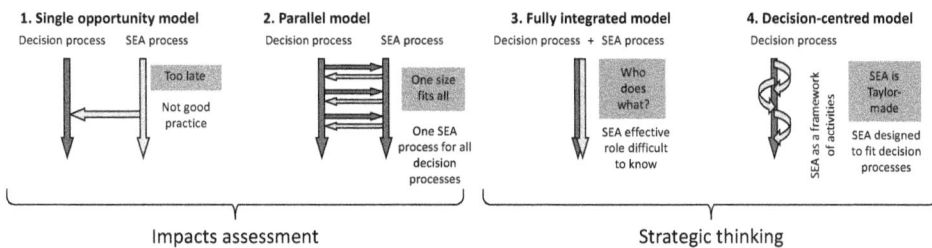

Figure 2.3 Design SEA to fit decision-making – four models (adapted from Partidário 2004, 2007).

decision process. This way, SEA increases its flexibility, can be made fit-for-purpose, and will be able to act as a facilitator of the strategic decision-making process.

The business case for SEA

Practice shows that across the world, with few exceptions, SEA seems to be touching the limits of its credibility. Increasingly professionals ask the question: what is SEA for after all? With SEA being so similar to EIA, what are differences other than the level to which it applies? Sometimes SEA may be seen only as "an easy taking" instrument when compared to EIA, with no real consequences, the perception being that it may be rather useless. So, what is the point of doing SEA?

Currently, this appears to be the dominant perception about SEA, particularly in Europe, but also wherever the impact-based and mitigation model has been preponderant (Noble et al. 2019; Partidário and Monteiro 2019). Isolated discussions with planners and EIA practitioners have revealed that, on the one hand, SEA delivers nothing because it does not get down to the actual details that require attention. But on the other hand, it is useless as it does not implicate any hard decisions with serious investment implications. The conclusion is that SEA does not serve short-term interests, and therefore decision-makers are not interested. Ultimately SEA serves for legal compliance; it does not implicate development decisions, while strategic decisions "are too strategic" to be assessed. At best, SEA is easy doing because it can get satisfied with much less (detailed) information and (often) it serves to overcome the need to do project's EIA.

It can be argued that the absence, or the lack, of strategic decision-making in general has been responsible for the difficulty in applying a truly strategic SEA. Largely, decisions that are taken under the environmental assessment realm are short term. Where there are strategic decisions, decision-makers do not want to discuss their intentions or have them challenged.

As argued, the value of SEA is a function of the extent it influences, and adds-value, to decision-making (Partidário 2000). Hence, the dissociation between the term and the function of SEA is not helping. One form of making SEA more useful is to adopt it as a framework to facilitate decision-making, with core elements as building blocks strategically placed in the policy, planning, and programmatic decision-making process, at strategic moments, as in model 4 (Figure 2.3) (Partidário 2007, 2012). Acting as a facilitator, with a flexible framework that molds to different governance contexts, SEA will increase its capacity to influence decision-making and will perform mostly as a governance exercise (Rodriguez and Partidário 2021). Like this, SEA will increase the chances of making a difference, becoming an added-value to sustainable decision-making.

Looking ahead – innovative forms of using SEA

The future of SEA is being challenged by the emergence of persistent problems that threaten societal systems (Partidário 2020). The Rockström et al. (2009) nine planetary boundaries are a notable reference, but also the world megatrends, of which the 11 European Environmental Agency megatrends (EEA 2015) are one, perhaps more comprehensive, example. Addressing these planetary trends and their major challenges require multi-scaling, -level, -actor and -sector, integrated and strategic, assessment approaches.

Emerging themes

In the last decade, scientific and Grey literature on SEA has acknowledged the potential role SEA can play in addressing global challenges, in particular those generated by climate change

(EC 2013; Larsen et al. 2012; among others), but also in promoting territorial-based ecosystem services in multi-actor contexts (Slootweg and Beukering 2008; EC 2013; Geneletti 2015; Rosas-Vasquez et al. 2019; among others) and in placing a greater focus on improved health conditions (WHO 2009; Fischer et al. 2010). The recently published *Handbook on SEA* (Fischer and González 2021) reserves different chapters to climate considerations, ecosystem services and health as part of the integrative scope of SEA.

The broad time and spatial scale, and strategic focus, of SEA makes it an ideal instrument to proactively advance climate-proof policy and planning actions, as well as creating policy and planning conditions that enhance healthy environments and the promotion of ecosystem services. Practice is still evolving, but if (and when) well adopted and implemented, with a truly strategic nature, SEA can become a very useful, value-added instrument in addressing associated concerns. Situational problems related to these themes can become triggers to the application of SEA, without depending on the existence of any policy, plan, or program; instead, it will be SEA generating the need for subsequent policies, plans and programs. This will bring significant change to the practice of SEA.

Another theme that is making its way into the SEA literature concerns the role of SEA in the achievement of sustainable development goals (SDG) (Nilsson and Persson 2017; IAIA 2019; Hacking 2019; González del Campo et al. 2020; Kørnøv et al. 2020). Nilsson and Person (2017) clearly recognized SDGs role in SEA "as a reference framework to define assessment criteria [...] so that SDGs and targets form an internationally harmonized and universal basis for impact assessment criteria" (Nilsson and Person 2017, 38).

Kørnøv et al. (2020) developed a framework to evaluate the extent to which SDG was being considered in environmental assessment in general, whether simply mentioned or if actually being part of the assessment. But while the role that SEA can, and should, play in relation to SDG is acknowledged, SEA needs to be more than a "mere mechanic measurement of indicators in assessing performances [...] and consider the interconnectedness of the 17 SDGs in a systemic way" (Partidário 2020, 3) enhancing the integration of the indivisible nature of SDG in policy and planning.

Strategic thinking-driven methodologies in SEA

Innovative forms of using SEA include strategic thinking approaches in SEA. A methodology for using strategic thinking for sustainability (ST4S) in SEA was proposed in Partidario (2007) and recently updated (Partidário 2021). This methodology offers key ingredients to enable coping with the complexity and the uncertainty brought forward by the emerging planetary trends: a strategic focus associated with a long-term vision, built over collaborative dialogues; a strategic discussion around future development options seeking to build futures, creating new conditions for more sustainable development; an ongoing learning process, well interconnected to policy and planning, aimed at creating technical and governance capacities to enable a sociopolitical process, with quick adjustments to emerging and unexpected events.

Strategic thinking SEA aims to enhance the transformative role of SEA toward sustainability, for which it requires recognition as a continuous learning and collaborative process based on dialogues, a governance exercise to assist strategic decision-making. In short, a good SEA, to act effectively, requires:

- that its added-value be recognized by committed top-level decision-makers, engaging responsible policy and planning authorities to take ownership of SEA, from congress or parliamentary politicians, to local, regional, and national government officers, and also corporate decision-makers – they need to be convinced and committed that SEA is an instrument that can bring benefits by enabling environment and sustainability priorities in development processes;

- that SEA flows continuously, aligned with strategic decision-making from initial stages, such as objectives and agenda-setting, throughout the implementation of strategies, including when these are revisited or re-aligned to be launched again, exactly with the purpose of supporting strategic decision-making;
- professional capacities for long-term, strategic and systems thinking, detached from the traditional EIA and project-driven mindsets, with social and political concerns targeting broad, strategic, resources planning and management, and cumulative processes;
- SEA as a governance exercise, more concerned with enabling agile decisions, reflection and re-alignment, considering options for sound and sustainable decisions, through coherent and timely advice.

Eventually, several concepts initially proposed by strategic thinking SEA have been swiftly adopted and are becoming part of SEA practice:

- the notion of SEA as a framework that facilitates decision-making, as opposed to formalized procedures;
- the notion of strategic options, as opposed to more operational alternatives;
- the notion of assessment of risks and opportunities, as opposed to the assessment of effects.

Strategic thinking SEA, with the ST4S methodology, has been used in several countries (e.g., Chile, Indonesia, Perú, Portugal) and is part of academic curricula in countries like Portugal and Sweden. A review of its application in Perú revealed the following benefits (Rodriguez and Partidário 2021):

- it improves the strategic dimension and content of plans;
- it amplifies the scope of considered themes toward sustainability and toward greater integration;
- the methodology leads to a broader focus beyond effects and impacts (as in EIA-based SEA);
- institutional coordination between public decisional organizations results improves, and creates new practices, opposed to silos thinking.

The future of SEA

This chapter presented SEA as a complex instrument, with multiple understandings and forms of application. Influenced by two dominant inspirations, SEA is now seen to unfold in different ways along a spectrum of multiple forms and functions, ranging from operational and management to strategic driven approaches. This spectrum also suggests a mix of different purposes with SEA, sometimes apparently contradictory and generating some confusion. This chapter intended to clarify ways of distinguishing the possible uses of different types of SEA but without the purpose of inventorying all acknowledged forms of SEA.

It appears that the earlier reasons that justified the rise of SEA are still valid – to enable earlier integration of environmental issues and set conditions that can improve the assessment of development projects. But in doing so, SEA is often bordering the line, and eventually getting rather mixed up, with EIA. It seems clear that the assessment of the environmental effects of individual development projects, or of sets of projects, should be kept outside of the SEA periscope. For one, because EIA exists as a powerful, concrete, well-tested instrument to play that role. And second, because SEA is needed to support development with broader, forward-looking perspectives, enhancing policy commitments to sustainable development.

The spectrum of possible SEA types reveals that an assessment instrument in the lines of the earlier notion of an EIA-type system appears to continue to make sense. Such traditional impact assessment-based SEA type instrument is more adequate to assess the positive and negative environmental and social effects of operational plans and programs when concrete actions are designed, anticipating development projects. This type of assessment can apply to different geographical scales, as local or regional assessments (see Chapter 10 in this book), also enabling the assessment of the cumulative effects of sets of projects (see Chapter 3 in this book). In this type of SEA, the strategic dimension is limited or totally absent. Traditional impact assessment-based SEA is more coherent with the process, methods, and expected outcomes (mitigation and documented information) that have been used since early days (it could even be named EIA in plan-making, or Programme EIA).

But a SEA that has a full strategic nature, and is more forward-looking and proactive, is also claimed as being needed. Such SEA works with policy and planning within strategic decision-making to help build futures and provide strategic direction. Using strategic thinking, SEA enables the creation of more sustainable development contexts, this way enhancing a strategic integration of environmental and sustainability priorities in development processes, before concrete actions are formulated in operational plans and programs. Such type of strategic thinking SEA searches priorities instead of effects, strategic options toward solutions, and is based on learning processes, constructive dialogues, and other governance mechanisms, engaging multi-stakeholders from early moments of the decision process, following a long-term vision.

The challenges created by global trends, such as climate change, biodiversity loss, food crisis, or poverty eradication, as well as the need to meet SDG and support transitions to sustainability, demand a strategic thinking SEA that can cope with the geographic and time scales of these grand societal problems. SEA needs to look largely beyond the assessment of effects and act more as an instrument that helps to pave the way for leading transformations, assuming the role of change agent. Traditional forms of SEA that persist in being essentially palliative, with the leading purpose of alleviating projects' negative impacts, assisting relatively short-term decisions and limited visions, will hardly have the capacity to embrace the increased complexity that permeates decisions on future actions.

SEA strategic hidden potential is still waiting to be unleashed. SEA needs to be recognized as an ongoing learning process that flows with, and throughout, policy and planning continued processes. Such SEA is concerned with paths and not with places. The strategic nature and capacity of SEA should be applied to governance processes that lead toward decisions and not to a decision per se. In short and finally, SEA should act as a framework to advance knowledge and practice, to help create contexts for sustainability, facilitating and being integral to decision processes through good governance.

References

Ahmed, Kulsum, and Ernesto Sánchez-Triana (Eds). 2008. *Strategic Environmental Assessment for Policies: An Instrument for Good Governance.* Washington D.C.: World Bank.

Azcárate, Juan. 2015. *Beyond Impacts: Contextualizing Strategic Environmental Assessment to Foster the Inclusion of Multiple Values in Strategic Planning.* PhD Thesis. Royal Institute of Technology, Sweden.

Bidstrup, Morten, and Anne M. Hansen. 2014. "The paradox of strategic environmental assessment". *Environmental Impact Assessment Review* 47: 29–35.

Bina, Olivia. 2007. "A critical review of the dominant lines of argumentation on the need for strategic environmental assessment". *Environmental Impact Assessment Review* 27(7): 585–606.

Bina, Olivia. 2003. *Re-Thinking the Purpose of SEA*. PhD thesis, Department of Geography, Cambridge University. Cambridge.

Boothroyd, Peter. 1995. "Policy assessment". In *Environmental and Social Impact Assessment*, edited by Frank Vanclay and Daniel Bronstein, 83–126. Chichester: John Wiley.

Caldwell, Lynton K. 1991. "Analysis-assessment-decision: The anatomy of rational policy making". *Impact Assessment Bulletin* 9: 81–92.

Cherp, A., A. Watt, and V. Vinichenko. 2007. "SEA and strategy formation theories – From three Ps to five Ps". *Environmental Impact Assessment Review* 27: 624–644.

Clark, Ray. 2000. "Making EIA count in decision-making". In *Perspectives in Strategic Environmental Assessment*, edited by Partidário and Clark, 15–27. New York: Lewis Publishers.

DETR (Department of the Environment, Transports and the Regions). 1999. *Proposals for a Good Practice Guide on Sustainability Appraisal of Regional Planning Guidance*. London: DETR.

EC (European Commission). 2013. *Guidance on Integrating Climate Change and Biodiversity into Strategic Environmental Assessment*. Brussels: European Commission.

EEA (European Environmental Agency). 2015. *The European Environment State and Outlook 2015 – Assessment of Global Megatrends*. Copenhagen: European Environmental Agency.

Feldman, Martha S., and Anne M. Khademian. 2008. "The continuous process of policy formation". In *Strategic Environmental Assessment for Policies*, edited by Kulsum Ahmed and Ernesto Sanchéz-Triana, 37–59. Washington, DC: The World Bank.

Fischer, Thomas. 2003. "Strategic environmental assessment in post-modern times". *Environmental Impact Assessment Review* 23(2): 155–170.

Fischer, Thomas, and Ainhoa Gonzalez. 2021. *Handbook of Strategic Environmental Assessment*. Cheltenham: Edward Elgar.

Fischer Thomas, B., Marco Matuzzi, and Julia Nowacki. 2010. "The consideration of health in strategic environmental assessment (SEA)". *Environmental Impact Assessment Review* 30(3): 200–210.

Geneletti, Davide. 2015. "A conceptual approach to promote the integration of ecosystem services in strategic environmental assessment". *Journal of Environmental Assessment, Policy and Management* 17: 1550035.

Gibson, R. B. 2006. "Sustainability assessment: Basic components of a practical approach". *Impact Assessment and Project Appraisal* 24(3): 170–182.

Gibson, Robert B., S. Hassan, S. Holtz, J. Tansey, and G. Whitelaw. 2005. *Sustainability Assessment: Criteria, Processes and Applications*. London: Earthscan.

González Del Campo, Ainhoa, Paola Gazzola, and Vincent Onyango. 2020. "The mutualism of strategic environmental assessment and sustainable development goals". *Environmental Impact Assessment Review* 82: 106383. https://doi.org/10.1016/j.eiar.2020.106383.

Hacking, Theo. 2019. "The SDGs and the sustainability assessment of private-sector projects: Theoretical conceptualisation and comparison with current practice using the case study of the Asian Development Bank". *Impact Assessment and Project Appraisal* 37(1): 2–16. https://doi.org/10.1080/14615517.2018.1477469.

Hacking, Theo, and Peter Guthrie. 2008. "A framework for clarifying the meaning of Triple Bottom-Line, Integrated, and Sustainability Assessment". *Environmental Impact Assessment Review* 28(2–3): 73–89.

Hayes, Stephen. 2019. "It's good to talk: Dialogue between strategic environmental assessment and plan-making". *Town Planning Review* 90(1): 57–79. https://doi.org/10.3828/tpr.2019.5.

IAIA. 2019. *Impact Assessment and the Sustainable Development Goals (SDGs) – IAIA fasTips Series Number 19*. Maria Partidário, Rob Verheem (Eds). International Association for Impact Assessment. https://www.iaia.org/uploads/pdf/Fastips_19%20SDGs.pdf.

Kørnøv, Lone. 2020. "SEA as a change agent: Still relevant and how to stay relevant?" *Impact Assessment and Project Appraisal* 39(1): 63–66.

Kørnøv, Lone, Ivar Lyhne, and Juanita G. Davila. 2020. "Linking the UN SDGs and environmental assessment: Towards a conceptual framework". *Environmental Impact Assessment Review* 85: 106463. https://doi.org/10.1016/j.eiar.2020.106463.

Kørnøv, Lone, and Wil Thissen. 2000. "Rationality in decision- and policy-making: Implications for strategic environmental assessment". *Impact Assessment and Project Appraisal* 18(3): 191–200.

Larsen, Sanne V., Lone Kørnøv, and A. Wejs. 2012. "Mind the gap in SEA: An institutional perspective on why assessment of synergies among climate change mitigation, adaptation and other policy areas are missing". *Environmental Impact Assessment Review* 33: 32–40.

Lee, Norman, and Fiona Walsh. 1992. "Strategic environmental assessment: An overview". *Project Appraisal* 7(3): 126–136.

Lee, Norman, and Christopher Wood. 1978. "EIA – A European perspective". *Built Environment* 4(2): 101–110.

Lobos, Víctor, and Maria Partidário. 2014. "Theory versus practice in Strategic Environmental Assessment (SEA)". *Environmental Impact Assessment Review* 48: 34–46.

Lyhne, Ivar, Maria R. Partidário, and Lone Kørnøv. 2021. "Just so that we don't miss it: A critical view on the meaning of decision in IA". *Environmental Impact Assessment Review* 86: 106500. https://doi.org/10.1016/j.eiar.2020.106500.

McCluskey, Daniel and Elsa João. 2011. "The promotion of environmental enhancement in Strategic Environmental Assessment". *Environmental Impact Assessment Review* 31(3): 344–351.

Ministério de Energía (Gobierno de Chile). 2015. *La evaluación ambiental estratégica en la política energética 2050.* http://www.minenergia.cl/archivos_bajar/ucom/publicaciones/EAE4_web.pdf (Accessed on May 2021).

Monteiro, Margarida B., Maria R. Partidário, and Louis Meuleman. 2018. "A comparative analysis on how different governance contexts may influence Strategic Environmental Assessment". *Environmental Impact Assessment Review* 72: 79–87.

Nilsson, M., and A. Persson. 2017. "Policy note: Lessons from environmental policy integration for the implementation of the 2030 Agenda". *Environmental Science and Policy* 78: 36–39. https://doi.org/10.1016/j.envsci.2017.09.003.

Nilsson, Mans, and Holger Dalkmann. 2001. "Decision making and strategic environmental assessment". *Journal of Environmental Assessment, Policy and Management* 3(3): 305–327.

Nitz, Tracey, and A. Lex Brown. 2001. "SEA must learn how policy making works". *Journal of Environmental Assessment, Policy and Management* 3(3): 329–342.

Noble, Bram, Robert Gibson, Lisa White, Jill Blakley, Kelechi Nwanekezie, and Peter Croal. 2019. "Effectiveness of strategic environmental assessment in Canada under directive-based and informal practice". *Impact Assessment and Project Appraisal* 37(3–4): 344–355.

Noble, Bram, and Kelechi Nwanekezie. 2017. "Conceptualizing strategic environmental assessment". *Environmental Impact Assessment Review* 62: 165–173.

Noble, Bram F. 2000. "Strategic environmental assessment: What is it and what makes it strategic?" *Journal of Environmental Assessment, Policy and Management* 2(2): 203–224.

Noble, Bram F. 2009. "Promise and dismay: The state of strategic environmental assessment systems and practices in Canada". *Environmental Impact Assessment Review* 29: 66–75.

Noble, Bram F. and Jill Gunn. 2015. "Strategic environmental assessment". In *Environmental Assessment in Canada: Practice and Participation,* edited by Kevin Hanna. pp 96–121. Don Mills: Oxford University Press.

Nooteboom, Sibout. 2006. *Adaptive Networks. The Governance for Sustainable Development.* Rotterdam: Erasmus University Rotterdam/DHV.

OECD-DAC. 2006. *Applying Strategic Environmental Assessment. Good Practice Guidance for Development Co-operation.* Paris: OECD.

O'Riordan, Timothy, and W. R. Derrick Sewell. 1981. *Project Appraisal and Policy Review.* Chichester: John Wiley & Sons.

Partidário, M. R., and W. R. Sheate. 2013. "Knowledge brokerage – potential for increased capacities and shared power in impact assessment". *Environmental Impact Assessment Review* 39: 26–36.

Partidário, Maria R. 1993. "Anticipation in environmental assessment: Recent trends at the policy and planning levels". *Impact Assessment* 11(1): 27–44.

Partidário, Maria R. 1996. "Strategic environmental assessment: Key issues emerging from recent practice". *Environmental Impact Assessment Review* 16: 31–55.

Partidário, Maria R. 1999. "Strategic environmental assessment – principles and potential". In *Handbook of Environmental Impact Assessment,* edited by Judith Petts, vol. 1., 60–73. Oxford: Blackwell.

Partidário, Maria R. 2000. "Elements of an SEA framework – improving the added-value of SEA". *Environmental Impact Assessment Review* 20: 647–663.

Partidário, Maria R. 2004. "Designing SEA to fit decision-making". Paper presented at the *24th Annual conference of the International Association for Impact Assessment*, Vancouver, BC.

Partidário, Maria R. 2005. "Future challenges of strategic environmental assessment". In *III Congreso Nacional de Evaluación de Impacto Ambiental*. Pamplona: Associación Española de Evaluación de Impacto Ambiental (informal paper).

Partidário, Maria R. 2007. "Scales and associated data - What is enough for SEA needs?". *Environmental Impact Assessment Review* 27(5): 460–478.

Partidário, Maria R. 2009. "Does SEA change outcomes?" Discussion Paper Number 2009-31. Joint Transport Research Centre, OECD. December 2009. https://www.oecd-ilibrary.org/transport/does-sea-change-outcomes_5kmmnc5ln3r0-en.

Partidário, Maria R. 2012. *Strategic Environmental Assessment Better Practice Guide - Methodological Guidance for Strategic Thinking in SEA*. Lisboa: Agência Portuguesa do Ambiente e Redes Energéticas Nacionais.

Partidário, Maria R. 2015. "A strategic advocacy role in SEA for sustainability". *Journal of Environmental Assessment, Policy and Management* 17(1): 1550015.

Partidário, Maria R. 2016. "Using strategic thinking and critical decision factors to achieve sustainability". In *Sustainability Assessment: Applications and Opportunities*, edited by Robert B. Gibson, 169–193. London: Earthscan.

Partidário, Maria R. 2020. "Transforming the capacity of impact assessment to address persistent global problems". *Impact Assessment and Project Appraisal* 38(2): 146–150. https://doi.org/10.1080/14615517.2020.1724005.

Partidário, Maria R. 2021. "Strategic thinking for sustainability in SEA". In *Handbook on Strategic Environmental Assessment*, edited by Thomas Fischer and Ainhoa Gonzalez, 41–57. Cheltenham: Edward Elgar Research Handbooks of Impact Assessment Series.

Partidário, Maria R., and Margarida B. Monteiro. 2019. "Strategic environmental assessment effectiveness in Portugal". *Impact Assessment and Project Appraisal*, 37(3–4): 247–265. https://doi.org.10.1080/14615517.2018.1558746.

Partidário, Maria R., and Filipe Moura. 2000. "Strategic sustainability appraisal – One way of using SEA in the move towards sustainability". In *Perspectives on Strategic Environmental Assessment*, edited by Maria R. Partidário and Ray Clark, 29–43. New York: CRC Press/Lewis Publishers.

Polido, Alexandra, and Tomás Ramos. 2015. "Towards effective scoping in strategic environmental assessment". *Impact Assessment and Project Appraisal* 33: 1–13. https://doi.org/10.1080/14615517.2014.993155.

Pope, Jenny, David Annandale, and Angus Morrison-Saunders. 2004. "Conceptualising sustainability assessment". *Environmental Impact Assessment Review* 24: 595–616.

REN/IST (Rede Eléctrica Nacional, S.A. / Instituto Superior Técnico). 2008. *Avaliação Ambiental Estratégica do Plano de Desenvolvimento e Investimento da Rede Nacional de Transporte de Electricidade 2009–2014 (2019) – Relatório Ambiental (Strategic Environmental Assessment of the National Transmission Grid Development and Investment Plan 2009-2014 (2019)*. Environmental Report. www.ren.pt, and at http://sensuist.pt.vu/, only in Portuguese.

Rockström, J., W. Steffen, K. Noone, Å. Persson, F. S. I. Chapin, E. Lambin, T. M. Lenton, M. Scheffer, C. Folke, H. J. Schellnhuber et al. 2009. "Planetary boundaries: Exploring the safe operating space for humanity". *Ecologic Society*, 14(2): 32.

Rodriguez, Juan J., and Maria R. Partidário. 2021. "Strategic thinking SEA in Perú – Results from recent pilot cases". In *IAIA21 Smartening Impact Assessment in Challenging Times Virtual Event*. May 18–21, 2021. https://conferences.iaia.org/2021/draft-papers.php.

Rozas-Vásquez, Daniel, Christine Fürst, Davide Geneletti, and O. Almendra. 2018. "Integration of ecosystem services in strategic environmental assessment across spatial planning scales". *Land Use Policy* 71: 303–310.

Sadler, Barry. 1994. "Environmental assessment and development policy-making". In *Environmental Assessment and Development*, World Bank Symposium, edited by Robert Goodland and V. Edmundson. Washington, DC: The World Bank

Sadler, Barry. 1999. "Environmental sustainability assessment and assurance". In *Handbook of Environmental Impact Assessment*, vol. 1, edited by Judith Petts, 12–32. Oxford: Blackwell.

Sadler, Barry, and Jiri Dusik (Eds). 2016. *European and International Experiences of Strategic Environmental Assessment: Recent Progress and Future Prospects*. London: Routledge.

Sadler, Barry, and Rob Verheem. 1996. *Strategic Environmental Assessment – Status, Challenges and Future Directions*. The Hague: Ministry of Housing, Spatial Planning and the Environment of The Netherlands.

Sheate, William R., Suzan Dagg, Jeremey Richardson, Ralf Aschemann, Juan Palerm, and Ulla Steen. 2003. "Integrating the environment into strategic decision-making: Conceptualizing policy SEA". *European Environment* 13: 1–18.

Sheate, William R., and Maria R. Partidário. 2010. "Strategic approaches and assessment techniques – Potential for knowledge brokerage towards sustainability". *Environmental Impact Assessment Review* 30: 278–288. https://doi.org/10.1016/j.eiar.2009.10.003.

Silva, A. W. L., P. M. Selig, A. A. Lerípio, and C. V. Viegas. 2014. "Strategic environmental assessment: One concept, multiple definitions". *International Journal Innovation Sustainable Development* 8(1): 53–76.

Stoeglehner, Gernot. 2019. "Strategicness – The core issue of environmental planning and assessment of the 21st century". *Impact Assessment and Project Appraisal* 38(2): 1–5. https://doi.org/10.1080/14615517.2019.1678969.

Tetlow, Monica, and Marie Hanusch. 2012. "Strategic environmental assessment: The state of the art". *Impact Assessment and Project Appraisal* 30: 15–24.

Thérivel, Riki. 1993. "Systems of strategic environmental assessment". *Environmental Impact Assessment Review* 13: 145–168.

Thérivel, Riki. 1998. "Strategic environmental assessment of development plans in Great Britain". *Environmental Impact Assessment Review* 18: 39–57.

Thérivel, Riki. 2004. *Strategic Environmental Assessment in Action*. London: Earthscan.

Thérivel, Riki, E. Wilson, S. Thompson, D. Heany, and D. Pritchard. 1992. *Strategic Environmental Assessment*. London: Earthscan Publications Ltd.

Thissen, W., et al. 2001. "Strategic environmental assessment and policy: Developments and challenges". In *Environmental Assessment Yearbook 2001*, edited by L. Billing. Manchester: The EIA Centre, University of Manchester.

Tonk, Jap, and Rob Verheem. 1998. "Integrating the environment in strategic decision-making – one concept, multiple forms". *Paper presented to the 18th Annual Conference of IAIA*. Christchurch: IAIA.

van Beukering, Peter J. H., Roel Slootweg, and D. Immerzeel. 2008. *Valuation of Ecosystem Services and Strategic Environmental Assessment – Influential Case Studies*. Utrecht, The Netherlands: Commission for Environmental Assessment.

Verheem, Rob, and Jiri Dusik. 2011. "A hitchhiker's guide to SEA: Are we on the same planet?" Opening plenary. *IAIA Special Conference on SEA*. September 21–23, Prague.

Wallington, Tabatha. 2002. *Civic Environmental Pragmatism – A Dialogical Framework for Strategic Environmental Assessment*, PhD Thesis. Murdoch University, Murdoch, Australia.

Wallington, Tabatha, Olivia Bina, and Wil Thissen. 2007. "Theorising strategic environmental assessment: Fresh perspectives and future challenges". *Environmental Impact Assessment Review* 27(7): 569–584.

White, Lisa, and Bram F. Noble. 2013. "Strategic environmental assessment for sustainability: A review of a decade of academic research". *Environmental Impact Assessment Review* 42: 60–66.

WHO (World Health Organization). 2009. *Health and Strategic Environmental Assessment*. Rome: WHO Consultation Meeting.

Wood, Christopher. 1988. "EIA in plan making". In *Environmental Impact Assessment – Theory and Practice*, edited by Peter Wathern, 98–114. Londres: Unwin Hyman.

Wood, Christopher, and Mohammed Djeddour. 1989. *Environmental Assessment of Policies, Plans and Programmes*. Interim report to the Commission of European Communities. EIA Centre, University of Manchester (final report submitted 1990, Contract N.° B6617-571-572-89). http://ec.europa.eu/environment/eia/.

Zhu, T., and K. C. Lam (Eds). 2009. *Environmental Impact Assessment in China*. Research Center for Strategic Environmental Assessment, Nankai University, China and Centre of Strategic Environmental Assessment for China, The Chinese University of Hong Kong. Available from: https://www.researchgate.net/publication/268503532_Environmental_Impact_Assessment_in_China (accessed May 4, 2021).

3

CUMULATIVE EFFECTS ASSESSMENT

Bram Noble

Introduction

Cumulative effects (CEs) are one of the most pervasive and challenging aspects of environmental assessment and management (Cronmiller and Noble 2018). Much of the concern about cumulative effects is triggered when large resource development projects are proposed, such as mining operations, hydroelectric dams, or major energy pipelines; yet, the cumulative effects issues that emerge usually extend beyond the project at hand, and concern the legacy effects of past developments, or the implications of the proposal for such matters as biodiversity, climate change, and sustainable futures. Increasingly, project proponents are expected to address the cumulative effects of their proposed undertakings – but understanding and managing cumulative effects are not tasks that can be easily achieved within the scope of a single project. Cumulative effects problems by nature are complex and, as Canter and Ross (2010) note, they require cumulative solutions.

Notwithstanding the rich history of cumulative effects research and evolving legislation, there remains a dissatisfaction with the state of cumulative effects assessment (CEA) practice globally (Bidstrup, Kørnøv, and Partidário 2016; Foley et al. 2017). Multiple reasons have been offered, including weak legislation and guidance (Connelly 2011; Wärnbäck and Hilding-Rydevik 2009), the lack of science-based considerations in decision-making (Schindler 2006; Greig and Duinker 2011), the provision of cumulative effects science that does not meet the needs of regulatory decision-makers (Wong, Noble, and Hanna 2019), and the project-focused environment in which cumulative effects are typically assessed (Duinker and Greig 2006). These enduring challenges to CEA are coupled with diverse understandings of what it means to assess cumulative effects, and thus the expectations for their consideration in impact management and decision processes (Arnold, Hanna, and Noble 2019; Hegmann and Yarranton 2011; Jones 2016).

This chapter explores the nature of cumulative effects and the fundamental principles of good CEA. The characteristics of cumulative effects are first explored, followed by foundational principles for assessing cumulative effects. A framework for CEA is then presented, along with an in-depth case study of CEA in action. This chapter is intended to be instructional, but not prescriptive, and to reinforce several good-practice principles for conducting CEA. It also highlights some of the enduring challenges to CEA, and opportunities for improvement.

DOI: 10.4324/9780429282492-4

Understanding cumulative effects

An effect is a change in the condition (e.g., structure, function) of a system, or a specific indicator, relative to some determined benchmark, target, range of natural variation, or other point of reference (Dubé et al. 2006; Kilgour et al. 2007). Effects arise from the stress of human-caused disturbances and natural processes. Effects can accumulate, or become cumulative, due to the repeated actions of a single stressor because of interactions between stressors over time or across space (Jones 2016).

The effects caused by an individual stressor, such as a single project or land use, can seem inconsequential when considered in isolation, or when compared to the impacts of other, much larger projects and disturbances. However, important to understanding cumulative effects is recognizing that the effects caused by a single project or activity do not occur in isolation – they add to, interact with, or amplify the effects of other disturbances and accumulate over multiple spatial scales and time frames (Noble 2020; Squires and Dubé 2013). The most devastating effects in ecological systems do not always result from a single, identifiable project or action, but from the combination of past effects, existing stresses, and the individually minor effects of disturbances over time (Clarke 1994; IFC 2013).

There is no universal definition of cumulative effects (Wärnbäck and Hilding-Rydevik 2008), and legislative and regulatory descriptions of cumulative effects vary (Jones 2016). The first formal reference to "cumulative effects" in environmental assessment regulations was under the United States National Environmental Policy Act, whereby the United States' Council on Environmental Quality (US-CEQ 1978) guidelines defined a cumulative effect (referred to as cumulative *impact*) as an "impact on the environment which results from the incremental impact of the action when added to other past, present, and reasonably foreseeable future actions" and that "cumulative impacts can result from individually minor, but collectively significant, actions taking place over a period of time". This definition has influenced most definitions of cumulative effects that have emerged over the last 40-plus years (Table 3.1).

Recognizing cumulative effects in ecological and social systems

Cumulative effects can encompass a spectrum of stressors and responses. The IFC (2013) explains that cumulative effects sometimes occur because of a series of stressors of a similar type, such as multiple mine sites in an important wildlife area, or multiple pulp and paper mills discharging to a river. In other instances, cumulative effects result from a diversity of stressors or activities on the landscape, for example, forest clearing, agricultural activity, hydroelectric power generation, urban development, and recreational use all occurring within the same watershed and adversely impacting aquatic health.

Consider the Athabasca River basin in northern Alberta, Canada. The Athabasca River is 1,400 km long with a drainage basin of approximately 150,000 km^2 and is the only major river in the province that does not have a dam to regulate flow. Since the 1960s, the Athabasca River has been subjected to increasing pressures from multiple land uses. The Athabasca region is home to the largest known deposits of crude bitumen in the world. Oil sands mining is the largest consumer of water from the Athabasca River. However, the Athabasca River basin has also experienced significant agricultural land conversion, forest harvesting, and pulp and paper mill development (Seitz, Westbrook, and Noble 2011). Between 1966 to 1976, and 1996 to 2006, Squires, Westbrook, and Dubé (2010) report that the total agricultural land area in the region increased by nearly 5 million acres and the number of pulp and paper mills discharging into the Athabasca River system increased from one to five. During this same period, oil sands

Table 3.1 Definitions of cumulative effects

The result of additive and aggregative actions producing impacts that accumulate incrementally or synergistically over time and space (Contant 1984).

Changes to the environment that are caused by an action in combination with other past, present, and future actions (Hegmann et al. 1999).

Effects that are likely to result from a project in combination with other projects or activities that have been or will be carried out (Canada Port Authority Environmental Assessment Regulations 1999).

Effects caused by the combined results of past, current, and future activities on a landscape (MacDonald 2000).

When two or more actions result in impacts that overlap in either time or space, affecting environmental, social, cultural, economic, or human health values. While cumulative impacts can occur in any place, they are more pronounced in areas that are rapidly undergoing change as a result of human development (AXYS Environmental Consulting Ltd and Salmo Consulting Inc. 2003).

The net result of the environmental impacts from multiple projects and activities (Cooper 2004).

Aggregated, collective, accruing, and (or) combined ecosystem changes that result from a combination of human activities and natural processes (Scherer 2011).

Effects on the environment which a project may produce if considered jointly with other projects (European Commission 2013).

Effects that result from the additive impacts caused by other past, present, or reasonably foreseeable actions, together with the proposed plan, program, or project itself, and the synergistic impacts that arise from the reaction between impacts of a development plan, program, or project on different aspects of the environment (Renewable UK 2013).

Changes in the environment are caused by multiple interactions among human activities and natural processes that accumulate across space and time (CCME 2014).

When two or more stimuli act together to influence the condition of a valued component of the environment (Sinclair, Doelle, and Duinker 2017).

leases increased from 2 to over 3,300, and water withdrawals increased from 12 million m³/yr to over 595 million m³/yr. The cumulative annual flow in the Athabasca River decreased by more than 500 m³/s, coupled with increases in temperature, turbidity, conductivity, and phosphorous, and an overall deterioration of the aquatic health of the Athabasca River system (Athabasca Watershed Council 2018).

Such adverse cumulative effects over time are not necessarily intentional. The problem emerges when the effects of each development, land use, or activity are considered individually insignificant, measured *against* (versus in *addition* to) the total effects of all other activities and land uses, or deemed too small to trigger regulatory impact assessment and management processes (Noble 2020). No matter how small an effect, the additional stress placed on an environmental system that is already stressed can still result in significant adverse ecological effects.

But cumulative effects are not unique to ecological systems – they also occur in social systems; they are just more difficult to understand and measure. The Hunter Valley, New South Wales, Australia, for example, is a high-density coal mining region with a mining history dating to the early 1800s. The valley is home to more than 40 open-cut coal mining operations, largely producing thermal and soft coking coal, and six coal-fired power stations (McArtney 2019). Mining leases in the Upper Hunter Valley extend over 60% of the valley floor. Most references to cumulative impacts in the Hunter Valley relate to biophysical concerns, health effects, and employment, but over time there have emerged significant social cumulative impacts (Franks, Brereton, and Moran 2010). The town of Muswellbrook, a once-rural farm town in the Hunter Valley, is now surrounded by several coal mining operations. Franks, Brereton, and Moran report that cumulative issues of concern in the community often revolve around increasing "social dis-location and changing sense of place" (303). These types of cumulative effects on social systems can be described as successive or incremental – i.e., effects that build up over time. According to Hackett, Liu, and Noble (2018), cumulative effects to social systems often involve the historical build-up of effects, or legacy effects, which may manifest as the loss of traditional lands caused by successive ecological disturbances (e.g., species habitat loss; successive flooding for hydroelectric development), shifts in livelihoods, or changes in identity, culture, and sense of place. What is clear from examples like the Athabasca and Hunter Valley is that cumulative effects are complex, extensive, pervasive, and usually unintentional – but they should always be expected.

Characterizing cumulative effects

The nature of cumulative effects has been explained in various ways, often based on different *typologies* that describe how such effects are triggered or the pathways that lead to cumulative change (Contant and Wiggins 1991). Baskerville (1986), for example, proposed that cumu-lative effects manifest in natural systems in three ways: (i) incremental insults to the system, whereby each increment adds to the previous over space and time; (ii) a single action or set of actions results in a significant shift in system structure or function, owing to the seemingly benign effects of previous actions with responses delayed in time; and (iii) the accumulation of impacts by shifting natural cycles over time, such as consecutive disturbances before the system has rebounded or recovered to the initial pre-disturbance state. Building on this classification, Beanlands and Duinker (1984) proposed that cumulative effects result from space-crowding or time-crowing, synergistic or indirect effects, and progressive nibbling – giving rise to the com-mon characterization of cumulative effects as *death by a thousand cuts*.

While definitions and typologies differ (see Table 3.1), most conceptualizations of cumula-tive effects emphasize several common principles and features (e.g., Baskerville 1986; Beanlands and Duinker 1984; Blakley et al. 2017; Contant and Wiggins 1991; Daman, Cressman, and Sadar 1995; Jones 2016; Kilgour et al. 2007; Noble 2020; Ziemer 1994):

- Cumulative effects are not always easily traced to their origins, especially in highly devel-oped landscapes.
- In complex ecological and social systems, cause and effect can be quite distant in time and space.
- Cumulative effects can aggregate linearly or exponentially and reach limits or tipping points, after which major changes can occur, which may be both unpredictable and irreversible.
- There may not be a single causal pathway that leads to cumulative effects; rather, cumula-tive effects may be the result of the interaction or aggregation of several unrelated sources.
- Even the most minor or seemingly insignificant actions or effects can lead to adverse cumulative environmental change.
- Cumulative effects caused by past actions can be delayed and go unnoticed, only to be trig-gered by a new action or disturbance – whether human-induced or a natural event.
- Cumulative effects can be positive effects, such as the cumulative response of ecological systems to mitigation actions or gains in social welfare because of employment and com-munity development.

The art and science of assessing cumulative effects

The International Finance Corporation (IFC) of the World Bank defines CEA as the process of analyzing the potential impacts and risks of proposed developments in the context of the potential effects of other human activities and natural environmental and social external drivers over time, and proposing measures to avoid, reduce, or mitigate such cumulative impacts and risks (IFC 2013, 21). Or, simply put, CEA is the assessment of changes that accumulate from multiple stressors, both human-induced and natural, over space and/or time (Dubé 2003; Reid 1993; Spaling and Smit 1993).

Legislated requirements or provisions for CEA are not new. The United States' National Environmental Policy Act of 1969 was the first national environmental assessment legislation to implement CEA requirements, requiring federal agencies to analyze project-specific and cumulative impacts of proposed federal undertakings before they are implemented (Ma, Becker, and Kilgore 2009). Guidelines for incorporating cumulative effects into the federal review process were subsequently introduced by the United States' Council on Environmental Quality in 1978 and revised in 1997. Canada also has a long history of CEA (Beanlands and Duinker 1984), with legislated requirements for CEA currently implemented under a national *Impact Assessment Act*. European Union directives for the content of both environmental impact assessments and strategic environmental assessments stipulate that cumulative effects should be considered (Council Directive 85/33/ EEC, 1985; Council Directive 97/11/EC, 1997; Council Directive 2001/ 42/EC, 2001), with several national legislations mirroring these requirements and identifying cumulative effects as a mandatory element of assessment – e.g., Germany, the Netherlands, the United Kingdom, and Denmark (Wärnbäck and Hilding-Rydevik 2009; Bidstrup, Kørnøv, and Partidário 2016). Of course, CEA requirements are not even across all jurisdictions. In Sweden, for example, Larsen et al. (2017) report Swedish Environmental Code does *not* include clear requirements for CEA, and that the expectations of permitting authorities and developers to consider cumulative effects are vague (Wärnbäck and Hilding-Rydevik 2009).

Thinking cumulatively

Sinclair, Doelle, and Duinker (2017, 184) suggest that good CEA requires adopting a "cumulative effects mindset". This means thinking cumulatively (Figure 3.1) and assuming that most (if not all) effects are cumulative. Operationally, thinking cumulatively orients CEA away from a *project-focused approach*, whereby emphasis is placed on minimizing the individual effects of a stressor (i.e., a single project) to an acceptable level, toward a *valued component-focused approach*, whereby emphasis is placed on the total effects of all stressors on the receiving environmental system or valued component (VC) of concern. Valued components are those aspects of environmental or social systems that are considered important and are the focus of CEA. Though often individual components, such as a wildlife species, fish, habitat, or air quality, VCs can also include more holistic concepts, such as ecosystem services, biodiversity, or well-being. The VC-focused approach considers the impacts from all sources of disturbances, whether natural or human-induced, regardless of whether or not the disturbance is caused by an action subject to legislated environmental assessment (Noble 2020).

The IFC (2013) explains that VCs are "integrators of the stressors that affect them" and are immersed in an "ever-changing environment that affects their condition and resilience". For example, the health of an aquatic species is affected by a variety of stressors, including extreme flood events, land uses (e.g., forest harvesting, agriculture), river crossings or riparian disturbances, and industrial discharge, to name a few. Thus, when assessing the cumulative effects of a newly proposed project, such as a hydroelectric generating facility, consideration must be given to how the health of the river system or aquatic species of concern has been influenced by the accumulation of all stressors, both past and into the future, in addition to (or in interaction with) the stress caused by the proposed development.

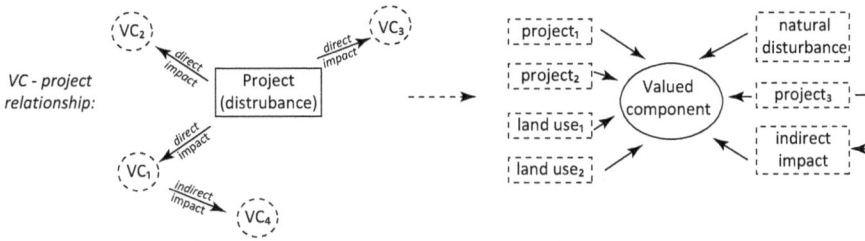

Concepts	Status quo thinking	----➤	Thinking cumulatively
Assumptions:	abundance	----➤	limits and thresholds
Systems:	simple, linear	----➤	complex, non-linear
Understanding:	certainty	----➤	uncertainty
Stressors:	single	----➤	multiple and interacting
Receptors:	single media	----➤	environmental system
Spatial scale:	local	----➤	multi-scaled
Temporal scale:	present	----➤	past, present, future
Scope:	regulated activities	----➤	all activities
Responsibility:	proponent	----➤	collective VC-centered

Figure 3.1 Status quo versus thinking cumulatively. Source: Adapted from Noble (2020) and IFC (2013).

The interactions between stressors and the implications for VC sustainability can be complex. For example, assume a watershed characterized by a legacy of industrial activity. Each industrial site may be operating within the regulatory standard for site operations, but slowly, over time, heavy metals accumulate in the aquatic system via surface runoff. The impacts on fish health may be concerning, but still tolerable and within an acceptable limit. Now assume that forest harvest licenses are issued in the same watershed, resulting in increased access roads and river crossings. As the forested area is cleared, coupled with disturbances to riparian habitat, the rate of introduction of heavy metals into the aquatic system increases due to increased surface runoff. Fish health starts to decline, but the population may still be sustainable. A hydroelectric project is now proposed, which will involve the construction of a large reservoir, resulting in further loss of riparian habitat. The creation of a large reservoir is also predicted to increase water temperature and lower dissolved oxygen levels – but such levels will remain within the range of tolerability for the fish species. However, under increased temperature and lower dissolved oxygen, the toxicity of heavy metals in the aquatic system is now amplified, and a healthy fish population is no longer sustainable. Each of these actions or disturbances (heavy industry, forestry, access roads, hydroelectric generation) may seem independent, and the effects even tolerable for the VC of concern, but the cumulative effects to aquatic health are significant. Thinking cumulatively means that the emphasis of CEA is on the resilience of the VC and its response to all stress – not only the stress of the development under consideration.

In practice, however, most jurisdictions continue to struggle with the cumulative effects mindset (Bidstrup, Kørnøv, and Partidário 2016; Duinker and Greig 2006). In some instances, CEA is considered "an irritant to the completion of a project environmental assessment […] [and] cumulative effects are assessed as a purely legal obligation without practical merit" (Sinclair, Doelle, and Duinker 2017, 183). The conventional practice is often to focus on identifying a project's individual impacts and reducing them to a proportionally acceptable level – versus considering how a project's impacts may interact with other stressors in a region and the implications for social and ecological sustainability. Duinker and Greig (2006) argue that this type of approach can do more harm than good, in that effects are rarely considered significant and cumulative effects go unchecked.

Effects- and stressor-based approaches

Although cumulative effects are the total effects to a VC, both effects- and stressor-based approaches are important for identifying, assessing, and understanding those effects. Effects-based approaches focus on measuring responses in ecological or social indicators relative to some benchmark or reference condition. Effects-based assessment is retrospective; attention is on measuring what is happening, or has happened, and quantifying change in response indicators (Dubé 2003). The objective is to understand the total effects of a VC from all sources, the accumulated state of the VC, and whether indicators are performing as desired (Dubé and Munkittrick 2001). In an aquatic system, for example, an effects-based assessment for fish might focus on liver weight or benthic invertebrate abundance (Hewitt et al. 2003), or on dissolved nitrogen or trace metals in the water as proxies for fish health (Seitz, Westbrook, and Noble 2011). The premise is that effects-based assessments can inform the identification of thresholds and risk-based assessments. Determining the cause or drivers of the observed change is often secondary (Dubé et al. 2013).

Stressor-based assessment, in contrast, is focused on the drivers of change or patterns of human activities or disturbances. Stressor-based assessments typically focus on identifying and quantifying trends (i.e., rates, patterns, distributions) of disturbances or land uses, and then projecting those trends or patterns into the future. The focus is on measuring and predicting the cumulative stress associated with agents of change, such as industrial footprints or fragmentation metrics (Noble 2020). The premise is that statistical associations exist between different types of disturbances and ecological responses, and stressors can be a suitable proxy for understanding and predicting cumulative threats or risks to individual ecological components. Stressor-based assessment adopts surrogate indicators, such as land-use and land-cover metrics (e.g., riparian habitat disturbance, stream crossing density, upslope cleared forest area), which serve as indicators for responses in, or risks to, fish health (Vos et al. 2001; Gergel et al. 2002). The association between changes in stressor indicators and ecological effects is descriptive of change, not diagnostic (Seitz, Westbrook, and Noble 2011).

The debate should *not* be which approach is better for assessing and managing cumulative effects. Each approach is valuable, and both are necessary. Effects-based approaches are essential to understanding the accumulated state and condition or health of a VC or ecological system and how conditions have changed over time. These changes can be related to changes in stressors on the landscape, such as pollution discharge locations, industrial concessions, or land-use and land-cover metrics, to build models for predicting the implications of future disturbances on VC condition. Monitoring actual outcomes can then validate or refine the association between stressors and responses, improve predictive capabilities, and inform land-use and management decisions (Dubé et al. 2013).

Basic components of a CEA framework

Multiple frameworks and step-by-step guidance for CEA have emerged over the years. Some of these are designed to meet the needs of project proponents who must conduct CEAs for their projects under legislative requirements (e.g., Hegmann et al. 1999); others are more tailored to regional and strategic environmental assessment approaches (e.g., CCME 2009; Dubé et al. 2013). There is no universal, one-size-fits-all model for CEA (Blakley et al. 2017), but most frameworks consist of several foundational components – regardless of whether the focus of application is an individual project or a regional land use. These core components are synthesized below, each comprising a main step in the CEA process. This chapter does not address supporting methods or tools for CEA, such as specific models or change detection methods, since the choice of methods or tools is specific to the cumulative effects problem, VC or indicator, and the nature and availability of data and resources.

Context setting

Context setting or scoping serves to establish the VCs of interest, identify potential cumulative effects concerns, and determine the scale of analyses. Critical decisions are also made about public participation and meaningful Indigenous engagement (see Chapters 13 and 14 in this book). Ideally, scoping occurs through an open and participatory process engaging affected communities and interests.

Valued components: The VCs scoped into CEA should reflect social, economic, cultural, and ecological values and cumulative effects concerns, in addition to ecological principles (Beanlands and Duinker 1984). Consideration must also be given to whether certain components need to be included because of regulatory requirements, such as threatened or rare species, or because they are identified in existing policies or regional or local land-use plans. In an analysis of federal impact assessments in western Canada, for example, Ball, Noble, and Dubé (2013) found regulatory compliance to be an important factor influencing VC selection, as well as government agency mandates and existing permitting or licensing arrangements. Such considerations may dominate the scope of CEAs when conducted by a single project proponent under regulatory project assessment requirements; however, good practices for VC selection in CEA suggests that several factors be considered, including VCs must reflect ecological, social, cultural, and/or economic values; there must be a reason to believe that the VC has been, is, or will be affected by human-induced or natural disturbances in the region of interest; there must be sufficient data or information available to be able to discern changes in VC conditions; and understanding change in VC conditions, or potential threats, must be informative to decision-making processes, such as land-use permitting, project licensing, or the triggering of management strategies (Noble 2020). As Parkins (2011) suggests, if there is nobody to tell, then the results of CEA initiatives are unlikely to have meaningful influence.

Valued component indicators: VCs can often be abstract concepts, such as "ecosystem health" or "well-being", and not directly measurable or indicative of change. The selection of indicators for VCs is thus important to understanding and predicting cumulative change. Indicators provide insight into changes in VC conditions or are indicative of potential risks or threats to VCs, meaning they must be responsive to disturbances caused by a project, land use, or natural events (Harriman and Noble 2008). Indicators for VCs can reflect effects-based (e.g., species abundance or condition) or stressor-based (e.g., habitat fragmentation metrics) understandings of cumulative effects, or both (Seitz, Westbrook, and Noble 2011). It is important that the indicators selected are useful for tracking change in baseline conditions over time and space, can be associated with human-induced disturbances or stressors (whether causative or correlative), and provide early detection of risk or signaling of change in VC condition (Wong, Noble, and Hanna 2019).

Assessment boundaries: There is no predetermined spatial scale for CEA that is best; however, there is consensus in the scientific literature that the spatial scale for understanding cumulative effects and changes in VC conditions must extend beyond the local, direct impact area of single development projects (Bidstrup, Kørnøv, and Partidário 2016; Duinker and Greig 2006; Noble 2008). Blakley et al. (2017) in their IAIA Fastips on CEA, suggest that adopting a regional scale of analysis is necessary to identify and assess cumulative effects on VCs (see Chapter 10 in this book). In practice, this often translates to the administrative or planning unit scale (Bidstrup, Kørnøv, and Partidário 2016; Scientific Advisory Committee 2007) or to the ecoregion or watershed scale (Dubé et al. 2013; Elk Valley Cumulative Effects Working Group 2017). For very large planning units or watersheds, however, such as the Athabasca River basin (discussed above), the spatial scale may prove impractical for a single project proponent to meet CEA obligations under regulatory requirements, *and* to provide information that is useful within the timeframe of a project-focused review. In such cases, nested impact zones may be delineated based on the pathways that influence how project-induced impacts may interact with other sources of stress to VCs of concern (Buttle 2002; Nadorozny 2009; Noble 2020). In a watershed setting, for example, Seitz, Westbrook, and Noble (2011) suggest focusing on the river reach where development is most concentrated as the first unit of analysis for CEA, where VCs are likely to be most responsive or at risk to the range of stressors interacting with the proposed project.

Analyses

Blakley et al. (2017) emphasize that CEA is more than a snapshot of VC and disturbance conditions; it involves identifying change over time and understanding longer-term trends. There are two important parts to the analysis stage: (i) accumulated state assessment, and (ii) forecasting or futures assessment (Dubé et al. 2013).

Accumulated state: Accumulated state assessments provide an understanding of the accumulated change in a VC (or indicator) over time and an analysis of the importance or significance of that change relative to some benchmark, such as a pre-disturbance condition, range of natural variability, or spatial reference point. Accumulated state assessments are typically effects-based and focused on measuring the change in VCs from historical conditions and identifying changes that are outside of *normal* conditions (Dubé et al. 2013). This provides insight to VCs that may be at higher risk or more vulnerable to any future disturbance or stress and identifies VCs that may already be impacted beyond an acceptable level and warrant restoration or management prior to any future development being approved. Changes in VCs are then related to key stressors, disturbances, or development patterns and trends on the landscape, and used to build conceptual or statistical models for assessing future conditions.

In some cases, however, data on VCs are simply not available (or accessible) to adequately conduct an effects-based assessment of cumulative change in VC condition. In such cases, accumulated state assessments may focus on establishing relationships between current conditions and disturbances using spatial reference points, and then backcasting VC conditions based on how stressors (e.g., development footprints, fragmentation metrics) have changed historically. This was the approach used in southwest Saskatchewan, Canada, for understanding the cumulative effects of development on biodiversity in a native prairie ecosystem (Noble 2008). Associations between range health indicators, species presence, and species diversity were established for different types of disturbances on the landscape (e.g., roads and petroleum and natural gas well sites) using spatially referenced gradient-to-background field assessments. Trends in landscape disturbances were assessed retrospectively using aerial photography, land-use plans and, where available, land-use and vegetation databases. Statistical models associating current land uses with ecological conditions were backcasted using historical land uses and disturbance patterns to understand how VC conditions may have changed over time (Scientific Advisory Committee 2007).

Forecasting or futures assessment: Whether they are effects- or stressor-based approaches, the outputs of accumulated state assessments form the core input to forecasting potential cumulative effects (Dubé et al. 2013). Associations between VC conditions and stressors (e.g., industrial footprint, fragmentation metrics, linear features, cleared area) are used to explore future cumulative effects or risk to VCs (Noble 2008). Scenario-based approaches are preferred, exploring different spatial configurations or intensities and rates of land uses and disturbances into the future, including alternative mitigation actions, the influence of natural disturbances and climate change, and, in the case of regulatory project-based assessment, the contributions of the project under review. This often requires the use of dynamic systems-based models, capable of capturing the complexity of pathways and relationships between multiple stressors and VC response (e.g., Duinker and Greig 2006; Salmo Consulting et al. 2004; Scientific Advisory Committee 2007). "Good" CEA is inherently about the future and ensuring that development decisions lead to desirable, versus most likely, outcomes and VC conditions.

Evaluation and management

When assessing cumulative effects, both retrospectively and prospectively, multiple scales of evaluation should be considered (Therivel and Ross 2007). Not all cumulative effects occur, or are recognizable, at all spatial scales. The growing tendency in CEA is to focus on broad, regional-scale stressors and effects. Although broad-scale approaches are necessary to identify cumulative effects pathways and to understand ecological processes, some cumulative effects may be concentrated or localized, and thus more responsive or susceptible to locally induced stress, such as a project-specific action, versus larger landscape-based disturbances or metrics. A multi-scale analysis is thus required to capture the distribution and potential cumulative effects and risks to VCs, and to understand the significance of those effects (Noble 2008).

Benchmarks or thresholds: Not all cumulative change is significant. The IFC (2013) suggests that the importance or significance of cumulative effects, or in some cases cumulative stress, must be judged based on thresholds. Thresholds are limits of change beyond which the VC condition, or the level of risk, is considered unacceptable – i.e., the sustainability of the VC is compromised. Thresholds can be ecological in nature, whereby even small changes can result in large, non-linear responses (Weber, Krogman, and Antoniuk 2012), or they can be management thresholds or limits of acceptable change in VC or land-use conditions (Antoniuk et al. 2009). For effects-based assessment, such thresholds are often based on VC response indicators, such as phosphorus or biological metrics in the case of water quality, and the significance of cumulative change assessed based on indicator change outside the range of natural variation (Munkittrick et al. 2009). Stressor-based indicators are often expressed as some form of land-use condition, such as upslope cleared vegetation or road–stream crossing density, and thresholds based on what is an acceptable level of landscape change or level of risk to the VC given known associations between certain drivers (e.g., stream crossing density and sedimentation) and VC conditions (e.g., water quality and fish health) (Dubé et al. 2013; Salmo Consulting et al. 2004).

Collaborative management: Under regulatory-based CEA, where the responsibility for assessment typically lies with the project proponent, managing cumulative effects can be challenging – if not sometimes impossible. Project proponents cannot be responsible for managing the adverse impacts of other land uses or stressors that are causing cumulative effects, but they are responsible for managing their project's contribution. The problem is that even the smallest of project-induced impacts can be cumulatively significant in already heavily developed or stressed environments. There are possible mitigation trade-offs – for example, a project proponent may be required to mitigate or reclaim previously abandoned sites (e.g., road reclamation, oil and gas site reclamation) or to contribute financially to such efforts, as a means to offset any additional habitat disturbance caused by their project. However, even for governments, it is legally (and politically) difficult to enforce more stringent mitigation requirements on projects that have *already* been approved because of the potential cumulative effects of a new development (Noble 2020). In almost all scenarios, effective management of cumulative effects requires collaboration among developers and land users, with clear government leadership and coordination (Sheelanere, Noble, and Patrick 2013).

Monitoring

Monitoring is the engine of CEA. In the absence of monitoring, the science informing accumulated state assessments is limited, cumulative effects models and forecasts are highly uncertain, and appropriate triggers for management responses are largely unknown. Cronmiller and Noble (2018) emphasize that monitoring is essential for understanding cumulative effects, recognizing them when they occur, and for supporting both longer- and shorter-term decisions about land and resource use and allocation. Monitoring is not unique to CEA, but cumulative effects science and information needs require that monitoring programs to support CEA reflect certain principles or characteristics (Wong, Noble, and Hanna 2019):

* Two types of monitoring are important to CEA and management – localized monitoring conducted by project proponents under the regulatory impact assessment, and regional monitoring conducted by government agencies and research programs. These programs must be complimentary. This requires coordination of the indicators selected for regional monitoring and the indicators required to be monitored under project approval conditions. A minimum set of common indicators is desirable, including consistent data collection and analytical methods, to track VC conditions at multiple scales.
* Monitoring data must be of sufficient resolution to detect changes in VC conditions as stressors change. This also means selecting indicators that are responsive to stress and that allow for early detection of risk or adverse effects to VC condition.
* Indicators used in CEA monitoring should be responsive to different types of stress or drivers of change, such that the cumulative contributions of multiple actions or disturbances can be understood.
* Monitoring data must be accessible to end-users. This requires that project proponents be willing, if not required, to share monitoring data on a select set of indicators that are valuable for tracking cumulative change. It also means that government-les or research-based monitoring data are available in a common and usable format.
* Monitoring must generate information that is useful for decision-makers and the types of day to day (i.e., project permitting and approvals) and longer-term (i.e., land-use planning and allocation) decisions that need to be made. This requires a conceptual model to guide monitoring efforts across regions and projects, including the science and management questions to be asked and answered, in a time frame that supports both longer-term cumulative change assessment and more immediate regulatory decision needs (Arnold, Hanna, and Noble 2019; Wong, Noble, and Hanna 2019).

Case study: Elk Valley cumulative effects assessment and management framework

The Elk Valley is located in southeastern British Columbia, Canada. Coal mining has persisted in the region for more than 100 years. There are currently five operating metallurgical coal mines in the valley (Fording River, Elkview, Line Creek, Coal Mountain, Greenhills), all within about a 100-kilometer-long area, producing more than 20 million tons of coal in 2015 (Powell 2016). The mines are owned and operated by Teck Resources. The effects of selenium pollution from mine waste rock on water quality in the Elk River and impacts to valued fish species, specifically Westslope cutthroat trout (WCT) (*Oncorhynchus clarkii lewisi*), have been a long-standing concern in the Elk Valley (Linnitt 2018). WCT are a sentinel species to the Elk River and are of high value to the Ktunaxa First Nations and to recreational fisheries. However, coal mining is not the only land use in the Elk Valley affecting aquatic conditions in the Elk River. Forestry is also a key natural resource industry in the region, and domestic and international tourism have increased substantially in recent years – in large part due to the region's world-class skiing. Other stressors on the landscape include agriculture and residential development. Between 1950 and 2014, total human disturbance, based on aerial imagery of landscape change, increased by over 850% (Golder 2015) – this includes over 5,400 kilometers of roads (Elk Valley Cumulative Effects Working Group 2017).

In 2012, an approval condition for a coal mine expansion required Teck to engage in broader discussions about the cumulative effects of development in the Elk Valley. A working group was formed, led by Teck and the Ktunaxa Nation Council, to undertake a CEA and develop a management framework for the Elk Valley. The approach was comprised of four stages. First, a scoping exercise was done to identify the components that would be the focus of CEA, representing key values in the Elk Valley. Five VCs were selected through an open and participatory scoping process, including WCT. Next, a retrospective analysis examined past land use and change in the Elk Valley to establish estimates for the range of natural variation in VCs, which would serve as a benchmark for understanding and evaluating cumulative effects over time and for assessing the significance of the future change. The analysis adopted a largely stressor-based approach, using past data to establish relationships between VC conditions and to characterize the potential risks or threats to VCs. For example, to understand cumulative effects or risks to WCT, attention focused on several indicators, including selenium concentrations, road densities in sub-watersheds (e.g., effects on water flow, temperature, sediment delivery, angler access), stream crossings (e.g., effects on sediment delivery and connectivity), riparian disturbance (e.g. effects on channel morphology, woody debris), and equivalent cleared area (e.g., effects on peak streamflow). Changes in indicator values for landscape-based metrics were converted to risk scores using benchmarks from various government policies and management plans, and communicated as hazard ratings (i.e., low, medium, high) for WCT in sub-units of the watershed. For example, British Columbia aquatic ecosystem value assessment standards were used to establish thresholds for road density on steep slopes (i.e., < 0.06 km/km^2, 0.06–0.12 km/km^2, > 0.12 km/km^2) and resulting hazards (i.e., low, medium, high) to WCT due to rates and volumes of sediment delivery to streams and impacts on peak streamflow (Elk Valley Cumulative Effects Working Group 2017).

In the third phase of the assessment, changes in stressors were explored using a cumulative effects simulator, ALCES (www.alces.ca), to examine how VC indicators might respond under future land uses and the subsequent risks posed to those VCs. Emphasis was placed on several development scenarios, land-use change, and natural disturbances in the Elk Valley to 2065. The scenarios included: business as usual, projecting recent rates of change and disturbances into the future; minimum growth, based on slower rates of mining, forestry and residential development than in the recent past; maximum growth, based on increased rates of mining, forestry, residential development, and other land uses; high natural disturbance, which modeled increased fire and forest pest outbreaks combined with the maximum growth scenario; intensive mitigation, which was combined with each growth scenario and included widespread mitigation of the impacts of development in old-growth forest and riparian areas, such as site reclamation and restoration of roads and river crossings; moderate mitigation, which was comprised of less stringent and more site-specific mitigation; and two climate scenarios (Elk Valley Cumulative Effects Working Group 2017).

The results showed that past land-use changes, especially the expansive road network in the Elk Valley, had resulted in a high risk of adverse cumulative effects to VCs, including WCT, that will persist into the future. Sixty percent of sub-watershed assessment units in the valley were found to present a high hazard to WCT due to high road densities, and over 90% of assessment units were found to present a high hazard due to road densities near streams (Elk Valley Cumulative Effects Working Group 2017). Future development under all scenarios was found to have only marginal, additional effects since most of the disturbance caused by road development in the valley has already happened due to the network of stream crossings and riparian disturbances, and the limited land available for future road network development. However, high selenium concentrations from past and future mining activity, and increased natural disturbances, were found to pose additional future risks to WCT. Mitigation scenarios were found to have significant positive effects on WCT – specifically mitigation actions that involved reducing the impacts of disturbances caused by roads and river crossings.

Included among the management recommendations in the final phase of the assessment were targeted management strategies for reducing stream crossing densities and riparian habitat reclamation in high-risk areas of the Elk Valley, such as removal of non-essential roads and repairing hanging road culverts. This means that in the future, when new roads are proposed as part of a development proposal, it may be necessary to consider the closure or reclamation of an existing road as a condition of project approval (Noble 2020). Reducing road densities within 100 meters of streams was also found to reduce access, and thus reduce the cumulative pressure on WCT from recreational fishing. Although results indicate that future coal mining activities are unlikely to contribute to increased road density in areas of high cumulative risk to WCT, the risks posed by high selenium concentrations persist. Additional recommendations for mitigation thus include retention of riparian areas at mine sites and rapid or enhanced reclamation to disturbed areas at mine sites to reduce overall cumulative effects.

It can be challenging for project proponents to assess anything other than the direct, immediate effects of their project, given the constraints to project assessment processes. Tight timelines, limited budgets, limited data, and other factors – not the least of which being proponents' vested interest in project approval – can work against the assessment of cumulative effects. There are also limits to what proponents can realistically accomplish in any one project assessment (Noble 2020). The Elk Valley CEA is an example of an assessment that was led by industry, in collaboration with First Nations and other interests, suggesting that project proponents *can* tackle complex cumulative effects issues that extend well beyond their project operations when done so in a collaborative setting. However, the Elk Valley case also illustrates that solving cumulative effects challenges, and appropriately managing cumulative risk to VCs, requires collaborative action. The Elk Valley CEA was limited in time and resources, and largely stressor-based. It did not include a long-term monitoring program, but a key recommendation emerging from the CEA was the integration of the assessment results with existing regional and proponent-led monitoring initiatives in the Elk Valley, including Teck Resources' (2014) area-wide effects-based water quality management program for selenium accumulation from historic, current and future mining activity (Teck Resources 2014). To facilitate the integration of CEA with broader management plans and regulatory decision-making, in 2019 the Elk Valley cumulative effects framework was adopted as part of the province's area-wide cumulative effects management framework.

Conclusions

This chapter explored the foundational principles of cumulative effects and CEA. The science to support CEA has evolved considerably, along with systems-based modeling and spatial analytical tools, for exploring the relationships between stressors and effects, and for projecting into the future VC responses and risks associated with human and natural disturbances. However, Larsen et al. (2017, 67) note that the impact assessment literature "is full of observations on how countries with well-established CEA-regimes continue to struggle to enact their ambitions". Part of the challenge lies in the siloed approach to how cumulative effects are typically monitored, assessed, and managed (Noble 2015) – coupled with the need for improved education and training on the fundamentals of CEA. In most jurisdictions, CEA is firmly embedded in project-based environmental assessment legislation, with the primary responsibility for CEA placed on project proponents. Minimizing the effects of individual projects is important, but it is insufficient for understanding, assessing, and appropriately managing cumulative effects (Arnold, Hanna, and Noble 2019; Dubé 2003; Foley et al. 2017; Jones 2016).

Solving complex cumulative effects challenges requires that project and regional CEA efforts, alongside long-term environmental monitoring, function in tandem (Cronmiller and Noble 2018; Parkins 2011; Schindler 2006) – whereby CEAs are conducted regionally, inform project-based reviews and decisions, and are supported by long-term monitoring programs. Science-based and technical challenges to CEA remain, but arguably the biggest challenges facing CEA are institutional and governance challenges (Cronmiller and Noble 2018; Jones 2016). Thinking cumulatively involves much more than adopting a VC-centered perspective; it requires embedding CEA into all planning, project review, monitoring, and decision-making processes (Larsen et al. 2017).

References

Antoniuk, T., S. Kennett, C. Aumann, M. Weber, S. Davis Schuetz, R. McManus, K. McKinnon, and K. Manuel. 2009. "Valued Component Thresholds (Management Objectives) Project." Retrieved from http://www.esrfunds.org/pdf/172.pdf.

Arnold, L.M., K. Hanna, and B. Noble 2019. "Freshwater Cumulative Effects and Environmental Assessment in the Mackenzie Valley, Northwest Territories: Challenges and Decision Maker Needs." *Impact Assessment and Project Appraisal* 6(37): 516-525.

Athabasca Watershed Council. 2018. "State of the Athabasca Watershed." Athabasca AB: Athabasca Watershed Council. Retrieved from https://awc-wpac.ca/resources/awc-reports/

AXYS Environmental Consulting Ltd and Salmo Consulting Inc. 2003. "Approaching Cumulative Impact Management in Northeast British Columbia." Vancouver BC: BC Oil and Gas Commission, The Muskwa-Kechika Advisory Board.

Ball, M., B. Noble, and M. Dubé. 2013. "Valued Ecosystem Components for Watershed Cumulative Effects: An Analysis of Environmental Impact Assessments in the South Saskatchewan River Watershed, Canada." *Integrated Environmental Assessment and Management* 9(3): 469–79.

Baskerville, G. 1986. "Some Scientific Issues in Cumulative Environmental Impact Assessments." In *Cumulative Environmental Effects: A Binational Perspective*, edited by G. E. Beanlands, W. J. Erckmann, G. H. Orians, J. 'Riordan, D. Policansky, M.H. Sadar, and B. Sadler. Ottawa ON: The Canadian Environmental Assessment Research Council and The United States National Research Council, Canada: Minister of Supply and Services.

Beanlands, G.E., and P.N. Duinker. 1984. "An Ecological Framework for Environmental Impact Assessment." *Journal of Environmental Management* 18: 267–77.

Bidstrup, M., L. Kørnøv, and M.R. Partidário. 2016. "Cumulative Effects in Strategic Environmental Assessment: The Influence of Plan Boundaries." *Environmental Impact Assessment Review* 57: 151–58.

Blakley, J., P. Duinker, L. Greig, G. Hegmann, and B. Noble. 2017. "Cumulative Effects Assessment." *International Association for Impact Assessment Fastips Series, No. 16.*

Buttle, J.M. 2002. "Rethinking the Donut: The case for Hydrologically Relevant Buffer Zones." *Hydrological Processes* 16: 3093–6.

Canada Port Authority Environmental Assessment Regulations. 1999. Retrieved from http://laws.justice.gc.ca/PDF/SOR-99-318.pdf

Canter, L.W., and B. Ross. 2010. "State of Practice of Cumulative Effects Assessment and Management: The Good, the Bad and the Ugly." *Impact Assessment and Project Appraisal* 28(4): 261–68.

CCME (Canadian Council of Ministers of the Environment). 2009. *Regional Strategic Environmental Assessment in Canada.* Winnipeg MB: CCME. 27p.

CCME (Canadian Council of Ministers of the Environment). 2014. *Canada-wide Definitions and Principles for Cumulative Effects.* Retrieved from https://www.ccme.ca/files/Resources/enviro_assessment/CE%20Definitions%20and%20Principles%201.0%20EN.pdf.

Clarke, R. 1994. "Cumulative Effects Assessment: A Tool for Sustainable Development." *Impact Assessment* 12: 319–31.

Connelly, R. 2011. "Canadian and International EIA Frameworks as They Apply to Cumulative Effects." *Environmental Impact Assessment Review* 31(5): 453–56.

Contant, C. K. 1984. *Cumulative Impact Assessment: Design and Evaluation of an Approach for the Corps Permit Program at the San Francisco District.* Ph.D. thesis. Stanford CA: Stanford University.

Contant, C. K. & Wiggins, L. L. (1991) Defining and analyzing cumulative environmental impacts. *Environmental Impact Assessment Review* 11(4): 297–309.

Cooper, L. M. 2004. *Guidelines for Cumulative Effects Assessment in SEA of Plans.* Retrieved from http://www.imperial.ac.uk/pls/portallive/docs/1/21559696.PDF.

Cronmiller, J., and B. Noble. 2018. "Integrating Environmental Monitoring with Cumulative Effects Management and Decision Making." *Integrated Environmental Assessment and Management* 14(3): 407–17.

Damman, D.C., D. Cressman, and M. Sadar. 1995. "Cumulative Effects Assessment: The Development of Practical Frameworks." *Impact Assessment* 13(4): 433–54.

Dubé, M.G. 2003. "Cumulative Effects Assessment in Canada: A Regional Framework for Aquatic Ecosystems." *Environmental Impact Assessment Review* 23: 723–45.

Dubé, M., and K. Munkittrick. 2001. "Integration of Effects-based and Stressor-based Approaches into a Holistic Framework for Cumulative Assessment in Aquatic Ecosystems." *Human and Ecological Risk Assessment* 7(2): 247–58.

Dubé, M., B. Johnson, G. Dunn, J. Culp, K. Cash, K. Munkittrick, I. Wong, K. Hedley, W. Booty, D. Lam, et al. 2006. "Development of a New Approach to Cumulative Effects Assessment: A Northern River Ecosystem Example." *Environmental Monitoring and Assessment* 113: 87–115.

Dubé, M.G., P. Duinker, L. Greig, M. Carver, M. Servos, M. McMaster, B. Noble, H. Schreier, L. Jackson, and K. Munkittrick. 2013. "A Framework for Assessing Cumulative Effects in Watersheds: An Introduction to Canadian Case Studies." *Integrated Environmental Assessment and Management* 9(3): 363–69.

Duinker, P., & Greig, L. (2006). The impotence of cumulative effects assessment in Canada: ailments and ideas for redeployment. *Environmental Management* 37(2): 153–161.

Elk Valley Cumulative Effects Working Group. 2017. *Elk Valley Cumulative Effects Assessment and Management Report.* Submitted to the Elk Valley Cumulative Effects Workshop Group, Cranbrook BC: Elk Valley Cumulative Effects Working Group.

European Commission. 2013. "Environmental Impact Assessment of Projects – Rulings of the Court of Justice." Retrieved from http://ec.europa.eu/environment/eia/pdf/eia_case_law.pdf.

Franks, D.M., D. Brereton, and C. Moran. 2010. "Managing the Cumulative Impacts of Coal Mining on Regional Communities and Environments in Australia." *Impact Assessment and Project Appraisal* 28(4): 299–312.

Foley, M., L. Mease, R. Martone, E. Prahler, T. Morrison, C. Murray, and D. Wojcik. 2017. "The Challenges and Opportunities in Cumulative Effects Assessment." *Environmental Impact Assessment Review* 62: 122–34.

Gergel, S.E., M.G. Turner, J.R. Miller, J.M. Melack, and E.H. Stanley. 2002. "Landscape Indicators of Human Impacts to Riverine Systems." *Aquatic Science* 64: 118–28.

Golder Associates. 2015. "Phase II – Quantitative Study, Pre-Development Study." Report Number: 1113460043/R02. Golder Associates. Submitted to Teck Coal Limited.

Greig, L., and P. Duinker. 2011. "A Proposal for Further Strengthening Science in Environmental Impact Assessment in Canada." *Impact Assessment and Project Appraisal* 29(2): 159–65.

Hackett, P., J. Liu, and B. Noble. 2018. "Human Health, Development Legacies, and Cumulative Effects: Environmental Assessments of Hydroelectric Projects in the Nelson River Watershed, Canada." *Impact Assessment and Project Appraisal* 36(5): 413–24.

Harriman, J.A.E., and B.F. Noble. 2008. "Characterizing Project and Strategic Approaches to Regional Cumulative Effects Assessment in Canada." *Journal of Environmental Assessment Policy and Management* 10(1): 25–50.

Hegmann, G., C. Cocklin, R. Creasey, S. Dupuis, A. Kennedy, L. Kingsley, W. Ross, H. Spaling, and D. Stalker. 1999. *Cumulative Effects Assessment Practitioners Guide*. Hull QC: XYS Environmental Consulting Ltd. and CEA Working Group for the Canadian Environmental Assessment Agency.

Hegmann, G., and G.A. Yarranton. 2011. "Alchemy to Reason: Effective use of Cumulative Effects Assessment in Resource Management." *Environmental Impact Assessment Review* 31(5): 484–90.

Hewitt, L.M., M.G. Dubé, J. Culp, D. McLatchy, and K. Munkittrick. 2003. "A Proposed Framework for Investigation of Cause for Environmental Effects Monitoring." *Human and Ecological Risk Assessment* 9(1): 195–211.

IFC [International Finance Corporation]. 2013. *Cumulative Impact Assessment and Management: Guidance for the Private Sector in Emerging Markets. Good Practice Handbook*. Washington DC: International Finance Corporation.

Jones, F.C. 2016. "Cumulative Effects Assessment: Theoretical Underpinnings and Big Problems." *Environmental Reviews* 24(2): 187–204.

Kilgour, B.W., M.G. Dubé, K. Hedly, C. Portt, and K. Munkittrick. 2007. "Aquatic Environmental Effects Monitoring Guidance for Environmental Assessment Practitioners." *Environmental Monitoring and Assessment* 130: 423–36.

Larsen, R.K., K. Raitio, M. Stinnerborm, and J. Wik-Karlsson. 2017. "Sami-State Collaboration in the Governance of Cumulative Effects Assessment: A Critical Action Research Approach." *Environmental Impact Assessment Review* 64: 67–76.

Linnitt, C. 2018. "For Decades B.C. Failed to Address Selenium Pollution in the Elk Valley – Now no one Knows how to Stop it." *The Narwhal* 14 December. Retrieved from https://thenarwhal.ca/for-decades-b-c-failed-to-address-selenium-pollution-in-the-elk-valley-now-no-one-knows-how-to-stop-it/

Ma, Z., D. Becker, and M. Kilgore. 2009. "Assessing Cumulative Impacts within State Environmental Review Frameworks in the United States." *Environmental Impact Assessment Review* 29: 390–98.

MacDonald, L. 2000. "Evaluating and Managing Cumulative Impacts: Process and Constraints." *Environmental Management* 26(3): 299–315.

McArtney, G. 2019. "Mining in the Hunter Valley: The Black Star." *The Australian Mining Review* 15 May 2019. Retrieved from https://australianminingreview.com.au/features/mining-in-the-hunter-valley-the-black-star/.

Munkittrick, K.R., C.J. Arens, R.B. Lowell, and G. Kaminski. 2009. "A Review of Potential Methods for Determining Critical Effect Size for Designing Environmental Monitoring Programs." *Environmental Toxicology and Chemistry* 28:1361–71.

Nadorozny, N. 2009. *Land Cover, Spatial and Temporal Scale, and Water Quality Patterns in the South Saskatchewan River Basin*. PhD Dissertation. Calgary AB: University of Calgary.

Noble, B. 2008. "Strategic Approaches to Regional Cumulative Effects Assessment: A Case Study of the Great Sand Hills, Canada." *Impact Assessment and Project Appraisal* 26(2): 78–90.

Noble, B.F. 2020. *Introduction to Environmental Assessment: Guide to Principles and Practice*. Don Mills, ON: Oxford University Press.

Parkins, J.R. 2011. "Deliberative Democracy, Institution Building, and the Pragmatics of Cumulative Effects Assessment." *Ecology and Society* 16(3): 20.

Powell, K. 2016. "Five of the Biggest Mines in BC are in the Elk Valley, Generate 65% of Mining Revenue" *Canadian Mining and Energy* 25 October. Retrieved from https://www.miningandenergy.ca/mininginsider/article/five_of_the_biggest_mines_in_bc_are_in_the_elk_valley_generate_64_of_mining/

Reid, L.M. 1993. "Research and Cumulative Watershed Effects." Gen. Tech. Rep. PSWGTR-141. Albany CA: Pacific Southwest Research Station, Forest Service, US Department of Agriculture. 118 p.

RenewableUK. 2013. *Cumulative Impact Assessment Guidelines*. Retrieved from https://ke.services.nerc.ac .uk/Marine/Members/Documents/Guidance documents/Cumulative Impact Assessment Guidelines .pdf.

Salmo Consulting, Axys Environmental Consulting, Forem Technologies, Wildlife and Company. 2004. *Deh Cho Cumulative Effects Study Phase 1: Management Indicators and Thresholds*. Calgary AB: Deh Cho Land Use Planning Committee, Fort Providence, Northwest Territories. 172 p.

Scherer, R.A. 2011. "Cumulative Effects: A Primer for Watershed Managers." *Streamline Watershed Management Bulletin* 14(2): 14–20.

Schindler, D.W., and W.F. Donahue. 2006. "An Impending Water Crisis in Canada's Western Prairie Provinces." *Proceedings of the National Academy of Sciences* 103(19): 7210–16.

Scientific Advisory Committee. 2007. *The Great Sand Hills Regional Environmental Study*. Regina SK: Canada Plains Research Centre.

Seitz, N., C. Westbrook, and B. Noble. 2011. "Bringing Science into River Systems Cumulative Effects Assessment Practice." *Environmental Impact Assessment Review* 31: 172–79.

Sheelanere, P., B. Noble, and R. Patrick. 2013. "Institutional Requirements for Watershed Cumulative Effects Assessment and Management: Lessons from a Canadian Trans-boundary Watershed." *Land Use Policy* 30(1): 67–75.

Sinclair, A.J., M. Doelle, and P. Duinker. 2017. "Looking up, Down, and Sideways: Reconceiving Cumulative Effects Assessment as a Mindset." *Environmental Impact Assessment Review* 62: 183–94.

Spaling, H., and B. Smit. 1993. "Cumulative Environmental Change: Conceptual Frameworks, Evaluation Approaches, and Institutional Perspectives." *Environmental Management* 17(5): 587–600.

Squires, A.J., and M.G. Dubé. 2013. "Development of an Effects-based Approach for Watershed Scale Aquatic Cumulative Effects Assessment." *Integrated Environmental Assessment and Management* 9(3): 380–91.

Squires, A.J., C.J. Westbrook, and M.G. Dubé. 2010. "An Approach for Assessing Cumulative Effects in a Model River, the Athabasca River Basin." *Integrated Environmental Assessment and Management* 6: 119–34.

Teck Resources Limited. 2014. *Elk Valley Water Quality Plan*. Sparwood BC: Teck Resources Limited. xxxii + 256 pp.

Therivel, R., & Ross, W. (2007). Cumulative effects assessment: does scale matter? *Environmental Impact Assessment Review* 27(5): 365–385.

US-CEQ. 1978. *Regulations Implementing the National Environmental Policy Act: 40, CFR 1508.7, 1508.8, 1508.25*. Retrieved from http://energy.gov/sites/prod/files/NEPA-40CFR1500_1508.pdf

Vos, C.C., J. Verboom, P. Opdam, and C.F.J. Ter Braak. 2001. "Toward Ecologically Scaled Landscape Indices." *American Naturalist* 183(1): 24–41.

Wärnbeck, A., and T. Hilding-Rydevik 2009. "Cumulative Impacts in Swedish EIA Practice – Difficulties and Obstacles." *Environmental Impact Assessment Review* 29: 107–15.

Weber, M., N. Krogman, and T. Antoniuk. 2012. "Cumulative Effects Assessment: Linking Social, Ecological, and Governance Dimensions." *Ecology and Society* 17(2): 22.

Wong, L., B.F. Noble, and K. Hanna. 2019. "Water Quality Monitoring to Support Cumulative Effects Assessment and Decision Making in the Mackenzie Valley, Northwest Territories, Canada." *Integrated Environmental Assessment and Management* 15(6): 988–99.

Ziemer, R.R. 1994. "Cumulative Effects Assessment Impact Thresholds: Myths and Realities." In *Cumulative Effects Assessment in Canada: From Concept to Practice*, edited by A.J. Kennedy. Calgary, AB, Canada: The Fifteenth Symposium of Alberta Society of Professional Biologists. pp. 319–26.

4

ASSESSING SOCIAL IMPACTS AND PROMOTING SUSTAINABILITY

Anne Merrild Hansen

Social impact assessment principles

Development can cause a wide range of social impacts within communities. While large-scale projects may offer the opportunity for economic growth, undesired impacts, such as pressures on local services, cultural impacts, increased inequality, and community conflict, can also be expe-

---

Box 4.1 What are social impacts?

Social impacts can be changes to any of the following:

- A way of life – that is, how they live, work, play, and interact with one another;
- Culture – that is, their shared beliefs, customs, values, and language or dialect;
- Community – cohesion, stability, character, services, and infrastructure;
- Political systems – the extent to which people are able to participate in decisions that affect their lives;
- Democratization and the resources provided for democracy;
- The environment within which people live – the quality of the air and water people use; the availability and quality of the food they eat; the level of hazard or risk, dust, and noise they are exposed to; the adequacy of sanitation; their physical safety; and their access to and control over resources;
- Human health and well-being – health is a state of complete physical, mental, social, and spiritual well-being and not merely the absence of disease or infirmity (Chapter 8 in this book provides an overview of health assessment);
- Personal and property rights – particularly whether people are economically affected, or experience personal disadvantage which may include a violation of their civil liberties;
- Fears and aspirations – their perceptions about their safety, their fears about the future of their community, and their aspirations for their future and the future of their children.

    (Adapted from Vanclay, F. 2003 International Principles for Social Impact Assessment. Impact Assessment & Project Appraisal 21(1), 5–11. http://dx.doi.org/10.3152/14715460378176649)

DOI: 10.4324/9780429282492-5

rienced. Like other aspects of assessment practice, *Social Impact Assessment* (SIA) is a tool aimed at anticipating and mitigating such impacts and supporting social sustainability through the management of social impacts related to human actions (Hansen et al. 2016;Vanclay 2002). SIA also provides an opportunity for enhancing the sustainability benefits of development (Chapter 6 in this book provides a review of sustainability assessment).

When we think about sustainable development and social impacts, we might be reminded to focus on the dynamic processes of societal changes, to consider the diverse paths that social progress and well-being can take, and to acknowledge the interdependency of cultural, political, economic qualities, and their ecological setting. Understanding and recognizing this interconnectedness can support a conceptual shift from a basic conservation and preservation perspective to more active and complex thinking about the realities of change and transformation inherent in development, and ensure such recognition is reflected in our planning processes (Petrova and Marinova 2015).

SIA is an opportunity for promoting social sustainability, understanding what people need from the places where they live and work, and understanding their aspirations and fears and how these may affect their readiness to respond or adapt to change brought on by different projects (Esteves et al. 2012). Assessing social impacts should also contribute to sustainable development by supporting the well-being of communities, and ensuring access to benefits while promoting development that is environmental and economically and socially sustainable. All this can seem ambitious, but it is not incompatible with supporting growth and development.

Rationale and objective

The overall assumption, or rationale, behind SIA is that social sustainability is advanced if social issues are identified and taken into consideration when development decisions are made – decisions that may affect the lives and futures of people living in an area subject to change (Vanclay 2003). Social sustainability should be broadly understood as the situation when "the formal and informal processes, systems, structures and relationships actively support the capacity of current and future generations to create healthy and livable communities" (McKenzie 2004, 18). There are criticisms of SIA, such as using it as a tool to legitimize projects rather than promoting sustainability, not integrating it well enough into overall assessment processes, and not using the information and knowledge SIA produces to inform decisions.

SIA is meant to contribute to the knowledge basis of the *carrying capacity* of peoples. Carrying capacity here can be understood to be their ability to adapt to and benefit from change, inform and identify relevant social thresholds, and understand the social and economic objectives of planning and management at the earliest phases of planning (for new projects). The knowledge produced in the overall assessment process informs decisions about how and if resources should be used, under which conditions, and to whose benefit. Like other assessment tools, SIA is a structured approach to considerations on social impacts and their management. By identifying potential costs and benefits for impacted communities in relation to planned activities, mitigation measures may also be identified and enhanced, which can serve to promote better-informed decisions (Vanclay et al. 2015; João et al. 2011). As with other aspects of assessment practice, SIA should be anticipatory, predictive, preventative, mitigative, and ultimately help identify and manage risks.

There is broad recognition of the utility and functional nature of SIA. Conducting an SIA involves knowledge creation and sharing and interactions between the involved parties. While communication between stakeholders is an inherent part of assessment best practice, it also

offers a platform for dialogue and mutual exchange of information and the opportunity to identify critical social factors, which can then be taken into consideration early in the planning and
design phase of activities. This can reduce risks – for communities (risk of not understanding a
project's implications), proponents (risk of conflict or rejection), and regulators (risk of making
a decision without good information or without full knowledge of impacts, or even political
risk). The SIA process ideally allows for knowledge, opinions, and perceptions to be shared,
thereby avoiding the spread of myths and mistakes. It can also be used as a platform to inform
a potentially impacted community about a proposed activity and provide the opportunity for
them to form and share opinions through "respectful" engagement (Wong and Ho 2015). Such
opinions, or knowledge, may also be used to help make decisions, and for communities to adapt
to the changes a project or activity may bring.

With all this in mind, a particularly important part of a SIA process involves engaging potentially impacted communities not only early but throughout the assessment process. Collaboration
with communities and local partners can take place in various ways. It is a prerequisite for
addressing local aspirations and fears that will be key considerations determining whether or
not a project will cause harm, or if it can support local development in ways that reflect local
wishes and needs (Olsen and Hansen 2014). Local support can be important, indeed essential.
If a project causes conflict, or creates community trauma, it can be time-consuming and costly
to companies, trust in political decision-makers may be weakened, and trust in future project
opportunities can be lost (Larsen et al. 2015; Prno and Slocombe 2012).

The social contexts covered in SIA refer to the present state of physical and social settings
in which people live, or in which something happens or a project would be developed. It also
includes the culture that individuals are situated within, including educational settings, individual relationships, and the people and institutions with whom they interact. Social context
is a broader concept than social class or social circle. It involves broad sociocultural aspects and
characteristics. The *International Principles for Social Impact Assessment* and the guidelines for social
impact assessment produced by the International Association for Impact Assessment commonly
define SIA as being:

> the processes of analyzing, monitoring and managing the intended and unintended
> social consequences, both positive and negative, of planned interventions (policies,
> programs, plans, projects) and any social change processes invoked by those interven-
> tions.
>
> *(Vanclay 2003)*

An impact is understood here as the difference between what would happen if an action were
implemented and what would happen without it. This definition is consistent with how an
impact is defined by the International Association for Impact Assessment (Vanclay 2003). Box
4.1 provides a description of how social impacts can be broadly framed. In order for a SIA to
be meaningful and add value to the assessment process, it should answer some key questions:

1. What is presently going on in the potentially impacted communities?
2. What will happen in the communities if the proposed activity is implemented (and if not)?
3. Which potential impacts are considered positive/negative by those impacted?
4. Who would gain from the development and who would lose?
5. What can be done to secure and enhance local benefits and mitigate negative impacts?

(Based on Flyvbjerg 2014)

Aspects of these value-based questions are also present in the framework developed by Vanclay et al. (2015). This is an approach that outlines a four-phase process for conducting an SIA (see Box 4.2).

Actions subject to SIA

There are many reasons to conduct an SIA. It can serve as a tool for evaluating past projects (a learning function); it can be an academic research approach to produce knowledge about a community or parts of a community (a baseline function), or it can be used as a strategic tool to identify actions and responses needed by NGO's, the proponent, governments, or other stakeholders. SIA can also be used as a tool to identify impacts on communities from natural hazards and climate change, and to inform prevention plans (a technical, risk management, or environment–human interactions identification function). And of course it can be a regulatory requirement.

SIA may have a broad or narrow scope. It depends on the focus and purpose of the assessment, the project, and the social or regulatory setting. Most often, SIA is applied according to the regulations of the jurisdiction, or international conventions, or even requirements by financial institutions such as the World Bank and the European Investment Bank. Independent social impact analysis reports with a thorough treatment of social impacts are increasingly being required. However, this is still far from standard practice. It is just as common for social impacts analysis to be absent from EIAs; social issues are not uncommonly given perfunctory attention in assessments (Larsen et al. 2018).

Regulatory SIAs follow the same basic steps, and the procedures are similar across various guidelines and legislation. Regulatory SIAs are typically conducted as a part of an application process for a license, permit, or other types of acceptance of a planned intervention and most often related to resource extraction projects.

In the impact assessment literature, actions are often used as a joint description of the concepts of policies, plans, programs, and projects. Actions in general can be explained as the things done to meet an objective and the needs and activities undertaken to implement it (Therivel and Brown 1999). The concept of "an action" covers a broad variety of activities. A policy is an *inspiration and guidance* for action. A plan is a set of coordinated and timed objectives for the implementation of the policy, and a program is a set of projects in a particular area (Wood and Djeddour 1991). A *project*, in the context of impact assessment, usually refers to physical undertakings. For example, the exploration and then development of a mineral or energy resource and the associated infrastructure. The different levels of action do not refer to the level of detail or the resources used, but only to the strategic level of actions or activities to which the impact assessment relates. For our purposes, it helps to reinforce that SIA is most commonly required and carried out on the project level.

There are many good and substantive guidelines on how to conduct SIA for individual sectors and in particular parts of the world (e.g., State of Queensland 2013; Arctic Council 2019), and for general best practice (e.g., Vanclay et al. 2015). Individual countries can also have detailed descriptions of how they expect the process to be carried out. In this chapter, the focus is not so much on the specific SIA requirements, but rather on important lessons learned in relation to ensuring that an SIA is fulfilling its purpose or potential, and providing a foundation for understanding SIA practice and opportunities.

Social science and knowledge

While the natural sciences that support other members of the impact assessment family are typically based on measurable parameters where thresholds can be determined, SIA methodologies are, in general, qualitative. Thinking of the social sciences as a set of tools that produce value-free knowledge is misguided (Flyvbjerg 2001). Rather, SIA can connect different kinds of knowledge, including technical and scientific, and community, local, and Indigenous. The objective is to provide input to dialogues and decision-making about the problems and risks communities face and how things may be done differently. Through engagement, the perceived weaknesses of social science approaches, such as the difficulty in building cumulative and predictive theory, can actually be turned into strengths, such as embracing and using practice-based and situated knowledge. The contribution of social science will not necessarily be explanatory or predictive models, as may be the case for the natural sciences, but it can provide a broader understanding of connections between the human dimensions of the environment and natural systems that support them (Flyvbjerg 2001).

In many countries, populations can include Indigenous peoples and other communities, who have lived in their areas for many generations. There is a growing recognition that Indigenous peoples, and indeed local people generally, possess important experience and knowledge of social and environmental relationships; SIA can bring this knowledge forward and highlight its value to assessment (Dahl and Hansen 2019). Accessing Indigenous knowledge and local knowledge, however, can be a challenge. Sometimes such knowledge is not immediately available or accessible in a way that can be directly applied to formal impact assessments (Dahl and Hansen 2019). It requires a special effort, and innovative methodological considerations, when Indigenous knowledge and local knowledge is presented or heard, for example, through storytelling, songs, works of art, or other approaches that do not fit neatly into other, common, forms of consultation or engagement. It may also be that people simply do not trust the assessment process, the proponent, or governments, and they may be wary of sharing knowledge with processes that they might see as unresponsive, colonial, or even oppressive. Knowledge sharing can require that proponents build trust and demonstrate respect, and it may require unique arrangements to protect the confidentiality of Indigenous or local knowledge. The last quality can also be difficult to accommodate within the transparency and open information requirements of some assessment processes.

There may be a need to draw on a wide range of data and methods when conducting a SIA. However, it is important to be aware that a SIA does not necessarily have to be based on an extensive or detailed database. Of course, an assessment needs to be based on valid data and knowledge, but depending on what is being investigated, and the strategic level of the action subject to investigation, different types and levels of detail of data will be needed. Finding such in social settings can be tough, and assumptions or requirements about detail may need to be adjusted for complex settings.

SIA approach and procedure

Along with a number of other impact assessment tools, as described in other chapters in this book, SIAs are based on a common basic procedure involving the steps of screening, baseline studies, scoping, analysis, mitigation, documentation, and later monitoring if a decision is made to approve the project. Each step covers a range of activities, but there will be differences depending on the legislation of a particular country or other contextual factors. An overview of the process is provided in Figure 4.1. This gives the impression that it is a linear process. In reality, this is seldom the case. The individual phases tend to overlap, and there are iterations between them. This is a process that reflects typical or best practices, meaning that it would be adapted to the specific decision-making process or contexts. This section illustrates how SIA is integrated into the assessment process, while Box 4.2 provides an illustration of specific SIA steps, or actions, that can be woven throughout the EIA process.

```
┌─────────────────────────┐        ┌─────────────────────────┐
│        Screening        │        │        Baseline         │
│  Investigating whether or│   ──▷  │   Understanding the     │
│  not an action is subject to│     │    impacted area and    │
│      mandatory SIA      │        │      communities        │
└─────────────────────────┘        └─────────────────────────┘
              │
              ▽
┌─────────────────────────┐        ┌─────────────────────────┐
│        Scoping          │        │        Analysis         │
│  Determining the relevant│   ──▷  │  Predicting impacts and │
│   scope of the assessment│        │     assessing their     │
│     and identifying     │        │      significanse       │
│      alternatives       │        │                         │
└─────────────────────────┘        └─────────────────────────┘
              │
              ▽
┌─────────────────────────┐        ┌─────────────────────────┐
│     Strategic Actions   │        │        Reporting        │
│  Identifying mitigation and│  ──▷ │  Documenting the process│
│  enhancement measures   │        │  Informing final decision -│
│   and indicators. Making a│       │        making           │
│      follow up plan     │        │                         │
└─────────────────────────┘        └─────────────────────────┘
```

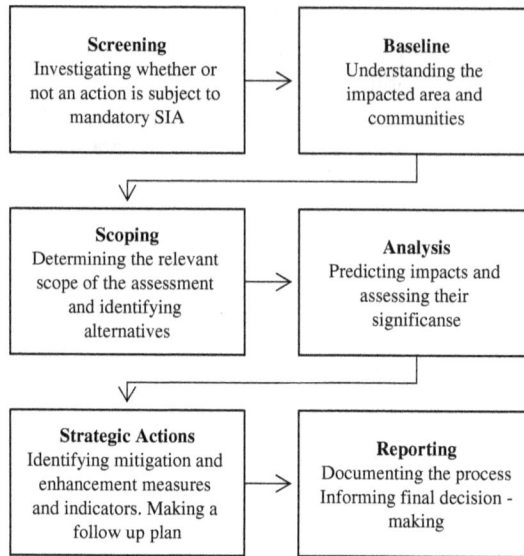

Figure 4.1 Overview of the SIA procedure (each step can inform decisions throughout the process on project design and lead to adjustments of the original plan).

Box 4.2 Four phases for doing a social impact assessment

Phase 1: Understand the issues

1. Understand the proposed project, including all phases and their activities.
2. Know the responsibilities and roles of all involved in or associated with the SIA, including connections and relevance to other studies being undertaken; and identify what laws, regulations, guidance, and standards are applicable.
3. Identify the preliminary "social area of influence" of the project, those likely which may be impacted and those who will benefit. Other stakeholders.
4. Understand the baseline conditions (profiling). Tools can include: (a) stakeholder analysis; (b) outline of the sociopolitical setting; (c) an assessment of the differing needs, interests, values, and objectives of the various parts of the affected communities, including a gender+ analysis (see Chapter 11 in this book); (d) an assessment of the experience of past projects and other historical impacts; (e) trends occurring in those communities; (f) outline the assets, strengths, and weaknesses of the communities; and (g) use surveys or similar instruments to learn about the communities.
5. Engage communities about: (a) the project; (b) similar projects elsewhere and potential interactions; (c) how they can be involved in the SIA; (d) their rights in the regulatory and social impacts for the project; and (e) their access to objection or consent, and feedback mechanisms.
6. Use participatory methods and deliberative processes (see Chapter 14 in this book) to help community members: (a) understand how they will be impacted; (b) determine the acceptability of likely impacts, risks (see Chapter 5 in this book), and projected (or promised) benefits; (c) make informed decisions about the project; (d) facilitate community visioning about desired futures; (e) contribute to mitigation and monitoring plans; (f) understand the worries and concerns of communities; and (g) outline for communities the changes that can occur.

7. Identify the social and human rights issues (part of scoping).
8. Assemble and organize relevant baseline data for key social issues.

Phase 2: Predict, analyze, and assess the likely impacts

9. Through analysis, define the social changes and impacts that will likely result from the project and its alternatives.
10. Consider the indirect impacts.
11. Consider how the project will contribute to any cumulative impacts (see Chapter 3 in this book).
12. Determine how the various affected groups and communities will likely respond.
13. Establish the significance of the predicted changes (i.e., prioritize or rank them, based on engagement and community input).
14. Actively contribute to the design and assessment of project alternatives, including no-go and other options.

Phase 3: Develop and implement strategies

15. Identify ways of addressing potential negative impacts (e.g., by using the mitigation hierarchy).
16. Develop and implement ways of enhancing benefits and project-related opportunities.
17. Develop strategies to support communities in managing change.
18. Develop and implement appropriate feedback, objection, or even potentially consent or agreement mechanisms.
19. Facilitate an agreement-making process between the communities and the developer leading to the drafting of an Impacts and Benefits Agreement (IBA).
20. Assist the proponent in facilitating stakeholder input and drafting a plan which puts into operation the benefits, mitigation measures, monitoring arrangements, and governance arrangements that were agreed to in the IBA, as well as plans for dealing with any ongoing unanticipated issues as they may arise.
21. Put processes in place to assist proponents, governments, Indigenous organizations, and civil society in implementing the measures identified an IBA or plan, and help such stakeholders develop their own respective management action plans for their organizations, outline and confirm roles and responsibilities throughout the implementation of those action plans, and maintain an ongoing role in monitoring.
22. Assist the proponent in developing and implementing ongoing social performance plans that address any obligations outlined during planning or in the assessment.

Phase 4: Monitoring

23. Develop a participatory monitoring plan.
24. Develop indicators to monitor change, performance, and meeting objectives (these may be defined during project planning, the assessment, or they may be conditions of approval).
25. Think about how flexibility, or adaptive management approaches, can be used when implementing a social impact monitoring system.
26. Evaluate, review, audit (the impacts, outcomes, benefits, performance of the project).
27. Use the results of follow-up and monitoring to inform future assessments, and to identify any needed adjustments to the operation of the project.

(Adapted from Vanclay et al. 2015)

Screening

Screening typically refers to the process of investigating whether or not an action is subject to SIA requirements in the given jurisdiction, while the baseline and the scoping steps are the first steps of the actual assessment process. Screening may lead to an early awareness of potential impacts that should be investigated further if it is determined that a SIA should be carried out. The screening stage involves getting an overview of relevant legislation and regulation in the given area as well as getting an overview of national and international standards and guidelines within the field. To get an idea of the potential social impacts an action may have, it is important to consider the entire life cycle of the proposed activity, and the related activities are undertaken to support the project's development and function. This includes all stages of development, from the early investigations, construction, production, and decommissioning, as well as activities after decommissioning (if the project is of a non-permanent character). During the screening, a socially defined *no-go* may already be identified. This is expanded below. A screening can lead to a draft overview of potentially significant impacts throughout the lifecycle that should be investigated in a full SIA process.

Baseline and scoping

If a full SIA is required, then the first step of the SIA begins. This involves baseline studies with the purpose of creating an understanding of the area and the communities that can be impacted. It also requires identifying relevant stakeholders to involve in the process. The scoping stage defines the framework for the SIA and the delimitations in relation to the area subject to investigation, social parameters/conditions, the level of detail for the study, the methods used, and data on which the analysis is based.

The baseline studies carried out as a part of an SIA are essential as they form the foundation on which the rest of the assessment is built. Scoping aims to ensure a good understanding of the potentially impacted location(s) and to identify focus areas to be investigated and addressed later in the process. In SIA, a baseline study can beneficially be conducted prior to the scoping phase. In this way, the baseline informs scoping. In environmental impact assessments, it is typically the other way around as baseline studies are carried out to close knowledge gaps after the scoping phase. The baseline study is the basis for identification and eventual mitigation of potential negative impacts and strengthens potentially positive impacts of the activities planned in the area. The description of baseline conditions provides authorities, proponents, and other relevant actors with an understanding of the starting point for the assessment.

As part of the baseline study, a systematic community profile can be created. The community profile can include a description and discussion of the sociopolitical context, the economy, who makes decisions and who has access to influence development, and an assessment of the different needs, interests, values, and wishes of different groups in a society.

The scoping phase aims to define the extent, content, method and timing of the assessment. Scoping builds on the baseline, but can also be somewhat overlapping. Scoping is about creating an early, and initial, overview of the significant possible impacts a particular project may have. This identifies the conditions (parameters) that must be analyzed and outlined in depth later in the impact assessment. The focus is also on identifying and defining the area that can be affected and the people who live there. In connection with scoping, the process, timeline, and methods used in the impact assessment are also defined. In many places, a scoping document, and perhaps even Terms of Reference (ToR), are published before the actual impact assessment is initiated, so that people have the opportunity to comment on the planned process and the content of the impact assessment before it is carried out.

Vanclay (2003, 2015) notes that social impacts are all impacts that influence peoples' ways of life (Box 4.1). Vanclay proposes a broad number of impact categories to be taken into consideration during the scoping, which can, in general, be grouped into the following categories of changes people may experience from a project. While acknowledging that it is difficult to cover everything of importance for every individual, we can add to these categories:

- Culture can be understood as shared beliefs, customs, norms, values, and language.
- Community covers cohesion, stability, character, services and facilities, including services related to the built environment such as sanitation systems, energy and water supply systems as well as political systems including access to influence regarding peoples' own lives and livelihoods, and the resources provided for this purpose.
- The environment is often understood as the natural environment, which may be subject to various ecosystem services dependent on the quality of the air and water people use; the availability and quality of available fresh foods; the hazard or risk, dust, and noise people are exposed to; their physical safety and their access to and control over resources; their health and well-being; their personal and property rights – particularly whether people are economically affected, or experience personal disadvantage (which may include a violation of their civil liberties); their fears and aspirations (perceptions about their safety, their fears about the future of their community, and their aspirations for their future and the future of their families).

Some types of impacts are best (or only) understood by looking at the relative differences between alternatives to the project, alternative ways of achieving a project's objectives, or even alternatives for implementing a policy, plan, or program (Glasson et al. 2013). By identifying and describing alternative ways of carrying out an activity and evaluating the alternatives in relation to each other, it is possible to understand and compare influences and mitigation measures. The recognition, discussion, and consideration of alternatives are an integral part of any impact assessment process (de Jesus et al. 2005). Consideration of alternatives is an assessment best practice, and alternatives may often be identified as a part of the scoping process. But consideration of alternatives can vary depending on the jurisdiction.

As was noted, SIA may most often be carried out in relation to physical projects. Ideally, the screening phase would exclude no-go sites from a social sustainability perspective. The potential alternatives to a proposed extractive project, which companies may consider in their assessments, are therefore not necessarily alternatives to the proposed activity in general or the location, but rather alternative ways of undertaking the proposed activity (Glasson et al. 2013; Vanclay 2003). The alternatives which are relevant to include in an impact assessment are determined by what is being assessed and what the objectives are. For physical projects, the alternatives to the project might be limited. The more strategic actions that are assessed (policies, plans, programs), the more strategic alternatives may be relevant to bring into play.

For many extractive activities, alternative sites are not usually available within a region as many of the activities related to a resource are very site-specific, particularly the location of the mineral/oil reserve. Some activities, such as shipping, often depend on deep-water access or proximity-related facilities, such as oil refineries, chemical treatment of minerals, or housing of workers. So, one alternative that should always be considered is the "no development" alternative or "zero alternative". As was mentioned, an impact is defined as the difference of what would happen in a community with or without the proposed project being implemented and is therefore always needed to describe the expected development without the project. From an efficiency perspective, the investigation and definition of the zero alternative could already begin during the baseline studies. In addition to the zero alternative, assessments can include, for example, technical and technological alternatives, such as choices on scale, appearance, timing, waste discharges, processing methods, and traffic management.

The alternatives to the proposed project should preferably be described in sufficient detail to "identify potential direct and indirect impacts, including cumulative effects" (see Chapter 3 in this book) and risks. The guidelines from the International Association for Impact Assessment recommend that impact assessment processes involve the affected communities and other stakeholders in identifying alternatives and selection of the preferred option or options (de Jesus 2005).

Analysis

Once the scoping is concluded, and the relevant issues to cover in the analysis are identified, the proposed activity is analyzed in order to predict the potential impacts and assess their significance. In general, a distinction is made as in other impact assessments between three impact categories: direct, indirect, and cumulative impacts. A direct impact is an effect that takes place at the same time and in the same place as the activity. For example, if a large number of temporary workers move into an area during the construction of a project, there may be an increase in demand on social services, housing, or a disproportionate impact on women (e.g., health, safety, loss of economic opportunity, rights) or marginalized populations. Direct impacts can take place at different times in the life cycle of activities. Indirect impacts are understood as influences that may occur at a different time or location than the activity that caused it. Indirect influences can follow complex paths. Cumulative impacts are the impacts to which a particular "recipient" is exposed cumulatively when the influence is seen over time and in connection with other influences of the same type. It may be that in an area, there is a particularly vulnerable group or community. For example, a group with fewer education or economic opportunities, or a lower school completion rate (than a regional or national average), could be particularly sensitive and vulnerable if a new initiative is implemented and other groups increase their earnings due to the project. Inflation can ensue, and this could exacerbate social issues if benefits do not also accrue to vulnerable communities. The potential for increased cumulative impacts on vulnerable groups could also rise. These impacts and their interactions can be difficult to measure and predict.

Cumulative influences can, however, also be about the total influences on a specific group, where the influences are of a different nature, and impact communities and groups differently (Dales 2011). For example, some forms of housing may be lost, but it may have little, if any, environmental significance. But if there are rent increases at the same time, it can again lead to more people becoming homeless, or reduced disposable income. Or if the number of people in the area increases, thereby increasing demand for homes, then prices may rise or other impacts may follow. Cumulative impacts can, however, also be influences that counteract each other. A housing shortage may be solved in part by new home construction spurred by the project, but the higher home prices may create new or intensified affordability challenges.

Strategic actions

The next step in the SIA process concerns the identification of measures/actions that can either mitigate (avoid, minimize, or compensate) unwanted impacts or strengthen desired impacts or the possibility that the desired impacts occur. In addition, strategies to empower society in the assessment process can be provided so that residents in impacted communities can better deal with and adapt to the changes that a project might cause.

Monitoring impacts that occur during the implementation of a project (if it is approved) will support follow-up or adjustment to the project at all stages. The type of indicators needed may vary depending on the project stages. For example, those used during the construction phase can be different from those applied during the operation or decommissioning stages. This allows tracking of impacts from development over the lifecycle of the project. Indicators can include population growth, health and safety indicators, social indicators (e.g., a rise in access needs for women's health services, crime rates, a rise in youth at risk), household income, employment levels, and other factors identified during the assessment related to the project. Once indicators have been identified and considered, there must be a description of how they can be monitored, who is responsible for the monitoring, and what efforts must be initiated if an unexpected impact occurs. Dealing with uncertainties, even by recognizing their potential, is important and can help plan for mitigation and follow-up or responses as might be needed.

Once the actual analysis of the individual factors has been carried out and possible initiatives have been identified to mitigate or respond to the impacts that have been identified, a monitoring plan is developed. Monitoring is about ensuring that it is detected in time and that action is taken if something unexpected happens or if, for example, an impact occurs, which was initially assessed as being very unlikely to occur. It allows regulators to ensure that a proponent meets the requirements of the assessment and any conditions of approvals. Monitoring can also help gauge the performance of projects and whether or not they meet the stated social, economic, or cultural objectives (e.g., benefits, impact mitigation, or distributional qualities). This would include a description of objectives, or strategic actions and outcomes, including mitigating measures and measures to increase the likelihood of desired objectives being achieved, as well as a proposal for a plan for monitoring and follow-up. As part of a monitoring plan, ongoing evaluation of the impact assessment and periodic reviews can be used. An action plan can not only outline the strategy and needs; it may prioritize them. But it is also important to assess the consequences of an activity after mitigation measures have been initiated, so that the SIA offers a sense of how effective the various mitigation efforts are not only expected to be, but how they actually appear once the project is underway.

In some jurisdictions, this phase also has a special purpose, specifically to help identify issues that can be included in a potential Impact Benefit Agreement (IBAs are discussed further below). This occurs in Greenland, parts of Canada, and Alaska, where some statutory impact assessments associated with natural resources projects have led to the IBAs with Indigenous communities (Hansen et al. 2019).

Reporting

As was noted above, an impact assessment is a comprehensive process, many choices are made during the process, and decisions are made about relevance, materiality, and prioritization. The purpose of the SIA report is not solely to state what conclusions have been reached. The report must also explain the process and describe how conclusions or decisions are made during each step. The methods must be clear and reproducible, and the data transparent.

In connection with the reporting, it is also relevant to reflect on whether there are conditions that emerged during the process, which may have a bearing on future assessments. For example, it may emerge that there is a group or part of a community that is particularly vulnerable, not only because of the project, but perhaps more generally. This can mean that special studies and/or efforts are needed to understand impacts to, and even opportunities for, this group. The way in which actors have been involved should also be described and any uncertainties in relation to data and results and, for example, risk assessments can be explained (see Chapter 5 for a discussion of risk assessment). The conclusion of the report will be a summary in non-technical language that explains the area studied, the main impacts and opportunities, the key proposals for improvements/adaptations, mitigating measures, and, potentially, recommendations for an Impact Benefit Agreement (discussed below).

Social thresholds and no-go factors

While SIA is to a large extent already mandatory in various jurisdictions and there is a general recognition of SIA potential for promoting sustainable development in communities, the tool is still critiqued as being carried out too late in the process when it comes to decision-making about approval and implementation, especially for large-scale projects (Hansen and Johnstone 2019). Mining provides an illustration of this. Mining begins with extensive exploration activities before a mine is developed, which can cause impacts on local communities well before a decision is made to pursue an actual mining license, or before extraction begins. This means that potential significant and irreversible social impacts that should, from an ethical perspective, lead to the decision not to propose an activity in a certain area (social no-go factors) are likely to be overlooked until later in the process, after a company may already have invested in the project prior to applying for permits or licenses. A "social no-go factor" can be defined as: *A social no-go factor is an expected adverse social impact imposed by a project, plan, or policy on local communities that violates human rights and for which no effective mitigation measures can be determined* (Aaen et al. 2021, 4).

For example, when mining companies consider potential mine sites, the focus is typically on factors such as geology (the key quality for sure), mine design (e.g., pit or underground), infrastructure, logistics (such as road access or power supplies or water availability), environmental conditions, and even politics (Aaen et al. 2021). But each category may involve "no-go factors", which might make the company modify or avoid specific activities in an area, or design mitigation measures. Social factors should also always be taken into consideration at an early stage when considering the desirability of investing in an area. Some projects are also more flexible in the choice of site, technology, or other factors, while other (such as mining) are much less so.

To ensure that social factors are considered early in the planning process, public and private sector organizations can use screening to evaluate different alternatives (for sites or other project characteristics) to make sure that human rights can be protected (Aaen et al. 2021). This can help define social issues that might inform *go* or *no-go* decisions. This suggests a form of human rights assessment, which may be important for proponents working in jurisdictions where social/political issues and related challenges are particularly difficult. It can help mitigate a range of political, social, financial, and even reputational risks. Such advance work during screening can inform a potential future full SIA and help define social thresholds and indicators to be included during the eventual monitoring of a project, if approval is obtained.

Making SIA meaningful

Good SIA practice should lead to improved local community development outcomes and thereby contribute to social sustainability. Even though SIA is generally recognized as a tool for promoting social sustainability, its rationale has, and is, still being contested because the social benefits that proponents and governments may ascribe to projects are not always as clear to affected communities as the project's adverse impacts can be (Gulakov 2020; Esteves et al. 2012). In addition, there are challenges to SIA practices, including climate change, awareness of the gendered nature of development, and increasing inequalities (Parsons 2020). Recent studies have found that a lack of regulation (the conditions for requirement) and guidelines for assessing social issues, even in countries where SIA is an integrated part of an EIA, results in social issues being addressed in a superficial manner and can lead to conflict and a lack of confidence in assessment processes (e.g., Kamran 2020; Larsen et al. 2018).

If SIA is to be meaningful to all involved in the process, and to potentially impacted communities, it is important that SIA does not merely become an academic exercise leading to legitimizing the approval of a project. Engagement of potentially impacted communities is requisite in SIA. Without such, it will not provide a thorough understanding of communities or be able to identify and address significant social impacts. The *meaningfulness* of SIA therefore relates directly to the engagement of citizens in the process. Meaningful engagement implies that those communities potentially affected by development are enlisted into the project planning and SIA process, and that project proponents are open to adjusting proposed projects and developing new plans, or even discarding existing ones, based on the knowledge and values of those affected (Noble and Udofia 2015; Prno and Slocombe 2012).

One tool that can be helpful in engaging the public in a meaningful manner is a Community Protocol. This provides a community profile (as mentioned in the baseline section of this chapter) outlined in an objective way, and based on the voices of community members. Community protocols serve to articulate the values, governance processes, and priorities of communities. Meaningful engagement not only means providing opportunities for those affected by development to become engaged in planning and decision-making, it also ensures that they have the capacity to do so (Noble and Hanna 2015). Kwiatkowski et al. (2009) also note that some Indigenous communities can lack the financial and other capacities (e.g., internal expertise, human resources, infrastructure) to become meaningfully engaged and to remain engaged throughout an assessment process, due in part, to the very size and complexity of the major project assessments and the timelines involved in review processes. This can lead to community *participation fatigue*, which in time weakens the efficacy of participation and engagement.

If it is to be applied and be effective, SIA needs to be meaningful from a procedural perspective – it needs to be is connected to the decision-making process; have a strong content perspective that ensures that new information and issues are taken into consideration; and supports engagement by ensuring that the SIA is open and meaningful to all involved parties, and that they have the capacity and resources to participate and engage.

Strategic action and benefit-sharing

At the national level, benefits from the development of new projects are often expected to derive from taxation and revenue distribution, job creation, project-related infrastructure and social services, and various multiplier effects such as the growth of secondary industries and increased purchasing power (Wilson 2019; Fjaertoft 2015). Governments are responsible for setting legislation and regulations related to the above. Governments are also responsible for land agreements and treaties, which may determine local and Indigenous rights to negotiate benefits. In some countries, governments establish sovereign wealth funds (Hansen et al. 2016). Unfortunately, the people who experience most of the direct social costs of development projects in various cases might not benefit from the "ripple effect" of revenue distribution and job creation (Wilson 2019; Hansen et al. 2018). Benefits from projects are not always as evident to affected communities as the project's adverse impacts can be (Gulakov 2020; Esteves et al. 2012). For example, impacted communities may not receive many of the tax payments, resource royalties, or other revenues generated by projects. Employment and procurement opportunities may exist, but local workers and businesses may not have the skills or experience to compete for jobs and contracts. Addressing such issues is a persistent challenge in assessment practice. It points to the need to identify ways that projects might also build capacity and skills that can have a lasting impact long after a project is decommissioned, and will have a social and economic sustainability benefit.

In order to enhance the potential local benefits from a project, strategic actions related to mitigation of negative outcomes and enhancement of potential positive ones are often identified as a part of the SIA process (Hansen et al. 2016). In some countries, IBAs are used to help communities define and gain benefits from projects. IBAs are also sometimes known as Community Development Agreements, Community Benefits Agreements, or Indigenous Land Use Agreements (O'Faircheallaigh 2013). An IBA is entered into between varying parties, but most often between the developer and a community organization, an Indigenous government, or a local government. An IBA regulates and defines processes and mechanisms for benefit-sharing, or in some cases even providing compensation. They are also used as a platform for negotiating social investments that can enhance the positive impacts of a project, or help ensure that some benefits accrue to people living in the project area.

The IBA is usually understood as a formal contract between an impacted party and the developer; however, governments are sometimes also involved, depending on the jurisdiction (Hansen et al. 2018). IBAs between companies and Indigenous and local communities are common in some countries, including Australia, Canada, and Greenland (Sosa and Keenan 2001). But globally, the experience is quite varied (Wilson 2019).

An IBA can include an outline of the expected impacts of the project and the related commitments and responsibilities of all relevant parties (Vanclay 2002). An overview of different types of benefits is provided in Table 4.1. IBAs can have different forms and purposes, and the parties involved can be quite different, depending on the jurisdiction. In Canada, for example, government involvement is rare. IBAs can be negotiated on an individual level between companies and communities or individuals, such as a hunter using an area disturbed by mining. In Alaska, oil companies have established mitigation funds that *boards* representing local communities can use in order to mitigate impacts identified before or while oil production is taking place in an area (Hansen and Ipalook 2020). While in Greenland, governments are involved – an IBA is negotiated between three parties: the Self-Government, the Municipality, and the proponent, together as an inherent part of the SIA process (Hansen et al. 2018).

An IBA describes how the local community will share in the benefits of the activity through, for example, taxation and revenue distribution, job creation, ownership of companies and shares, negotiated agreements, community development programs, and social investment expenditure (Wilson 2019). But agreements such as IBA's, however, do not by themselves guarantee successful outcomes. Some IBAs in Northern Canada have been found to be inequitable or simply not achieving what communities had hoped for, and some may have the effect of drawing attention away from broader industrial impacts and different, more desirable development opportunities

Table 4.1 Elements of benefit-sharing models for local and Indigenous communities (adapted from Wilson 2019)

Type of model	Description
A. State-controlled benefit-sharing	
Taxation, revenue payments, and revenue distribution	Governments have the responsibility to establish and enforce regulations, and there is increasing pressure for greater transparency of revenues paid to governments and how these are used. Companies near their own responsibility for paying taxes, not avoiding them, and where required, e.g., under the Extractive Industries Transparency Initiative (EITI), to report on what they pay to governments.
Local content obligations	Targets for the hiring of local workers and procurements of local goods and services may be included in host government agreements with companies, and in cases is legislated. Government-mandated local content is frequently interpreted as "national" content, rather than targeting local and Indigenous communities.
Mandatory social investment	Social investment spending can be mandatory as part of a host government agreement or national legislation, whereby companies are required to invest in infrastructure programs, such as road construction or health facilities, as a condition of their license.
B. Voluntary company-led initiatives	
Philanthropy	Companies may voluntarily engage in community spending in addition to their mandatory obligations under contracts and licenses. Philanthropic support might include medical facilities, cultural or sports programs, scholarships, and environmental projects.
Strategic social investment	Increasingly companies seek to target their social investment spending on programs designed to survive beyond the life of the industrial project and/or to create value for the industrial project. These might include micro-credit programs, local livelihoods support programs, skills training, enterprise development support, or conservation programs.
C. Partnership model	
Voluntary local content initiatives	Companies may develop partnership programs based on voluntary targets and initiatives to train and bring in the local and Indigenous workforce to a project, with training and enterprise support linked to opportunities to secure employment or contracts, often with an element of preferential contracting. This may or may not form part of a wider benefit-sharing agreement.
Benefit-sharing agreements	Benefit-sharing agreements are negotiated directly with communities and may include cash payments, profit sharing, local hiring, skills development, education, cultural support and environmental protection. These are likely to be closely related to impact assessments and may also provide the basis for a process of free, prior, and informed consent (FPIC).
D. Indigenous ownership and control	
Indigenous ownership	Indigenous ownership might include Indigenous peoples' ownership of companies or equity shares in the enterprise involved in extracting or processing resources or enterprises providing services to the industry. Opportunities can be enhanced through government support and preferential hiring and contracting.
Indigenous control	Indigenous control relates to Indigenous peoples' right to determine their own development priorities and strategies and includes participation in strategic-level decision-making on resource-related policies, programs and regulations, including resource mapping, zoning and land allocations, and processes of FPIC where appropriate.

(Caine and Krogman 2010). There is a need to broaden the understanding of what *benefits* entail and to be clearer about what benefits are actually going to accrue to communities (Wilson 2019). Benefits can mean more than jobs or royalties; they can also mean longer-term opportunities or improvements to a range of well-being factors (social, economic, and even ecological) that are important for communities and their sustainability, but not necessarily apparent at first glance to proponents.

Concluding remarks

Social Impact Assessment is a tool that can promote social (and environmental) sustainability in the planning and decision-making processes for projects that could impact the well-being of communities. It is a key part of environmental impact assessment. SIA offers a systematic approach to the management of social change processes. But in order for SIA to be meaningful and effective, it is important that the SIA provides early information about communities and emerging or potential issues to create an understanding of community values and the values and qualities that may be affected by a project. It is also important to outline and understand the desired directions and capacities of the affected communities and to understand economic and other opportunities.

Often, questions about projects will remain even after an SIA is done. And it may not always be clear to those affected how the final decision will be made, and whether communities or local institutions will be part of that discussion. To ensure that benefits and costs are accounted for and distributed fairly, IBAs can be used to negotiate and enter into formal agreements on benefit-sharing. But SIAs and IBAs cannot alone create or secure sustainability. That may require changes or other actions beyond the scope of the project, or the abilities and capacities of a proponent. As part of good assessment practice, SIA needs to be seen and applied as a part of an ongoing dialogue, and not as just as a source of information that supports a one-time decision about a project. SIA can be an important tool for realizing sustainability, and for making projects, planning, and decision-making better.

References

Aaen, S. B., Hansen, A. M., & Kladis, A. 2021. "Social no-go Factors in Mine Site Selection". *Extractive Industries and Society*, 8:2. https://doi.org/10.1016/j.exis.2021.100896

Caine, K. J., & Krogman, N. (2010). Powerful or Just Plain Power-Full? A Power Analysis of Impact and Benefit Agreements in Canada's North. *Organization & Environment*, 23(1): 76–98. https://www.jstor.org/stable/27068674

Dahl, P. P. E., & Hansen, A. M. 2019. "Does Indigenous Knowledge Occur in and Influence Impact Assessment Reports? Exploring Consultation Remarks in Three Cases of Mining Projects in Greenland". *Arctic Review*, 10, 165–189. https://doi.org/10.23865/arctic.v10.1344

Dales, J. T. 2011. "Death by a Thousand Cuts: Incorporating Cumulative Effects in Australia's Environmental Protection and Biodiversity Conservation Act". *Pacific Rim Law and Policy Journal*, 20:1, 149–178

de Jesus, J., Bingham, C., Croal, P., & Fuggle, R. 2005. "Alternatives in Project EIA". *International Association for Impact Assessment*, FASTIPS No. 11. Retrieved from http://www.iaia.org/uploads/pdf/FasTips_11_AlternativesinProjectEIA.pdf

Esteves, A. M., Franks, D., & Vanclay, F. 2012. "Social Impact Assessment: The State of the Art". *Impact Assessment & Project Appraisal*, 30:1, 34–42. https://doi.org/10.1080/14615517.2012.660356

Fjaertoft, D., & Modeling. 2015. "Russian Regional Economic Ripple Effects of the Oil and Gas Industry: Case Study of the Republic of Komi". *Regional Research of Russia*, 5, 109–121.

Flyvbjerg, B. 2001. *Making Social Science Matter: Why Social Inquiry Fails and How it Can Succeed Again* (S. Sampson, Trans.). Cambridge: Cambridge University Press. https://doi.org/10.1017/CBO9780511810503

Glasson, J., Therivel, R., & Chadwick, A. 2013. *Introduction to Environmental Impact Assessment*. London, New York: Routledge.

Gulakov, I., Vanclay, F., & Arts, J. 2020. "Modifying Social Impact Assessment to Enhance the Effectiveness of Company Social Investment Strategies in Contributing to Local Community Development". *Impact Assessment and Project Appraisal*, 38:5, 382–396, https://doi.org/10.1080/14615517.2020.1765302

Hansen, A., Vanclay, F., Croal, P., & Skjervedal, A. S. 2016. "Managing the Social Impacts of the Rapidly Expanding Extractive Industries in Greenland". *Extractive Industries and Society*, 3:1, 2016-01, 25–33.

Hansen, A. M., Larsen, S. V., & Noble, B. (2018). "Social and Environmental Impact Assessments in the Arctic". In M. Nuttall, T. Christensen, & M. Siegert (eds.), *The Routledge Handbook of the Polar Regions*. Routledge International Handbooks.

Hansen, A.M., & Ipalook, P. (2020). "Local views on oil development in a village on the North Slope of Alaska". In R.L. Johnstone & A.M. Hansen (eds), *Regulation of Extractive Industries: Community Engagement in the Arctic*. Routledge Research in Polar Law.

IWG (Interagency Working Group on Environmental Justice). 2011. "Agency Responses to Comments Received During the 2011 Alaska Forum on the Environment". *EJ IWG Community Dialogue*, 7–11.

João, E., Vanclay, F., & den Broeder, L. 2011. "Emphasising Enhancement in all Forms of Impact Assessment". *Impact Assessment & Project Appraisal*, 29:3, 170–180.

Khan, I. 2020. "Critiquing Social Impact Assessments: Ornamentation or Reality in the Bangladeshi Electricity Infrastructure Sector?". *Energy Research & Social Science*, 60.

Koivurova, T. 2008. "Transboundary Environmental Assessment in the Arctic". *Impact Assessment and Project Appraisals* 26:4. 265–275 https://doi.org/10.3152/146155108X366031

Kwiatkowski, R. E., Tikhonov, C., Peace, D. M. & Bourassa, C. (2009) "Canadian Indigenous engagement and capacity building in health impact assessment", *Impact Assessment and Project Appraisal*, 27(1): 57–67, DOI: 10.3152/146155109X413046

Larsen, S.V., Hansen, A. M., & Nielsen, H. 2018. "The Role of EIA and Weak Assessments of Social Impacts in Conflicts Over Implementation of Renewable Energy Policies". *Energy Policy*, 115, 43–53. https://doi.org/10.1016/j.enpol.2018.01.002

Larsen, S., Hansen, A., Lyhne, I., Aaen, S., Ritter, E., & Nielsen, H. 2015. "Social Impact Assessment in Europe: A Study of Social Impacts in Three Danish Cases". *Journal of Environmental Assessment Policy and Management*, 17:4. https://doi.org/10.1142/S1464333215500386

McKenzie, S. (2004). "Social Sustainability: Towards Some Definitions". *Working Paper Series, 27*. Magill, Australia: University of South Australia, Hawke Research Institute.

Noble, F., & Udofia, A. 2015. *Protectors of the Land: Toward an EA Process that Works for Aboriginal Communities and Developers*. Ottawa, ON: MacDonald-Laurier Institute.

Noble, B. N., & Hanna, K.S. 2015. "Environmental Assessment in the Arctic: A Gap Analysis and Research Agenda". *Arctic*, 68:3, 341–355.

Faircheallaigh, C. 2013. "Community Development Agreements in the Mining Industry: An Emerging Global Phenomenon". *Community Development*, 44:2, 222–238. https://doi.org/10.1080/15575330.2012.705872

Olsen, A. H. S.., & Hansen, A. M. 2014. "Perceptions of Public Participation in Environmental Impact Assessment: A Case Study of Offshore Oil Exploration Industry in Northwest Greenland". *Impact Assessment & Project Appraisal*. 32:1, 72–80.

Parsons R. 2020. "Forces for Change in Social Impact Assessment". *Impact Assessment and Project Appraisal*, 38:4, 278–286. https://doi.org/10.1080/14615517.2019.1692585

Petrova, S., & Marinova, D. 2015. "Using 'Soft' and 'Hard' Social Impact Indicators to Understand Societal Change Caused by Mining: A Western Australia Case Study". *Impact Assessment and Project Appraisal*, 33:1, 16–27. https://doi.org/10.1080/14615517.2014.967987

Prno, J., & Slocombe, S. 2012. "Exploring the Origins of 'Social License to Operate' in the Mining Sector: Perspectives from Governance and Sustainability Theories". *Resources Policy*, 37:3, 346–357.

Sosa, I., & Keenan, K. 2001. "Impact Benefit Agreements Between Aboriginal Communities and Mining Companies: Their Use in Canada". *Canadian Environmental Law Association, Environmental Mining Council of British Columbia, CooperAcción*.

State of Queensland, Department State Development, Infrastructure and Planning, July 2013, 100 George Street, Brisbane Qld 4000. (Australia)

Therivel, R. & Brown, A. L. (1999). Methods of strategic environmental assessment. In: J. Petts (ed). Handbook of Environmental Impact Assessment, vol 1, pp. 441–464.

Vanclay, F., Esteves, A., Aucamp, I., & Franks, D. 2015. "Social Impact Assessment: Guidance for Assessing and Managing the Social Impacts of Projects". IAIA (International Association for Impact Assessment). Retrieved from https://www.iaia.org/uploads/pdf/SIA_Guidance_Document_IAIA.pdf

Vanclay, F. 2003. "International Principles for Social Impact Assessment". *Impact Assessment and Project Appraisal*, 21:1, 5–12.

Vanclay, F. 2002. "Conceptualising Social Impacts". *Environmental Impact Assessment Review*, 22:3, 183–211.

Wilson, E. 2019. "What is Benefit Sharing? Respecting Indigenous Rights and Addressing Inequities in Arctic Resource Projects". *Resources*, 8:2, 74.

Wong, C. H. M., & Ho, W. 2015. "Roles of Social Impact Assessment Practitioners". *Environmental Impact Assessment Review*, 50, 124–133. https://doi.org/10.1016/j.eiar.2014.09.008

5

RISK ASSESSMENT AND RISK ANALYSIS

Ayla De Grandpré and Karaline Reimer

Introduction

Risk is part of life. It is present in almost every decision we make, even when our perception of risk is unconscious. A large part of environmental impact assessment (EIA) practice is about identifying the potential negative outcomes, or risks, of a project and managing or mitigating them, and ultimately making decisions with knowledge of risks and their potential outcomes. Perceived risks – to people, buildings, ecosystems, or environmental values – can form the core of what EIA examines, and what participation asks proponents to account for and mitigate. In EIA practice, public participation can come down to the following questions about risks:

- If you build it, what are the chances it will leak?
- If a leak happens, what is the likelihood it will be at a water crossing?
- What are the risks to fish if that happens?
- Could it blow up?
- What types of accidents can happen?
- What are the risks to human health?
- What risks does your project pose to our rivers and lakes?
- What are the risks to wildlife?
- What are the risks during construction, and what are ones during operation?
- There was an accident at a similar facility, and people were hurt. What is the chance it will happen here, and what are you doing to make sure it doesn't happen here?
- How often do such accidents happen?
- Can you tell me the risk will never happen?

Risk assessment is a fundamental part of effective EIA. While at its core a risk assessment can seem to be a simple calculation of a problematic consequence multiplied by the probability or likelihood of it occurring, there are many factors that complicate this basic formula. Uncertainties for calculating and assessing risk come in many forms. For example, data discrepancies, complicated biological systems, personal or institutional biases, and perceptions and tolerances, all factor into risk assessment. There is also no one template for conducting a risk assessment. The type of project, its location, and the data and information we have determine which qualitative and/or

DOI: 10.4324/9780429282492-6

quantitative methodologies are appropriate, and ultimately how to communicate those results. This chapter outlines concepts, terminology, and methods for risk assessment, and provides a fundamental guide to the risk assessment process as it can be applied to environmental impact assessment practice. The focus is on quantitative and qualitative approaches common in project design and engineering. Natural events and disaster risk management in EIA are discussed in Chapter 9.

Risk analysis and EIA?

In the 1970s, along with the creation of the Environmental Protection Agency (EPA), the US federal government enacted several new environmental laws, including the 1977 Clean Water Act (Suter 2009). The implementation and enforcement of these laws required regulators to assess risks to the environment and public, and to use such assessments as the foundation for decision-making. Early EIA practitioners employed risk analysis to understand the effects of projects on human and ecosystem health. In EIA, risk functions as a driver of caution and proactive planning. Risk is what causes society to worry, regulators to be vigilant, and requires processes such as EIA, which require proponents to analyze and mitigate the risks associated with a project. In this way, risk analysis is and has long been an essential constituent of the EIA process (Morgan 1999).

Risk analysis in EIA has become more complex and multidisciplinary. While the evolution of EIA has varied among various jurisdictions, there has been a general broadening of scope from individual effects toward cumulative social, health, and ecological effects (Suter 2009). Today, the objectives of EIA include identification, prediction, evaluation, and mitigation of any risks (or "possible adverse effects"), associated with the proposed activity, and ultimately support planning and decision-making about projects (Hanna 2016; Therivel 2018). Risk assessment is woven throughout the EIA process (see Chapter 1 in this book).

Within EIA, risk analysis is a flexible tool (or set of tools) used for understanding and predicting the likely outcomes of risks (e.g., an accident), and whether or not they can be mitigated, in order to understand a project's impacts and decide if it is worth doing (MELP 2000). Knowing and understanding risks and their characteristics helps EIA practitioners reduce uncertainty and increase the quality of decisions. In turn, this creates a more robust rationality for EIA decisions and reduces the reliance on predicative worst-case scenarios (Suter et al. 1987). Awareness of risks creates an opportunity for the proponent to establish practices and strategies that mitigate them, ideally leading to better decisions and project outcomes (MELP 2000). The most effective risk assessments for EIA include clear management goals created in collaboration with decision-makers, assessors, scientists, and stakeholders – an interdisciplinary approach that leads to the clear communication of objectives to all parties (Dale et al. 2008.

Risk analysis in EIAs generally lacks standardization, particularly in terms of the methodologies and depth to which the analysis is performed (Ostrom 2019). While this allows for flexible and contextually appropriate approaches, a lack of standardization can make cross-comparisons difficult and may create ambiguity in terms of risk interpretation and the methodological rigor involved. Another issue that arises is the tendency to adopt an overly reductionist approach to risk. Given the environment is a complex system involving endless linkages and feedback loops at multiple scales, adopting a reductionist lens and isolating risks could lead to misrepresentations and poor-quality data (MELP 2000). Moving forward, risk analysis and management in EIA may need to better balance standardization with the creativity of methods, and reductionism with holism, or systems thinking, in terms of the approach to risk.

In this chapter, risk analysis is defined as the consolidation of the assessment, evaluation, and management of risks (see "The risk analysis process" section below for further detail). Risk assessment, risk evaluation, and risk management are therefore distinct but complementary stages involved in the risk analysis framework are established. A proponent may use different methods, and not all stages of analysis may be included. Thus, the stages of risk analysis will likely look different depending on the context and application of the analysis. Risk analysis, or the information it provides, is used at all stages of the EIA process (see Chapter 1 in this book).

Defining and understanding risk and related concepts

Risk is inherently subjective; it can be understood in different ways depending on who is defining the term and for what purpose. While standardized definitions of risk have been established by subject-matter authorities, such as the Society for Risk Analysis, to create consensus, there is still considerable debate regarding how it should be interpreted (Aven 2016). In broad terms, risk is understood as the potential for an adverse natural and/or human-caused event, or "hazard", that produces undesirable consequences for humans (Hewitt 1983; Ostrom 2019). For example, a mudslide, hurricane, or earthquake might be categorized as natural hazards, while human hazards refer to events such as social unrest, political instability, poverty, and food insecurity (see Chapter 9 in this book for a discussion of disasters). To illustrate, climate change is commonly considered an environmental hazard. However, given that the environment both impacts and is impacted by humans, and that human culture, activities, and decision-making drive it, climate change can also be considered a human-caused hazard (Adger et al. 2013; Tanner et al. 2014).

Because risk can also be difficult to capture and measure, it is often defined as the product of two key elements:

1. The *consequence*, or the seriousness of the harm caused by an impact.
2. The *likelihood* or frequency (used instead of probability) of exposure to the impact, which helps reflect uncertainty.

(Haroon and McMillan 2018)

The context and characteristics of a particular risk also provide various ways to understand the term. Risk can be characterized based on the type, likelihood, severity, timing and spatial distribution of adverse outcomes, and the size and characteristics of the population exposed to risk (MacDonell et al. 2018). For example, a risk may be categorized based on *what* is being affected (e.g., project, product, operational, environmental, and health risks) but also *who* is being affected (e.g., individual risks which relate to individual concerns and impacts that relate to individual values, or societal risks, referring to widespread impacts that have implications for general society) (Therivel 2018). So, when defining risk or interpreting the use of the term, we need to understand both the context in which it is being used and the intended purpose.

Defining risk and its vocabulary is an important first step in approaching risk analysis. The language used can have an impact on the methods used and the effectiveness of the assessment (Aven 2016). For example, if risk is defined as the "exposure to a proposition", the consequence is highlighted, and the probability of consequence(s) may be de-emphasized (Aven et al. 2018, 4). As a result, the assessment may not employ the needed predictive tools and/or models, or it may not explicitly communicate the probability of the unwanted occurrence. This makes the analysis inaccurate or misleading because of the lack of probability represented in the results. While many interpretations of risk exist, effective risk analyses recognize, utilize, and communicate consequence(s) and likelihood (or probability may be used), but there are many variations to this basic formula (see Figure 5.1 for two examples).

In statistics, *probability* and *likelihood* are not the same thing.[1] But *likelihood* is used throughout risk analysis, including in standards and guidelines (Rausand and Haugen 2020, 44). While determining the *likelihood of something happening* is based on ideas from probability theory, in risk analysis it is not true-to-life to assume that events are repeatable under the same conditions. In other words, we cannot have the same landslide repeatedly happen under the same conditions. For example, a proponent or regulator will think about routing options for a pipeline, the related geotechnical hazards, any special provisions for construction, and ultimately the likelihood of a landslide happening at certain points. This limits the utility of a strict or classical (frequentist) approach to probability.

In risk analysis, analysts prefer a Bayesian approach. It considers probability as subjective. Now, the word *subjective* may have a negative connotation, so some practitioners prefer to use the term *personal probability*, since the probability is a personal judgment about whether an event will occur, or not, based on the best knowledge and information available to the analyst (Rausand and Haugen 2020).

Establishing an understanding of risk and its associated terms is also critical for avoiding confusion and misunderstandings, especially in risk communication and decision-making. The difference between risk and impact, for example, is often poorly understood. Though interrelated, risk describes the abstract idea or probability (or likelihood) of an adverse event occurring, and the impact occurs only if/when the risk is realized and becomes materialized (Haroon and McMillan 2018; Therivel 2018). Similarly, the concepts of the incident versus the accident can be mistakenly used as interchangeable terms. Hazards, or a source of danger that could increase the effect or likelihood of a risk, can initiate both *incidents* and *accidents* (Therivel 2018). An *incident* refers to the sequence of actions or events leading up to the impact of a "near-miss", and an *accident* refers to the harmful consequence(s) that may occur because of these actions or events (Drupsteen, Groenewg, and Zwetsloot 2013). So, not all incidents are accidents, but all accidents are *incidents*. These are important terms, not only in understanding possible outcomes but also for communicating risk.

Finally, we need to recognize that all risk involves uncertainty, because the likelihood and consequence of an impact are usually not understood or predicted with absolute confidence (Park and Shapira 2018). You cannot effectively analyze risk without acknowledging such uncertainty (Suter et al. 1987). Addressing uncertainty in risk assessment is critical to the process. Unaddressed and unidentified uncertainties affect the accuracy of evaluation and are highly consequential for risk management (Thompson 2002). Uncertainty is particularly endemic to environmental risk assessments due to the difficulty in obtaining complete knowledge of complex systems and the inherent randomness of nature (Skinner et al. 2014). But we may have little information or understanding of how complex natural systems will react to cumulative effects of development such as landscape change, climate change, or incremental increase in a pollutant.

Risk = Likelihood x Consequence

or

Risk = Probability x Losses

Figure 5.1 Two risk equations.

With respect to EIA, Suter et al. (1987) identify two types of uncertainty relevant to risk assessments: (1) defined uncertainty, which describes what is known to be unknown; and (2) undefined uncertainty, which refers to what is inherently unknowable about the state of the world (the "unknown unknowns"). High-quality risk assessments will identify and address defined uncertainties, such as model errors or oversimplifications of likelihood and consequence estimates, parameter uncertainty, or natural stochasticity (natural irregularity or randomness that makes it impossible to capture and measure complex ecological variables precisely); however, a margin of error should be maintained for undefined uncertainties (Suter et al. 1987).

This conceptualization, however, does not create an explicit distinction between uncertainty and variability. The National Research Council's (NRC) report *Science and Judgment in Risk Assessment* highlighted the importance of creating the distinction between uncertainty and variability (NRC 1994). The NRC work argues that uncertainty and variability have different functions in risk analysis, and thus should be treated as distinct but interrelated variables. For instance, uncertainty occurs due to imperfect knowledge or the impossibility of prediction, imploring practitioners to examine confidence limits, possible standard deviation, and probability of over- or underestimation (Thompson 2002). But variability occurs due to tangible and inherent variation that cannot be removed by better knowledge or measurement (e.g., climatic variability), and therefore requires coping or adaptation strategies for management (Thompson 2002). Establishing a clear distinction between uncertainty and variation helps create a more transparent and comprehensive analysis of risk and the limitations of a risk analysis.

The risk analysis process

How are risks analyzed in EIA?

All projects carry risk. Risk is not something that can be wholly eliminated. Even forgoing a project or action may entail risks. Understanding risk involves systematically analyzing and managing it, and then using this information to support the EIA process. There are four stages in risk analysis.

1. Identification: establish the context and characterize the hazards

In preparation for the risk assessment, the proponent establishes the assessment context of the project by defining the purpose, scope, and scale of the assessment, and then identifying the methods to be employed and the resources and reporting that are required (MELP 2000). The scope of the risk assessment (and the EIA) is determined by the regulatory requirements of the jurisdiction in which it is being completed and can include the construction, operation, and decommissioning phases of a project. It should include the breakdown of all activities that fall under those phases and the oversight of the EIA laws in the region.

Project scale is particularly important when establishing risk assessments for an EIA. To readily identify patterns, a risk assessment must be scaled according to the stressors and resources being evaluated, ecological receptors, repetitive events such as spawning or migrations, and the response time of systems which can be incredibly variable (Dale et al. 2008). Due to the interconnected character of natural systems, risks and effects can have impacts well beyond an immediate project, both spatially and temporally. These concerns are increasingly being addressed through cumulative effects assessment. Originally devised to describe the relative impacts of multiple human pressures on the marine environment (Hammar et al. 2020), cumulative effects assessment analyzes environmental and social risks beyond the spatial and temporal scope of traditional EIA. This acknowledgment is essential for risks caused by projects that are in themselves considered small, such as a forestry operation, a road, or a single gas well. On their own, the impacts may be considered minimal, but their accumulation can carry significant risk for species and habitat fragmentation (see Chapter 3 in this book for a review of cumulative effects assessment).

Potential hazards will also be identified and characterized in this stage, including their source, area of impact, potential initiating events, and causes (Aven 2016). A risk can only be effectively dealt with if it has been identified and acknowledged. Once identified, risks may be grouped or categorized (Sharma 2013) to streamline the assessment phase. This requires qualified professionals with the use of specialized tools and techniques, such as checklists or hazard and operability studies. For example, an approach for identifying potential hazards within operational systems, such as Hazard and Operability Analysis (HAZOP), can provide a systematized technique for risk management (Aven 2016). An example of how hazard identification and categorization could work in an EIA would be baseline information that provides wildlife counts for species that forage, hunt, transit, or reproduce in an area and then identifying their respective level of sensitivity to disturbances such as roads, land clearing, noise, or more people in an area.

2. Risk assessment

Once a list of hazards has been compiled, practitioners will examine the consequences and likelihood of these hazards (Therivel 2018). They will ask: what could go wrong? And what is the likelihood that something will go wrong? This assessment can be done using either qualitative risk assessment or quantitative risk assessment, or a combination of the two. Using both might be considered best practice, but the applicability can vary depending on the setting and what risks we are looking at.

Qualitative risk assessment

Qualitative risk assessment is a descriptive and categorical approach to understanding risk(s) (Thievel 2008). It is deductive and normally relies on the perception of experts and affected stakeholders, elicited through qualitative methods such as interviews, focus groups, and surveys, to identify and assess the likelihood and consequences of impacts (Animah and Shafiee 2020). Furthermore, qualifying risks involve a more subjective understanding of risk(s). This is because

		Likelihood				
		Highly unlikely	Unlikely	Possible	Likely	Almost certain
Consequence	Negligible	Low				
	Minor		Moderate			
	Moderate			High		
	Major				Extreme	
	Catastrophic					

Figure 5.2 Example of a risk matrix.

Box 5.1 Qualitative risk assessment in the management of an Australian fishery

Astles et al. (2006) provide a good illustration of environmental risk assessment for a wild capture fishery in New South Wales, Australia. Institutional reforms for the wild catch fishing industry required an update of the existing fishery management plans to provide a more ecosystem-based management system. But this meant that an understanding of the fisheries' ecological, social, and economic risk factors was needed. The lack of extensive data about the target species and how they interact with the ecosystem led practitioners to adopt a qualitative approach.

1. **Identify the sources of risk**: Sources of risk to the fishery were identified and then subdivided into individual activities and tasks. For example, ocean trawling was broken down into trawling, harvesting, discarding (by-catch), loss of fishing gear, travel, marketing, maintenance, and general disturbance. The ecosystem was also divided and then subdivided into categories. These divisions were based on government guidelines, expert opinion, literature, and historical records. This provided a set of integrated classifications that linked science, institutional structures, and historic knowledge.
2. **Characterize the risk**: Risk levels were assigned to the fishing activities and the vulnerability of ecosystems based on knowledge from existing literature. The components that had a negligible risk were eliminated from further assessment.
3. **Evaluate the risk**: A qualitative risk matrix was developed. The *y*-axis represented the fisheries impact and considered how susceptible a species was to capture and mortality. This information was gathered from the research literature and fisheries reports. The *x*-axis represented species resilience and how capable it was of recovering after a depletion event.

The outcome of this qualitative analysis was to clearly identify the specific issues surrounding the fisheries that aid decision-makers when creating the management plans. Fishing tactics can be adjusted based on the most vulnerable ecosystem components to bring the fishery into alignment with sustainability principles. The method also allows for regular re-assessment if changes are made to fishing tactics. An additional benefit of this sort of analysis is how it highlights knowledge or data gaps that prevent a more robust quantitative analysis from being completed.

In the context of an environmental assessment, this risk analysis identified the key hazards and vulnerabilities of specific fish stocks, which allows EA practitioners to suggest amendments or mitigation strategies to developers that would promote the sustainability of the resource.

---

the risk is understood largely based on the perception of informants, which will differ according to the values, experiences, and worldviews of the individual (or group) who perceives the risk (Ostrom et al. 2019). In situations where there are not readily available data sets or existing data has problematic gaps, qualitative risk assessment may be preferable. Qualitative methods are better able to incorporate local or traditional knowledge (Mantyka-Pringle et al. 2017) that has the added benefit of increasing stakeholder involvement in the EIA process.

The results of qualitative risk analysis are commonly communicated using a risk matrix. In these matrices, the consequence and likelihood of risk are expressed, as the name of this technique suggests, qualitatively, as descriptive levels and/or colors on a qualitative scale (Haroon and McMillan 2018). The cells of these matrices can be divided into risk classes – for example, low, medium, and high – to represent approximate levels of risk (MELP 2000). Events that have both a high likelihood and a significant consequence rank higher and will be afforded greater priority.

Qualitative assessments are well suited to understanding society-environment relations, or risks that involve or depend upon societal order and social perspectives. For example, these assessments are better suited to analyzing social risks, such as labor issues and shortages, where risk perspectives are relevant and quantification may be difficult. Some examples of where a qualitative risk assessment has been utilized in EIA include the incorporation of perceived risks and values in parks management in Australia (Carey et al. 2005), using expert opinions to determine the risk of non-indigenous species in eastern Canadian waters (Therriault and Herborg 2008), and the inclusion of community/local knowledge in government assessments of desertification in North Africa (Davis 2005). The last is particularly relevant as EIA agencies increasingly strive to meet their obligations under UNDRIP. Qualitative risk assessment methodologies are likely better equipped to incorporate Indigenous knowledge into the risk assessment process (see Chapter 13 in this book).

While these are valuable tools, there are many limitations to consider when employing qualitative methodologies in risk assessment. Qualitative risk assessments have limitations:

- They are often not comprehensive and do not provide the level of detail required for risk prevention (Tiusanen 2017).
- May be less rigorous. They are subjective and based on a relative and less verifiable perspective of risk and rely on the expertise of potentially biased analysts (Therivel 2018; Tiusanen 2017).
- They often are not able to capture complex and technical risks or account for all the variables in complex systems (Therivel 2018).
- Can employ risk matrices (e.g., Figure 5.2) that must be interpreted cautiously, as establishing management decisions and priorities require greater information and complexity than a rating of severity and likelihood (Tiusanen 2017).

Quantitative risk assessment

This involves probabilistic analysis, which is a formal systematic approach that understands risk as something that is objective and can be quantified (Therivel 2018). Quantitative risk assessment often relies on modeling software to establish the numerical probability (likelihood) of a detrimental event occurring, as well as the possible outcomes and consequences (see Box 5.2 for a quantitative case study) (Smith and Simpson 2020).

Other tools and techniques used by QRA practitioners to establish numerical risk values include:

- **Failure Mode and Effects Analysis (FMEA)**: FMEA is commonly used across industry to identify and eliminate known sources of failure and to improve the safety of complex systems (Zhou and Thai 2016). It is a systematic process where any *potential* issues are identified, evaluated, and prioritized in advance of a process commencing, thereby offering an opportunity to enact positive controls. This technique is intended to be applied in advance of a task.
- **Event Tree Analysis (ETA)**: ETA is an intuitive probabilistic analysis that visually lays out the possible outcomes of an event. It begins with a single initiating event, followed by a series of reactionary events, each mutually exclusive of each other (and usually binary, or

an either/or) that provide conditional probabilities. The accumulation of these probabilities represents the overall likelihood of a final outcome. An Event Tree Analysis results in several potential outcomes, which can be analyzed or ranked.

- **Fault Tree Analysis (FTA)**: FTA is very similar to Event Tree Analysis, and they are often used in conjunction. Fault trees demonstrate the relationships among events, while event trees show a sequence of events linked by conditional probabilities (Paté-Cornell 1984). It is a top-down analysis of failures that uses Boolean logic (true or false) where the undesired outcome is the "top event" and the instances or decisions that lead to it branch out from there. Fault Tree Analysis can be used both prior to a task or in incident investigations after the fact and can be qualitative, quantitative, or both, depending on the scope of the analysis (Lee et al. 1985).

- **Bow-Tie Analysis (BTA): Event Tree Analysis + Fault Tree Analysis)**: Bow-Tie Analysis is a graphical representation of the relationship between hazards, threats, controls, and consequences (Cockshott 2005). The process creates a two-sided diagram centered over the critical event where a fault tree on the left side identifies the possible events causing the incident, and the right side is an event tree showing the possible consequences caused by the breakdown of control measures (Khakzad, Khan, and Amyotte 2012). See Figure 5.3 for an example. Bow-Tie Analysis is both a qualitative and quantitative analysis; it shows the relationship between events in a logical and ranked (or able to be ranked) form, and by assigning probabilities, it can also be used to conduct a quantitative analysis.

Probability assessment models and methodologies are not incontrovertible. Many researchers have adapted these various models over the years, adjusting and combining them to account for uncertainty (e.g., fuzzy probability) and dynamic environments, while using faster computer processing. For example, risk scenarios can be used as a tool to identify hazards and associated risks, and Fault Tree Analysis and Event Tree Analysis (or Bow-Tie Analysis) can help to establish a logical relationship between the events that led up to an accident and the associated risk (Ferdous et al. 2011). While Fault Tree Analysis helps us identify the events leading up to an accident, Event Tree Analysis is used to understand and measure the consequences and likelihood of

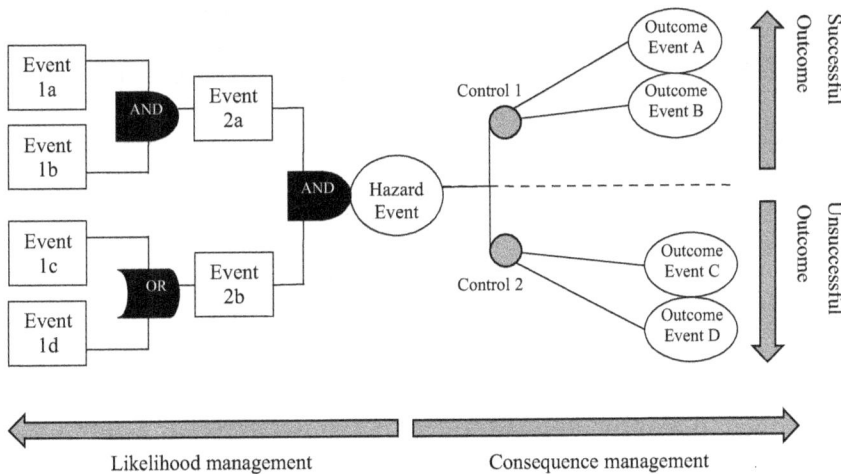

Figure 5.3 A basic Bow-Tie Analysis.

risk. The evaluation of a fault tree can be a probabilistic risk analysis (PRA) and a probabilistic safety assessment (PSA); both are also important parts of this approach. They are used primarily to understand potential accidents and impacts on human health (Therivel 2018) (see Chapter 4 in this book for a review of human health impact assessment).

Quantitative assessments can capture and incorporate many variables and provide an objective and replicable understanding of risks that can be difficult for people to envision and evaluate (Ostrom et al. 2019). For example, quantitative risk assessment is especially useful for engineering, where a high level of expertise and precision can produce risk estimates that can provide the public and regulators with a better understanding of specific risks, including which ones to worry about, which ones are perhaps less important, and which technologies or design options might address concerns best (Bowles 2020). Examples of where quantitative risk assessment has been applied in an EIA setting include the incorporation of in-situ measurements to analyze the effect of green spaces on urban temperatures and air quality (Cohen et al. 2014), the quantification of the greenhouse gas emissions of a highway over 20 years through the expenditure of energy in manufacturing and maintenance (Park et al. 2003), and a risk assessment of industrial spills from onshore pipelines on water quality (Bonvicini et al. 2015).

All of these projects included measurable data with large datasets making them excellent candidates for a quantitative risk assessment. While the calculated outcomes of quantitative risk assessment may seem intimidating for small projects, the results can be communicated clearly to different audiences. Results can be presented as statistical probabilities (e.g., there is a 25% chance that risk X will occur), as charts or graphs depicting the probability (y-axis) over time (x-axis), or as risk contour maps, where each contour line is assigned a numerical value, based on the results of the risk equation (Therivel 2018).

Box 5.2 Case study: Quantitative risk assessment in a liquefied natural gas terminal project

In the liquefied natural gas (LNG) industry, quantitative risk assessments are standard practice. Liquefying natural gas requires gradual chilling and compression. This makes it more space-efficient for long-distance transportation (e.g., by ship). LNG facilities are complex and have many components and operational steps. Dan et al. (2014) developed a case study for a quantitative risk analysis of fire and explosion for an offshore LNG Floating Production Storage and Offloading facility (FPSO). Their example provides an illustration of how quantitative risk assessment can be conducted for just one component operation (in this case, a gas valve system), that supports a larger analysis of the FPSO. A quantitative risk assessment was done because of previous incidents/accidents that have occurred in similar component operations at other facilities, notably those related to fire and explosion risk factors.

1. **Identification of hazards and their consequences**: A secondary literature was used to identify the primary hazards of concern, fire and explosion, which both pose a risk to human health and to the facility. The literature was also used to select accident scenarios, which included three different accident profiles representing accidents where leakages in a specific gas valve system occurred. Based on this initial knowledge, leakage points were

assumed at five expansion valves because these have the most probable failure rate in the overall liquefaction process. The size of a leak would be linked to the different hole sizes at the expansion valves (5, 50, 100 mm), and the effect would also vary with the leak size. For the accident scenarios, immediate ignition may result in a jet fire, while delayed ignition would result in a flash fire and explosion. Because leaked LNG turns into gas, a pool fire was not considered (a pool fire is caused when a flammable liquid is released on the ground or water and ignites).

2. **Assessment of the risk**: *Consequence analysis* – Computational modeling was used to determine the consequence of different scenarios by simulating and calculating the severity of impacts over distance (e.g., extent and intensity – leading to fatalities). The output from this step was the creation of consequence maps, which plotted the number of fatalities (y-axis) over distance in meters (x-axis) for each scenario. In this stage, it was found that leakage from the largest hole (100 mm), with fire or explosion resulting, produced the most severe consequence, which may not be surprising.

Frequency analysis– ETA was used to calculate the probability of a valve leakage occurring and leading to unwanted consequences. The initiating event on this branch of the ETA was a gas release, which then initiated a chain of events leading to the outcome events (jet fire, flash fire, etc.). The probability of these outcome events occurring was calculated by multiplying the modeled probability of each of the events in the chain leading up to the outcome. What the ETA tells us are the possible consequences that could occur from a single initiating event and their frequency (presented statistically as a numerical or percentage value). But the analysis was limited by uncertainty stemming from having an early stage schematic of the facility. But such techniques could also be used to examine different configurations of a facility. And frequency is often used instead of probability in calculations (Rausand and Haugen 2020).

3. **Evaluation of the risk**: At this point, the results from the consequence and frequency analysis were input into another computational model, which produced a risk contour map. These maps visualize the risk information (consequence and likelihood) in relation to space, using colored concentric rings or contours. In this way, risk was expressed spatially by overlaying colored contours, representing different risk levels, on top of a blueprint of the project site. On the map, the red contours (highest risk) occur closest to the expansion valve, and risk is diffused in concentric rings until the white zone (zero risk, safe zone) is reached near the bow of the FPSO.

4. **Risk management**: The results of this analysis were used to inform the next stage of risk management, in which certain controls and procedures would be put in place to reduce the operational risks associated with an expansion valve leak. In this case, safety-level zones were implemented according to the risk levels defined in the risk map.

The risk analysis followed key steps, including the identification of key hazards and their impacts, a quantitative risk assessment, and an evaluation of the risk using a risk contour map, which also served as a communication device by translating complex quantitative data into a more accessible format.

Like qualitative risk assessment, quantitative methods need to be applied carefully due to potential constraints:

- They are often more time-consuming and costly to complete (Bowles 2020).
- It can require a high level of expertise, very high-quality data, and technological supports that may not be readily available in many circumstances (Bowles 2020).
- They may underestimate uncertainty by emphasizing predictability and quantifiability of risks (Altenbach 1995).
- The results may not account for less quantifiable, human elements of risk, such as "irrational" decision-making and human error (Altenbach 1995).
- The results are often challenging to communicate to decision-makers, project managers, stakeholders, and the public who are unfamiliar with the methods and terminology (Bowles 2020; Dale et al. 2008).

There is no one method that is necessarily "better" overall than another – some are simply better at answering different questions and are used based on the availability and quality of data, expertise, money, and time. In an EIA, there is likely to be a combination of tools used: a hydroelectric dam project may utilize quantitative risk assessment for river flow and geotechnical data to estimate engineering risks associated with the project, but rely on a qualitative risk assessment and expert community ranking of aquatic species risks due to a lack of data for a remote area.

3. Risk evaluation

The risk presented in the proponent's proposal is evaluated by the regulator, who may produce a risk profile that highlights different classes of risk (e.g., high, medium, low) (Aven 2016; MELP 2000). Risk profiles weigh the potential benefits against the risks associated with the proposal and decide whether the results of the assessment meet a predefined criterion for "tolerable" or "acceptable risk". Risks can fall into one of three key categories: (1) acceptable; (2) tolerable; or (3) unacceptable (Therivel 2018).

- **Acceptable risks**: are those where the perceived benefits of the action outweigh the trade-offs (potential harms).
- **Tolerable risks**: are those not always acceptable to a population but are seen by experts, and sometimes society, as a worthwhile trade-off given the benefits.
- **Unacceptable risks**: those that exceed some critical threshold of acceptability for both society and its individuals as well as experts.

Determining the appropriate qualitative criteria or quantitative ranges for each of these levels of risk is a complex and often highly politicized task. Policymakers, risk practitioners, and general society, for example, might interpret levels of acceptable risk differently. This lack of consensus complicates both the definition of acceptable risk and the communication of these risks. Different jurisdictions may interpret acceptable risk differently, and assign different criteria, further complicating comparisons across regions. In addition to acceptable risk, some jurisdictions may also include criteria for tolerable risk levels, which describes the level of risk a society is willing to live with, in exchange for the benefits gained by taking the risk (Therivel 2018). These variations can be observed in many global regions and can be reflected by regulatory qualities such as health standards or pollution discharge limits, or design and operating standards for roads, buildings, waste management, or pipelines.

Acceptable risks are determined using both objective criteria (e.g., a quantified risk threshold based on the risk equation) and subjective criteria (e.g., a risk threshold that is established subjectively by experts, policymakers, regulators, and/or implicated stakeholders, based on standards informed by their knowledge, experience, beliefs, values, and interests) (Therivel 2018). While it is tempting to see objective metrics as being preferable, subjective measures may also be appropriate given the inherent subjectivity of risk. It's fair to say that risk assessments that involve communities in identifying risks, and acceptable risk levels, will see greater public confidence in the assessment outcomes, confidence in the measure taken to manage risk, and the project will ultimately have a better chance of acceptance. This highlights the growing need to integrate risk analysis methods, to account for both objective and subjective perspectives of acceptable risk. An example of an acceptable risk in EIA would be one where the probability of the occurrence falls below a predefined limit, or the benefits of the project are seen to outweigh a low probability of a negative outcome. But some risks cannot be mitigated and may not be able to meet the standards of acceptable risk, regardless of any intervention.

4. Risk management

Risk assessments are not a form of decision-making but rather inform decisions. Early approaches to EIA did not necessarily distinguish between the stages of risk assessment and risk management. However, this began to change after a report from the *US Committee on the Institutional Means for Assessment of Risks to Public Health* was released in 1983. The report recommended the separate treatment of risk assessment and risk management based on the ambiguity and a lack of transparency and credibility that occurred when these stages were not differentiated (Therivel 2018). The approach holds that *differentiation* aids in the delineation of analytic (assessment) and decision and policymaking functions (management) and renders the analysis clearer and more credible, given that management decisions are not necessarily made by those who assess the risk.

Risk management involves interpreting the results of the evaluation stage and then choosing interventions and mitigation strategies to help reduce risk to a level that is as low as reasonably possible (ALARP). There are three best practices to manage risk:

1. **Risk-informed management (mitigation)**: This includes mitigation and management measures that deal with risk by avoiding, minimizing, transferring, or compensating for an undesirable impact. An example in EIA would be a project proponent creating new salmon spawning habitat downstream for a project to mitigate the change to channel banks from development.
2. **Cautionary or precautionary risk management**: This includes mitigation strategies that contain and diffuse risk, as well as the development of alternative strategies, designs and contingency plans (e.g., safety plans). The key principle for this form of risk management is to retain system flexibility and adaptive capacity. In EIA, this could be a requirement for the proponent to remove vegetation from the reservoir floor of a hydroelectric dam to prevent methyl-mercury pollution.
3. **Discursive risk management**: This involves using strategies that remove or minimize defined uncertainties and ambiguities to build confidence. Discursive management often involves consultation with the affected populations. For example, an agreement between

communities, governments, and the proponent for a long-term monitoring plan during operation, decommissioning, and reclamation (Aven 2016).

Once mitigation measures and management plans have been established, a report of the assessment is produced, including an overview of the key impacts, their likelihood and consequence(s), levels of uncertainty, and the limitations of analysis (MELP 2000).

Box 5.3 As low as reasonably practicable (ALARP)

ALARP originates from a UK court decision in 1949 that held that the risk of harm must be weighed against the cost of the measures needed to avert it (Taylor and Israni 2014). The meaning of the term and application of the concept has evolved from the health and safety sector to wider applications of risk assessment. ALARP is a guiding principle for risk assessment rather than a prescriptive risk management approach (Pike, Khan, and Amyotte 2020). The key elements, "low", "reasonably", and "practicable", are also the most problematic, being undefined and subjective in nature (Melchers 2001). ALARP is based on the concept of a "maximum tolerable risk" and a "broadly acceptable risk"; where any activities or projects that exceed that predefined maximum risk level must be brought back down (mitigated) to an acceptable tolerance level before the project can continue (Baybutt 2013). The resulting analysis will always include an element of cost–benefit thinking, which has its own limitations and challenges. The use, or status, of ALARP varies from country to country but is increasingly being adopted across Europe as a risk management decision-making support tool (Jones-Lee and Aven 2011).

How are risks communicated?

Perhaps the most important part of analyzing risk is communication. Without effective communication, the risk analysis lacks applicability and will be of little use to practitioners, the public, and decision-makers who need this information. Risk communication is threaded throughout the risk analysis process, and the ways that risk is communicated is important, as it affects who can access the results and how they are disseminated and integrated into risk management (Lundgren and McMakin 2018). For risk analysis to be effective in guiding decision-making, it needs to be presented clearly, in a way that can be used to help assess the costs and benefits of a project (Pike, Khan, and Amyotte 2020).

There are different approaches to risk communication – such as mental models (an explanation of how something works) different written communication styles, or even modes of communication (e.g., graphic, digital, printed). Choosing a method normally reflects who is communicating findings and who the targeted audience is (Lundgren and McMakin 2018). Risk communication that is more technical, such as event trees or risk contour maps, reflect the qualitative background or training of the researcher, and are chosen based on the desire to communicate the results of the study to a select audience who have particular expertise. The use of Geographical Information System (GIS) technologies to visualize risk represents a strong tool to communicate the results of complex environmental risk assessments (see also Chapter 12 in this book) (Lahr and Kooistra 2010).

As with other communication and engagement aspects of the EIA process, risk communication can be challenging. The more diverse stakeholders who are involved in an assessment are, the more demanding it can become. Successful risk communication depends on trust and credibility, both of which are largely seen to be in diminishing capacity between government, corporations, and the public (Trettin and Musham 2000). Risk perception will vary from person to person. Regulators, scientists, engineers, or the public will perceive certainty and uncertainty differently, but these are the foundation of any risk analysis (Garvin 2001). Risk perception is inherently subjective. It is affected by a wide range of factors such as education, political ideology, and locational distance from the project (Clarke et al. 2016).

Risk perception can influence an EIA in many ways. There is likely to be increased pressure both publicly and politically on practitioners over developments like those that have involved perceived "unjustifiable" risks and have drawn intense scrutiny or protest in recent memory. An EIA practitioner's own perception of the risks will be shaped by their background and knowledge, even where a conscious effort is being made to remain unbiased. In many cases, the public perception of risk may not align with the numerical reality of a calculated risk assessment. An example of a difference in cultural and geographic perception of risk toward a project can be seen in Canada when looking at proposed pipeline projects across adjacent provinces of Alberta (where oil is produced) and British Columbia (BC) (westernmost province which the pipelines would cross to reach coastal oil terminals). In Alberta, the economic importance of the oil industry may significantly reduce the negative risk perception of pipelines. In BC, on the other hand, the perceptions are different. BC residents may feel they are accepting undue risks to their coastline (from oil tankers) and lands and waterways (pipeline risks) while realizing few if any economic or social benefits from an industry located in another jurisdiction. So, for many BC resident pipelines are an environmental issue, whereas for Albertans they are an economic issue. Powers (2007) notes that faced with public uncertainty about risk-related issues, regulators can be inclined toward simple disclosure, rather than further dialogue and explanation, which does not really help consultation throughout the EIA process.

Communicating risk is a process of early engagement, listening, answering, and taking time.

Risk analysis innovations for EIA

Improvements in data, communication, modeling, and communication, coupled with faster and more powerful computing technologies, has increased the complexity and accuracy of the quantitative analyses we can do. This is particularly important for large, complicated systems and projects with high levels of variability and uncertainty (Ward et al. 2012). Improved computation has also made running quantitative analysis on smaller projects much more efficient, which has increased its use across a wide variety of industries and disciplines. The strength of a model's output is dependent on the quality of data and access to it. The ability to store data sets on cloud servers and share them allows both researchers and professionals to direct their resources to the analysis itself rather than the individual burden of information gathering.

Increased data and computing power have led to the recent development of some Dynamic Risk Assessment Methodologies (DRA) that go a step further than traditional methods to address the challenges of environments that are constantly changing. Some DRA models are designed to use real-time data derived from monitoring (see Box 5.4) (Abimbola and Khakzad 2014), which provide a constantly updating decision-making tool. The DRA framework, constantly analyzing emerging risks, has a much better chance of foreseeing incidents as it incorporates all the early warnings and minor related risks that may be missed by traditional assessment methods (Villa et al. 2016). DRA has the potential to become a tool that is applied throughout a project, from the design to construction and operation, and for support monitoring and follow-up (Villa et al. 2016).

GIS has not only improved the spatial analysis and representation of data, but it has also helped risk communication. The use of risk mapping to communicate the results of probabilistic quantitative analysis to stakeholders and the public in a clear and concise way and also helps build better-informed participation in EIA (see Chapter 12 in this book) (Lahr and Kooistra 2010).

Box 5.4 Real Time Mine Monitoring Real Time Mine Monitoring

Real-time mine monitoring

Innovation equipment, data management, and analytical software means that instruments can be put in place at the mine site to provide *real-time* geotechnical monitoring for mine managers and engineers. Different instruments can be used to observe and report changes in geotechnical constraints such as ground movements and pore pressure. Early warnings allow managers to respond quickly to observed changes. Real-time monitoring systems also allow engineers to access data immediately for rapid decision-making and reduce the time needed to collect and analyze data.

- Open pit mines – Movement of slope walls and groundwater levels can be monitored to watch for areas of instability or aid in pit dewatering efforts.
- Tailings dams and waste dumps – Stability can be monitored to determine vertical or horizontal movement, increases in pore water pressure, or seepage. Monitoring programs can be designed for site-specific characteristics (e.g., foundation materials, dam materials, height, and construction style) and to monitor specific or overall risks.
- Underground mines – Monitoring programs can use instruments to monitor tunnel roof subsidence/convergence and shaft instability.

(Adapted from: https://rstinstruments.com/mines/)

Current issues and challenges

Risk analysis and management in EIA, despite its widespread acceptance, still faces many challenges. Risk as a concept can be challenging to communicate to EIA stakeholders. The word *risk* itself has become fractious and heavily used in the media to create antagonism and polarize discussions (Pike, Khan, and Amyotte 2020). Practitioners must increasingly contend with preconceived notions about projects driven by online news or social media that complicate rational discussions based on calculated probabilities. The subjective nature of risk makes complete agreement on risk factors unlikely even within similar groups. While risk assessments attempt to incorporate and accommodate uncertainty and variability into models, the lack of unequivocal answers can frustrate decision-makers and the public (Garvin 2001). Similar to risk being considered subjective, so is cost–benefit analysis, such as the application of principles. For example, when an EIA is required because some sort of development project has been proposed, the perceived benefit of that project can bias the risk tolerance of some stakeholders and decision-makers.

EIAs will often include a combination of qualitative and quantitative risk assessment methods. While the qualitative assessment relies on institutional knowledge or the recruitment of subject-matter experts, quantitative assessment requires an existing data set. Finding or accessing a data set with sufficient entries can be difficult, and further challenged by uncertainties surrounding data quality, quality assurance, or indeterminate collection methods. While increasing digitization has made more government datasets freely available (Melchers 2001), there is unlikely to be any standardization between assessment agencies, making the comparing and contrasting of information flawed, if not useless. Industry groups meanwhile may consider their data proprietary and may be unwilling to make it available at all. Where data is unavailable, either time and/or money must be spent to acquire it, or the analysis may run with such a high degree of uncertainty that it becomes difficult to justify the outcomes to stakeholders.

EIA and risk analysis are inherently multidisciplinary in nature. Yet, there is still insufficient integration of risk assessments across all types of EIAs (Cormier and Suter II 2011). The combination of methodologies like risk assessment with monitoring-based assessments increases the opportunities for finding links between different perspectives on the process.

Selecting what risks are significant and what is a subjective and value-laden judgment can reflect the role and values of the decision-makers. As mentioned above, risk tolerance is highly variable, and social perceptions of risk may differ from the standardized criteria established by experts and policymakers. Risk evaluation is also dependent on the jurisdictional practices and norms prescribed in that area. Inconsistencies, contradictions, and the subjective nature of risk make it challenging to conduct risk analyses that speak to everyone in the EIA process. A common or semi-standardized approach could help to improve the credibility of such assessments, although the issue of negotiation of divergent perspectives on acceptable levels of risk will always be there and can only be addressed through consultation and negotiations with communities. Regardless of the degree of technicality and attempted objectivity involved, risk analyses will always be contested by multiple parties with divergent interests, values, and worldviews.

Best practices for risk analysis in EIA

With everything above in mind, what are the best practices for EIA? We know that ultimately they will be contextual, but the following points highlight some of the key practices identified in the risk analysis literature, which can be applied to EIA.

1. **Risk analysis provides "tools" not absolutes**

   Risk analysis tools, such as the qualitative or quantitative risk assessment tools presented in the case studies, should not be applied without acknowledging their appropriate uses and limitations (Aven 2016). Even when appropriate, a tool should never be treated as "truth", given the limitations imposed by unavoidable and inherent uncertainties, variabilities, and risk subjectivity. In the case of EIA, any defined (knowable) uncertainties should be acknowledged and disclosed by the proponent to the regulating body as part of their application. Risk subjectivity should also be considered by both the proponent and regulating bodies, who must understand and negotiate how their perceptions, official standards, and public perceptions might differ and affect risk tolerance and public acceptance of the project.

2. **Apply qualitative and quantitative assessments where they are strongest and most suitable to the context, scale, and purpose of the project**

    Given the inherent strengths and weaknesses of both qualitative and quantitative risk assessments, EIA practitioners need to focus time and resources on which methods are suitable to the context, scale, and purpose of the project in question. For example, qualitative methods would be inappropriate for evaluating the complex risks of complex projects (e.g., LNG facilities or nuclear plants) that require a high level of precision (Animah and Shafiee 2020). But quantitative methods might not be best when considering the sociocultural and economic impacts of a project on a local population, such as Indigenous Peoples and their way of life. In EIA, employing the appropriate assessment methods is vital to conveying risk in an accurate and meaningful way to regulators and the public.

3. **Where possible, apply an integrated approach**

    Given the strengths and weaknesses of Qualitative and Quantitative Risk Assessments, they can be used together as *mutually informing approaches* to answer different questions or in different applications (Ostrom et al. 2019). Integrated risk analysis is a growing area of research; however, more work is needed to test hybrid methodologies (Animah and Shafiee 2020). Rather than trying to eliminate uncertainty, integrated methods may help EIA practitioners and proponents manage uncertainties, as they consider a greater number and type of variables (Therivel 2018).

4. **Communicate**

    Risk communication might be the most overlooked component of risk analysis. For EIA, risk analyses must be communicated effectively, meaning that the results (and limitations) must be accessible to the target audience (e.g., by using an appropriate method, vocabulary, and level of jargon) and presented in a way that presents the costs and benefits and other information required to make decisions about future actions and the need for mitigation. Successful risk communication also depends on trust and credibility, which can be a challenge given the subjective nature of how risk is experienced and perceived by the stakeholders involved in an EIA process and the contentious nature of some projects. Regardless, dialogue about risk and building a *risk literacy* can strengthen trust and understanding of decisions made in the EIA process.

Concluding remarks

The concept of risk is a fundamental characteristic of effective EIA. Risks cannot be avoided, deferred, or eliminated; we need to acknowledge, understand, and address them. Risks may be what people care or worry about most when projects are proposed. Risk encourages caution, proactive planning, and the implementation of mitigation strategies. The quality of a risk analysis – the identification, assessment, evaluation, and management of risks – is an important part of effective EIA. And the ways in which risks are communicated can be key to the way a project is received by the regulators and society.

Better data, improvement in modeling, and technology innovations are increasing the capacity of quantitative risk analyses in complex systems, where variability and uncertainty are highest. Dynamic Risk Assessment Methodologies are one example of an efficient and capacity-building tool, which can help researchers to iteratively monitor and assess data in real-time to observe risks. Data storage advances are increasing our capacity to gather and store massive amounts of data. GIS technologies support analysis and representation of spatial data and provide new ways for EIA practitioners and risk analysts to communicate their work in more accessible ways.

Risk analysis is an integral part of EIA practice. It is present in some form at every stage of the EIA process. Risk can define what people worry about when projects are proposed, it can frame the information the public needs for effective participation and engagement, and it will be a significant part of what regulators are interested in when making decisions.

Note

1 For probability, the hypothesis is treated as a given and the data are free to vary, but for likelihood, it is the data that are a given while the hypotheses vary.

References

Adger, W. N., Barnett, J., Brown, K., Marshall, N., & O'Brien, K. (2013). Cultural dimensions of climate change impacts and adaptation. *Nature Climate Change, 3*(2): 112–117. https://doi.org/10.1038/nclimate1666

Animah, I., & Shafiee, M. (2020). Application of risk analysis in the liquefied natural gas (LNG) sector: An overview. *Journal of Loss Prevention in the Process Industries, 63,* 103980. doi:10.1016/j.jlp.2019.103980

Astles, K. L., Holloway, M. G., Steffe, A., Green, M., Ganassin, C., & Gibbs, P. J. (2006). An ecological method for qualitative risk assessment and its use in the management of fisheries in New South Wales, Australia. *Fisheries Research, 82*(1–3), 290–303.

Aven, T. (2016). Risk assessment and risk management: Review of recent advances on their foundation. *European Journal of Operational Research, 253*(1), 1–13. doi:10.1016/j.ejor.2015.12.023

Aven, T., Ben-Haim, Y., Andersen, H. J., Cox, T., Lopez Droguett, E., Greenberg, M., Guikema, S., Kroger, W., Renn, O., Thompson, K. M. & Zio, E. (2018). *Society For Risk Analysis Glossary. Society for Risk Analysis (SRA).* Retrieved from: https://www.sra.org/wp-content/uploads/2020/04/SRA-Glossary-FINAL.pdf

Bonvicini, S., Antonioni, G., Morra, P., & Cozzani, V. (2015). Quantitative assessment of environmental risk due to accidental spills from onshore pipelines. *Process Safety and Environmental Protection, 93,* 31–49.

Bowles, D. (2020). *Summary of USSD emerging issues white paper on dam safety assessment: What it is? Who's using it and why? Where should we be going with it?* Report prepared for The Working Group on Risk Assessment USSD Committee on Dam Safety.

Carey, J. M., Burgman, M. A., Miller, C., & Chee, Y. E. (2005). An application of qualitative risk assessment in park management. *Australasian Journal of Environmental Management, 12*(1), 6–15.

Cockshott, J. E. (2005). Probability bow-ties: A transparent risk management tool. *Process Safety and Environmental Protection, 83*(4), 307–316.

Cohen, P., Potchter, O., & Schnell, I. (2014). A methodological approach to the environmental quantitative assessment of urban parks. *Applied Geography, 48,* 87–101.

Cormier, S. M., & Suter II, G. W. (2008). A framework for fully integrating environmental assessment. *Environmental Management (New York), 42*(4), 543–556. doi:10.1007/s00267-008-9138-y

Dan, S., Lee, C. J., Park, J., Shin, D., & Yoon, E. S. (2014). Quantitative risk analysis of fire and explosion on the top-side LNG-liquefaction process of LNG-FPSO. *Process Safety and Environmental Protection, 92*(5), 430–441. doi:10.1016/j.psep.2014.04.011

Dale, V. H., Biddinger, G. R., Newman, M. C., Oris, J. T., Suter, G. W., Thompson, T.,... & van Heerden, I. L. (2008). Enhancing the ecological risk assessment process. *Integrated environmental Assessment and Management, 4*(3), 306–313. doi:10.1897/IEAM_2007-066.1

Davis, D. K. (2005). Indigenous knowledge and the desertification debate: problematising expert knowledge in North Africa. *Geoforum, 36*(4), 509–524.

Drupsteen, L., Groeneweg, J., & Zwetsloot, G. I. J. M. (2013). Critical steps in learning from incidents: Using learning potential in the process from reporting an incident to accident prevention. *International Journal of Occupational Safety and Ergonomics, 19*(1), 63–77. doi:10.1080/10803548.2013.11076966

Garvin, T. (2001). Analytical paradigms: The epistemological distances between scientists, policy makers, and the public. *Risk Analysis, 21*(3), 443–456. doi:10.1111/0272-4332.213124

Haroon, R., & McMillan, T. (2018). Understanding and Managing Risk [PDF]. *Info Series prepared for and published by The Centre for Environmental Assessment Research at UBC.*

Hewitt, K. (1983). *Interpretations of calamity from the viewpoint of human ecology.* Boston: Allen & Unwin.

Jones-Lee, M., & Aven, T. (2011). ALARP – What does it really mean?. *Reliability Engineering & System Safety, 96*(8), 877–882. doi:10.1016/j.ress.2011.02.006

Khakzad, N., Khan, F., & Amyotte, P. (2012). Dynamic risk analysis using bow-tie approach. *Reliability Engineering & System Safety, 104*, 36–44.

Lee, W. S., Grosh, D. L., Tillman, F. A., & Lie, C. H. (1985). Fault tree analysis, methods, and applications: A review. *IEEE Transactions on Reliability, 34*(3), 194–203.

Lundgren, R. E., & McMakin, A. H. (2018). *Risk communication: A handbook for communicating environmental, safety, and health risks* (Fifth ed.). New Jersey: IEEE Press. doi:10.1002/9780470480120

Mantyka-Pringle, C. S., Jardine, T. D., Bradford, L., Bharadwaj, L., Kythreotis, A. P., Fresque-Baxter, J.,... & Lindenschmidt, K. E. (2017). Bridging science and traditional knowledge to assess cumulative impacts of stressors on ecosystem health. *Environment international, 102*, 125–137. doi:10.1016/j.envint.2017.02.008

Ministry of the Environment, Lands & Parks (MELP). (2000). *Environmental Risk Assessment (ERA): An approach for assessing and reporting environmental conditions.* Retrieved from: http://www.env.gov.bc.ca/wld/documents/era.pdf

Morgan, R. K. (1999). *Environmental impact assessment: A methodological approach.* Norwell, MA, USA: Springer Science & Business Media. Retrieved from: https://books.google.ca/books?hl=en&lr=&id=2ehQHkg6iloC&oi=fnd&pg=PR7&dq=risk+environmental+impact+assessment&ots=skmdwUm-Mjb&sig=Gz_RUpPGxax1mYvclHfhJopvCZI&redir_esc=y#v=onepage&q=risk&f=false

National Research Council (NRC), Commission on Life Sciences, Division on Earth and Life Studies, Committee on Risk Assessment of Hazardous Air Pollutants, Commission on Life Sciences, & Board on Environmental Studies and Toxicology. (1994). *Science and judgment in risk assessment.* National Academies Press. doi:10.17226/2125

Ostrom, L. T., Wilhelmsen, C. A, & Knovel (2019). *Risk assessment: Tools, techniques, and their applications* (Second ed.). NJ: John Wiley & Sons.

Park, K. F., & Shapira, Z. (2018). Risk and uncertainty. In *The palgrave encyclopedia of strategic management*, 1479–1485. London: Palgrave Macmillan UK. doi:10.1057/978-1-137-00772-8_250

Park, K., Hwang, Y., Seo, S., & Seo, H. (2003). Quantitative assessment of environmental impacts on life cycle of highways. *Journal of Construction Engineering and Management, 129*(1), 25–31.

Paté-Cornell, M. E. (1984). Fault trees vs. event trees in reliability analysis. *Risk Analysis, 4*(3), 177–186.

Pike, H., Khan, F., & Amyotte, P. (2020). Precautionary principle (PP) versus as low as reasonably practicable (ALARP): Which one to use and when. *Process Safety and Environmental Protection, 137*, 158–168. doi:10.1016/j.psep.2020.02.026

Rausand, M., & Haugen, S. (2020). *Risk assessment: Theory, methods, and applications.* Hoboken, NJ, USA: John Wiley & Sons.

Sharma, S. K. (2013). Risk management in construction projects using combined analytic hierarchy process and risk map framework. *IUP Journal of Operations Management, 12*(4), 23–53.

Skinner, D. J., Rocks, S. A., Pollard, S. J., & Drew, G. H. (2014). Identifying uncertainty in environmental risk assessments: The development of a novel typology and its implications for risk characterization. *Human and Ecological Risk Assessment, 20*(3), 607–640. doi:10.1080/10807039.2013.779899

Suter II, G. W., & Cormier, S. M. (2011). Why and how to combine evidence in environmental assessments: Weighing evidence and building cases. *Science of the Total Environment, 409*(8), 1406–1417. doi:10.1016/j.scitotenv.2010.12.029

Suter, G. W. (2008). Ecological risk assessment in the United States Environmental Protection Agency: A historical overview. *Integrated environmental assessment and management, 4*(3), 285–289. doi:10.1897/IEAM_2007-062.1

Suter, G. W., Barnthouse, L. W., & O'Neill, R. V. (1987). Treatment of risk in environmental impact assessment. *Environmental Management (New York), 11*(3), 295–303. doi:10.1007/BF01867157

Tanner, T., Lewis, D., Wrathall, D., Bronen, R., Cradock-Henry, N., Huq, S.,..., & Stockholm Environment Institute. (2014). Livelihood resilience in the face of climate change. *Nature Climate Change, 5*(1), 23–26. doi:10.1038/nclimate2431

Taylor, M., & Israni, C. (2014, March). Understanding the ALARP Concept: Its Origin and Application. In *SPE International Conference on Health, Safety, and Environment.* Society of Petroleum Engineers.

Therivel, R. (2018). *Methods of environmental and social impact assessment* (Fourth ed.). New York: Routledge. doi:10.4324/9781315626932

Therriault, T. W., & Herborg, L. M. (2008). A qualitative biological risk assessment for vase tunicate Ciona intestinalis in Canadian waters: Using expert knowledge. *ICES Journal of Marine Science, 65*(5), 781–787.

Thompson, K. M. (2002). Variability and uncertainty meet risk management and risk communication. *Risk Analysis*, *22*(3), 647–654. doi:10.1111/0272-4332.0004

Tiusanen, R. (2017). Qualitative risk analysis. In N. Moller, S. Ove Hansson, J. Holmberg & C. Rollenhagen (Eds.), *Handbook of Safety Principles*, (pp. 463–492). John Wiley & Sons, doi:10.1002/9781119443070.ch21

Trettin, L., & Musham, C. (2000). Is trust a realistic goal of environmental risk communication?. *Environment and Behavior*, *32*(3), 410–426. doi:10.1177/00139160021972595

Villa, V., Paltrinieri, N., Khan, F., & Cozzani, V. (2016). Towards dynamic risk analysis: A review of the risk assessment approach and its limitations in the chemical process industry. *Safety Science*, *89*, 77–93.

Ward, J. D., Mohr, S. H., Myers, B. R., & Nel, W. P. (2012). High estimates of supply constrained emissions scenarios for long-term climate risk assessment. *Energy Policy*, *51*, 598–604.

Zhou, Q., & Thai, V.V. (2016). Fuzzy and grey theories in failure mode and effect analysis for tanker equipment failure prediction. *Safety Science*, *83*, 74–79.

6

SUSTAINABILITY ASSESSMENT PRINCIPLES AND PRACTICES

Angus Morrison-Saunders, Alan Bond, Jenny Pope, and Francois Retief

Introduction

Environmental impact assessment (EIA), which is arguably the most widely used environmental protection and management instrument in use globally (e.g., Morgan 2012; Yang 2019), has long been framed as a 'tool for promoting sustainable development' (Morrison-Saunders 2018, 7). This objective of EIA is explicitly identified in best practice principles and other guidance materials published by the International Association for Impact Assessment (e.g., IAIA and IEA 1999; Partidário et al. 2012). This focus can be traced back to earlier accounts, such as the writing of van Pelt et al. (1992) and Gibson (1993), which link EIA with the sustainable development agenda arising from the publication of *Our Common Future* in 1987. Principle 17 of the Rio Declaration on Environment and Development (United Nations Conference on Environment and Development 1992) explicitly stated that EIA should be used as the tool for assessing projects with potentially significant impacts, leading to the formalization of the sustainable development focus for EIA in national sustainable development strategies worldwide. While the terminology of 'sustainable development' or 'sustainability' has only been in use since the late 1980s, the sentiment is implicit in the world's first EIA legislation (Bond et al. 2010), whereby the provisions for 'environmental impact statements' in the *National Environmental Policy Act* 1969 from the United States of America expect consideration of:

> The relationship between local short-term uses of man's [sic] environment and the maintenance and enhancement of long-term productivity.
>
> *(s. 102(2)(c))*

In recent times, EIA has been promoted as an important means for realizing the United Nations Sustainable Development Goals (SDGs) (e.g., Hacking 2018; UNEP 2018; Morrison-Saunders et al. 2020). A sustainability focus is also central to the notion of 'next generation' EIA, as reflected in a set of principles proposed to better equip impact assessment to address the complex challenges of 21st-century decision-making (Sinclair et al. 2018).

EIA is also argued to have other goals, such as Environmental Policy Integration (EPI) (Rega and Bonifazi 2014; Runhaar et al. 2014) and/or democratic governance (Kidd and Fischer 2007). In this chapter, however, we focus on sustainable development as the goal. The term

DOI: 10.4324/9780429282492-7

'sustainability assessment' has been coined to describe 'any process that directs decision-making towards sustainability' (Bond et al. 2015a, 3). This definition is broad enough to embrace an extensive range of processes and practices (Pope et al. 2017), not all of which relate to the practice of EIA as targeted in this Handbook. Yet even within the domain of sustainability assessment as a form of EIA, there is considerable variation in the application, process, and indeed in interpretations of the term 'sustainability', widely acknowledged to be a pluralistic and contested concept (Bond et al. 2013; Jacobs 1999).

We first deal with the matter of applications of sustainability assessment. Like any form of impact assessment, sustainability assessment informs decision-making. This decision-making can be at any level, from policies, plans, and programs associated with the practice of strategic environmental assessment to projects typically associated with EIA practice, including the modification of existing projects. It is useful to consider the nature of decisions that might be informed by sustainability assessment. In the impact assessment field, there are broadly two kinds of decisions: choice and threshold (Pope et al. 2017). Choice decisions involve asking which of a range of alternatives or options is best; in the case of sustainability assessment, 'best' means the 'most sustainable'. Choice decisions are therefore, about the comparison. Choices can be made across the entire hierarchy of alternatives utilized in EIA (e.g., commencing with need or demand for development, and continuing to alternative modes, scale, location, and timing of development) and extending on through the hierarchy of mitigation measures to enhance positive outcomes and avoid, minimize, rectify, and offset negative impacts (Morrison-Saunders 2018). Importantly, the 'higher' up the hierarchy of alternatives that sustainability assessment is applied and thus more strategic level of decision-making, the greater the potential contribution to sustainability (Morrison-Saunders and Therivel 2006; Hacking and Guthrie 2008; Gibson 2013). For each step further down the hierarchy, 'the options become progressively narrower in scope' and offer less 'room to move' (Morrison-Saunders and Pope 2013, 57) in this regard. Threshold decisions, in contrast, involve a single alternative that is compared with an established benchmark of some kind, which may be a minimum acceptability limit or an aspirational target. Much EIA practice revolves around threshold questions, asking whether a proposed development is environmentally acceptable in a given location or not. Because of the difficulty in defining these benchmarks with respect to sustainability, threshold questions in sustainability assessment are far more difficult to answer than choice decisions.

Sustainability assessment processes can also be categorized according to who undertakes the assessment (Pope et al. 2017), which might variously be a proponent, a regulator, or third party such as a community group. EIA and sustainability assessment processes are typically complex and take place over many months or years, and thus there may be opportunities for different stakeholders to pursue sustainability assessment thinking at different moments within the overall proposal lifecycle. For example, a proponent might undertake an evaluation of development options or locations well in advance of formally submitting a preferred development proposal subject to mandatory EIA to a regulator and utilize sustainability assessment to support the choice decision. As repeatedly stressed in Morrison-Saunders (2018), deciding to adopt a sustainability assessment approach is a *choice* of individual practitioners. This choice can be exercised notwithstanding that EIA regulations in a given jurisdiction may seemingly provide limited scope for sustainability assessment to be realized, since the option of operating 'beyond compliance' (Morrison-Saunders 2018, 35) remains open for exploration.

In the remainder of this chapter, we outline the principles and practices of sustainability assessment, highlighting some of the points of variation, debate, and contention. Our content is derived from literature review, drawing largely upon distillation and synthesis of our own thinking and writing on the topic along with the work of others. As such, our account might be construed as personal and pointed; we have not attempted to carry out a systematic or comprehensive review of the field. We start in the next section with an exploration of the concept of 'sustainable development' or 'sustainability' (for the purposes of this chapter using the two terms interchangeably), first in general terms and then more specifically related to sustainability assessment. Next, we summarize some of the key challenges practitioners of sustainability assessment face, before concluding with a proposed set of principles for best practice sustainability assessment in the last section.

Exploring sustainability

Understanding sustainability is an essential first step in understanding sustainability assessment (Morrison-Saunders and Pope 2013; Bond et al. 2015a; Pope et al. 2017). We explore theoretical conceptualizations of sustainability in three ways. We commence by positing sustainability as 'one of a number of environmental governance discourses' (Bond and Morrison-Saunders 2009, 327), which has consequences for the legitimacy of environmental decision-making. We then consider two common interpretations of sustainability, whereby sustainability is conceptualized in terms of the maintenance and management of different forms of capital, or alternatively in terms of intra- and inter-generational equity. We then highlight four distinct discourses that can be identified within the theory and practice of sustainability assessment, where a discourse is 'a specific ensemble of ideas, concepts, and categorizations that are produced, reproduced, and transformed in a particular set of practices and through which meaning is given to physical and social realities' (Hajer 1995, 44). Throughout the discussion, we identify some key issues for sustainability assessment practice.

Sustainable development as a contested environmental governance discourse

Bond and Morrison-Saunders (2009), drawing on Svarstad et al. (2008), introduced four main types of environmental governance discourse, making it clear that these were simply common discourses and that there are potentially many more:

1. Traditionalist;
2. Sustainable development;
3. Ethical management; and
4. Promethean.

The traditionalist discourse understands that local actors are best placed to manage the environment. This is founded on evidence for indigenous people living in harmony with the environment for many thousands of years in some cases, without degrading the environment.

The sustainable development discourse is grounded in the Brundtland Commission definition of the term as: 'development that meets the needs of the present without compromising the ability of future generations to meet their own needs' (WCED 1987, 9). This definition brings together the twin imperatives of development and environmental protection, the nexus of the two also being the point at which EIA functions. This discourse was legitimized by the Earth Summit in 1992 in terms of being the appropriate goal framing EIA practice, as mentioned previously. The implication is therefore that EIA with a sustainability imperative should seek to ensure that future generations are not disadvantaged by development decisions today.

Ethical management is based on an understanding that people should have an ethical relationship with the environment (Clarke 2002). The stark differences between environmental ethics and sustainability discourses on biodiversity outcomes were highlighted by Bond et al. (2021), who suggested that even strong sustainability corresponds with a weak position in relation to environmental ethics, whereby few species enjoy rights on a par with human beings. The consequence of this is inevitable incremental biodiversity loss.

A Promethean discourse reflects confidence in human ingenuity, and the potential for technology to resolve emerging issues. In the context of environmental governance, this discourse is inherently positive in the face of environmental losses, arguing that solutions will be found.

These discourses are introduced to illustrate the conflict that can potentially arise through the use of a tool like EIA, since a stated policy goal of sustainable development is a contested position in itself; those who disagree that sustainable development is an appropriate environmental governance framing will not accept the validity of sustainability assessment. Such views matter given increasing provisions for public involvement in environmental decision-making, as they can undermine the legitimacy of sustainability assessment recommendations as valid bases of decision-making. With this in mind, given the focus of this chapter on sustainability assessment, the focus for the remainder of this section will be on alternative framings of sustainability, which also raise legitimacy concerns. The legitimacy of an impact assessment process has been defined as 'one which all stakeholders agree is fair and which delivers an acceptable outcome for all parties' (Bond et al. 2016, 188).

Sustainability in terms of maintaining and managing capitals

A simple internet search for dictionary definitions of sustainability provides interpretations and synonyms such as: keeping in existence, prolonging, maintaining, continuing, and lasting. Discussion of what is to be sustained, and the implications for future generations, is typically framed in terms of forms of capital. Here, three capitals are often considered, also known as the triple-bottom-line (TBL) or three pillars of environmental, social, and economic capital (Elkington 1997; Pope et al. 2004). Elsewhere, as many as six capitals have been mooted (e.g., IIRC 2013) through the addition of manufactured, intellectual, and human capital in addition to the TBL. Apart from environmental (or natural) capital, the other forms (regardless of how many are distinguished) refer specifically to human-related activity or considerations; thus, we simply differentiate only between natural and socioeconomic capital in this chapter.

EIA is triggered by development proposals likely to have a significant adverse effect on the environment. By definition, adverse impacts represent an erosion of capital, and thus a tension immediately comes to the fore when seeking to implement sustainability assessment. It gives rise to consideration of the spectrum of possible sustainability positions ranging from weak to strong sustainability. A weak sustainability position enables trade-offs between different forms of capital with the expectation that total capital continues to increase, while a strong sustainability stance expects there to be no net loss in any individual form of capital (Cabeza Gutés 1996; Neumayer 2010). Weak sustainability is commonly associated with EIA, especially when evaluation of new development proposals is 'reduced to the "jobs versus the environment" dilemma' (Glasson et al. 2012, 206), meaning that natural or environmental capital is substituted for immediate or short-term socioeconomic gain.

Yet even the strong sustainability position poses a challenge for sustaining natural capital in the face of an ever-expanding human population and the pursuit of economic growth. In this context, immutably sustaining natural capital without some form of change or trade-off arising in the practice of EIA or sustainability assessment is not possible. In an examination of the implications for biodiversity (i.e., a sub-set of natural capital), using an ethical, rather than sustainability framing, Bond et al. (2021, 5–6) concluded that while 'EIA will continue to consider the implications for biodiversity of human development [...] incremental loss of biodiversity is inevitable'. They further noted that if EIA were to give full protection of biodiversity without any losses, 'then only brownfield development would be possible' (1). Similarly, for non-living parts of the environment, such as mining, no new resource development could be contemplated, only recycling of currently used materials. This scenario is untenable. Only an 'absurdly strong sustainability' position can live up to the generic definition of sustaining, and this has been dismissed as being 'absurd' to expect from a new development associated with EIA practice (Daly 1995; Bond et al. 2021). A pragmatic form of strong sustainability is instead advocated in the context of sustainability assessment whereby critical thresholds are identified to delineate non-negotiable capital that cannot be traded whether across the pillars of sustainability through substituting one form of capital for another, or within individual pillars through offsetting (Morrison-Saunders and Pope 2013). There are already many long-established thresholds applied in relation to pollution controls (e.g., air and water quality standards that must be upheld) as well as more implicit expectations for protecting human lives (e.g., it is unlikely a proponent would put forward a development proposal that would intentionally cause human death or serious injury). In relation to biodiversity, an example of a critical threshold might be the presence of a protected species, meaning that development cannot take place unless some means can be found of preventing the loss of that species (for example, the Birds Directive within the European Union: Council of the European Communities 1979). While mitigation and offsetting measures can in principle deliver no net loss outcomes whereby the adverse impacts of development are truly counterbalanced, Pope et al. (2021) note that current theory and practice is 'conceptually murky' and go on to specify the conditions to be met for such improvement measures 'to be considered offsets rather than compensations'.

Sustainability in terms of intra- and inter-generational equity

The adverse and beneficial impacts of development are typically not evenly distributed in space or time, making it imperative to address intra-generational and inter-generational equity in sustainability assessment. It was suggested by George (1999) when discussing the potential role of EIA in delivering sustainable development that 'only two tests are needed for whether or not a proposed development is sustainable development: is it equitable for future generations, and is it equitable for the present generation?' (180). George (1999) subsequently points out that: 'if intragenerational equity is achieved, it will automatically become inter-generational, provided people are aware of development's likely effects on future generations and are in a position to make rational decisions about them' (183). Hermans and Knippengerg (2006) likewise focus on these two forms of equity, further noting that 'thinking about justice for future generations [...] [also necessitates thinking] about resilience' (312), a point to which we return later.

Thinking about the long term, especially extending into consideration of future generations, is challenging for sustainability assessment practice. It is a point of difference relative to traditional EIA, which focuses on the life cycle of development activities (i.e., from design through to decommissioning where applicable), but which is nevertheless criticized for having an undue emphasis on the immediate or short-term, especially the stages leading up to the approval decision (Glasson 1994; Weston 2000; Morrison-Saunders 2018). Hermans and Knippenberg (2006) point out that 'in the weak sustainability view [...] the present generation's wants and needs clearly outweigh the wants and needs of future generations' (304). There is implicit bias as the value judgments made in sustainability assessment 'can only be made by the present generation' (George 1999, 185). By default, new development activity seeks to enhance socioeconomic conditions in the short-term while conservation of natural capital 'relates specifically to the interests of future generations' (George 1999, 185). In the face of current trends of declining biodiversity and natural capital (Scholes and Biggs 2005), the trajectory is for the short-term benefit for current generations and long-term cost for future generations. It is also important to acknowledge here the converse argument that an emphasis on protecting natural resources comes to uphold inter-generational equity ideals at the expense of today's poor and disadvantaged (Barrett and Grizzle 1999). A further complicating factor is that even if current generations can think and act in terms of inter-generational needs, there is ambiguity as to exactly what time horizon this might represent given how much longevity varies between different human populations on the planet (Bond and Morrison-Saunders 2011), a point to which we return in the next section.

There are thus inherent challenges associated with addressing intra- and inter-generational equity in sustainability assessment practice and no guarantee that the two concepts are compatible. With the aim of guiding practice, Lamorgese and Geneletti (2013) derived examples of equity criteria, posing a suite of yes/no questions to apply during assessment for intra- and inter-generational equity alike. In subsequent work, noting that the sustainability assessment literature often treats each type of equity separately, they proposed an approach that 'allows for concurrent consideration of conditions of inter- and intra-generational equity' (Lamorgese and Geneletti 2015, 71). To this end, they defined four equity perspectives relating to (i) 'equity of opportunity for everyone' to have an 'acceptable quality and standard of living'; (ii) 'distributional fairness', meaning a 'fair share or proportionate distribution of benefits and dis-benefits' from development; (iii) 'distributional fairness across generations'; and (iv) 'justice for an imperfect world' meaning that a sustainability assessment process gives 'critical scrutiny and a comparative approach in assessment for identifying the major synergies, conflicts or trade-offs' (Lamorgese and Geneletti 2015, 65). We are not aware of any attempt in practice to apply this unique approach to conducting sustainability assessment.

Sustainability framing within sustainability assessments

The previous two sections have highlighted how two different ways of framing sustainability influence EIA (or sustainability assessment) practice. Building on this, in this section, we summarize the four sustainability discourses outlined in Pope et al. (2017), noting that this approach drew heavily on the work of Hugé et al. (2013) in taking an approach that 'sought to move away from the dichotomy of weak and strong sustainability' (Pope et al. 2017, 211). These discourses could all potentially be reflected in sustainability assessment practice, though some are more common than others.

The first and most commonly applied sustainability discourse, termed 'the pragmatic integration of development and environmental goals' (Pope et al. 2017, 210), is characterized by the notion of balancing and trading off between sustainability capitals. Traditional approaches to EIA with an emphasis on mitigation of significant impacts fall into this category along with more objectives-led approaches associated with strategic environmental assessment, and both lack clarity as to what sustainability goals or outcomes will be achieved (Pope et al. 2004). The challenges for practice discussed in the section 'Sustainability in terms of maintaining and managing capitals' are inherent within this discourse.

The second sustainability discourse, termed the 'idea of limitations on human activities' (Pope et al. 2017, 210), is based on assessment taking place in the context of clearly established limits or boundaries at the global scale that must not be crossed. By ensuring that the impacts of development stay within the limits of acceptable change, an acceptable sustainability outcome will be met. This might sound easy, but in practice understanding or translating global limits to inform local sustainability for individual development activities undergoing assessment is extremely challenging. This discourse embodies the notion of 'non-negotiable capital' mentioned in the section 'Sustainability in terms of maintaining and managing capitals'.

The third sustainability discourse termed 'a process of directed change/transition' (Pope et al. 2017, 210) recognizes that current conditions may actually be unsatisfactory and not something that should be sustained. This might variously be specific cases of environmental pollution and degradation, or social deprivation, injustices, or inequalities ranging through to global conditions and undesirable trajectories occurring at national and international scales. Applying this discourse in sustainability assessment means actively seeking improvements to current circumstances through processes that 'encourage positive steps towards greater community and ecological sustainability, towards a future that is more viable, pleasant and secure' (Gibson 2006, 172). Rather than mitigating development activity in relation to the baseline conditions, this discourse actively seeks to shift that baseline in a positive direction, even if the desired end goal is not well defined.

The fourth sustainability discourse termed 'promotion of resilience and justice' by Pope et al. (2017, 211) was originally mooted by Hermans and Knippenberg (2006). As discussed in the section 'Sustainability in terms of intra- and inter-generational equity', the notion of justice is considered in terms of intra-generational equity (i.e., justice for those present now and affected in some way by the development activity under consideration in a sustainability assessment) and inter-generational equity (i.e., justice for those to come in the future). Resilience is defined as the ability of the system to maintain functionality or maintain the elements needed to renew and reorganize in response to a large perturbation (Walker et al. 2002).

By way of conclusion regarding the four sustainability discourses apparent within the practice of sustainability assessment, the key point is that the very nature of a given sustainability assessment will be determined by the discourse that underpins key decisions and actions arising. Each discourse points to a different purpose and type of outcome to be pursued. While it is likely that individual practitioners and stakeholders may favor a particular discourse, what is applicable in any given individual assessment is a product of the context at stake. Discourse analysis is rare in research studies, given the time-demanding requirements for interviews, and extensive analysis required. Nevertheless, some examples exist that demonstrate that discourses do influence the interpretation of sustainability in the EIA context (e.g., Rozema et al. 2012), or that different discourses exist as to the extent to which EIA has sustainable development as a goal (Runhaar et al. 2013). Assessing and evaluating the future consequences of human actions requires consideration of the nature of development activities and the characteristics of the receiving environment, which will bear the brunt of the consequences of development (Morrison-Saunders 2018).

We do not advocate a 'one-size-fits-all' approach to interpreting sustainability for the purposes of sustainability assessment. Rather, different discourses or combinations of them may be brought into play for individual applications and importantly 'each sustainability assessment process should be tailor-made for context' (Bond et al. 2012, 59). What is essential for effective practice is having a clear understanding of which discourse(s) are going to be applied. This might involve a visioning exercise carried out with stakeholders, some other explicit process to 'facilitate debate on appropriate discourses and representations of sustainability within a given decision-making context' (Pope et al. 2017, 214), or it might simply reflect prevailing government policy for a region. For example, the *Impact Assessment Act* of Canada[1] (2019) provides a definition of sustainability:

sustainability means the ability to protect the environment, contribute to the social and economic well-being of the people of Canada and preserve their health in a manner that benefits present and future generations.

(Section 2)

A factor that must be taken into account during impact assessment is 'the extent to which the designated project contributes to sustainability' (Section 22(1)(h)). Thus, this framing of sustainability is aligned with the process of directed change discourse along with the pragmatic integration of development and environmental goals discourse.

Regardless of how sustainability is defined or interpreted for a given sustainability assessment, it is essential that the discourse is shared or at least clearly communicated to all involved in the sustainability assessment because it is a determinant of the 'goals and criteria that establish and operationalize the vision for sustainability in the context of the proposed activity. And it provides the framework against which alternatives will be compared and the preferred option determined' (Morrison-Saunders and Pope 2013, 56). As advocated in Bond and Morrison-Saunders (2011), sustainability assessment 'needs to be seen as a vehicle for deliberation in order to address the policy controversies' (5) that arise from the application of competing sustainability discourses.

Practical challenges in sustainability assessment

There are many practical challenges in the conduct of sustainability assessment, some of which are common to impact assessment generally (for example, prediction of impacts, managing uncertainty etc.) and others that relate to the contested concept of sustainability itself. We focus on two of those here: the challenge of operationalizing sustainability for the purpose of the assessment and the challenges associated with time horizons alluded to above.

Operationalizing sustainability

The sustainability discourse being applied in a sustainability assessment needs to be operationalized in a format to enable the practical evaluation of a proposed development activity to proceed. As Bond and Morrison-Saunders (2009) put it, the 'key component of any sustainability assessment is the sustainability indicator' (325). Here, the notion of an indicator is that 'it should provide a simplified, but still sufficient, representation of sustainability' (Pope et al. 2017, 211). A long-standing body of literature is devoted to the development and selection of sustainability indicators which we do not seek to repeat. Here, we simply reiterate the point made by Pope et al. (2017):

> Sustainability indicators are developed in the sustainability assessment process by ascribing values to these variables and comparing these with relative values (reflecting a particular discourse) in order to assess the sustainability of the proposal at hand.
>
> *(212)*

With respect to carrying out sustainability assessment in practice, determination of the sustainability discourse and the appropriate indicators can take place simultaneously through the involvement of relevant stakeholders (Bond and Morrison-Saunders, 2009). Failure to involve a broad range of stakeholders, for example, by having developers and their consultants alone perform this role, will result in sustainability assessment frameworks that 'favor their own discourses' (Bond and Morrison-Saunders 2009, 326). In the absence of sets of indicators having been established by regulators in individual jurisdictions, the 17 SDGs established by the United Nations along with the 169 sub-goals may provide a foundational starting point for practice (Hacking 2018). There is, however, considerable scope for interpretation of those for the purposes of an individual assessment. In addition, sustainability assessment is regarded as being context-specific, with different goals being more or less relevant in different contexts. For example, it is clear that the sustainability considerations for a decision on a mining project will be very different from those for a land-use plan or other more strategic types of decision. Consequently, an important first step is to define the meaning of the term in the context of the decision at hand. This is in line with Bina (2008), who argued that a strategic environmental assessment system should be context-specific, in that it needs to be flexible and adapt to the different dimensions of context (which she indicates are values, cultural; political; and social). We agree that the consideration of context is all important, but would suggest that it is not just the sustainability assessment method that needs to accommodate the context; it is also the framing of sustainability.

Whether the selected sustainability indicators take the form of disaggregated triple-bottom-line measures or composite sustainability variables (Pope et al. 2017), they must be appropriately holistic in scope to capture the sustainability discourse at hand. It is also important that they are compatible with each other so as to be mutually reinforcing, since conflicting sustainability indicators will give rise to substantive trade-offs irrespective of other characteristics of a sustainability assessment process (Morrison-Saunders and Pope 2013). The 'sustainability appraisal' guidance of the UK Office of the Deputy Prime Minister (OPDM) (2005, 120–121) advocates carrying out a test of the internal compatibility of sustainability assessment objectives to highlight any points of tension or conflict between them prior to embarking on evaluating development activities. In this approach, each sustainability assessment objective established for application within a given assessment is systematically compared with each of the others. A similar approach was put forward by Nilsson et al. (2016) for mapping the interactions between the SDGs and individual sub-goals or targets, but in this case with a seven-point scale of 'indivisible, reinforcing, enabling, consistent, constraining, counteracting, cancelling' (321) being used to denote the nature of the relationship between any two components. Such tools provide a means of 'managing (and in this case, avoiding) potential trade-offs in sustainability assessments' (Morrison-Saunders and Pope 2013, 57).

A problem associated with the development of indicators is the debate over the extent to which sustainability assessment should be reductionist and the degree to which it should be holistic (Bell and Morse 2008). Reductionism we define as breaking down complex processes into simple terms or component parts (i.e., selecting a few sustainability indicators to represent the sustainability of a whole system). Evidence currently suggests that the emphasis in sustainability assessment is very much on reductionism, but that the degree of reductionism varies a great deal within particular systems (e.g., illustrated in Bond and Morrison-Saunders, 2011 using examples of practice from England and Western Australia, see Box 6.1). These sustainability assessments can be criticized by observers for using the wrong indicators, or too few indicators, or too many indicators. From a pragmatic point of view, a large number of indicators leads to an unwieldy, time consuming and expensive sustainability appraisal (SA) exercise. In this context, the indicators available through the UN SDGs clearly present problems in terms of the practicality of their use. Bell and Morse (2008) provide the example of Maximum Sustainable Yield (MSY) as a pervasive policy indicator, which has led to significant environmental impacts through its use. MSY is the extreme of reductionism whereby the understanding of the environmental system is reduced to a single measurement. It is typically used (although being phased out) in management of marine fish stocks. Fish stocks can be fished up to the tonnage specified, after which further fishing is banned to preserve the stocks. Bell and Morse (2008) cite the use of MSY as the cause of the collapse of the Peruvian anchovy stock in 1972, given that the calculations did not understand the implications of, or take into account, El Niño events.

Box 6.1. Number of indicators used in sustainability appraisal (SA) reports associated with the local develop plan process of a sample of English local planning authorities in 2005–2007

Local authority	SA report publication year	Number of indicators in SA framework
Ashford	2006	233
Blaby	2006	101
Blackburn	2007	112
Charnwood	2006	70
Chelmsford	2006	60
Doncaster	2005	150
Great Yarmouth	2006	106
Guildford	2006	137
Scarborough	2006	133

Source: Adapted from Bond and Morrison-Saunders (2011).

Steinemann (2000, 640) defines a holistic approach as one which facilitates 'moving away from analyses of isolated risks and toward a broader understanding'. In this vein, Grace and Pope (2015) recommend a systems approach to sustainability assessment, in which the sustainability variables or indicators are defined such that the relationships between them, and the feedback loops they create can be visually represented and system behavior analyzed. There are examples of systems approaches to sustainability assessment being applied in practice; see, for example, Audouin et al. (2015), who describes the development of a causal loop diagram of a socio-ecological system as the basis for the sustainability assessment of a platinum mine in South Africa.

To conclude, much has been written about the development of appropriate indicators within impact assessment (e.g., Donnelly et al. 2007; Laedre et al. 2015), with some consensus that broad engagement is necessary to develop an appropriate set of indicators. Such engagement tends to implicitly include a range of sustainability discourses, thereby increasing the likelihood that the sustainability assessment will be regarded as legitimate.

Factoring in long-term time horizons

As we have already discussed above, despite the acknowledgment of inter-generational equity as a core principle of sustainability, in practice, impact assessment processes consider only the lifetime of the plan or project being assessed and not its long-term implications (Bond and Morrison-Saunders 2011). Box 2 introduces the particular case of radioactive waste in the UK, where the timescale associated with potential impacts extends for a period longer than human beings have existed as a species.

Box 6.2 Time frames for the UK Managing Radioactive Waste Safely (MRWS) program

The UK has the longest history of civil nuclear power of any nation, dating back to the opening of the Windscale nuclear power plant in 1956. In 2008, the UK adopted a policy to develop new nuclear power stations to replace those due to be decommissioned at the end of their life (Department for Business Enterprise & Regulatory Reform, 2008). This policy decision continues to increase the inventory of radioactive waste currently stored in various sites around the UK. The inventory at the time of writing is 133,000 m³ of waste of all activities, with expected future waste arising totaling an additional 4,420,000 m³ (https://ukinventory.nda.gov.uk/the-2019-inventory /2019-uk-data/).

Recognizing the need for a long-term solution for the management of radioactive waste, and following a failed attempt to develop a geological repository for radioactive waste in 1997, the UK government embarked on a *Managing Radioactive Waste Safely* program (Department for Environment Food and Rural Affairs, Department of the Environment, National Assembly for Wales and Scottish Executive 2001), in order to identify a publicly acceptable solution to the growing problem of radioactive waste storage. A Committee on Radioactive Waste Management (CoRWM) was established to advise government, and concluded its investigations in 2006, culminating in recommendations to government (CoRWM 2006). The Committee grappled, in particular, with the need to consider impacts over more than 100,000 years, as this is the timescale over which the waste will remain a potential hazard. However, regulators refused to accept a safety case made for radioactive waste disposal for a period greater than 300 years because that is the longest period they have confidence that institutional control can be guaranteed (CoRWM 2006). The timescale required for radioactive waste management is so long that there are even debates about how to warn future generations of the existence of buried, hazardous waste, when there are no guarantees that languages spoken will be the same, and no expectation that any signage would last even a fraction of that time (https://www.bbc.com/future /article/20200731-how-to-build-a-nuclear-warning-for-10000-years-time). What is to stop a future generation from digging at the site of an underground repository for radioactive waste?

Gee and Stirling (2004) distinguish between *risk* (where impacts and their probabilities are known), *uncertainty* (where impacts are known but their probabilities are not), and *ignorance* (where neither impacts nor their probabilities are known). Over very long timescales, predictions in sustainability assessment are likely to be based on both uncertainty and ignorance. There is little practice on which to draw for such predictions and certainly no follow-up studies. Some preliminary ideas for dealing with uncertainty, ambiguity and ignorance in impact assessment are put forward in Bond et al. (2015b) based on using scenario methods and embedding resilience thinking in procedures. The key point we wish to make here is simply that accommodating long-term time horizons in sustainability assessment will require new ways of thinking and carrying out practice relative to long-standing project-based EIA approaches. At the very least, there ought to be the explicit acknowledgment of the time scales underpinning any given sustainability assessment.

Conclusions: principles for sustainability assessment practice

In this chapter, we have shown that the practice of EIA is inextricably linked with ideas for sustainability assessment. We have demonstrated that sustainable development is by no means universally accepted as the appropriate framing for environmental governance and that this will continue to have implications for the perceived legitimacy of EIA. Assuming that sustainable development is desired by stakeholders, then the variety of interpretations of sustainability adds further complexity to assessment; the very goal the assessment is striving to achieve is contested. This remains an unresolved problem that only broad dialogue can currently address. It is also clear that these many sustainability discourses are reflected in the variety of discourses associated with sustainability assessment itself, along with more obfuscating factors associated with the role of impact assessment more generally as a decision-support tool. Together, these many complications do not stop sustainability assessment from being a valuable tool. They mean that sustainability assessment has to embrace and reflect human diversity, and the better it does so, the more equitable the outcome will be.

Traditional EIA practice is clearly a form of sustainability assessment (Hacking and Guthrie 2008; Pope et al. 2004), even if it is limited in terms of the full spectrum of possibilities. As such then, all of the well-established principles for EIA best practice (e.g., IAIA and IEA 1999) are applicable for sustainability assessment, and we do not duplicate that content here. Instead, we distill key points arising from our previous discussion that add three principles that should also apply specifically to the practice of sustainability assessment:

Principle 1

Broad engagement is required to define an understanding of the meaning of sustainability in the context of the proposed action to be assessed. This principle is key to the consideration of discourses, and supports Runhaar et al. (2010), who previously argued that EIA could act as a vehicle for discourse reflection to prevent environmental knowledge inconsistent with dominant discourses from being ignored.

Principle 2

Timescales need to be explicitly stated, and need to consider a number of human generations into the future, appropriate to the context for the assessment. This principle underpins the focus of sustainable development on future (as well as current) generations (WCED 1987), acknowledging that the implications of decisions made now could be felt by thousands of future generations. It also recognizes that it is not feasible to know or understand what decisions future generations would opt for if they were able to have a stake in current impact assessment processes.

Principle 3

Ethics should form part of the assessment in that the future implications of adopted sustainability discourses on natural capital should be considered. This principle reflects the prevalence of economic discourses on sustainable development that consider the implications of change in financial terms, which tend to favor the one species that employs a monetary system: humans. As argued in Bond et al. (2021), an ethics-based assessment can act as a counterweight to the situation where the rights of all other species on the planet are subordinate to the rights of humans.

Sustainability remains a very vague and contested concept, and this fact makes the practice of sustainability assessment extremely challenging. Upholding the first two principles will help to ensure that the pitfalls associated with a process that purports to drive development toward sustainability are at least clear on what this means. The final principle seeks further reflection on whether the outcomes for future generations are really what those generations might expect.

Note

1   https://www.parl.ca/Content/Bills/421/Government/C-69/C-69_4/C-69_4.PDF (accessed December 4, 2020)

References

Audouin, M., M. Burns, A. Weaver, D. le Maitre, P. O'Farrell, R. du Toit and J. Nel (2015), 'Chapter 14: An introduction to sustainability science and its links to sustainability assessment', in A. Morrison-Saunders, J. Pope and A. Bond (eds), *Handbook of Sustainability Assessment*, Cheltenham, UK: Edward Elgar Publishing Ltd., pp321–345

Barrett, C. B. and R. E. Grizzle (1999), 'A holistic approach to sustainability based on pluralistic steward-ship', *Environmental Ethics*, **21**: 23–42.

Bell, S. and S. Morse (2008), *Sustainability Indicators: Measuring the Immeasurable?* London, Sterling, VA: Earthscan.

Bina, O. (2008), 'Context and systems: Thinking more broadly about effectiveness in strategic environmental assessment in China', *Environmental Management*, **42**: 717–733.

Bond, A. and A. Morrison-Saunders (2009), 'Sustainability appraisal: Jack of all trades, master of none?', *Impact Assessment and Project Appraisal*, **27**(4): 321–329.

Bond, A., C. V. Viegas, C. Coelho de Souza Reinisch Coelho and P. M. Selig (2010), 'Informal knowledge processes: The underpinning for sustainability outcomes in EIA?', *Journal of Cleaner Production*, **18**(1): 6–13.

Bond, A. and A. Morrison-Saunders (2011), 'Re-evaluating sustainability assessment: Aligning the vision and the practice', *Environmental Impact Assessment Review*, **31**(1): 1–7.

Bond, A., A. Morrison-Saunders and J. Pope (2012), 'Sustainability assessment: The state of the art', *Impact Assessment and Project Appraisal*, **30**(1): 56–66.

Bond, A., A. Morrison-Saunders and R. Howitt (eds) (2013), *Sustainability Assessment: Pluralism, Practice and Progress*, Abingdon: Routledge.

Bond, A., J. Pope and A. Morrison-Saunders (2015a), 'Introducing the roots, evolution and effectiveness of sustainability assessment', in A. Morrison-Saunders, J. Pope and A. Bond (eds), *Handbook of Sustainability Assessment*, Cheltenham: Edward Elgar, pp. 3–19.

Bond, A., A. Morrison-Saunders, J. Gunn, J. Pope and F. Retief (2015b), 'Managing uncertainty, ambiguity and ignorance in impact assessment by embedding evolutionary resilience, participatory modelling and adaptive management', *Journal of Environmental Management*, **151**: 97–104.

Bond, A., J. Pope, A. Morrison-Saunders and F. Retief (2016), 'A game theory perspective on environmental assessment: What games are played and what does this tell us about decision making rationality and legitimacy?', *Environmental Impact Assessment Review*, **57**: 187–194.

Bond, A., J. Pope, A. Morrison-Saunders and F. Retief (2021), 'Taking an environmental ethics perspective to understand what we should expect from EIA in terms of biodiversity protection', *Environmental Impact Assessment Review*. https://doi.org/10.1016/j.eiar.2020.106508.

Cabeza Gutés, M. (1996), 'The concept of weak sustainability', *Ecological Economics*, **17**(3): 147–156.

Clarke, A. H. (2002), 'Understanding sustainable development in the context of other emergent environmental perspectives', *Policy Sciences*, **35**(1): 69–90.

CoRWM (2006), 'Managing our radioactive waste safely, CoRWM's recommendations to government', available at: https://www.gov.uk/government/publications/managing-our-radioactive-waste-safely-corwm-doc-700 (accessed October 28, 2020).

Council of the European Communities (1979), 'Council directive 79/409/EEC of 2 April 1979 on the conservation of wild birds', *Official Journal of the European Communities*, **L103**: 1–18.

Daly, H. E. (1995), 'On Wilfred Beckerman's critique of sustainable development', *Environmental Values*, **4**(1): 49–55.

Department for Business Enterprise & Regulatory Reform (2008), 'Meeting the energy challenge: A white paper on nuclear power', available at: https://webarchive.nationalarchives.gov.uk/+/http://www.berr.gov.uk/files/file43006.pdf (accessed December 14, 2020).

Department for Environment Food and Rural Affairs, Department of the Environment, National Assembly for Wales and Scottish Executive (2001), 'Managing radioactive waste safely. Proposals for developing a policy for managing solid radioactive waste in the UK', available at: http://www.ni-environment.gov.uk/ra_waste.pdf (accessed March 17, 2009).

Donnelly, A., Jones, M., O'Mahony, T. and Byrne, G. (2007), 'Selecting environmental indicators for use in strategic environmental assessment', *Environmental Impact Assessment Review*, **27**(2): 161–175.

Elkington, J. (1997), *Cannibals with Forks: The Triple Bottom Line of 21st Century Business*, Oxford: Capstone Publishing Limited.

Gee, D. and A. Stirling (2004), 'Late lessons from early warnings: Improving science and governance under uncertainty and ignorance', in M. Martuzzi and J. A. Tickner (eds), *The Precautionary Principle: Protecting Public Health, the Environment and the Future of Our Children*, Copenhagen: WHO Regional Office for Europe, pp. 93–120.

George, C. (1999), 'Testing for sustainable development through environmental assessment', *Environmental Impact Assessment Review*, **19**: 175–200.

Gibson, R. (1993), 'Environmental assessment design: Lessons from the Canadian experience', *The Environmental Professional*, **15**: 12–24.

Gibson, R. (2006), 'Sustainability assessment: Basic components of a practical approach', *Impact Assessment and Project Appraisal*, **24**(3): 170–182.

Gibson, R. (2013), 'Avoiding sustainability trade-offs in environmental assessment', *Impact Assessment and Project Appraisal*, **31**(1): 2–12.

Glasson, J. (1994), 'Life after the decision: The importance of monitoring in EIA', *Built Environment*, **20**(4): 309–320.

Glasson, J., R. Therivel and A. Chadwick (2012), *Introduction to Environmental Impact Assessment*, 4th edition, London: Routledge.

Grace, W. and Pope (2015), 'Chapter 13: A systems approach to sustainability assessment', in Morrison-Saunders, A., Pope, J., and Bond, A. (eds), *Handbook of Sustainability Assessment*, Cheltenham: Edward Elgar, pp. 285–320.

Hacking, T. and P. Guthrie (2008), 'A framework for clarifying the meaning of triple bottom-line, integrated, and sustainability assessment', *Environmental Impact Assessment Review*, **28**(2–3): 73–89.

Hacking, T. (2018), 'The SDGs and the sustainability assessment of private sector projects: Theoretical conceptualisation and comparison with current practice using the case study of the Asian Development Bank', *Impact Assessment and Project Appraisal*, **37**(1): 2–16.

Hajer, M. A. (1995), *The Politics of Environmental Discourse: Ecological Modernization and the Policy Process*, New York: Oxford University Press.

Hermans, F. and L. Knippenberg (2006), 'A principle-based approach for the evaluation of sustainable development', *Journal of Environmental Assessment Policy and Management*, **8**: 299–319.

Hugé, J., T. Waas, F. Dahdouh-Guebas, N. Koedam and T. Block (2013), 'A discourse-analytical perspective on sustainability assessment: Interpreting sustainable development in practice'. *Sustainability Science*, **8**: 187–198.

IAIA and IEA – International Association for Impact Assessment and Institute for Environmental Assessment UK (1999), 'Principles of environmental impact assessment best practice', available at: www.iaia.org/uploads/pdf/principlesEA_1.pdf (accessed October 28, 2020).

IIRC – International Integrated Reporting Council (2013), 'The international IR framework', IIRC, available at: http://www.theiirc.org/wp-content/uploads/2013/12/13-12-08-THE-INTERNATIONAL-IR-FRAMEWORK-2-1.pdf (accessed October 28, 2020).

Jacobs, M. (1999), 'Sustainable development as a contested concept', in A. Dobson (ed.), *Fairness and Futurity: Essays on Environmental Sustainability and Social Justice*, Oxford and New York: Oxford University Press, pp. 21–45.

Kidd, S. and T. B. Fischer (2007), 'Towards sustainability: Is integrated appraisal a step in the right direction?', *Environment and Planning C*, **25**: 233–249.

Laedre, Ö., T. Haavaldsen, R. A. Bohne, J. Kallaos and J. Lohne (2015), 'Determining sustainability impact assessment indicators', *Impact Assessment and Project Appraisal*, **33**(2): 98–107.

Lamorgese, L. and D. Geneletti (2013), 'Sustainability principles in strategic environmental assessment: A framework for analysis and examples from Italian urban planning', *Environmental Impact Assessment Review*, **42**: 116–126.

Lamorgese, L. and D. Geneletti (2015), 'Equity in sustainability assessment: A conceptual framework', in A. Morrison-Saunders, J. Pope and A. Bond (eds), *Handbook of Sustainability Assessment*, Cheltenham: Edward Elgar, pp. 57–76.

Morgan, R. K. (2012), 'Environmental impact assessment: The state of the art', *Impact Assessment and Project Appraisal*, **30**(1): 5–14.

Morrison-Saunders, A. and R. Therivel (2006), 'Sustainability integration and assessment', *Journal of Environmental Assessment Policy and Management*, **8**(3): 281–298.

Morrison-Saunders, A. and J. Pope (2013), 'Conceptualising and managing trade-offs in sustainability assessment', *Environmental Impact Assessment Review*, **38**: 54–63.

Morrison-Saunders, A. (2018), *Advanced Introduction to Environmental Impact Assessment*, Cheltenham: Edward Elgar.

Morrison-Saunders, A. and M. Hughes (2018), 'Overcoming sustainability displacement – The challenge of making sustainability accessible in the here and now', in M. Brueckner, R. Spencer and M. Paull (eds), *Disciplining the Undisciplined?: Perspectives from Business, Society and Politics on Responsible Citizenship, Corporate Social Responsibility and Sustainability*, Cham: Springer, pp. 39–53.

Morrison-Saunders, A, L. E. Sánchez, F. Retief, J. Sinclair, M. Doelle, M. Jones, J.-A. Wessels and J. Pope (2020), 'Gearing up impact assessment as a vehicle for achieving the UN sustainable development goals', *Impact Assessment and Project Appraisal*, **38**(2): 113–117.

Neumayer, E. (2010), *Weak versus Strong Sustainability: Exploring the Limits of Two Opposing Paradigms*, Cheltenham: Edward Elgar.

Nilsson, M., D. Griggs and M. Visbeck (2016), 'Map the interactions between sustainable development goals', *Nature*, **534**: 320–322.

ODPM – Office of the Deputy Prime Minister (UK) (2005), *Sustainability Appraisal of Regional Spatial Strategies and Local Development Documents: Guidance for Regional Planning Bodies and Local Planning Authorities*, London: ODPM, 156 pp.

Partidário, M. (2012), 'Impact assessment', Fastips No. 1, Fargo: International Association for Impact Assessment, available at: www.iaia.org/uploads/pdf/Fastips_1-Impact Assessment.pdf (accessed December 15, 2020).

Pope, J., D. Annandale and A. Morrison-Saunders (2004), 'Conceptualising sustainability assessment', *Environmental Impact Assessment Review*, **24**: 595–616.

Pope, J., A. Bond and A. Morrison-Saunders (2015), 'A conceptual framework for sustainability assessment', in Morrison-Saunders, A., J. Pope and A. Bond (eds), *Handbook of Sustainability Assessment*, Research Handbooks on Impact Assessment, Cheltenham: Edward Elgar, pp. 20–42.

Pope, J., A. Bond, J. Hugé and A. Morrison-Saunders (2017), 'Reconceptualising sustainability assessment', *Environmental Impact Assessment Review*, **62**: 205–215.

Pope, J., A. Morrison-Saunders, A. Bond and F. Retief (2021), 'When is an offset not an offset? A framework of necessary conditions for biodiversity offsets', *Environmental Management*, **67**: 424–435.

Rega, C. and A. Bonifazi (2014), 'Strategic environmental assessment and spatial planning in Italy: Sustainability, integration and democracy', *Journal of Environmental Planning and Management*, **57**(9): 1333–1358.

Rozema, J. G., A, J. Bond, M. Cashmore and J. Chilvers (2012), 'An investigation of environmental and sustainability discourses associated with the substantive purposes of environmental assessment', *Environmental Impact Assessment Review*, **33**(1): 80–90.

Runhaar, H., P. R. Runhaar and T. Oegema (2010), 'Food for thought: Conditions for discourse reflection in the light of environmental assessment', *Environmental Impact Assessment Review*, **30**(6): 339–346.

Runhaar, H., F. van Laerhoven, P. Driessen and J. Arts (2013), 'Environmental assessment in The Netherlands: Effectively governing environmental protection? A discourse analysis', *Environmental Impact Assessment Review*, **39**: 13–25.

Runhaar, H., P. Driessen and C. Uittenbroek (2014), 'Towards a systematic framework for the analysis of environmental policy Integration', *Environmental Policy and Governance*, **24**(4): 233–246.

Scholes, R. J. and R. Biggs (2005), 'A biodiversity intactness index', *Nature*, **434**(7029): 45–49.

Sinclair, J. A., M. Doelle and R. B. Gibson (2018), 'Implementing next generation assessment: A case example of a global challenge', *Environmental Impact Assessment Review*, **72**: 16–176.

Steinemann, A. (2000), 'Rethinking human health impact assessment', *Environmental Impact Assessment Review*, **20**: 627–645.

Svarstad, H., L. K. Petersen, D. Rothman, H. Siepel and F. Wätzold (2008), 'Discursive biases of the environmental research framework DPSIR', *Land Use Policy*, **25**(1): 116–125.

UNEP (2018), *Assessing Environmental Impacts- A Global Review of Legislation*, Nairobi: United Nations Environment Program.

United Nations Conference on Environment and Development (1992), *Earth Summit '92*, London: Regency Press.

van Pelt, M. J. F., A. Kuyvenhoven and P. Nijkamp (1992), 'Sustainability, efficiency and equity: Project appraisal in economic development strategies', in A. G. Colombo (ed.), *Environmental Impact Assessment*, Dordrecht: Kluwer Academic Publishers, pp. 287–309.

Walker, B., S. Carpenter, J. Anderies, N. Abel, G. S. Cumming, M. Janssen, L. Lebel, J. Norberg, G. D. Peterson and R. Pritchard. (2002), 'Resilience management in social-ecological systems: A working hypothesis for a participatory approach', *Conservation Ecology*, **6**(1): 14..

WCED – World Commission on Environment and Development (1987), *Our Common Future*, Oxford, UK: Oxford University Press.

Weston, J. (2000), 'EIA, decision-making theory and screening and scoping in UK practice', *Journal of Environmental Assessment Policy and Management*, **43**(2): 185–203.

Yang, T. (2019), 'The emergence of the environmental impact assessment duty as a global legal norm and general principle of law', *Hastings Law Journal*, **70**(2): 525–572.

7

CLIMATE CHANGE IN ENVIRONMENTAL ASSESSMENT IN EUROPE

A lot of potential and a lot to do

Alexandra Jiricka-Pürrer and Thomas B. Fischer

Introduction

Identifying impacts arising from climate change and finding ways for adapting to them constitute complex challenges for policymakers and planners. In this context, decision-support instruments such as strategic environmental assessment and environmental impact assessment (together referred to in this chapter as "environmental assessment" – EA) offer opportunities for better integrating climate change early on in policy, plan, program, and project processes. EA can therefore support climate-friendly and more sustainable socioeconomic development.

That climate change is an important issue for EA was first suggested over two decades ago and has been discussed regularly since then (Fischer 1999; Birkmann and Fleischhauer 2009; Fischer and Sykes 2009; Agrawala et al. 2010; Posas 2011; Byer et al. 2012; Wende et al. 2013). In this context, while issues of climate change mitigation were covered early on, adaptation became part of the discussion only more recently. However, it is now attracting a considerable amount of attention in many countries throughout the world.

There are opportunities for considering climate change mitigation and adaptation throughout at the various stages of EA, and information on climate change can be integrated into the assessment of environmental impacts in different ways. EA reports continue to be based primarily on legal requirements and standards in most systems. Despite some initial attempts (EBA 2015), climate change is rarely integrated into these existing standards. However, insights from climate scenarios and climate impact models can be used in the future as additional information becomes available for the development of mitigation and compensation measures. This implies that validated data (used in climate impact models) are established by authorities and are made available by, e.g., climate service centers.

Issues that are frequently raised and discussed in the EA and climate change literature include capacity building, uncertainty (related to, e.g., the role of different planning levels for investigating climate change impacts and developing measures to deal with them at the appropriate level) and the availability of climate projections (particular at the regional and local level). Larsen et al. (2013) were the first to discuss the implications of lack of certainty, suggesting that climate

DOI: 10.4324/9780429282492-8

change should not be ignored in EA just because of uncertainty. However, focusing on SEA practices in Denmark, they observed that climate change was poorly addressed. Subsequently, the need for planning to embrace EIA and SEA for better addressing climate change mitigation and adaptation was stressed by, e.g., Jiricka-Pürrer et al. (2016, 2018).

In the European Union, based on the recent amendment of the EIA Directive (2014/52/EC), consideration of climate change impacts – especially of potential climate change-related accidents and disasters – has become mandatory at the project level. Strategies for coping with climate risks are often prepared at higher tiers though, i.e., through policies, plans, and programs (see, e.g., Marshall and Fischer 2006; Wende et al. 2013; Jiricka-Pürrer et al. 2016) and therefore need to be covered by SEA. However, currently, the EU's SEA Directive (dating back to 2001) does not explicitly mention climate change, even if guidance was released in 2013, dealing with how climate change should be integrated with SEA (e.g., EC 2013a).

In this chapter, we discuss the legal requirements as well as entry points for the consideration of climate change mitigation and adaptation through EA. In this context, the interaction between EIA and SEA in meeting challenges is discussed. How to maximize potential benefits from EA application is also elaborated on.

The role of EIA for climate change mitigation and adaptation

This section consists of three parts. These revolve around the identification of key impact factors and thematic entry points for the consideration of climate change as well as methodological implications.

Key impact factors

Previous publications on barriers for considering climate change in EA have highlighted the tension between uncertainties on the one hand and the application of regulatory and standards-based processes on the other (Larsen et al. 2013; Jiricka-Pürrer et al. 2016). Overall, key factors impacting on how climate change is considered can be attributed to one of the following three dimensions (adapted from Jiricka-Pürrer et al. 2019a):

1. **Framing conditions**: These include legislation at national level (at times framed at international level), guidance, regulations, standards, as well as procedural and methodological provisions;
2. **Data and information**: These require field-specific expertise of climate change impacts and options for adaptation, including climate change scenarios, impact models, and downscaling at multiple spatial levels; and
3. **Capacities of relevant actors (particularly authorities, project developers, and consultants but also partly NGOs and others)**: These include knowledge of climate change impacts, values, and responsibilities.

Thematic entry points for the consideration of climate change

Climate change mitigation

Reduction of greenhouse gas emissions (GHG) for meeting the targets of the Paris Agreement from 2015 and after 2021 those of the Glasgow COP26 is a continuous effort throughout the world (EEA 2018). Furthermore, and depending on the specific situation of application, other area-specific objectives are also relevant, including, e.g., the Alpine Climate Target System 2050 (Alpine Convention 2019) for the European Alps that are particularly affected by climate change.

EA can highlight implications of policies, plans, programs, and projects for global climate change and can support the early consideration of system and technological alternatives (Stöglehner 2020; Wende et al. 2013; Fischer 2006). Furthermore, EA can help foster the integrative consideration of climate change mitigation targets.

If approached as a framework, (see e.g. Fischer and Gonzalez 2021) EA can help to streamline the consideration of specific issues and their objectives, and the consideration of associated alternatives as well, e.g., mitigation measures at different decision tiers (policies, plans, programs, and projects). Reduction options of GHGs can be considered at all phases of a project's lifetime, including context setting (in, e.g., policies, plans, and programs), project planning, construction, and operation phases (depending on project type). Some countries, such as Austria, introduced requirements for considering a project's CO_2 emissions (impacts on global climate change) in EIA earlier than the EU Directive 2014/52/EC. However, to date, in practice, "carbon footprint statements" (CFSs) have been largely ignored and tend to result in only marginal changes, if any (Margelik and McCallum 2014). Similarly, for England, Hands and Hudson (2016) showed that mitigation targets were rarely considered in transport planning EIAs, despite their strong impact on global climate change. Importantly for EA, Fischer (2006) showed that overall measures for climate change were most effectively addressed in visions and policies (and their associated SEA), while measures for adaptation were found to be of particular importance for EA at the levels of programs and projects.

Climate change adaptation

Due to the complexity of the interrelationships from impacts of different environmental issues, the need for a precautionary consideration of climate change in planning has been advocated (Agrawala et al. 2010; Runge et al. 2010; Byer et al. 2012; McCallum et al. 2013). In 2013, the EIA climate change adaptation guidance from the European Commission (EC 2013a) put a focus on thematic entry points of adaptation into EIA. Following the revised European EIA Directive (2014/52/EU), since May 2017, both CC mitigation and adaptation need to be considered in EIA prepared in EU member states, as EU regulations set the frame for national implementation. However, and importantly, the interpretation and integration in the member states' legal settings vary, despite the support provided by the European Commission through guidance and evaluation reports with specific feedback on implementation in the different member states. With regards to climate change, overall, concrete messages for action have remained vague, in particular on, e.g., the integration of impact models and the consideration of "climate resilient measures" at diverse planning levels. The Institute of Environmental Management and Assessment (IEMA) 2015 Guidance on Climate Resilience and Adaptation in EIA (revised in 2020) was one of the first internationally to provide comprehensive support for concrete methodological entry points, suggesting that "sensitivity of topic-specific environmental receptors to climate change" should be established, asking to focus on those environmental issues that are "reliant on specific climatic conditions". In 2013, the European Commission (EC 2013b) outlined challenges and chances for the consideration of climate change in SEA. National guidance followed (e.g., by the Irish EPA 2015), outlining a range of thematic aspects for various sectoral applications of SEA as well as for diverse planning levels.

There is also guidance available focusing on specific aspects, for example, UNECE Water (2009). An overarching theme in guidance is the use of climate scenarios (climate projections) in impact models, supporting the assessment of the vulnerability of environmental aspects in a given country or region. For example, mountainous territories, sea-shorelines, and wetland areas that are already affected by multiple impacts, and weather patterns that are predicted to continue to change with shifts toward, e.g., more extreme precipitation and temperatures (Gobiet et al. 2014). Ensuing impacts include, e.g., erosion, flooding, heatwaves, droughts, wind erosion, and wind throw, to only name a few of the most prominent (IPCC 2019).

Effects of intense localized rainfall can include landslides, mudslides, and unstable slopes (Stoffel and Huggel 2012). These, in turn, can lead to considerable costs due to reconstruction measures, blockage of strategically important routes, network failures, or even physical injury to persons (Haurie et al. 2009; Altvater et al. 2011; Birkmann et al. 2012). In this context, information on the susceptibility of "soil", particularly its water retention capacity, including aspects such as the permeability and consolidation of sealed surfaces and vegetation cover, are increasingly considered. In some regions, in particular in low- and mid-range altitudes, and depending on the climatic conditions, a rise in winter temperatures could lead to an increase in precipitation on unfrozen ground. This, in turn, can increase the risk of, e.g., landslides. Increased soil sensitivity should therefore be taken into account when considering climate change adaptation in EIA, including issues such as site selection, depth of foundations for buildings, and slope stability (EC 2013a).

Wind throw caused by more frequent and powerful storm events can severely impact projects subject to EIA, such as road and rail infrastructure or high-voltage power lines (Jiricka-Pürrer et al. 2018). In particular, the combination of drought and pests can affect the resilience of forests and vegetation. A rise in the frequency of fires on embankments and in nearby (protective) forests (Leidinger et al. 2013; Birkmann et al. 2012) are possible indirect effects. They can lead to an increased likelihood of hazards to planned projects.

Next to thematic aspects relevant for the climate proofing of projects and affecting issues such as soils, water, and vegetation, changed susceptibilities of other environmental aspects also need to receive some attention (EC 2013a; Jiricka et al. 2014; IEMA 2015). Global climate change is, above all, one of the major threats to biodiversity and an important contributor to its decline (Bellard et al. 2012; Lambers 2015). Climate-induced stress and biological invasion as a result of global change are posing pressure on existing green spaces and their biodiversity (Martinson and Raupp 2013). In this context, cold-adapted species are expected to be increasingly negatively affected (Dullinger et al. 2012), whereas warm-adapted species are likely to profit (Gottfried et al. 2012; Gentili et al. 2015; Ferrarini et al. 2017). Amplification of negative impacts of projects subject to EIA can deteriorate habitat conditions and affect a range of species (Jiricka-Pürrer et al. 2016).

Climate change is observed to have a strong impact on both water quality and quantity, as well as on water ecosystems. Due to an increase in mean annual temperatures, a rise in water temperatures and the associated change in oxygen levels have already been observed (Payne et al. 2014). As early as 2009, the European Environment Agency summarized the impacts of changing water resources and their implications for biodiversity (EEA 2009). These include shifts and loss of species in rivers, streams, and lakes. Cold-water fish species especially are stressed by warmer water temperatures, whereas other fish seem to profit (Burkhardt-Holm 2009; Melcher et al. 2012, 2013; Schmid et al. 2014; Pletterbauer et al. 2015). Among others, Pletterbauer et al. (2012) discussed climate change-related changes in river flow regimes. Seasonally altered runoff intensities (not only intense rainfall and glacial melt but also low water levels) amplify altered habitat conditions. Industrial projects and energy production can also affect water ecosystems (Schinegger et al. 2016).

These are only some of the key climate change issues for EA. Generally speaking, three main categories of impacts can be summarized:

1. Direct impacts on plans/programmes/projects and their proposals;
2. Indirect impacts on plans/programmes/projects due to altered sensitivities of environmental issues; and
3. Increased impacts of plans/programmes/projects due to altered sensitivities of environmental issues.

Figure 7.1 illustrates the potential interrelationships between projects and their environments being influenced by the multiple direct and indirect impacts of climate change.

Methodological implications

EA applied to a particular policy, plan, program or project is organized as an assessment process, and the different stages of that process are entry points for particular issues. To start with, scoping can be used to check what data and information on climate change and its impacts (climate scenarios, parameters, time horizons, impact models) are required and what climate change mitigation and adaptation objectives should be included and used. For example, the vulnerability of a project to climate change impacts is associated with its location and environment. Projects are increasingly exposed to potential climate change impacts, for example, near water bodies, in hillsides, or if they are located in areas affected by drought.

Project developers may have access to databases on past hazards, and it can be useful to refer to those for the assessment of potential impacts on development (climate change adaptation/climate proofing). In addition, it is advisable to obtain expert knowledge from climatologists or climate services already during the scoping phase and to integrate support for the interpretation of climate data or impact models right from the beginning. During scoping, environmental authorities together with climate service providers can help to clarify whether and to what extent potential climate change impacts are likely to occur more frequently and/or more

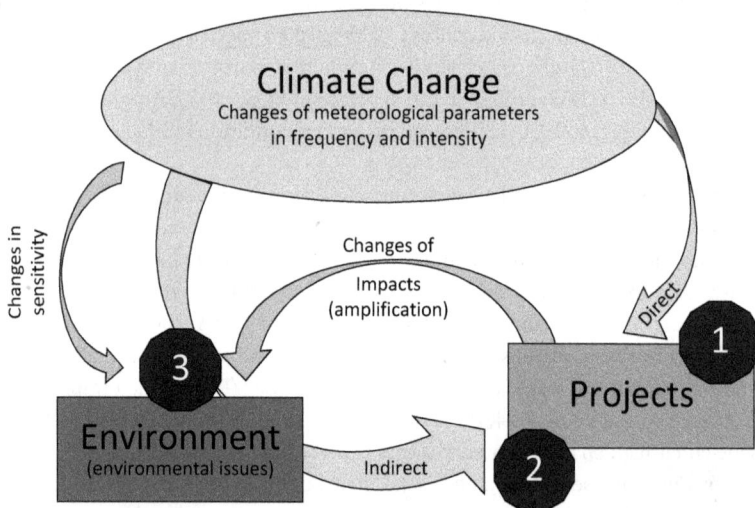

Figure 7.1 Direct and indirect impacts of climate change to be considered in EIA. Source: Adapted from Jiricka et al. (2014).

intensely and how, if necessary, the development of specific adaptation measures should have priority. Furthermore, the relevance of identifying alternatives to avoid severe climate change impacts or to reduce their negative effects can be discussed. In this context, relevant authorities can establish what climate change impacts and possible conflicts of interest and resources are already related to specific areas and which should be considered (scope and cumulative impacts reflecting the territorial relevance for each environmental issue arising).

Environmental goals referring to climate change mitigation and adaptation requirements and standards (e.g., for climate proofing issues such as soil and water) as well as uncertainties deriving from higher-level planning (e.g., spatial planning and landscape planning) should be documented. The zero option can support the assessment of environmental impacts if the description of the environment refers to climate projections in the event a development is not implemented. The zero option may also be used as a complementary benchmark for the assessment of environmental impacts and as background information for the identification of mitigation measures and their monitoring. Climate change adaptation measures can increase the resilience of a project and reduce negative impacts on environmental issues at the same time. They should be designed in such a way that even under projected changes of climatic conditions, in the long term, they are able to fulfill their intended functions effectively. This includes reducing risks for water quality or quantity in times of drought and heat or the maintenance of connectivity despite changes in habitat conditions and displacement of species. In this context, green (vegetation) and blue (water) Infrastructures can play an important role in supporting both climate change adaptation and mitigation of negative impacts. The European Commission introduced the concept of green infrastructure in 2013 (EC 2013c) as a policy instrument for connectivity of habitats, creating multiple other benefits for other sectors outside nature conservation. Following the primary target of the enhancement of green (and blue) infrastructure, the EC encourages the inclusion of ecological connectivity into biodiversity policies of EU member states.

Monitoring also needs to be increasingly adaptive, in particular for environmental issues that are climate-sensitive (EC 2013a; IEMA 2015). The desired outcomes of mitigation and compensation measures (in case significant environmental impacts cannot be mitigated appropriately, e.g., in cases of forest clearing) should serve as controlling parameters during implementation phases. If measures are at risk of being ineffective, consideration should be given to whether an improvement of climate-sensitive measures is necessary. Also, follow-up inspections should be a requirement in approval notices. This can include suggestions for risk management, particularly with regard to habitats and species covered by the Habitats Directive of the European Union. In summary, Table 7.1 presents key questions for the consideration of both climate change mitigation and adaptation in EA.

How to foster an integrative consideration of climate change in EA

This section consists of two parts. First, the potential ability of SEA to minimize or avoid conflicts of objectives and/or resources related to climate change adaptation or mitigation is considered. Second, opportunities to foster co-benefits for adaptation and mitigation in both SEA and EIA are discussed.

The potential ability of SEA to help minimizing conflicts likely to emerge from climate change

Potential areas of conflict between adaptation and mitigation pertain increasingly to the use of space and impacts on land and landscape (O'Mahony 2021). From the literature on adaptation

Table 7.1 Key questions to consider in climate change mitigation and adaptation in EA

Environmental IA process	Mitigation of climate change	Adaptation to climate change
1. Scoping	• Are targets for GHG reduction considered? • What trends exist for GHG emissions, what is the likely impact of the project/plan/program? • What adaptation objectives can imply synergies contributing to GHG reduction (early identification of likely co-benefits)?	• What are the objectives of adaptation strategies related to environmental issues? • Are there any conflicts of interest/objectives likely to result from adaptation measures? • What is the trend for the changed sensitivity of environmental issues?
2. Consideration of alternatives	• What alternatives exist to reduce GHGs? • Are system alternatives relevant? • What alternatives are likely to reduce/prevent conflicts of interest (e.g., locational, technical alternatives)	• What alternatives contribute to the reduction of impacts, and what conflicts (with regards to, e.g., scarcity of resources) are likely to occur/are aggravated when implementing adaptation measures? • What alternatives reduce impacts related to the changed sensitivity of environmental issues due to CC?
3. Assessment of significant environmental impacts	• Are significant negative impacts on environmental issues likely which contradict cc mitigation targets (e.g., degradation of moors, deforestation)? • What lock-in effects (e.g., destruction of carbon sinks) could result (directly or indirectly) from the plan/programs objectives/activities?	• What environmental impacts of the project/ plan/program could be amplified due to the changed environmental sensitivity? • What impacts <u>on</u> the project/plan/program could result from the changed sensitivity of environmental issues?
4. Mitigation and compensation measures	• What co-benefits for CC mitigation can result from mitigation and compensation measures?	• How far could CC affect the effectiveness of mitigation measures? • Is more flexibility in the implementation of mitigation measures needed/possible in order to adapt them in case they are not meeting the initial targets (e.g., in case of re-vegetation)? • How can conflicts of interest be avoided/ reduced when planning and implementing mitigation and compensation measures?

(Continued)

Table 7.1 (Continued)

| 5. Monitoring | • What objectives for GHG reduction were supported by the plan/program/project or the mitigation measures? | • Is a longer timeframe for monitoring needed to survey the effectiveness of some mitigation/compensation measures?
• What are the key challenges that need to be surveyed to review the impacts caused by the changed sensitivity of environmental issues?
• How can a dynamic response work in case of unexpected impacts due to climate change?
• What are key targets for the effective implementation of mitigation measures?
• How can monitoring uncover maladaptation (avoidance of lock-in effects)? |
| **6. Participation (Consultation of the public)** | • What institutions could contribute input on data and objectives regarding the reduction of GHGs (relevant already during scoping)?
• Are the appropriate institutions already involved at the stage of scoping to identify alternatives that contribute to GHG reduction early enough? | • What institutions could contribute input on data about adaptation targets and/or the changed sensitivity of the environmental issues and if possible also indirect effects on the project/plan/program (already relevant during scoping)? |

and mitigation strategies, reports, and studies (Biesbroek et al. 2010; Aguiar et al. 2018), three main categories can be established of potential conflicts related to climate change (see Jiricka-Pürrer and Wachter 2019):

- Conflicts related to emerging competition in the utilization of resources due to climate change impacts;
- Conflicts related to adaptation measures in response to climate change; and
- Conflicts related to mitigation measures (in some strategies, these are partly addressed as adaptation measures for peripheral energy supply).

SEA potentially allows for an integrated system analysis, which can help to prevent conflicts. To support cross-sectoral consideration of impacts, as indicated in some national and state climate change adaptation strategies (e.g., BMNT (formerly BMFLUW) 2012; BMU 2015; BAFU 2014), SEA can serve to identify areas of conflicts (Jiricka-Pürrer and Wachter 2019; Barker and Fischer 2003). It can thus support EIA to consider interrelationships between environmental issues and also of cumulative impacts (Bragagnolo et al. 2012).

When applied in a timely manner and at a suitable spatial level, SEA can play an important role due to its strategic orientation. Possible areas that should receive greater attention regarding preservation and minimization of conflicts are summarized in Table 7.2. In particular, the consideration of potential conflicts of objectives with nature conservation targets is highly relevant as the deterioration and reduction of green and blue infrastructures can have negative consequences for both adaptation and mitigation. To allow for cross-sectoral precautionary consideration of conflicting topics, the consideration of objectives of thematically related policies, plans, and programs during scoping is likely to be relevant.

The examination of alternatives and the consideration of a potential mitigation hierarchy is a particular SEA strength that can contribute to conflict prevention (Fischer et al. 2019). Furthermore, co-benefits for sustainable, climate-friendly planning can also be identified during the design of mitigation measures (further explained below). Hazard prevention could create synergies with nature conservation through, e.g., allowing agricultural land to flood (e.g., in flood-prone areas) or through the conservation of moors and marshlands. Additionally, forestry can consider appropriate reforestation (e.g., with marshland vegetation) in retention areas in case this is not contradictory to hazard protection targets as suggested in several adaptation strategies (e.g., Bayerisches Staatsministerium für Umwelt und Verbraucherschutz 2017, 41). Compensation measures are a further means to consider conflict minimization. The promotion of natural water retention areas can minimize conflicts and maximize synergies to meet climate mitigation targets. Through the consideration of interrelationships, SEA can consider both negative side effects and positive synergies across different environmental issues.

Potential areas of conflict can be considered at different spatial levels, e.g., in regional and local applications, especially when considering water resources and tree species for reforestation. In this context, preservation of fresh air corridors or impacts on groundwater resources – typically assessed in SEA – could gain increasing importance. Meanwhile, when it comes to an overall consideration of water resources, SEA at the national (or state) level can take a leading role if it is applied, for example, to water management plans/programs (Jiricka-Pürrer et al. 2019; Mustow 2021). Selection of the respective spatial and temporal application is important to prevent conflicts and lock-in effects. Monitoring will be of particular importance here. Longer time frames and adaptive approaches (Bulling and Köppel 2017) are required to reconsider conflicts and adjust measures to avoid and minimize them, e.g., in cases where water-saving measures are not sufficiently efficient, and water resources are more affected in terms of quantity and quality than expected.

When it comes to potential conflicts, balancing the needs of various EU Directives will be important (Marot et al. 2021). Article 9 of the European Flood Directive (2007/60/EC), for instance, implies the need for coordination with the Water Framework Directive (WFD 2000/60/EC). The measures of the Flood Directive, involving key adaptation targets to cope with climate change impacts, should be in line with the goals of the WFD for surface waters, groundwater, coastal, and estuarine water resources and their associated habitats and species, including fisheries. According to the EU Flood Directive, management plans also need to be in line with the targets of the Habitats Directive (92/43/EWG) and where appropriate, the EIA Directive (2014/52/EC). Primarily, SEA can, due to the variety of environmental issues covered and the obligation to assess interrelationships between them, serve as a tool of coordination between these Directives. For example, targets of the Habitat Directive can be considered in the course of the assessment of the environmental issues under flora/fauna/biodiversity. Uncertainties and further development of mitigation and compensation measures to reduce and avoid conflicts and balancing interests can be picked up in subsequent EIA processes, if applicable. This also allows to include the surveillance of conflicts of resources in mandatory monitoring processes carried out by authorities. Moreover, it would allow joining initiatives between different monitoring obligations, e.g., in the use of data and their overlay.

Table 7.2 Thematic entry points for SEA to prevent and minimize conflicts

Conflicting resources / interests	Thematic aspects	Environmental issue(s) likely to be affected	Plans / Programs which could reflect these topics (subject to SEA or relevant for the consideration of objectives / sources of information)	Sectors likely to be primarily involved
Changing water resources	Water quantity (scarcity, flooding)/quality (e.g., oxygen, temperature), alternation of species (introduction of neobiota), barrier effects through flood prevention	Groundwater/surface water/fauna/flora/ biodiversity/human health/population	Spatial plans/programs, PA management plans, i.e., Nature 2000 management plans, Water Framework Directive (2000/60/EC)management plans, hazard prevention plans, forest management/development plans	Human health, agriculture, forestry, energy production, tourism, industry, hazard prevention, spatial planning, nature conservation
Changing soil/ land resources	Sealing of soils, conversion of valuable soils, deterioration of soil quality/quantity, increased susceptibility to erosion	Soil/land/natural hazards (human health)	Spatial plans/programs (energy planning), hazard prevention plans/ programs	Agriculture, forestry, energy production, recreation (tourism), Industry, hazard prevention, spatial planning, traffic planning
New pests (reduction/ prevention)	alternation of species (introduction of neobiota), emission of pesticides	Flora/fauna/biodiversity/ human health/ groundwater	PA management plans i.e., Nature 2000 management plans, forest management/development plans/ WFD management plans	Agriculture, forestry, human health, nature conservation
Utilization/ deterioration of landscape resources	Functionality and aesthetics changed through installations (e.g, hazard prevention, renewable energy supply)	Landscape/human health/ population (recreation)/ cultural heritage	Spatial plans/programs (energy planning), hazard prevention plans/ programs,	Spatial planning, energy production, hazard prevention, tourism, nature conservation

Source: Jiricka-Pürrer and Wachter (2019).

Fostering co-benefits for adaptation and mitigation in EA

As early as 2013, the European Commission emphasized the potential role of EIA to foster co-benefits between objectives of climate change adaptation (e.g., to heatwaves, drought events, and heavy rainfall) and mitigation, as well as preservation and restoration of biodiversity (EC 2013a). Thanks to its interdisciplinary perspective, SEA can – even more than EIA – facilitate mainstreaming and capacity building of climate change mitigation and adaptation, and coordinate such processes.

Provided that climate change mitigation and adaptation are already taken into account during the examination of environmental objectives (in scoping) and the examination of alternatives (going as far as a consideration of system alternatives, if applicable), and that results from monitoring are reflected on and integrated, SEA can contribute to these targets. SEA co-benefits can be identified in particular during the systematic analysis of alternatives and the consideration of synergies of (environmental) objectives. Both SEA and the EIA – especially during the development of measures and monitoring – are particularly well suited to examine whether objectives for climate change mitigation and adaptation are recognized and achieved or if maladaptation is occurring (see Figure 7.2).

To achieve the goals of climate proofing and to create co-benefits for climate change mitigation, green infrastructure (GI) (e.g., green roofs or green bridges) and what has been called nature-based solutions (NBS) (e.g., filtration stripes, multi-functional spaces to increase water storage capacity) play key roles (Diaz et al. 2009). Appropriate planning of GI in urban and suburban areas is recognized as one of the most efficient ways to tackle the challenges of climate change by reducing negative impacts of heatwaves, but also droughts and heavy rainfall (Akbari et al. 2001; Gill et al. 2007; Rizwan et al. 2008). In this context, the long-term and strategic development of green networks and corridors as well as the preservation of carbon sinks (e.g., moors and forests) is highly important. Again, overarching strategies and full exploitation of the mitigation hierarchy are necessary (Mörtberg et al. 2007; Bigard et al. 2020). Effects of land scarcity and pressures on green areas/green belts, particularly in urban and peri-urban areas, deserve long-term strategies to foster cross-sectoral adaptation needs and maximize the potential for climate change mitigation at the same time. Depending on its level of application, SEA can encourage an integrative consideration of the multiple effects and support the long-term preservation of GI (Fischer et al. 2018). Concepts that allow foresight planning to identify complex problems and likely solutions are needed. In this context, Honeck et al. (2020) stressed the need to include strategic but flexible adaptive management options to achieve multiple benefits from GI, particularly for climate change mitigation and adaptation.

To achieve climate proofing targets, EIA can integrate a variety of NBS, depending on the project as well as the natural surrounding conditions and likely impacts by climate change. NBSs can serve in particular to reduce the impacts of heavy rainfalls on, e.g., traffic infrastructure or settlements, minimize impacts of wind and storm events to a certain extent, and, in addition, contribute to coping with the impacts of heat and drought for nearby areas (infrastructure/settlements) to support aims for the preservation of, e.g., (drinking) water resources. For example, NBS as mitigation measures in transport planning EIA comprise green bridges, water storage systems and retention areas, filtration stripes, re-vegetation of slopes, protection forests, and windbreakers. These measures can fulfill the minimization of negative environmental impacts and achieve co-benefits for mitigation and adaptation at the same time. In addition to climate change adaptation targets, NBS can fulfill core targets for conservation and enhancement of biodiversity and can serve to foster positive impacts on human health, such as the reduction of emissions and improvement of the scenery of landscapes for touristic purposes (Fischer et al.

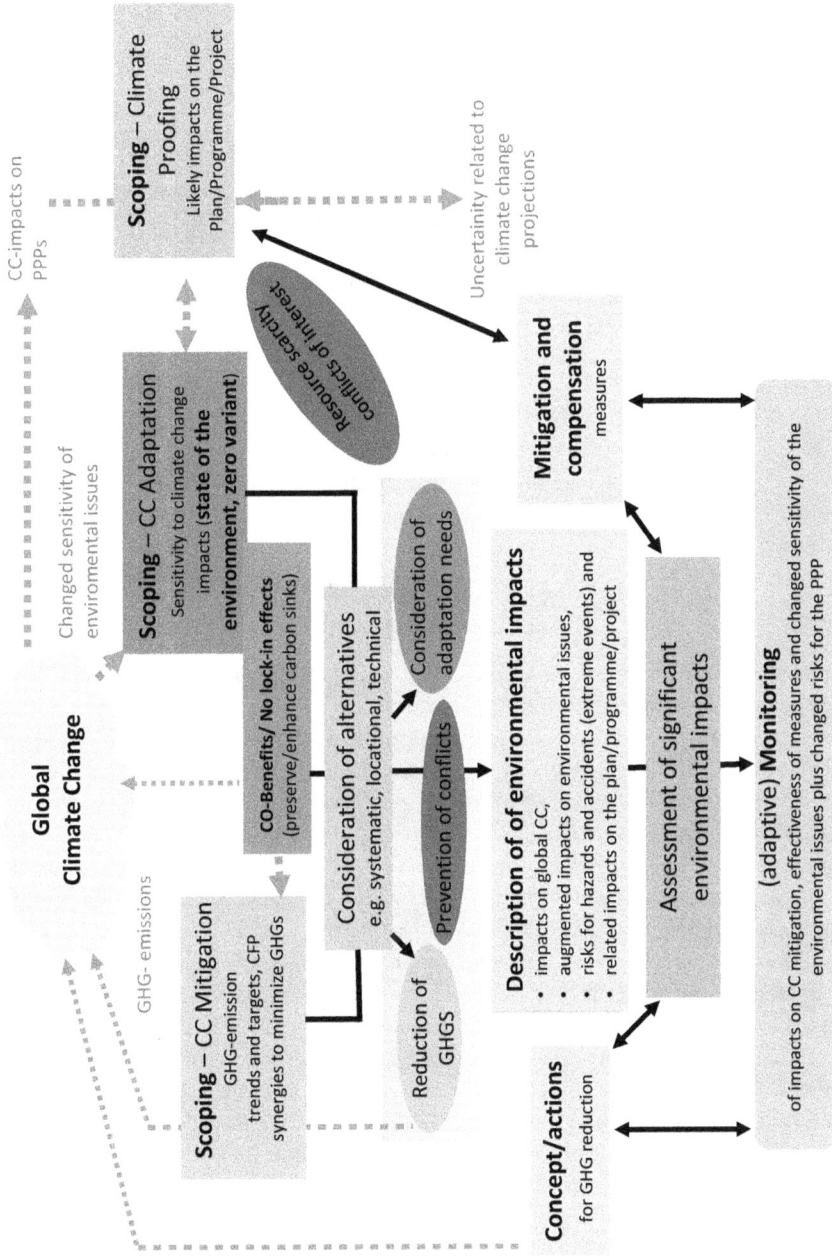

Figure 7.2 Options to consider climate change in an integrative manner. Source: Jiricka-Pürrer (2020).

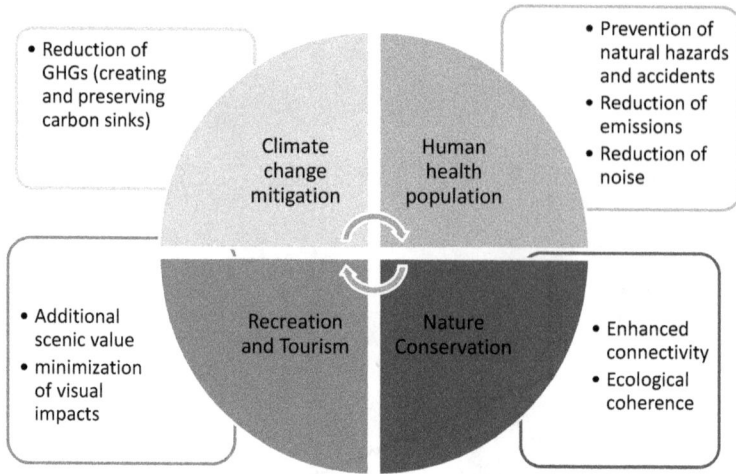

Figure 7.3   Key benefits of considering green infrastructure in EIA of transport planning.

2018). Figure 7.3 summarizes the benefits of considering GI in EIA of transport infrastructure. In order to adapt to climate change impacts, an adequate selection of species, e.g., for re-vegetation or reforestation, is essential in order to reduce the susceptibility to drought, pests, and wind throw.

Conclusion and outlook

Environmental assessment (consisting of project EIA and policy, plan and program SEA) is an important decision-support approach that can lead to a more effective consideration of objectives, targets, and actions of climate change mitigation and adaptation. It can be used to coordinate activities at different geographical levels, sectors, and at administrative tiers (i.e., across policies, plans, programs, and projects; Mörtberg et al. 2007; Bigard et al. 2020; Fischer and González 2021). While climate change adaptation measures can create positive synergies for achieving other environmental targets (for example, through GI measures), they are also likely to increase conflicts with, e.g., land and water resources. Similarly, climate change mitigation can – especially in the field of renewable energy production – interfere with, e.g., biodiversity measures. Effects of land scarcity and pressures on green areas (i.e., GI), particularly of urban and peri-urban areas, deserve long-term strategies to foster cross-sectoral adaptation and maximize the potential for climate change mitigation at the same time.

SEA can help minimize conflicts and maximize co-benefits for adaptation (e.g., prevention of erosion, water retention capacities, climate regulation) and mitigation measures (e.g., preservation and enhancement of carbon sinks). EIA can thus profit later, consider cumulative impacts and device-specific options for climate mitigation and adaptation.

Consideration of natural hazards and climate proofing is now strongly encouraged in the EU and can be derived from the EU Directive on EIA from 2014. However, the need to consider and mitigate changed susceptibility of environmental issues (such as flora/fauna/biodiversity/water) due to climate change and the effective consideration of the interrelationship with cumulative impacts of different projects (e.g., barrier effects for species with varying habitat conditions, water scarcity, or rise in water temperature) have not been achieved, yet. This deserves stronger attention throughout the EA process, starting with scoping to include impact models as well as consideration of climate impacts on environmental issues in the zero option. The selection of climate-robust mitigation and compensation measures can foster climate proofing for projects subject to EIA. They also reflect the changed susceptibility of environmental issues.

Bibliography

Agrawala, S., Matus Kramer, A., Prudent-Richard, G., Sainsbury, M. 2011. Incorporating climate change impacts and adaptat ion. In: OECD (ed.). *Environmental Impact Assessments, Opportunities and Challenges.*

Akbari, H., Pomerantz, M., Taha, H. 2001. Cool surfaces and shade trees to reduce energy use and improve air quality in urban areas. *Solar Energy*, 70(3), 295–310.

Aguiar, F.C., Bentz, J., Silva, J.M.N., Fonseca, A.L., Penha-Lopes, G., Santos, F.D., Penha-Lopes, G. 2018. Adaptation to climate change at local level in Europe: An overview. *Environmental Science and Policy*, 86, 38–63.

Altvater, S., Görlach, B., Osberghaus, D., McCallum, S., Dworak, T., Klostermann, J., van de Sandt, K., Tröltzsch, J. & Frelih Larsen, A. 2011. Recommendations on priority measures for EU policy mainstreaming on adaptation – task 3 report. Ecologic Institute, Berlin.

BAFU – Bundesamt für Umwelt 2014. Anpassung an den Klimawandel in der Schweiz, Aktionsplan 2014–2019. Zweiter Teil der Strategie des Bundesrates vom 9. April, Bern.

Bayrisches Staatsministerium für Umwelt und Verbraucherschutz 2017. Bayerische Klima -Anpassungsstrategie Ausgabe 2016, München.

Barker, A., Fischer, T.B. 2003. English regionalism and sustainability: Towards the development of an integrated approach to SEA. *European Planning Studies*, 11(6), 697–716.

Bellard, C., Bertelsmeier, C., Leadley, P., Thuiller, W., Courchamp, F. 2012. Impacts of climate change on the future of biodiversity. *Ecology Letters*, 15(4), 365–377.

Biesbroek, G.R., Swart, R.J., Carter, T.R., Cowan, C., Henrichs, T., Mela, H., Morecroft, M.D., Rey, D. 2010. Europe adapts to climate change: Comparing national adaptation strategies. *Global Environmental Change*, 20(3), 440–450.

Bigard, C., Thiriet, P., Pioch, S., Thompsona, J.D. 2020. Strategic landscape-scale planning to improve mitigation hierarchy implementation: An empirical case study in Mediterranean France. *Land Use Policy*, 90, 104286.

Bragagnolo, C., Fischer, T.B., Geneletti, D. 2012. Cumulative effects in strategic environmental assessment of spatial plans – Evidence from Italy and England. *Impact Assessment and Project Appraisal*, 30(2), 100–110.

Byer, P., Cestti, R., Croal, P., Fisher, W., Hazell, S., Kolhoff, A., Kørnøv, L. 2012. IAIA statement on climate change and impact assessment. In: *IAIA – Climate Change in Impact Assessment: International Best Practice Principles.* Special publication series no. 8.

Birkmann, J., Fleischhauer, M. 2009. Anpassungsstrategien der Raumentwicklung an den Klimawandel: "Climate Proofing" – Konturen eines neuen Instruments. *Raumforschung und Raumordnung*, 67(2), 114–127.

Birkmann, J., Schanze, J., Müller, P., Stock, M. 2012. Anpassung an den Klimawandel durch räumliche Planung. Grundlagen, Strategien, Instrumente, E-Paper der ARL Nr. 13, Hannover.

BMNT, vormals BMFLUW Bundesministerium für Land- und Forstwirtschaft, Umwelt und Wasserwirtschaft (Hrsg.). 2012. Die österreichische Strategie zur Anpassung an den Klimawandel. Teil 1 – Kontext. Wien.

BMU – Bundesministerium für Umwelt, Naturschutz und Reaktorsicherheit. 2015. Fortschrittsbericht zur Deutschen Anpassungsstrategie an den Klimawandel. Berlin, www.bmu.de/download/fortschritts-bericht-zur-klimaanpassung/.

Bulling, L., Köppel, J. 2017. Adaptive Management in der Windenergieplanung – Eine Chance für den Artenschutz in Deutschland? *Naturschutz und Landschaftsplanung*, 49(2), 73–79.

Burkhardt-Holm, P. 2009. Climate change and decline in abundance of brown trout – Is there a link? Results from Switzerland. *Umweltwissenschaften und schadstoff-Forschung*, 21(2), 177–185.

Byer, P., Cestti, R., Croal, P., Fisher, W., Hazell, S., Kolhoff, A., Kornov, L. 2012. Climate Change in Impact Assessment – International Best Practice Principles IAIA Special Publication Series No. 8. USA: International Association for Impact Assessment. Cornish E. Futuring: the exploration of the future. World Future Society, Bethesda, MD 2004. 313 pp.

Dullinger, S., Gattringer, A., Thuiller, W., Moser, D., Zimmermann, N.E., Guisan, A., Willner, W., Plutzar, C., Leitner, M., Mang, T., Caccianiga, M., Dirnböck, T., Ertl, S., Fischer, A., Lenoir, J., Svenning, J., Psomas, A., Schmatz, D.R., Silc, U., Vittoz, P., Hülber, K.-C., Psomas, A., Schmatz, D.R., Silc, U., Vittoz, P.,

Hülber, K. 2012. Extinction debt of high-mountain plants under twenty-first-century climate change. *Nature Climate Change*, 2(8), 619–622.

EC – European Commission 2013a. Guidance on integrating climate change and biodiversity into Environmental Impact Assessment, Brussels. http://ec.europa.eu/environment/eia/pdf/EIA%20Guidance .pdf.

EC – European Commission 2013b. Guidance on Integrating Climate Change and Biodiversity into Strategic Environmental Assessment, Brussels. https://ec.europa.eu/environment/eia/pdf/SEA %20Guidance.pdf

EC – European Commission 2013c. The EU Strategy on adaptation to climate change. Strengthening Europe's resilience to the impacts of climate change, Brussels. https://ec.europa.eu/clima/sites/clima/ files/docs/eu_strategy_en.pdf.

EEA – European Environment Agency 2018. Greenhouse gas emissions, (AIRS_PO2.5). (https://www .eea.europa.eu/airs/2018/resource-efficiency-and-lowcarbon-economy/greenhouse-gas-emission).

EPA-Environment Protection Agency. 2015. *Integrating Climate Change into Strategic Environmental Assessment in Ireland – A Guidance Note*, Environment Protection Agency, Washington, DC, USA.

Ferrarini, A., Alatalo, J., Gustin, M. 2017. Climate change will seriously impact bird species dwelling above the treeline: A prospective study for the Italian Alps. *Science of the Total Environment*, 590–591, 686–694.

Fischer, T.B. 2006. SEA and transport planning: Towards a generic framework for evaluating practice and developing guidance. *Impact Assessment and Project Appraisal*, 24(3), 183–197.

Fischer, T.B. 1999. The consideration of sustainability aspects within transport infrastructure related policies, plans and programmes, *Journal of Environmental Planning and Management*, 42(2): 189–219.

Fischer, T.B., González, A. 2021. Conclusions – Towards a theory of strategic environmental assessment? In: Fischer, T. B. and González, A. (eds.). *Handbook on Strategic Environmental Assessment*, Cheltenham: Edward Elgar(chapter 27): 425–437.

Fischer, T.B., Welsch, M., Jalal, I. 2019. Guidelines for strategic environmental assessment of nuclear power programmes – Preparation process, contents and consultation feedback. *Impact Assessment and Project Appraisal*, 37(2), 165–178.

Fischer, T.B., Jha-Thakur, U., Fawcett, P., Nowacki, J., Clement, S., Hayes, S. 2018. Consideration of urban green space in impact assessment for health. *Impact Assessment and Project Appraisal*, 36(1), 32–44.

Fischer, T.B., Dalkmann, H., Lowry, M., Tennøy, A. 2010. The dimensions and context of transport decision making. In: Joumard, Robert and Gudmundsson, Henrik (eds.). *Indicators of Environmental Sustainability in Transport, Les Collections de l'Inrets*, Paris, 79–102. http://hal.archives-ouvertes.fr/docs/00/49/28/23/ PDF/Indicators_EST_May_2010.pdf.

Fischer, T.B., Sykes, O. 2009. The new EU Territorial Agenda – Indicating progress for climate change mitigation and adaptation? In Davoudi, S. et al. (eds.). *Planning for Climate Change*, London: Earthscan, 111–124.

Gentili, R., Badola, H.K., Birks, H.J.B. 2015. Alpine biodiversity and refugia in a changing climate. *Biodiversity*, 16(4), 193–195.

Gill, S.E., Handley, J.F., Ennos, A.R., Pauleit, S. 2007. Adapting cities for climate change. The role of the green infrastructure. *Built Environment*, 33(1), 115–133.

Gobiet, A., Kotlarski, S., Beniston, M., Heinrich, G., Rajczak, J., Stoffel, M. 2014. 21st century climate change in the European Alps – A review. *Science of the Total Environment*, 493, 1138–1151.

Gottfried, M., Pauli, H., Futschik, A., Akhalkatsi, M., Barančok, P., Benito Alonso, J.L., Coldea, G., Dick, J., Erschbamer, B., Fernández Calzado, M.R., Kazakis, G., Krajči, J., Larsson, P., Mallaun, M., Michelsen, O., Moiseev, D., Moiseev, P., Molau, U., Merzouki, A., Nagy, L., Nakhutsrishvili, G., Pedersen, B., Pelino, G., Puscas, M., Rossi, G., Stanisci, A., Theurillat, J.-P., Tomaselli, M., Villar, L., Vittoz, P., Vogiatzakis, I., Grabherr, G. 2012. Continent-wide response of mountain vegetation to climate change. *Nature Climate Change*, 2(2), 111–115.

Grimm, M., Köppel, J., Geißler, G. 2019. A shift Towards landscape-scale approaches in compensation suitable mechanisms and open questions. *Impact Assessment and Project Appraisal*, 37(6), 491–502

Haurie L., Sceia, A. & Theni, J. 2009. Inland Transport and Climate Change. A Literature Review. 3. Nov. 2009. http://www.unece.org/ fileadmin/DAM/trans/doc/2009/wp29/WP 29-149-23e.pdf

Habib, T.J., Farr, D.R., Schneider, R.R., Boutin, S. 2013. Economic and ecological outcomes of flexible biodiversity offset systems. *Conservation Biology*, 27(6), 1313–1323.

Hands, S., Hudson, M.D. 2016. Incorporating climate change mitigation and adaptation into environmental impact assessment: A review of current practice within transport projects in England. *Impact Assessment and Project Appraisal*, 34(4), 330–345.

Honeck, E., Moilanen, A., Guinaudeau, B., Wyler, N., Schlaepfer, M.A., Martin, P., Sanguet, A., Urbina, L., von Arx, B., Massy, J., Fischer, C., Lehmann, A. 2020. Implementing green infrastructure for the spatial planning of peri-urban areas in Geneva, Switzerland. *Sustainability*, 12(4), 1387.

IEMA-Institute of Environmental Management and Assessment 2015. IEMA EIA Guide to: Climate Change Resilience and Adaptation, IEMA-Institute of Environmental Management and Assessment, Lincoln, UK. www.iema.net.

IPCC 2019. *IPCC Special Report on Climate Change, Desertification, Land Degradation, Sustainable Land Management, Food Security and Greenhouse Gas Fluxes in Terrestrial Ecosystems.*

Jiricka-Pürrer, A., Wachter, T. 2019. Coping with climate change-related conflicts – The first framework to identify and tackle these emerging topics. *Environmental Impact Assessment Review*, 79.

Jiricka-Pürrer, A., Wachter, T., Driscoll, P. 2019. Perspectives from 2037 – Can Environmental Impact Assessment be the solution for an early consideration of climate change related impacts? *Sustainability-Basel*, 11(15), 4002.

Jiricka-Pürrer, A., Czachs, C., Formayer, H., Wachter, T.F., Margelik, E., Leitner, M., Fischer, T.B. 2018. Climate change adaptation and EIA in Austria and Germany – Current consideration and potential future entry points. *Environmental Impact Assessment Review*, 71, 26–40.

Jiricka, A., Formayer, H., Schmidt, A., Völler, S., Leitner, M., Fischer, T.B., Wachter, T.F. 2016. Consideration of climate change impacts and adaptation in EIA practice – Perspectives of actors in Austria and Germany. *Environmental Impact Assessment Review*, 57, 78–88.

Jiricka, A., Völler, S., Leitner, M., Formayer, H., Fischer, T.B., Wachter, T.F. 2014. Herausforderungen bei der Integration von Klimawandelfolgen und – anpassung in Umweltverträglichkeitsprüfungen – Ein Blick auf die Planungspraxis in Österreich und Deutschland. *UVP-Report*, 28(3+4), 179–185.

Jiricka-Pürrer, A. 2020. The capacity of SEA and EIA for tackling complex global challenges Habilitation thesis at University of Natural Resources and Life Sciences, Vienna.

Kujala, H., Whitehead, A.H., Morris, W.K., Wintle, B.A. 2015. Towards strategic offsetting of biodiversity loss using spatial prioritization concepts and tools: A case study on mining impacts in Australia. *Biological Conservation*, 192, 513–521.

Lambers, J.H.R. 2015. Ecology. Extinction risks from climate change. *Science*, 348(6234), 501–502.

Larsen, S.V., Kørnøv, L., Christensen, P. 2018. The mitigation hierarchy upside down – A study of nature protection measures in Danish infrastructure projects. *Impact Assessment and Project Appraisal*, 36(4), 287–293.

Larsen, S.V., Kørnøv, L., Driscoll, P.A. 2013. Avoiding climate change uncertainties in Strategic Environmental Assessment. *Environmental Impact Assessment Review*, 43, 144–150.

Le Treut, H., Somerville, R., Cubasch, U., Ding, Y., Mauritzen, C., Mokssit, A., Peterson, T., Prather, M. 2007. Historical overview of climate change. In: Solomon, S., Qin, D., Manning, M., Chen, Z., Marquis, M., Averyt, K.B., Tignor, M. and Miller, H.L. (eds.): Climate Change 2007: The Physical Science Basis. Contribution of Working Group I to the Fourth Assessment Report of the Intergovernmental Panel on Climate Change. Cambridge University Press, Cambridge, United Kingdom and New York, NY, USA.

Leidinger, D., Formayer, H., Arpaci, A., (2013). Analysis of current and future fire weather risk in Tyrol. [Int. Conference on Alpine Meteorology (ICAM), Kranjska Gora, 3.–7. Juni 2013]. *ICAM, 32nd Conference on Alpine Meteorology*, 3–7 June 2013, Kranjska Gora, Slovenia, "Book of Abstracts".

Li, Y., Cohen, J.M., Rohr, J.R. 2013. A review and synthesis of the effects of climate change on Amphibians. *Integrative Zoology*, b, 145–161.

Margelik, E., McCallum, S. 2014. Umweltverträglichkeitsprüfung in Österreich – Einblicke in ein umfassendes Dokumentationssystem. *UVP-Rep*, 28, 3 und 4, 128–132.

Maron, M., Dunn, P.K., McAlpine, C.A., Apan, A. 2010. Can offsets really compensate for habitat removal? The case of the endangered red-tailed black-cockatoo. *Journal of Applied Ecology*, 47(2), 348–355.

Marot, N., Fischer, T.B., Sykes, O., Golobič, M., Muthoora, T., González, A. 2021. Territorial impact assessment. In: Fischer, T.B. and González, A. (eds.). *Handbook on Strategic Environmental Assessment*, Edward Elgar, Cheltenham (chapter 5).

Martinson, H.M., Raupp, M.J. 2013. A meta-analysis of the effects of urbanization on ground beetle communities. *Ecosphere*, 45(5), 60.

McCallum, S., Dworak, T., Prutsch, A., Kent, N., Mysiak, J., Bosello, F., Klostermann, J., Dlugolecki, A., Williams, E., König, M., Leitner, M., Miller, K., Harley, M., Smithers, R., Berglund, M., Glas, N., Romanovska, L., van de Sandt, K., Bachschmidt, R., Völler, S., Horrocks, L. 2013. *Support to the Development of the EU Strategy for Adaptation to Climate Change: Background Report to the Impact Assessment, Part I – Problem Definition, Policy Context and Assessment of Policy Options*, Environment Agency, Vienna, Austria.

Melcher, A.H., Kremser, H., Pletterbauer, F., Schmutz, S. 2012. Effects of climate change on fish assemblages in terms of lakes and their outlets in Alpine areas – Explained by the case study Traunsee. In: Schmidt-Kloiber, A., Hartmann, A., Strackbein, J., Feld, C. K. and Hering, D. (eds.). *Current Questions in Water Management. Book of Abstracts to the WISER Final Conference Tallinn*, 117–121.

Melcher, A.H., Pletterbauer, F., Kremser, H., Schmutz, S. 2013. Temperaturansprüche und Auswirkungen des Klimawandels auf die Fischfauna in Flüssen und unterhalb von Seen. *Österreichische Wasser- und Abfallwirtschaft*, 65(11–12), 11–12, 408–417.

Mörtberg, U.M., Balfors, B., Knol, W.C. 2007. Landscape ecological assessment: A tool for integrating bio-diversity issues in strategic environmental assessment and planning. *Journal of Environment Management*, 82(4), 457–470.

Mustow, S.E. 2020. SEA in the water sector. In: Fischer, T. B. and González, A. (eds.). *Handbook on Strategic Environmental Assessment*, Edward Elgar, Cheltenham (chapter 13).

O'Mahony, C. 2021. Integration of climatic factors into Strategic Environmental Assessments. In: Fischer, T.B. and González, A. (eds.). *Handbook on Strategic Environmental Assessment*, Edward Elgar: Cheltenham (chapter 16).

Payne, J.T., Wood, A.W., Palmer, R.N., Lettenmaier, D.P., Lettenmaier, D.P. 2004. "Mitigating the effects of climate change on the water resources of the Columbia River Basin". *Climatic Change*, 62(1–3), 233–256.

Pletterbauer, F., Melcher, A.H., Ferreira, T., Schmutz, S. 2015. Impact of climate change on the structure of fish assemblages in European rivers. *Hydrobiologia*, 744(1), 235–254.

Posas, P.J. 2011. Exploring climate change criteria for strategic environmental assessments. *Progress in Planning*, 75(3), 109–154.

Rehnus, M., Marconi, L., Hackländer, K., Filli, F. 2013. Seasonal changes in habitat use and feeding strategy of the mountain hare (Lepus timidus) in the Central Alps. *Hystrix*, 24(2), 161–165.

Rizwan, A.M., Dennis, L.Y., Chunho, L.I.U. 2008. "A review on the generation, determination and mitigation of urban heat island". *Journal of Environmental Sciences*, 20(1), 120–128.

Runge, K., Wachter, T., Rottgart, E. 2010. Klimaanpassung, climate proofing und Umweltfolgenprüfung. *UVP-Report*, 24(4), 165–169.

Schinegger, R., Palt, M., Segurado, P., Schmutz, S. 2016. Untangling the effects of multiple human stressors and their impacts on fish assemblages in European running waters. *Science of the Total Environment*, 573, 1079–1088.

Schmid, M., Hunziker, S., Wüst, A. 2014. Lake surface temperatures in a changing climate: A global sensitivity analysis. *Climatic Change*, 124(1–2), 301–315.

Stoeglehner, G. 2020. Strategicness – The core issue of environmental planning and assessment of the 21(st) century. *Impact Assessment and Project Appraisal*, 38(2), 141–145.

Wende, W., Bond, A., Bobylev, N., Stratman, L. 2013. Climate change mitigation and adaptation in strategic environmental assessment. *Environmental Impact Assessment Review*, 32(1), 88–93.

8

HEALTH IMPACT ASSESSMENT

Chris G. Buse

Introduction: health and environmental impact assessment

Over the course of the past several decades, the increasing acknowledgment of the health impacts of policies, programs, and projects (PPPs) led to health's incorporation into environmental impact assessment (EIA) policy and practice (Lock 2000; Wernham 2011). However, within the context of EIA, health is often scoped at the level of a single project and often focuses on direct, biophysical risks – such as industrial emissions and impacts on respiratory health – rather than the broader determinants of health (Buse et al. 2019a; Parkes 2016). In other words, to say that health is a mainstream inclusion in EIA would be an overstatement (Morgan 2011). Fortunately, there are other tools beyond EIA that can enable a more robust accounting of health in relation to PPPs.

Health impact assessment (HIA) is one such tool. HIA is a standalone methodology for understanding the direct and indirect health risks of PPPs. While their implementation has grown over the past 20 years, requirements for conducting HIAs are still largely absent from the international policy climate (Winkler et al. 2013) and the subordination of HIA within the EIA process may limit HIA's influence over decision-making (Davenport et al. 2006).

For example, in the Canadian context, "health is insufficiently incorporated into environmental assessment" (Peterson and Kosatsky 2016), "mandatory health impact assessments are long overdue" (Benusic 2014), and there is an established "need for HIA to be integrated into all federal EIA processes" (CAPE 2016). Internationally, similar calls have been made (Steinemann 2000), and increasingly, practitioners and researchers have questioned how attentive HIAs have been to the issue of health equity (the idea of what is fair or just in terms of health outcomes or the distribution of disease in a given population and which stands as a core value of public health practice) (Heller et al. 2014).

Increasingly, an assessment of the health impacts of PPPs is required for social license to operate, and EIA offices around the world are more often including HIA in EIA processes. Assessing health can help build trust between proponents, regulators, and impacted communities, mitigate healthcare costs that are typically unintended and may be inadequately accounted for in project proposals, and can help build a holistic picture of the "impact" of a project and connections between environments, communities, and health.

DOI: 10.4324/9780429282492-9

This chapter introduces the theory and practice of HIA. First, the notion of health and related terminology are introduced and defined before unpacking the history of HIA and its now established protocols for implementation. Next, two case studies are presented to demonstrate the challenges and opportunities of HIA in contemporary settings, and to animate the utility of HIA procedures in relation to EIA. The chapter ends with a discussion of challenges and opportunities for the integration of health into EIA policy and practice more generally, and future directions for improving assessment methodology and rigor.

HIA concepts and history

It is likely that at some point in your life, you have fallen ill or suffered an injury. These experiences make it immediately clear that our health has declined or changed relative to when we were "healthy". The process of becoming ill or injured may have precipitated the need to engage with the healthcare system. However, health is not simply just the absence of disease or an outcome to be measured, and it does not begin or end in healthcare settings. We all have our own understanding of what it means to be healthy. For some it might translate into individual health-promoting practices of being active and sourcing proper nutrition, and for others it might mean being surrounded by supportive friends, families, and community members.

In the context of this chapter, health refers to the suite of contexts in which an individual is situated (e.g., social, cultural, environmental, political, genetic), and how they work in tandem to influence our exposure to acute ill-health effects and longer-term health impacts of PPPs. This means that HIAs need to be attentive not only to direct health impacts of specific PPP, but also the indirect impacts to the determinants of health, such as those defined in Box 8.1.

The complexities of human exposures, sensitivities, and ability to adapt or cope mean that no one sector can effectively govern health impacts alone, raising significant challenges for assessing the attributive dimensions of a single PPP to health status over time. Recognition of this has built interest in how decision-making in other sectors might impact health and/or impact the health sector (Cole and Fielding 2007). If health is important and can be modified by PPPs, what then, is a HIA? HIA is an impact assessment process that combines qualitative or quantitative methods and tools to assess the health impacts of a PPP. The principal goal of any HIA is to provide information that can enable the minimization of negative impacts on health, while maximizing positive impacts over the life cycle of the PPP.

HIA largely arose out of developments in the field of health promotion and public health. The Ottawa Charter of Health Promotion (WHO 1986) was a significant public health milestone that not only exemplified the connections between supportive environments (e.g., ecosystems, social networks, care provision, and social safety nets) and health, but began a movement to address so-called health-in-all-policies (HiaP) and healthy public policy (HPP) (Collins and Koplan 2009; Kemm 2001). The monikers HiaP and HPP are often used interchangeably, and they refer simply to the incorporation of health in non-health sector decision-making. HIA was subsequently developed to enable HiaP and HPP (Harris et al. 2012), building on years of epidemiological and qualitative research tools (Lock 2000). As a result, HIA is most frequently utilized outside of healthcare and public health settings, although each may provide input and contribute expertise to an HIA in partnership with impact assessors.

While multiple definitions and approaches to HIA have been forwarded over the years, Harris-Roxas and Harris (2011) developed a typology of common approaches which include mandated, decision-support, advocacy, and community-led HIA:

Box 8.1. Key HIA terminology defined

Health is "a state of complete physical, mental and social well-being and not merely the absence of disease or infirmity" (WHO 2020a).

Healthcare is the provision of medical care to individuals or communities through family physicians, specialists, community care facilities, and/or hospitals. Healthcare is primarily responsible for the treatment of acute conditions and/or disease, although preventive health information is often delivered to individuals to reduce future health risks (e.g., changing diet to reduce the risks of diabetes).

The *biomedical model of health* is primarily concerned with biological or genetic factors that directly impact health status. The biomedical model largely ignores the psychological, environmental, or social determinants of health, and is what is largely practiced in acute care settings through healthcare provision.

Public/Population Health refers to the health of entire groups or populations, and is primarily concerned with disease prevention (as opposed to treatment). The field of public and population health typically utilizes epidemiological methods to undertake disease surveillance and monitoring, share public communications on known health risks, and was born out of concerns over clean air, drinking water, and food (see Buse et al. 2018).

Social determinants of health (SDoH) are the conditions or settings in which people are born, live, work, and age that are modified by social structures and institutions such as the distribution of financial resources and power at different levels of a society (WHO 2008). Social determinants are typically conceptualized as being related to genetic, behavioral, political, cultural, and social factors, and may also influence engagement or interaction with the healthcare system.

The *ecological determinants of health* (EDoH) refer to the ecological processes and natural resources that are essential to the survival and flourishing of humans and other species on our planet. Principal ecological determinants of health include air, water, and food, but also the ecosystem services that cycle nutrients, create a breathable atmosphere, and detoxify environments making them livable for a variety of species (Hancock et al. 2015).

- *Mandated HIA* typically occurs within the context of a formal EIA, and is done to meet some a policy or other statutory requirement;
- *Decision-support HIA* is typically a voluntary process with the aim of improving decision-making and the implementation of a PPP;
- *Advocacy-oriented HIA* is typically conducted not by project proponents or decision-makers, but by other individuals or organizations seeking to influence the decision-making process of a specific project and advocate for better or fairer health outcomes;
- Finally, *community-led HIA* refers to processes where affected communities lead an assessment on issues that are of relevance and concern for their jurisdiction.

HIA tends to follow similar steps to other forms of impact assessment described in this textbook (e.g., social impact assessment, environmental impact assessment, cumulative effects assessment). Box 8.2 synthesizes multiple existing frameworks guiding the implementation of HIA into several discrete phases of screening, scoping, assessment, recommendations and reporting, and monitoring and evaluation.

Box 8.2. Steps in completing a HIA

Like most forms of impact assessment, there are multiple frameworks that guide assessment protocols and implementation. HIA is no different and most practitioners familiar with EIA protocols will see overlaps and similarities. In the wake of concerns regarding the multidiscplinarity of HIA and its requirement to understand and represent different international contexts (see Krieger et al. 2003; Parry and Stevens 2001), the World Health Organization and US Centers for Disease Control developed similar guidance frameworks; however, it should be noted that guidelines and policies dictating HIA vary country by country, or even according to mandates of lower-tier governments (Hebert et al. 2012). Importantly, the steps provided below are not necessarily required to be completed in the prescribed order. Rather, HIAs should be seen as an iterative process where each phase is in dialogue with all other component parts of the assessment, including protocols that may be external to the HIA itself (e.g., a cumulative effects assessment; social impact assessment). Common phases of HIA include the following:

Screening establishes whether an HIA is required by determining the health relevance of the PPP.

Scoping identifies both perceived and actual health issues of relevance to the policy or project and any relevant public concerns. Spatial and temporal boundaries are then decided upon to move forward with the assessment.

Assessment or appraisal can be qualitative or quantitative in nature. Assessment uses the best available evidence on known health hazards and contextual baseline data to assess or predict relevant health impacts of concern, any vulnerable populations, and their significance as related to the program or policy to drive risk mitigation options.

Recommendations and reporting share relevant mitigation options or abatement strategies to lesson exposures to drivers of ill-health. Information is typically shared with the results of the assessment in the form of a synthesis report that may comment on opportunities to improve health status in relation to the PPP.

Monitoring and evaluation seeks to build a robust system for documenting changes to baseline conditions over the project through continued follow-up assessment of valued health components. Interventions to reduce health risks should also be evaluated and reported on to enhance the existing evidence base and understand emergent health benefits and detriments that may result from the PPP, but which were unanticipated at the time of the initial assessment. Monitoring and evaluation should also seek to engage with quality assurance protocols to ensure future evaluation of HIA's successes and failures, and to confirm issues raised through the assessment are met in accordance with existing best practices (Green et al. 2019).

(Adapted from the CDC 2016 and WHO 2007)

HIAs can be prospective or retrospective, and rapid (i.e., cross-sectional based on a desktop review of evidence and secondary data) or systematic (i.e., involving original data collection and longitudinal monitoring and analysis efforts). Irrespective of the phase of HIA being completed, HIAs are also methodologically pluralistic, drawing from established methods in knowledge synthesis, qualitative research approaches, and quantitative analytics. Example methods and associated opportunities for informing HIA are provided in Table 8.1.

HIAs are necessarily multidisciplinary, and require engaging with the determinants of health (e.g., income, education, housing, etc.) – not just biophysical health outcomes such as cancer,

Table 8.1 Example research methods and their strategic uses in HIA

Research method	Strategic uses
Literature review (scholarly or gray)	Source contextual evidence about the locality and impacted communities that are in scope of the assessment; clarify evidence on pathways between specific projects and health; review best practices in health risk mitigation
Quantitative analysis	Statistical analysis of primary or secondary quantitative datasets (e.g., census data, health administration databases, community surveys) to understand human health and its determinants over the life cycle of a PPP; cost-benefit analysis is a related quantitative tool that can quantify the cost of health impacts over time in relation to identified risk mitigation procedures
Qualitative analysis	General evaluative public engagement strategies (e.g., town halls), key informant interviews, focus groups, and/or arts-enabled methods that either (1) source information on valued health components that are of strategic interest to local stakeholders and rightsholders that have perceived or actual impacts in relation to the PPP; or (2) ground truth existing indicators and data through the lived experience(s) of local stakeholders and rightsholders

Box 8.3. The realized (and then missed) opportunity of HIA in British Columbia, Canada

In the 1990s, and in the wake of the first international conference on health promotion that recognized the importance of supportive environments for health and consideration of health-in-all-policies, the BC Ministry of Health established the Office for Health Promotion. The Office for Health Promotion was primarily responsible for leading the development of an HIA implementation guideline and province-wide capacity building workshops. The office recommended mandatory HIA for any new development project or provincial policy. However, a shifting political climate in the mid-1990s led to health being interpreted as a discretionary part of the province's impact assessment framework and HIA was largely lost in the mix.

Health was later incorporated as one of five pillars of sustainability (which include environmental, economic, social, economic, cultural, and health) in EIA by the BC Environmental Assessment Office under the 2018 BC Environmental Assessment Act. During the province's 2018 environmental assessment revitalization process, public comment indicated that despite being a named pillar, health was rarely ever formally included in EIA documents. Indeed, standalone HIAs run by project proponents or regulators are rare, and those that do exist (such as the HIA case study of the Mt. Polley Tailings Pond failure described later in this chapter) are typically commissioned independent of EIA review boards or panels.

Federally, Canada enacted the *Impact Assessment Act* in 2019 creating the Impact Assessment Agency of Canada which replaces the Canadian Environmental Assessment Agency. The *Impact Assessment Act* significantly broadens the consideration of valued components in EIA to include socioeconomic and health components, and thus the role of health in federally mandated impact assessments has been strengthened (NCCHP 2019). Consideration of various health criteria, Indigenous health, and the deployment of best practices in HIA is now seen as an integral component of federally mandated EIAs in Canada, although those that fall solely under the purview of the province of BC may still not be adequately scoped for health impacts or require the implementation of a formal HIA (see Buse et al. 2019a; Mahboubi et al. 2015).

respiratory illness, etc. This requires HIA practitioners to work across sectors and form partnerships with stakeholders and rightsholders to provide data, insight, and connections to other relevant considerations (Inmuong et al. 2011; Knol et al. 2010).

As health is increasingly seen as a relevant valued component in EIA policy and practice, HIA has a somewhat unique relationship to EIA process and procedure. For example, some jurisdictions may simply require health to be scoped, screened, and assessed as part of an existing EIA process (see Box 8.3 on the example of British Columbia, Canada). Other jurisdictions (such as Wales, see Box 8.4) have mandated HIA for all PPPs overseen by the Welsh government. However, even in cases where robust policy frameworks exist to assess health or mandate HIA, HIAs are still often scoped for a single PPP and may miss connections between past, present, future interactions between multiple stressors, and determinants of health (Mahboubi et al. 2015; Parkes 2016). This is to say that cumulative health impacts are as much of a concern as cumulative environmental and community effects and impacts that manifest as a result of interactions between existing land uses and new PPPs (Gillingham et al. 2016).

Box 8.4. Wales HIA support unit (WHIASU)

The WHIASU actualized some of the missed potential in British Columbia, Canada (see Box 8.3) by developing focused and sustained professional attention to support HIA. Created in 2004 to support health impact assessment practice in Wales, the WHIASU subsequently revised its role in 2017 in the wake of sweeping mandates by the Welsh government that made HIA statutory for public bodies implementing new PPPs. According to its mission statement, the WHIASU "is an all Wales service responsible to Public Health Wales and funded by Welsh government as part of a wider strategy to improve health and reduce inequalities and to assist organizations" in the pursuit of good health.

In addition to providing expert advice, guidance and support to multiple ministries within the Welsh government, the WHIASU additionally runs continuing professional development sessions that provide training capacity building in HIA. Staff also work to facilitate rapid project and policy appraisals, and share relevant resources to those who are conducting HIAs. A key aspect in achieving their goal(s) is to build multi-sectoral partnerships with statutory, voluntary, community, and private organizations in Wales, and to contribute to new research, build the evidence base, and improve judgments on the health impacts of policies, programs, and/or projects. For more information, and to access relevant resources and sample assessments, visit: https://whiasu.publichealthnetwork.cymru/en/

HIA's relationship to multiple exposures and the broader determinants of health

An increasing body of literature now recognizes that health is located at the confluence of complex webs of proximal and distal determinants, and that ecosystems and are essential supportive environments that directly contribute to planetary drivers of health and well-being (Barton and Grant 2013; Briggs 2008; Buse et al. 2018; Whitmee et al. 2015). Indeed, not only do PPPs pose risks to ecosystems, communities, and human health through a variety of potential exposures, but the cumulative impacts of multiple land uses can pose significant risks to health security given the intertwined nature of social-ecological systems and human health (Buse et al. 2020; Parkes et al. 2019). Accordingly, HIA has some way to go in order to account for multiple determinants of health and the challenges they pose for health equity (Buse et al. 2019b; Heller et al. 2014).

In spite of this complexity, a significant aspect of any impact assessment is the appropriate screening and analysis of indicators that represent health values that may be impacted (either perceived or actual) by a PPP. HIA is not different from other forms of impact assessment insofar as processes for establishing valued components that reflect core aspects of the assessment are still required to be defined and populated with relevant data. While a case could be made that social impact assessments may include health, like EIA, health is typically a secondary consideration rather than a principal focus in these assessment architectures, leading to missed opportunities to fully consider connections between valued components.

But how well have HIAs, EIAs, and social impact assessments engaged with the complexity of the determinants of health? Brisbois et al. (2018) completed a review of the literature analyzing the health impacts of mining and oil and gas extractive activities. In their analysis, they quantified the exposure pathways identified in the peer-review literature, noting that 58% assessed some form of direct toxic exposure (e.g., contaminated air/water/soil) on human health, and that 17% of articles assessed direct occupational exposures (e.g., workplace health and safety). By comparison, only 4% of identified studies included any consideration of the social determinants of health.

Similarly, Buse et al. (2019a) interrogated the availability and use of indicators of the social determinants of health in provincial EIA processes by conducting a review of EIA reports and comparing them to HIAs or related documents focused on extractive projects in northern British Columbia, Canada. Using a purposeful maximum variation sampling method, 11 EIA reports and 7 gray literature reports were selected for analysis. A total of 552 indicators were extracted from these documents, and coded according to ten broad themes representing the social determinants of health: demographics, housing, education, infrastructure and services, agriculture and food, health service delivery, work environments, economy and politics, Indigenous culture and identity, and community and social values. Of those indicators, nearly 80% had some form of publicly available data. However, only 74 indicators (13%) were obtained directly from provincial EIA documents, with the remaining indicators coming from HIA documents and extra-provincial reports. This led Buse et al. (2019a) to argue that EIA in British Columbia inadequately attends to the determinants of health given the limited indicator selection but broad availability of health data. Similar deficiencies in terms of health's incorporation into EIA have been documented in other jurisdictions as well (Harris et al. 2009).

These findings further suggest considerable variability in indicator selection to support EIAs in terms of their substantive focus and level of analysis (e.g., individual vs. community-level). Indicator and data availability can also be limited in rural and remote settings where population numbers are low, and sharing health data, especially for rare conditions, can be potentially identifying and breach confidentiality and anonymity of impacted community members. Moreover, EIA documents may be more likely to place greater emphasis on physical natural environments and direct health risks (e.g., contaminated air and water and their impacts on health) than indirect determinants of health that can be modified by PPPs.

Moving beyond the simple articulation of biophysical risks remains a challenging but not impossible task in EIA, and standalone HIAs enable a more robust accounting of impacts to the determinants of health by enabling the analysis of "upstream" (i.e., the determinants of health) and "downstream" (i.e., health outcomes) impacts over time (Parkes 2016). Integrating HIAs within EIAs and/or social impact assessments is also possible, and could enable a more robust and integrative accounting of health impacts of PPPs (Ahmad 2004; Bhatia and Wernham 2008; Parkes 2016), although some concern has been raised as to the degree to which HIAs enable long-term evaluation and monitoring (Schuchter et al. 2014).

Case examples of HIA in action

HIAs have been leveled at numerous health challenges posed by PPPs. For example, a review of 27 US case studies found that topics examined included living wages, after-school programs, power plants, public transport, housing and urban redevelopment, and home energy (Dannenberg et al. 2008). In other jurisdictions, HIAs have been utilized to understand the impacts of active transportation policies (Meuller et al. 2015), waste management facilities (Forastiere et al. 2011), large-scale energy projects (Wernham et al. 2007; Witter et al. 2013), industrially contaminated sites (Sarigiannis and Karakitsios 2018), and traffic-related air pollution (Khreis et al. 2018), among others.

But do HIAs work, and to what end? The 2016 Health Impact Project – a national research collaboration in the United States – analyzed 388 HIAs and interviewed 149 respondents with the goal of evaluating outcomes related to the implementation of HIAs. Results indicate that when HIAs were used to guide decisions, they can: (1) build trust and strengthen relationships between decision-makers and community members living in a (potentially) impacted area; (2) contribute to more equitable access to health-promoting resources such as healthy foods, safe places for physical activity, transit and healthcare; and (3) help protect vulnerable communities from disproportionate exposures to environmental health hazards (Dannenberg 2016; Sohn et al. 2018). In order to animate the "why" and the "how" of HIA, two case studies are presented in further detail, below.

HIA in the aftermath of the Mt. Polley tailings pond failure

On August 4, 2014, a dam in central British Columbia that contained tailings from the Mt. Polley gold and copper mine run by Imperial Metals suffered catastrophic failure. By August 8, the tailings pond had completely emptied, resulting in approximately 17 million cubic meters of tailings wastewater and 8 million cubic meters of tailings being deposited into Polley Lake, Quesnel Lake, and Hazeltine Creek (BC Newsroom 2014). It was later determined that engineers failed to account for glacial silt underneath the tailings pond, compromising its structural integrity and resulting in the collapse of its dam. Remediation activities have continued since 2014, and at the time of writing this chapter, no fines or formal charges have been administered in response to this disaster.

While authorities, the proponent, and the mining dependent towns of the region reeled at the potential environmental impacts of this event, others were concerned with what this meant for the health of the lake and the health of communities who relied on it for recreation and sustenance. Local Indigenous groups, supported by the BC First Nation's Health Authority (FNHA), contracted a HIA to assess communities impacted by this event and better understand how health and its determinants were impacted.

The resulting HIA took a participatory approach across two phases of work: a scoping phase and a full HIA. HIA practitioners worked with six community based coordinators to contact 46 Indigenous and 1 settler community, with 22 participating in the scoping phase based on their geographic location, experience of impacts to traditional lands from the dam failure, established relationships with FNHA and local leadership. The first phase of the assessment was utilized to identify potentially impacted communities, review available environmental, industry and community health data, identify probable community-level impacts on the determinants of health that were linked to the incident, undertake a gap analysis of existing literature and data to identify additional evidence to inform the HIA, and to identify interim measures.

This work documented multiple post-breach impacts experienced by Indigenous communities. First, there were significant impacts to the health of the Fraser River system to which Quesnel Lake drains. There were both perceived and actual impacts to the viability of salmon, direct impacts on treaty rights to fish on traditional and un-ceded territories due to total loss of or impacted access to sacred land or traditional sources of food and medicine, decreases in individual fishing practices, impacts on commercial fisheries, significant emotional stress during and after the event, increased administrative burdens on band offices responding to queries, impacts to local and traditional diets, concerns over the quality of drinking water, and changes to physical activity patterns (Shandro et al. 2016).

Indirect impacts included losses of livelihood and mental health impacts driven by perceived and real trauma which were exacerbated by a lack of transparent communication by the proponent following the event, which in some cases led to community conflict (Shandro et al. 2016). At the time of completing the chapter, the second phase of the full HIA was still underway. It is slated to scope the impacts to the cultural determinants of health through surveys, and to assess clinical health outcomes of self-reported health issues and biophysical markers of chemical exposure through toxicological analysis of blood and hair samples. Routinely collected health information will also assess pre- and post-dam failure on long-term health status. Finally, economic data related to commercial fisheries, and the systematic longitudinal sampling of fish, plankton, and wildlife for heavy metals is being undertaken to understand issues of biomagnification when consuming wild game.

Findings from the original scoping phase of the HIA demonstrate the strong links between Indigenous communities, land, natural resources, and culture, and how impacts to any one precipitate health outcomes. The importance of salmon fishing to support economic livelihoods, physical recreation, and shared cultural identity were also found to be strongly impacted, all of which are important factors influencing emotional and mental health (FNHA 2016). The important work that will follow, aims to establish appropriate risk mitigation activities in relation to identified health impacts, and to clarify impacts on treaty rights in a time where increased focus has been given to the rights of Indigenous people and Truth and Reconciliation over the ills of colonization in the Canadian context.

HIA of air pollution across 25 European cities

Air pollution is a persistent challenge for communities, and PPPs can influence air pollution, which leads to increased impacts on respiratory and circulatory health outcomes (Henschel et al. 2012; Makri and Stilianakis 2008). PPPs can alter traffic patterns and associated vehicle emissions, add new emissions into an existing airshed, or increase risks for certain events known to exacerbate air quality issues (e.g., forest fires).

The Aphekom Project was a collaborative project that aimed to provide new tools and information to guide more effective local and European policies, assist health practitioners advise vulnerable population groups, and empower individuals to make informed decisions that reduce circulatory and respiratory health risks associated with particulate matter and ozone exposure. The three-year project started in 2008, and involved more than 40 scientists across 25 European cities (Pascal et al. 2013). While not attending to a specific PPP, this HIA aimed to forecast different air quality scenarios to understand how changes to local and national policies would impact health and the economic impacts of poor air quality.

Specific attention was given to developing new health impact indicators of air quality, and cost-benefit analysis of air pollution mitigation strategies. The project utilized standardized health impact assessments of urban air pollution using exposure–response measurement to model the health impacts of air pollution and their monetary costs, and to use this information to strengthen policy. Ambient air quality and health records were analyzed, and two scenarios were modeled: one where air pollutants decreased by a fixed amount and another where they decreased according to World Health Organization air quality guidelines. Economic evaluation of each scenario assessed the cost of illness and death.

Results indicated that complying with the WHO guideline for particulate matter of 2.5 (i.e. particulate matter smaller than 2.5 micrometers is commonly occurring in the form of dust, smoke, pollen, etc.) would add 22 months of life expectancy on average after age 30 and would delay a corresponding 19,000 deaths amounting to more than €31 billion in annual savings on health expenditures, absenteeism from work, life expectancy, and quality of life (Pascal et al. 2013). Findings further demonstrated that many European cities have ambient air pollution levels higher than World Health Organization guidelines, and strengthening air quality regulations would be an economic and health benefit for all jurisdictions included in the analysis. This work resulted in significant advocacy and tool development to improve sulfur dioxide, ozone, and particulate matter regulations (Chanel et al. 2016).

Opportunities and future directions

The case studies above exemplify HIA in action, and the benefits of undertaking a focused analysis of health impacts. What then, are the future opportunities for applying HIA and strengthening existing methodologies?

Numerous policy exemplars exist that signify the increasing recognition of health in EIA processes. The World Health Organization and Europe's Health 2020 policy framework supports governments to understand the determinants of health and health impacts, with the goal of fostering intersectoral action (WHO 2020b). Numerous European Union policies have also been established to support health impact assessments in EU member countries (see Abrahams et al. 2004). The Welsh example serves as leading edge practice by mandating HIA for any and all new PPPs. Canada's new *Impact Assessment Act* falls short of mandating HIA, but increases the purview of the Impact Assessment Agency of Canada to assess health and its determinants in federal EIAs, with HIA being a named tool of relevance to such a process.

Irrespective of the policy climate of a given PPP, HIAs can still be implemented by trained professionals as part of an EIA or as a standalone methodology. However, more research and practice-based evidence is required to support initial successes of HIA. For example, Dannenberg et al. (2006) signal the need for more concerted workforce capacity and training, and conducting pilots of relevant HIA tools. Since that publication, numerous resources and pilot studies have been developed. However, there remains a significant need for concerted funding and attention to be directed toward evaluation and monitoring efforts after the initial assessment is completed (Schuchter et al. 2014).

Other opportunities are also present within the HIA literature. As environmental health challenges become more complex and driven by cumulative land uses and their implications for environments, communities, and health, there is a requirement to understand the health risks of large-scale emergencies such as climate change through comparative risk assessment approaches (see Gillingham et al. 2016; Patz et al. 2008). There is also a requirement to understand and be attentive to how climate change can be incorporated into EIA and HIA more generally (Turner et al. 2013). Moreover, HIAs will increasingly be required to not only assess health impacts at the scale of a specific project, but at larger regional scales that enable consideration of how trans-boundary flows impact the determinants of health across space and time (see Parkes 2016; and Chapter 10 on regional assessment in this book).

It has also been postulated that HIA is perhaps best leveled prospectively, prior to the construction of a project or implementation of a program or policy, and that it must be done in a manner that is tactical, insofar as it aligns with institutional rules and the values of impacted actors (Harris et al. 2014). Future research should seek to clarify the degree of influence that prospective HIAs have in relation to policy and regulatory changes, relative to retrospective or post-hoc analyses of health impacts that have accumulated over time, and to understand the mechanisms that influence policy change in contexts where HIAs have been utilized.

While positioned as a tool beyond the health sector, there are also questions about the health sector's involvement in HIA and the training required to undertake an HIA. It could be assumed that HIAs are but another form of impact assessment, and anyone with a background in impact assessment could complete an HIA. While this could certainly be the case, HIAs are likely to be more robust and rigorous when they involve researchers and practitioners with health credentials or expertise. The collaborative involvement of the health sector (notably public health professionals) is a boon to HIA practice, and lends institutional credibility to impact assessments. A review of collaborative HIAs with and without health sector involvement could be particularly helpful in determining the quality and impact that HIAs have over policy- and decision-making.

Critiques have also been leveled at HIA for not adequately incorporating considerations of equity, fairness, and justice into assessment protocols – despite being a central aspect of HIA (see Heller et al. 2014; Sohn et al. 2018). Equity is important in HIA because it asks about the conditions under which unequal health outcomes manifest and attempts to address their root issues. Buse et al. (2019b) reviewed several methodologies that could assist advancing equity in HIA, but found that equity was inconsistently utilized in HIA and often conflated with unequal health outcomes rather than their systemic driving forces. Other methodologies such as health equity impact assessment offer significant opportunities to tackle systematic issues that affect health status (e.g., poverty, racism/discrimination, sexism, ableism, homophobia, etc.).

Common to HIA methodology is an orientation toward participatory research, or the direct involvement of impacted communities in their assessment of health and well-being (Cameron and Wasacase 2017). Participatory approaches should not necessarily only focus on detriments orientations toward health and wellness (e.g., quantifying the things that are wrong with any given PPP and its negative impacts on health), but should also be assets-oriented and engage people in the aspects of their community that enable them to be healthy. Assets-oriented engagement approaches can empower individuals to maximize the health benefits of community assets and ensure their protection (Buse and Patrick 2020). These are key aspects to building trust and social capital at the community level, and ensure that HIAs influence decision-making by addressing the health needs of those who may be most impacted by a particular PPP.

Finally, there are significant opportunities to advance integrated impact assessment frameworks that more adequately understand the relationships between ecological, socioeconomic, sociocultural, and health impacts of PPPs. Health can be both positively and negatively impacted by PPPs, but also by ecological and social changes. The context of the PPP will largely determine the directionality and intensity of health impacts. Forwarding integrated methodologies that do not simply silo environmental, social, and health impact assessment methods, but actively work to understand how changes in one value may influence changes to another are increasingly commonplace, but rarely actualized in impact assessment practice (Buse et al. 2019b; Gillingham et al. 2016).

An integrated, conceptually neutral analysis model has obvious benefits. It does not privilege one domain of assessment over another, thereby overcoming a sense of asymmetry between sectors or disciplines involved in the assessment; it actively works to understand interrelationships among identified valued components; and it may actively assist with streamlining regulatory processes and reducing regulatory burdens (NCCHPP 2014a). Weak integration of values represents what could be argued as the status quo in impact assessment, where sectoral impact assessments are carried out independent of one another. Strong integration would see each domain folded into a single assessment architecture (Bond et al. 2001). Through a review of integrated impact assessments, the National Collaborating Centre for Healthy Public Policy found that:

> the effect of systematizing the decision-making process through the practice of integrated impact assessment leads to a strengthening of analytical rigour, greater transparency and, depending on the mode of practice, a social dialogue. In this manner, integrated impact assessment helps to improve the policy-making process.
>
> *(NCCHPP 2014b)*

Conclusion

Health impact assessment has come a long way since its inception. Over the past several decades, numerous HIAs have been completed for diverse PPPs as a means to reconcile changes to health values over time. While there are still opportunities to strengthen health's inclusion in EIA and develop integrated assessment methods, HIA represents a significant methodological and analytic leap for understanding the interrelationships between ecological and social change, and their implications for human health and well-being.

References

Abrahams, Debbie, Lea den Broeder, Cathal Doyle, Rainer Fehr, Fiona Haigh, Odile Mekel, Owen Metcalfe, Andrew Pennington, and Alex Scott-Samuel. 2004. "Policy Health Impact Assessment for the European Union: Final Project Report". Germany; Ireland; Netherlands; United Kingdom: European Union. http://ec.europa.eu/health/ph_projects/2001/monitoring/fp_monitoring_2001_a6_frep_11_en.

Ahmad, Balsam S. 2004. "Integrating Health into Impact Assessment: Challenges and Opportunities". *Impact Assessment and Project Appraisal* 22 (1): 2–4. https://doi.org/10.3152/147154604781766094.

Aphekom group, Olivier Chanel, Laura Perez, Nino Künzli, and Sylvia Medina. 2016. "The Hidden Economic Burden of Air Pollution-Related Morbidity: Evidence from the Aphekom Project". *The European Journal of Health Economics* 17 (9): 1101–1115. https://doi.org/10.1007/s10198-015-0748-z.

Barton, H. and Grant, M.. 2013. Urban planning for healthy cities. *Journal of urban health*, 90(1): 129–141..

BC Newsroom. 2014. "Mount Polley Tailings Pond Situation Update". August 8, 2014. https://web.archive.org/web/20140810113441/http://www.newsroom.gov.bc.ca/2014/08/friday-aug-8---mount-polley-tailings-pond-situation-update.html.

Benusic, Michael A. 2014. "Mandatory Health Impact Assessments Are Long Overdue | BC Medical Journal". *BC Medical Journal* 56 (5): 238–239.

Bhatia, Rajiv, and Wernham Aaron. 2008. "Integrating Human Health into Environmental Impact Assessment: An Unrealized Opportunity for Environmental Health and Justice". *Environmental Health Perspectives* 116 (8): 991–1000. https://doi.org/10.1289/ehp.11132.

Bond, R., J. Curran, C. Kirkpatrick, N. Lee, and P. Francis. 2001. "Integrated Impact Assessment for Sustainable Development". *World Development* 29 (6): 1011–1024.

Briggs, David J. 2008. "A Framework for Integrated Environmental Health Impact Assessment of Systemic Risks". *Environmental Health* 7 (1): 61. https://doi.org/10.1186/1476-069X-7-61.

Brisbois, Ben W., Jamie Reschny, Trina M. Fyfe, Henry G. Harder, Margot W. Parkes, Sandra Allison, Chris G. Buse, Raina Fumerton, and Barbara Oke. 2018. "Mapping Research on Resource Extraction and Health: A Scoping Review". *The Extractive Industries and Society*, October. https://doi.org/10.1016/j.exis.2018.10.017.

Buse, Chris G., Donald C Cole, and Margot W. Parkes. 2020. "Health Security in the Context of Social-Ecological Change". In *Human Security in World Affairs: Problems and Opportunities*, edited by A. Lautensach and S. Lautensach, (p. 20pp). BCCampus and UNBC. https://opentextbc.ca/humansecurity/chapter/social-ecological-change/

Buse, Chris and Rebecca Patrick. 2020. "Climate Change Glossary for Public Health Practice: From Vulnerability to Climate Justice". *Journal of Epidemiology and Community Health* 74: 867–871. http://dx.doi.org/10.1136/jech-2020-213889

Buse, Chris, Katie Cornish, Margot W. Parkes, Henry Harder, Raina Fumerton, Drona Rasali, Crystal Li, Barb Oke, David Loewen, and Melissa Aalhus. 2019a. "Towards More Robust and Locally Meaningful Indicators for Monitoring Health and the Social Determinants of Health Related to Resource Extraction and Development across Northern BC". Prince George, BC: University of Northern British Columbia. https://www.northernhealth.ca/sites/northern_health/files/services/office-health-resource-development/documents/nh-unbc-indicators-report.pdf.

Buse, Chris G., Valerie Lai, Katie Cornish, and Margot W. Parkes. 2019b. "Towards Environmental Health Equity in Health Impact Assessment: Innovations and Opportunities". *International Journal of Public Health* 64 (1): 15–26. https://doi.org/10.1007/s00038-018-1135-1.

Buse, Chris G, Jordan Sky Oestreicher, Neville R Ellis, Rebecca Patrick, Ben Brisbois, Aaron P Jenkins, Kaileah McKellar, et al. 2018. "Public Health Guide to Field Developments Linking Ecosystems, Environments and Health in the Anthropocene". *Journal of Epidemiology and Community Health* 72 (5): 420–425. https://doi.org/10.1136/jech-2017-210082.

Cameron, Colleen, and Tanya Wasacase. 2017. "Community-Driven Health Impact Assessment and Asset-Based Community Development: An Innovate Path to Community Well-Being". In *Handbook of Community Well-Being Research*, (pp. 239–259). Dordrecht: Springer.

Canadian Association of Physicians for the Environment. 2016. "The Need for Health Impact Assessments to Be Integrated into All Federal Environmental Assessment Processes: A Submission from Health Organizations and Health Professionals to the Expert Panel Established by the Minister of Environment and Climate Change to Review Federal Environmental Assessment Processes". CAPE. https://cape.ca/wp-content/uploads/2017/01/HIA-EA_final.pdf.

Centres for Disease Control. 2016. "Healthy Places – Health Impact Assessment (HIA)". 2016. https://www.cdc.gov/healthyplaces/hia.htm.

Cole, Brian L., and Jonathan E. Fielding. 2007. "Health Impact Assessment: A Tool to Help Policy Makers Understand Health Beyond Health Care". *Annual Review of Public Health* 28 (1): 393–412. https://doi.org/10.1146/annurev.publhealth.28.083006.131942.

Collins, Janet, and Jeffrey P. Koplan. 2009. "Health Impact Assessment: A Step Toward Health in All Policies". *JAMA* 302 (3): 315. https://doi.org/10.1001/jama.2009.1050.

Dannenberg, Andrew L. 2016. "A Brief History of Health Impact Assessment in the United States". *Chronicles of Health Impact Assessment* 1 (1): 1–8. https://doi.org/10.18060/21348.

Dannenberg, Andrew L., Rajiv Bhatia, Brian L. Cole, Carlos Dora, Jonathan E. Fielding, Katherine Kraft, Diane McClymont-Peace, et al. 2006. "Growing the Field of Health Impact Assessment in the United States: An Agenda for Research and Practice". *American Journal of Public Health* 96 (2): 262–270. https://doi.org/10.2105/AJPH.2005.069880.

Dannenberg, Andrew L., Rajiv Bhatia, Brian L. Cole, Sarah K. Heaton, Jason D. Feldman, and Candace D. Rutt. 2008. "Use of Health Impact Assessment in the U.S.". *American Journal of Preventive Medicine* 34 (3): 241–256. https://doi.org/10.1016/j.amepre.2007.11.015.

Davenport, C. 2006. "Use of Health Impact Assessment in Incorporating Health Considerations in Decision Making". *Journal of Epidemiology & Community Health* 60 (3): 196–201. https://doi.org/10.1136/jech.2005.040105.

First Nations Health Authority. 2014. *Mount Polley Health Impact Assessment*. https://www.fnha.ca/about/news-and-events/news/mount-polley-health-impact-assessment

First Nations Health Authority. 2016. "Mount Polley Health Impact Assessment". 2014. https://www.fnha.ca/about/news-and-events/news/mount-polley-health-impact-assessment

Forastiere, Francesco, Chiara Badaloni, Kees de Hoogh, Martin K von Kraus, Marco Martuzzi, Francesco Mitis, Lubica Palkovicova, et al. 2011. "Health Impact Assessment of Waste Management Facilities in Three European Countries". *Environmental Health* 10 (1): 53. https://doi.org/10.1186/1476-069X-10-53.

Gillingham, M. P., Halseth, G. R., Johnson, C. J., and Parkes, M. W. (Eds.). 2016. *The Integration Imperative—Cumulative Environmental, Community and Health Effects of Multiple Natural Resource Developments*. Springer. https://www.springer.com/gp/book/9783319221229

Green, Liz, Benjamin J. Gray, Nerys Edmonds, and Lee Parry-Williams. 2019. "Development of a Quality Assurance Review Framework for Health Impact Assessments". *Impact Assessment and Project Appraisal* 37 (2): 107–113. https://doi.org/10.1080/14615517.2018.1488535.

Harris, Patrick J., Elizabeth Harris, Susan Thompson, Ben Harris-Roxas, and Lynn Kemp. 2009. "Human Health and Wellbeing in Environmental Impact Assessment in New South Wales, Australia: Auditing Health Impacts within Environmental Assessments of Major Projects". *Environmental Impact Assessment Review* 29 (5): 310–318. https://doi.org/10.1016/j.eiar.2009.02.002.

Harris, Patrick John, Lynn Amanda Kemp, and Peter Sainsbury. 2012. "The Essential Elements of Health Impact Assessment and Healthy Public Policy: A Qualitative Study of Practitioner Perspectives". *BMJ Open* 2 (6): e001245. https://doi.org/10.1136/bmjopen-2012-001245.

Harris, Patrick, Peter Sainsbury, and Lynn Kemp. 2014. "The Fit between Health Impact Assessment and Public Policy: Practice Meets Theory". *Social Science & Medicine* 108 (May): 46–53. https://doi.org/10.1016/j.socscimed.2014.02.033.

Harris-Roxas, Ben, and Elizabeth Harris. 2011. "Differing Forms, Differing Purposes: A Typology of Health Impact Assessment". *Environmental Impact Assessment Review* 31 (4): 396–403. https://doi.org/10.1016/j.eiar.2010.03.003.

Hebert, Katherine A., Arthur M. Wendel, Sarah K. Kennedy, and Andrew L. Dannenberg. 2012. "Health Impact Assessment: A Comparison of 45 Local, National, and International Guidelines". *Environmental Impact Assessment Review* 34 (April): 74–82. https://doi.org/10.1016/j.eiar.2012.01.003.

Heller, Jonathan, Marjory Givens, Tina Yuen, Solange Gould, Maria Jandu, Emily Bourcier, and Tim Choi. 2014. "Advancing Efforts to Achieve Health Equity: Equity Metrics for Health Impact Assessment Practice". *International Journal of Environmental Research and Public Health* 11 (11): 11054–11064. https://doi.org/10.3390/ijerph111111054.

Henschel, Susann, Richard Atkinson, Ariana Zeka, Alain Le Tertre, Antonis Analitis, Klea Katsouyanni, Olivier Chanel, et al. 2012. "Air Pollution Interventions and Their Impact on Public Health". *International Journal of Public Health* 57 (5): 757–768. https://doi.org/10.1007/s00038-012-0369-6.

Inmuong, Uraiwan, Panee Rithmak, Soomol Srisookwatana, Nathathai Traithin, and Pornpun Maisuporn. 2011. "Participatory Health Impact Assessment for the Development of Local Government Regulation on Hazard Control". *Environmental Impact Assessment Review* 31 (4): 412–414. https://doi.org/10.1016/j.eiar.2010.03.008.

Hancock, T., Spady, D., and Soskolne, C. L. (2014). *Ecological Determinants of Health*. http://www.cpha.ca/uploads/policy/edh-brief.pdf

Kemm, J. 2001. "Health Impact Assessment: A Tool for Healthy Public Policy". *Health Promotion International* 16 (1): 79–85. https://doi.org/10.1093/heapro/16.1.79.

Khreis, Haneen, Kees de Hoogh, and Mark J. Nieuwenhuijsen. 2018. "Full-Chain Health Impact Assessment of Traffic-Related Air Pollution and Childhood Asthma". *Environment International* 114 (May): 365–375. https://doi.org/10.1016/j.envint.2018.03.008.

Knol, Anne B, Pauline Slottje, Jeroen P van der Sluijs, and Erik Lebret. 2010. "The Use of Expert Elicitation in Environmental Health Impact Assessment: A Seven Step Procedure". *Environmental Health* 9 (1): 19. https://doi.org/10.1186/1476-069X-9-19.

Krieger, N. 2003. "Assessing Health Impact Assessment: Multidisciplinary and International Perspectives". *Journal of Epidemiology & Community Health* 57 (9): 659–662. https://doi.org/10.1136/jech.57.9.659.

Lock, K. 2000. "Health Impact Assessment". *BMJ* 320 (7246): 1395–1398. https://doi.org/10.1136/bmj.320.7246.1395.

Mahboubi, Pouyan, Margot W. Parkes, and Hing Man Chan. 2015. "Challenges and Opportunities of Integrating Human Health into the Environmental Assessment Process: The Canadian Experience Contextualised to International Efforts". *Journal of Environmental Assessment Policy and Management* 17 (04): 1550034. https://doi.org/10.1142/S1464333215500349.

Makri, Anna, and Nikolaos I. Stilianakis. 2008. "Vulnerability to Air Pollution Health Effects". *International Journal of Hygiene and Environmental Health* 211 (3–4): 326–336. https://doi.org/10.1016/j.ijheh.2007.06.005.

Morgan, Richard K. 2011. "Health and Impact Assessment: Are We Seeing Closer Integration?" *Environmental Impact Assessment Review* 31 (4): 404–411. https://doi.org/10.1016/j.eiar.2010.03.009.

Mueller, Natalie, David Rojas-Rueda, Tom Cole-Hunter, Audrey de Nazelle, Evi Dons, Regine Gerike, Thomas Götschi, Luc Int Panis, Sonja Kahlmeier, and Mark Nieuwenhuijsen. 2015. "Health Impact Assessment of Active Transportation: A Systematic Review". *Preventive Medicine* 76 (July): 103–114. https://doi.org/10.1016/j.ypmed.2015.04.010.

National Collaborating Centre for Healthy Public Policy. 2014b. "Main Challenges and Issues Tied to IIA". 6. Series on Integrated Impact Assessment: Quebec, QB: National Collaborating Centre for Healthy Public Policy. http://www.ncchpp.ca/docs/2014_GovInt_IIANote1_En.pdf.

———. 2014a. "Overall Situation and Clarification of Concepts". 1. Series on Integrated Impact Assessment: Quebec, QB: National Collaborating Centre for Healthy Public Policy. http://www.ncchpp.ca/docs/2014_GovInt_IIANote1_En.pdf.

Parkes, Margot, Sandra Allison, Henry Harder, Dawn Hoogeveen, Diana Kutzner, Melissa Aalhus, Evan Adams, et al. 2019. "Addressing the Environmental, Community, and Health Impacts of Resource Development: Challenges across Scales, Sectors, and Sites". *Challenges* 10 (1): 22. https://doi.org/10.3390/challe10010022.

Parkes, Margot W. 2016. "Cumulative Determinants of Health Impacts in Rural, Remote, and Resource-Dependent Communities". In *The Integration Imperative: Cumulative Environmental, Community and Health Effects of Multiple Natural Resource Developments*, edited by Michael Gillingham, Greg Halseth, Chris Johnson, and Margot W. Parkes, 117–149. New York, NY: Springer International Publishing.

Parry, J., and A. Stevens. 2001. "Prospective Health Impact Assessment: Pitfalls, Problems, and Possible Ways Forward". *BMJ* 323 (7322): 1177–1182. https://doi.org/10.1136/bmj.323.7322.1177.

Pascal, M., M. Corso, O. Chanel, C. Declercq, C. Badaloni, G. Cesaroni, S. Henschel, et al. 2013. "Assessing the Public Health Impacts of Urban Air Pollution in 25 European Cities: Results of the Aphekom Project". *Science of The Total Environment* 449 (April): 390–400. https://doi.org/10.1016/j.scitotenv.2013.01.077.

Patz, Jonathan, Diarmid Campbell-Lendrum, Holly Gibbs, and Rosalie Woodruff. 2008. "Health Impact Assessment of Global Climate Change: Expanding on Comparative Risk Assessment Approaches for Policy Making". *Annual Review of Public Health* 29 (1): 27–39. https://doi.org/10.1146/annurev.publhealth.29.020907.090750.

Peterson, Emily, and Tom Kosatsky. 2016. "Incorporating Health into Environmental Assessments in Canada". *Environmental Health Review* 59 (1): 4–6. https://doi.org/10.5864/d2016-006.

Sarigiannis, Dimosthenis A., and Spyros P. Karakitsios. 2018. "Addressing Complexity of Health Impact Assessment in Industrially Contaminated Sites via the Exposome Paradigm". *Epidemiologia & Prevenzione* 42 (5–6S1): 37–48. https://doi.org/10.19191/EP18.5-6.S1.P037.086.

Schuchter, Joseph, Rajiv Bhatia, Jason Corburn, and Edmund Seto. 2014. "Health Impact Assessment in the United States: Has Practice Followed Standards?" *Environmental Impact Assessment Review* 47 (July): 47–53. https://doi.org/10.1016/j.eiar.2014.03.001.

Shandro, Janis A., M Winkler, L Jokinen, and A Stockwell. 2016. "Health Impact Assessment of the 2014 Mount Polley Mine Tailings Dam Breach: Screening and Scoping Phase Port". West Vancouver: First Nations Health Authority (80pp.). http://www.fnha.ca/Documents/FNHA-Mount-Polley-Mine-HIA-SSP-Report.pdf.

Sohn, Elizabeth Kelley, Lauren J. Stein, Allison Wolpoff, Ruth Lindberg, Abigail Baum, Arielle McInnis-Simoncelli, and Keshia M. Pollack. 2018. "Avenues of Influence: The Relationship between Health Impact Assessment and Determinants of Health and Health Equity". *Journal of Urban Health* 95 (5): 754–764. https://doi.org/10.1007/s11524-018-0263-5.

Steinemann, A. 2000. "Rethinking Human Health Impact Assessment". *Environmental Impact Assessment Review* 20 (6): 627–645. https://doi.org/10.1016/S0195-9255(00)00068-8.

Turner, Lyle, Katarzyna Alderman, Des Connell, and Shilu Tong. 2013. "Motivators and Barriers to Incorporating Climate Change-Related Health Risks in Environmental Health Impact Assessment". *International Journal of Environmental Research and Public Health* 10 (3): 1139–1151. https://doi.org/10.3390/ijerph10031139.

Wernham, Aaron. 2007. "Inupiat Health and Proposed Alaskan Oil Development: Results of the First Integrated Health Impact Assessment/Environmental Impact Statement for Proposed Oil Development on Alaska's North Slope". *EcoHealth* 4 (4): 514–514. https://doi.org/10.1007/s10393-007-0143-z.

———. 2011. "Health Impact Assessments Are Needed In Decision Making About Environmental And Land-Use Policy". *Health Affairs* 30 (5): 947–956. https://doi.org/10.1377/hlthaff.2011.0050.

Whitmee, S., Haines, A., Beyrer, C., Boltz, F., Capon, A. G., Dias, B. F. de S., Ezeh, A., Frumkin, H., Gong, P., Head, P., Horton, R., Mace, G. M., Marten, R., Myers, S. S., Nishtar, S., Osofsky, S. A., Pattanayak, S. K., Pongsiri, M. J., Romanelli, C., … Yach, D. 2015. Safeguarding human health in the Anthropocene epoch: Report of The Rockefeller Foundation–Lancet Commission on planetary health. *The Lancet*, 386(10007), 1973–2028. https://doi.org/10.1016/S0140-6736(15)60901-1

WHO. 2007. "HIA Procedure". 2007. https://www.who.int/teams/environment-climate-change-and -health/air-quality-and-health/hia-tools-and-methods/hia-procedure.

———. 2020. "HIA Policy". 2020. https://www.euro.who.int/en/health-topics/environment-and-health /health-impact-assessment/policy.

WHO Commission on Social Determinants of Health. 2008. *Closing the Gap in a Generation: Health Equity Through Action on the Social Determinants of Health : Commission on Social Determinants of Health Final Report*. World Health Organization.

Winkler, Mirko S., Gary R. Krieger, Mark J. Divall, Guéladio Cissé, Mark Wielga, Burton H. Singer, Marcel Tanner, and Jürg Utzinger. 2013. "Untapped Potential of Health Impact Assessment". *Bulletin of the World Health Organization* 91 (April): 298–305. https://doi.org/10.2471/BLT.12.112318.

Witter, Roxana Z., Lisa McKenzie, Kaylan E. Stinson, Kenneth Scott, Lee S. Newman, and John Adgate. 2013. "The Use of Health Impact Assessment for a Community Undergoing Natural Gas Development". *American Journal of Public Health* 103 (6): 1002–1010. https://doi.org/10.2105/AJPH.2012.301017.

World Health Organisation. 1986. Ottawa Charter for Health Promotion: First International Conference on Health Promotion Ottawa, 21 November 1986. https://www.healthpromotion.org.au/images/ ottawa_charter_hp.pdf

9

ENVIRONMENTAL IMPACT ASSESSMENT AND DISASTER RISK MANAGEMENT

Troy McMillan

On the surface, the management of disaster risk and environmental impact assessment (EIA) seem to be somewhat disconnected domains with differing purposes. Disaster risk assessment (DRA) aims to make communities and countries safer from disasters, while EIA aims to address the impacts of development. However, as each practice matures, there is increasing opportunity for data exchange, methodological advancement, and possible integration. In what was a practice predominantly focused on project-level environmental protection, environmental impact assessment has grown to look at cumulative effects, social and economic impacts, considerations of culture, varying geographical scales (e.g., local vs. regional), and more. This book details key facets of the evolving discipline.

In parallel, disaster risk assessment has followed a similar progression. DRA endeavors to develop evidence to inform disaster risk reduction efforts and to build resilience, in consideration of a diverse range of events, from severe inclement weather to significant seismic activity, to industrial accidents, to terrorism. It has matured to include many of the same dimensions examined by EIA, to better consider the social systems and environments impacted by physical forces of planet and people, at varying scales (local and national), and in consideration of future-oriented change such as demographics and climate change and the capacities of those potentially impacted. While DRA and EIA have different primary intents and the associated risk is considered from a different catalyst (hazard event vs. project impact), they draw upon many of the same sources of data, using similar approaches to examine that data, and produce recommendations to reduce risk. Risk assessment and management approaches often draw upon common concepts, standards, and best practices across disciplines; however, while similarities exist, it is valuable to understand how the examination, measurement and remediation of risk has been tailored by discipline in order to improve best practice and identify potential interdisciplinary opportunities. This chapter describes how risk is examined within the context of disasters and how logical linkages with impact assessment exist. EIA risk assessment methods and concepts are discussed in Chapter 5 in this book.

DOI: 10.4324/9780429282492-10

Why consider disaster risk within environmental impact assessment practice?

When we look at the impact of development activities, from resource extraction (e.g., mines, oil drilling) to the construction of transportation systems, there is an intersection between these activities and the natural and built environments in which they are conducted. We have seen substantial growth in the new lenses applied through impact assessment which considers the increasingly connected and complex environments in which we build and analyze, with increased focus on strategic environmental assessment and cumulative effects. These are discussed in other chapters of this textbook.

Much of these efforts increasingly consider how multiple development projects influence risk in aggregate, across various systems, variable geographical and time scales, and increasingly in consideration of a climate-impacted future; often in parallel, risk experts within disaster and emergency management (DEM) are working within many of the same variables, examining how natural hazards and human-caused threats (accidental and malicious) impact the same communities and countries. In examining the linkages between the two disciplines, asking ourselves why we do each makes for a sound starting point.

Why do we conduct environmental impact assessment?

Simply, we conduct these activities to see how a project may positively and/or negatively impact the planet (or country or community) and its dependent systems (i.e., social, environmental, and economic). EIA is integral to 'sound decision-making, serving both an information-gathering and an analytical component, used to inform decision-makers concerning the impacts and management of proposed developments' (Noble 2015).

In parallel, why do we practice disaster risk assessment?

Ultimately, we perform this work to see how the planet and its dependent systems, particularly its social systems, may be adversely impacted by the occurrence of a hazard or threat. DRA is meant to inform decision-makers, through disaster risk management practice, on what risks from natural and human-induced threats may overwhelm a society and what risk reduction actions may be applied to reduce them to an acceptable residual level (Coppola 2015).

So, both disciplines focus us on how significant event or change to an environment may adversely impact the established, existing systems of societies and the natural and built environments.

Unpacking this further, the notion of capacity is integral to both disciplines. Will this new project adversely impact the surrounding environment to such a degree that it will not be able to recover? Will it pollute a waterway to a degree that makes it harmful to dependent communities who use the water for consumption, for which existing purification systems cannot cope? Will the net positive economic and employment (permanent and temporary) result in a net positive benefit, in consideration of the adverse impacts of the project? These are some of the initial questions asked within EIA practice when determining if what is being put forth by the proponent supports an improved quality of life for the citizens impacted by the initiative, and if certain capacities – such as the absorption of pollution or environmental degradation – are sufficient or will be irreparably harmed. When we look at disasters, we are concerned with how the existing capacity of a community, region, or country can avoid, absorb, resist, and recover from an adverse event (Cardona et al. 2012). It is when this *capacity is exceeded that an incident becomes a disaster.*

This leads us to a natural intersection between the two areas or practice: *sustainability.* Broadly, sustainability speaks to the need to ensure that we do not take so much from the planet today that we don't leave enough for future generations – that we do not exceed the *capacity* of the planet to provide for healthy, quality life for today and tomorrow. Frequently, sustainability is described as having three core dimensions: the environment, the economy, and society (Lamorgese and Geneletti 2013). If we do not develop in a way that promotes environmental, economic, and social sustainability, we leave the planet worse tomorrow than it is today (see Chapter 6 in this book). By extension, if we continuously absorb destruction of the environment, economy, and social fabric of our communities because we do not implement appropriate safeguards or are ignorant to the hazards which exist where we develop, we erode the resources available to us which can be invested or used to develop more smartly. Worse yet, we may be placing communities, and the people that live within them, in harm's way by not understanding the disaster risk which may exist prior to development.

In substantive ways, the integration of disaster risk assessment with impact assessment is akin to robust cumulative effects assessment (see Chapter 3 in this book). With cumulative effects assessment (CEA), we consider how multiple projects in aggregate, over time, may change the environment and intersect with each other to increase damaging effects; appreciating the disaster risk within EIA practice as a potential change can also identify potential compounding risks within the development landscape that may not otherwise be considered. For example, the EIA examinations of a rail line development and water treatment facility as independent projects several kilometers apart may not appear to be related, until a disaster risk lens is applied: a landslide that causes rail cars to spill toxins upstream of the treatment facility, adversely impacting the potable water supply of one or more communities, highlights this relationship. If integrated, DRA-informed EIA likely would have resulted in increased setback of the line from the waterway, a relocation further from the landslide hazard, an alternate placement of the treatment facility, or a combination thereof.

In essence, development is not sustainable if societies are disaster prone (Zhou, Wang, and Wang 2016; Kim and Lim 2016; Collins 2018), and unsustainable development or resource exploitation may contribute to increased disaster risk (Cutter et al. 2008; Oliver-Smith et al., 2016). When considering sustainability, the linkages between EIA and disaster risk reduction strengthen; smart development choices, supported by sound EIA practice, can play an integral part in ensuring communities do not face increased disaster risk (Figure 9.1 provides a conceptual overview of this relationship). Additionally, the incorporation of disaster risk considerations into the EIA process can provide valuable context such as hazard and social information (e.g., vulnerabilities) to inform not only project-level EIA, but SIA and CEA as well. Sustainability of *how* we develop, and *where* we develop, highlights the nexus between EIA and disaster risk management. Both these disciplines directly support objectives of sustainability, aiming to better inform decision-makers and the public on how to develop and live smarter, while better coexisting with the natural environment. We must examine both how the project intersects with the hazards of the area and how it itself may impact sustainability, such that the correct location for development is identified, or, at minimum, that the necessary safeguards are built into the project to mitigate those risks to an acceptable level.

An overview of disasters, disaster risk, and key dimensions of disaster risk

Disasters

Disasters occur when the ability of a social system (e.g., a community, region) to cope with the impact of adverse events such as an earthquake, flood, or terrorist act. These events are frequently characterized by the loss of life and extensive damage to the built and natural environments,

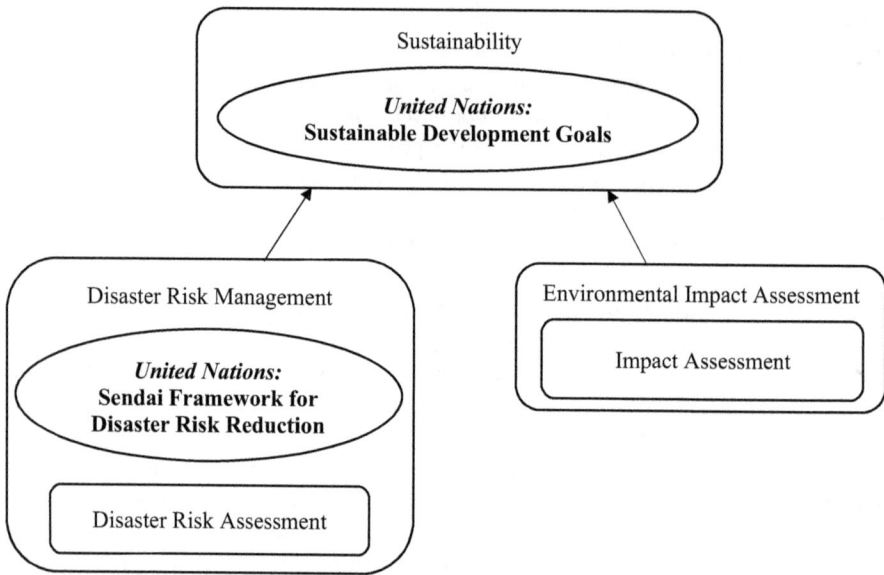

Figure 9.1 Linkages between sustainability, EIA, disaster risk, and United Nations initiatives

often resulting in substantial financial losses due to damage and the costs required to rebuild. Disasters may occur from rapid (e.g., earthquake) or slow (e.g., drought) onset events, and recovery may take extended periods of time – worse yet, some societies may never fully recover.

From a global perspective, the United Nations (UN) through its Office for Disaster Risk Reduction (UNDRR), defines a disaster to be a

> serious disruption of the functioning of a community or a society at any scale due to hazardous events interacting with conditions of exposure, vulnerability and capacity, leading to one or more of the following: human, material, economic and environmental losses and impacts.
>
> *(UNDRR n.d.a.)*

Drawing upon the climate change lens, the Intergovernmental Panel on Climate Change (IPCC) defines disaster as

> severe alterations in the normal functioning of a community or a society due to hazardous physical events interacting with vulnerable social conditions, leading to widespread adverse human, material, economic, or environmental effects that require immediate emergency response to satisfy critical human needs and that may require external support for recovery.
>
> *(Lavell et al. 2012)*

The inability to cope with events is driven by the capacity (or lack thereof) of the impacted social systems and the susceptibility of it to harm. Disasters are *socially constructed*. They do not exist in a vacuum but occur when social systems (or society, used here interchangeably) are harmed by events, whether natural or human-induced (e.g., accident or terrorism). This is why the discipline is moving away from the term *natural disaster*. Disasters are not natural phenomena, but unnatural events caused by human action to settle in hazard-prone areas or otherwise develop in unsustainable ways. To draw an analogy of filling a glass with water: like a given society, the glass has a fixed amount of capacity to be filled until water spills all over the table; an emergency can be seen as the moments where the water approaches the top of the glass with attempts made to stop it from overflowing (akin to how we respond to events with paramedics, police, and firefighters), but the tipping point to *disaster* occurs when the capacity to successfully absorb is exceeded and water spills over.

In part due to increased urbanization and continued development in risk areas, and in large part due to a changing climate, research from the Centre for Research on the Epidemiology of Disasters (CRED) has documented that disasters are increasing in both terms of severity and impact (CRED 2020). With a climate-impacted future, this trend will continue: more intense precipitation events, higher average temperatures, sea level rise, droughts, and heat waves will continue to adversely impact infrastructures and the people who rely upon them (Jabareen 2013). Many countries are already experiencing these adverse changes, including Canada and the United States (Public Safety Canada 2019a; Smith 2020).

This continues to highlight the importance of linking the disciplines of EIA and disaster risk assessment and management; it is increasingly important to understand the risk to our communities and social systems, in concert with the risk that may increase or decrease through development initiatives. In sum: we know that poor development drives disaster risk, and poor (or absent) disaster risk information may impede robust EIA. Therefore, we must maximize our understanding of both domains in concert, not in isolation, to fully gain the holistic picture of the risk that may erode sustainability.

Disaster risk: a basic overview

Risk, simply, is the likelihood of something negative (to someone or something) happening over a given time frame. By extension, disaster risk can be understood as the likelihood (within a specified time frame) of an adverse event occurring that is significant enough to overwhelm a society through negative impacts. However, disaster risk is more complex than just these two dimensions of consequence and likelihood. It is a composite of the hazard or threat (the source of harm), its likelihood of occurrence, the consequence of occurrence, and several other key factors including the vulnerability and resilience of people and the built and natural environments and their respective exposure to the hazard or threat.

The IPCC elaborates that disaster risk can be viewed as

> the likelihood over a specified time period of severe alterations in the normal functioning of a community or a society due to hazardous physical events interacting with vulnerable social conditions, leading to widespread adverse human, material, economic, or environmental effects that require immediate emergency response to satisfy critical human needs and that may require external support for recovery ... [and] derives from a combination of physical hazards and the vulnerabilities of exposed elements and will signify the potential for severe interruption of the normal functioning of the affected society once it materializes as disaster.
>
> *(Lavell et al. 2012)*

The concept of exposure considers the physical placement of people, their communities, and the things they depend on (infrastructures, services, assets) relative to the source of harm that they face. Exposure is a key concept when understanding disasters; a hazard (the source of harm) must intersect with someone (people, communities) or something (infrastructure, housing) in a way which disrupts or destroys the sound functioning of a social system. Simply, if a hazard occurs – an earthquake in uninhabited area, for example – where no one is harmed, there is no disaster. The UNDRR simply describes exposure as the 'situation of people, infrastructure, housing, production capacities and other tangible human assets located in hazard-prone areas' (UNDRR n.d.a.). The IPCC elaborates that it is the 'presence (location) of people, livelihoods, environmental services and resources, infrastructure, or economic, social, or cultural assets in places that could be adversely affected by physical events and which, thereby, are subject to potential future harm, loss, or damage' (Lavell et al. 2012).

It is important to understand that disaster is a socially constructed concept – that is, if something bad happens but people are not impacted by the event – so exposure considers what people and things may be adversely impacted should a threat or hazard materialize. A strong earthquake which occurs in a remote location where few people live is significantly different than an equivalent size of earthquake occurring in a densely populated area which collapses buildings, destroys infrastructure, and causes a multitude of structural fires; the hazard itself may be similar, but the *exposure* of people, the infrastructures on which they rely, and the social systems in which they exist drives whether the occurrence of hazard is disastrous.

In relating disaster risk to environmental impact assessment practice, risk within EIA focuses on the post-development risk associated with a proponent's project such as the potential negative impact to the environment or the health of communities within the area; disaster risk assessment, in contrast, considers the risk that may befall a community (or region or country) due to a natural hazard, or human-induced threat (e.g., terrorism). Both disciplines include the analysis of impacts to communities, and draw upon many of the same variables and questions to derive an approximation of the risk:

- Who and what can be impacted?
- How can people be impacted, and how badly?
- What is the likelihood of potential impacts occurring (over a given time frame)?
- What are the economics for loss?

Given that many jurisdictions mandate a form of disaster risk assessment to inform emergency management planning, there are opportunities for EIA practitioners to draw upon these existing analyses to inform EIA efforts within the respective jurisdiction. These data sources can expedite EIA-based risk analysis and are elaborated on further within this chapter.

A key difference is that EIA contrasts the risk of negative impact with considerations of the positive impact, if any, of the project (Arnold and Hanna 2017; Noble 2015). After all, a proponent has put the project together to advance something positive for someone – it is the role of EIA to determine its validity within the context of the environment and if that positive element is not to the detriment of others or the environment (or at least propose a way forward to address the negative elements). This positive lens is rarely a consideration when examining disaster risk given these situations are analyzed to effectively determine *how bad can or will it be?* Yet, when thinking through opportunities to reduce disaster risk and devising appropriate solutions – commonly described to as disaster prevention and mitigation – the net positive benefits of these interventions then follow a similar logical process as EIA, contrasting the potential positive value of the disaster risk reduction (DRR) effort with the disaster risk assessed.

Beyond hazard, threat, and likelihood: other key dimensions and drivers of disaster risk

Vulnerability

Broadly, disaster-related vulnerability is concerned with the susceptibility of something – typically people (and the communities in which they exist), infrastructures, and environments – to disaster risk. It is the potential for loss, as a function of exposure to hazards or threats, sensitivity to disturbances due to hazards or threats, and a limitation of adaptive capacity (i.e., the ability to change) (Frazier, Thompson, and Dezzani 2014). Vulnerability is inherent within societies before disasters materialize, and increases susceptibility to adverse events (Cutter et al. 2008; UNDRR n.d.a.). It is the result of a complex aggregate of 'historical, social, economic, political, cultural, institutional, natural resource, and environmental conditions and processes' (Lavell et al. 2012).

Additionally, depending on the specific hazard or threat, vulnerability may vary within a social system. A housing development that adheres to seismic standards from a robust building code but has a dry, thick cedar hedge immediately beside it is likely to survive an earthquake, but not fare well during a wildland–urban interface fire. A lack of access to resources also drives vulnerability, particularly as it relates to adaptive capacity. Often, this is a function of income or financial resources, but can also be a lack of access to information or education. For example, individuals in a developing country forced to live in slums within a flood plain are significantly more likely to experience disaster loss or harm during a flood than a wealthy individual who may be exposed to a similar flood hazard, but has access to financial resources to raise their home, waterproof its foundation, and purchase insurance. The marginalized within communities are often the most vulnerable to disaster.

It is important to consider vulnerability within the assessment of disaster risk, as it may significantly increase the impact a hazard or threat may have. If significant vulnerability erodes a community's ability to cope with adverse events, the likelihood of a disaster occurring also increases even if the hazard frequency remains unchanged. By extension, disaster risk management efforts should consider opportunities to reduce vulnerability to reduce overall disaster risk.

Resilience

Resilience as a concept continues to gain prominence as a means to address disaster risk, by building capacities that allow for social systems to adapt and cope with the increasing occurrence and severity of disasters. Resilience is often viewed in a positive contrast to vulnerability, and may be seen as a 'desirable condition, overall aim, and cultural attribute' (Fekete, Hufschmidt, and Kruse 2014). The highlight of resilience which separates it from other disaster risk reduction concepts, is its potential to cope with unknown risks in addition to known risks (Gunderson 2010). Given that not all risks can be identified and assessed, developing resilience can increase the survivability of a social system when facing disaster.

The UNDRR describes resilience, in relation to disaster risk, as the

> ability of a system, community or society exposed to hazards to resist, absorb, accommodate, adapt to, transform and recover from the effects of a hazard in a timely and efficient manner, including through the preservation and restoration of its essential basic structures and functions through risk management.
>
> *(UNDRR n.d.a.)*

In essence, this definition considers resilience as the ability to bounce back to a sound level of functioning from a disaster, drawing upon a collection of multidisciplinary capacities. However, there is an increasing orientation of resilience to not only return to a pre-event level of functioning, but to draw upon adaptive capacity more such that impacted society learns and is better than before. The Resilient Cities Network characterizes this as adaptation and growth, whereas the IPCC characterizes it as improvement (Lavell et al. 2012; Resilient Cities Network n.d.). This evolution of resilience integrates the notion of continuous learning into the disaster risk management process, with an aim to continuously reduce risk while augmenting adaptive capacity – reducing known risk while being able to better cope with the unknown.

Given the relationship of resilience with increased adaptive capacity and understanding that vulnerability is linked with a lack of adaptive capacity, we can look at vulnerability and resilience as related but different concepts which are both underpinned – and therefore linked – by the capacity to adapt (Joakim, Mortsch, and Oulahen 2015). This becomes useful for disaster risk management, as it may focus DRR efforts to support the augmentation of adaptive capacity, including access to funds, training, information, or other appropriate resources. However, as there are multiple factors which drive both resilience and vulnerability – skills, education, access to finances, location, and more – one must consider both dimensions separately and in the context of which the risk is being examined. For example, a senior citizen who has significant financial resources may purchase insurance to safeguard their home from fire-induced losses but lack mobility to readily evacuate during an emergency; the same individual is both resilient and vulnerable. It is also important to consider that exposure to hazard events intersects with resilience and vulnerability: the senior citizen may be resilient through insurance, and vulnerable due to limited mobility, but these considerations are most important when *exposed* to a fire or other hazard.

Risk drivers

As disasters are socially constructed, so too is much of disaster risk. Only a natural hazard can be considered *natural*; the notion of a *natural disaster* in the modern world is misleading. This undermines the fundamental social processes which underpin disaster risk, instead focusing on the physical aspect of disaster (Oliver-Smith et al. 2016). People frequently live where hazards exist, either willingly (i.e., 'it won't happen to me'), through ignorance (i.e., insufficient research or a lack of data), or without choice (i.e., vulnerable populations). Raising this is not meant to place blame on any single person or process, but to highlight that disaster risk is driven through human decision-making processes. Whether this is the choice to continue to contribute to anthropogenic climate change resulting in more, and more severe, climate-driven hazards or to expose more people and infrastructure to hazards through insufficient land use planning, human decisions amplify exposure, vulnerability, and even hazards themselves. It is therefore important to consider these drivers of risk when assessing the potential for disaster and consider them in mitigative solutions to reduce risk. The United Nations identifies that these processes or conditions are often related to development (UNDRR n.d.a.), further highlighting how good EIA can contribute to reduced disaster risk.

The Integrated Research on Disaster Risk program, funded in part by the UNDRR, has developed the Forensic Investigations of Disasters (FORIN) conceptual framework to aid practitioners and researchers examine these underpinnings of disaster risk (Oliver-Smith et al., 2016). FORIN describes that risk drivers, in recognition of increasing inequities and key social processes, are 'dynamic conditions that accentuate existing or create new forms of risk at all levels … [increasing] exposure and vulnerability, risk and disaster' (Oliver-Smith et al., 2016).

It is a valuable framework for practitioners to better understand the root causes of disaster risk, moving beyond traditional assessments focused on hazard and vulnerability.

In sum, disaster risk is a product of an adverse event (hazard or threat), the likelihood of that event over a time frame, and in consideration of who and what is exposed; it is increased by the vulnerability of the society impacted and underpinning risk drivers, and decreased by the resilience of the impacted society. Understanding the composite parts of disaster risk allow for a more wholesome refection on the activities and social processes that create and drive disaster risk, including those activities which we consider within EIA practice. A sound understanding of how disaster risk is manifested allows for improved assessment of the risk facing society and helps to create and inform the necessary evidence required to develop risk reduction interventions.

Assessing disaster risk to support disaster risk management

Environmental impact assessment as a practice typically not only reviews a project for impacts, but also examines risk reduction opportunities (mitigative actions, proposal of alternate project scales or sites, and so forth) for the project which is included in the recommendation to the governing body to which the impact assessment is submitted (Arnold and Hanna 2017). In addition, the organization responsible for the EIA practice, typically through a form of legal authority, also monitors the project as it is developed and implemented (Hanna 2016; Noble 2015). While this scope may differ between countries, it is common for the EIA process to not only assess risk, but to also support management of the risk whether directly, via recommendations, through a form of compliance, or a combination thereof.

In contrast, disaster risk assessment is typically looked at as an underpinning process which support disaster risk management, often enabled through emergency management plans and disaster risk reduction efforts (either embedded within plans or complementary to them). To draw comparisons between the fields of practice, disaster risk assessment is most akin to the assessment activities within the broader EIA process, and disaster risk management more closely parallels EIA writ large. That said, the legislative underpinnings of disaster risk management (across different level of government) are often focused on emergency response activities, and not necessarily the assessment or proactive reduction of risk. For example, Canada's Emergency Management Act identifies that each Minister must identify risks within their respective purview, but only insofar as to prepare emergency plans to be able to respond to those risks if realized (Government of Canada 2007).

The United Nations promotes that sound DRA 'considers all relevant hazards and vulnerabilities, both direct and indirect impacts, and a diagnosis of the sources of risk will support the design of policies and investments that are efficient and effective in reducing risk' (UNISDR 2017). The inclusion of an *all hazards* approach here is important. While a community may be familiar with a certain hazard (e.g., flooding) from a recent event, it is important to be as impartial as possible when scoping hazards and be driven by evidence. Risk assessment should focus on *what is likely to happen* in the future, so that efforts to reduce risk focus on the ones most relevant to the community. Historical context is important to draw upon for informing risk reduction efforts, but care should be given to ensure that bias to recent events does not reduce the impartiality of evidence-informed risk assessment.

Scales and intent of disaster risk assessment

Given that hazards are often geographically bound, legislation for emergency and disaster management typically focuses on a 'bottom-up' approach, where local governments are responsible for developing plans to reduce risk. Theses can be through proactive efforts (e.g., building in safeguards to reduce hazard impact or exposure) or through augmenting preparedness and response capacity (e.g., firefighting, evacuation protocols). It is therefore valuable to inform the local approach to risk reduction through a scale of assessment that is most relevant to the relevant communities.

Sound local risk assessment provides awareness of the hazards and threats facing a community, and the vulnerability and resilience of people, infrastructure, and environment exposed to those sources of harm. All components described earlier within this chapter should be considered. Additionally, risk assessment should include an evaluation component, which aims to prioritize risks within the jurisdictional context in which it is conducted. It looks at the collection of analyzed disaster risks, and endeavors to rank them in relative order of their seriousness within the context of the impacted society (Coppola 2015).

Akin to sound EIA practice, public participation in local risk assessment can solicit valuable insights from community members while facilitating awareness and empowerment; however, its ability to scale-up remains a challenge (Pelling 2007). Other methodological approaches can be used to develop national lenses on disaster risk; national risk assessment (NRA) is globally gaining traction to support national approaches to risk reduction, recognizing the continued growth of disaster losses. In the European Union (EU), Decision 1313/2013/EU makes the periodic development of NRAs mandatory to inform disaster risk management, with the intent to reduce European disaster risk (Casajus Valles et al. 2019).

Scale and *intent* of the assessment is important: a NRA developed to prioritize national risk reduction efforts is likely to miss important local context. In contrast, a local assessment to inform a regional governments emergency plans is often difficult to aggregate to a national level. This parallels some challenges facing EIA, where reconciling different assessment approaches can result in the downplaying of the key *intent* of some assessment types (Tajima and Fischer 2013). Balancing a desire for efficiency should not be at the expense of effectiveness (i.e., the intent). This context is important for EIA practitioners drawing upon disaster risk assessment for context into their assessment efforts. Whereas NRA results may provide valuable information for strategic environmental assessment (SEA), it is likely too high-level to be valuable to a local project EIA or cumulative effects assessment. Conversely, a local disaster risk assessment may be valuable for a project EIA, but too granular or region-specific to be valuable for SEA.

Broadly, disaster risk assessment is meant to provide evidence to DEM practitioners and decision-makers to select the most appropriate mechanisms to reduce risk and prepare for those that cannot be mitigated or avoided. However, like the plethora of EIA assessment types, the type or scale of DRA should be tailored for the type of evidence required: such as local DRA for local government and communities and NRA for national risk reduction initiatives.

Advancing disaster risk management through mitigation and prevention

As disasters increase both in terms of severity and impact, communities and countries must look at ways to increase the ability to understand, anticipate, respond to, and recover from disasters. Disaster risk management regimes must consider these dimensions, and examine pathways to reduce disaster risk before, during, and after they occur. DEM has a rich history of focusing on response activities to cope with emergencies and disasters. For EIA practice, many of these activities would parallel the safety plans required as part of the operational stage of a significant development project. Activities may include evacuation plans for a mine, fire suppression for forestry operations, and so forth. These are necessary capacities (the 'cups' we need from overflowing), whether for an operator of mine or a local or national government, to ensure that adverse events can be combated with an eye to saving lives.

However, from a broader disaster risk management perspective, the focus of the reactionary facet of disaster risk management is often at the detriment of proactive interventions to reduce disaster risk before the event occurs (Oliver-Smith et al., 2016). Given an increase in global disaster frequency and extent, opportunities to prevent or reduce the impact of disasters through mitigative or preventive activities – often referred to as risk treatment – is required. Unlike the use of the term *mitigation* within the climate change domain referring to the reduction of green house gas emissions, *mitigation* within DEM refers to the proactive, pre-event activities taken to attenuate the impact of disasters on society. The UNDRR identifies it as the 'lessening or minimizing of the adverse impacts of a hazardous event' (UNDRR n.d.a.). These activities may include improved building codes, engineering activities to divert the hazard (e.g., floodways, retaining walls), and the re-establishment of natural environment buffers (e.g., coastal marshlands). The National Institute of Building Sciences (NIBS) within the United States has identified that mitigation activities provide a benefit-cost rations between 4:1 and 11:1 depending on hazard and intervention (National Institute of Building Science 2019).

Disaster prevention activities refer to those which eliminate factors of disaster risk composition to avoid disasters (Lavell et al. 2012). Returning to the notion that disaster is socially constructed and driven by the exposure of people (and things) to sources of harm, robust land use planning which limits residential development in known hazardous areas can be seen to be an enabler of disaster prevention. To illustrate: a community that identifies a high-risk wildland-urban interface through disaster risk assessment, and subsequently develops these areas as a buffering linear park and recreational area, are likely to prevent disaster from occurring by proactively reducing exposure of homes and infrastructure from wildland fire.

Like climate change adaptation (CCA) efforts, opportunities to shift decision-makers' views to a longer-term horizon is valuable for proactive disaster risk reduction: it can be used as a catalyst to sustain interest among disaster risk stakeholders. Particularly when linked to sustainable development, concern with climate change has the benefit of refocusing disaster risk dialogue using a long-term horizon, moving away from reactionary approaches focused on short term benefit (Burns and Des Johansson 2017; Frame 2008). An increased focus within DEM on mitigation and prevention relative to natural hazards and human-induced events more appropriately parallels the proactive lens placed on development activities by EIA to reduce risk before the commencement of the initiative. *While the catalyst for risk differs – development vs. hazard – the intent to proactively reduce detrimental activities is shared between disaster risk reduction and EIA.*

Reviewing vulnerability, resilience, and risk drivers

As a social construct based on potential hazards or threats, disasters and associated risk is 'seriously and dominantly conditioned by societal perceptions, priorities, needs, demands, decisions and practices' (Oliver-Smith et al., 2016). As a result, it is imperative that when examining disaster risk assessment results, and devising risk reduction plans as part of the broader disaster risk management regime, scope is not myopically framed by hazard and exposure. Like increasing emphasis on the social dimension of EIA (see Chapter 4 in this book), disaster risk management can consider risk reduction efforts that reduce vulnerability and increase resilience within the impacted society. Targeted measures which improve access to resources and information increase adaptive capacity, which has the potential to both reduce vulnerability and build resilience. Beyond efforts to reduce inequity, the sharing of hazard information, improving access to training, and participation in emergency planning and exercises are easily facilitated examples by which disaster risk reduction can occur by empowering people to make better decisions.

Considering an 'all hazards' lens for risk treatment

Additionally, efforts to reduce risk should endeavor to scale in an all hazards manner, where appropriate. For example, the implementation of natural buffers to reduce riverine flood risk through land use planning are a positive step toward proactive risk reduction; however, a more comprehensive land use regime which also includes wildland-urban interface fires and urban flooding may be more difficult to implement, but safeguard the community more fully. Disaster risk assessment, with a sound evaluation of identified and assessed risks, serves to provide the evidence by which decision-makers can select which risks should be prioritized for reduction efforts. At a national level, a country updating a building code can also draw upon their national disaster risk assessment, where available, to incorporate proactive elements that move beyond a single hazard bias (e.g., earthquake) to one where flood proofing, fire resistance, and other hazards and threats are incorporated (UNISDR 2017).

Supporting EIA practice: pragmatic value from disaster risk assessment

At the heart of this chapter is to provide EIA practitioners resources from which they can leverage from the disaster risk assessment practice within disaster risk management. This section draws upon an adapted list of steps stages the EIA practice as outlined by Hanna (2016), and then identifies tools and data that EIA practitioners may find useful to inform impact assessment. The stages can be framed as follows:

1. Project proposal: what is it we want to do?

In the commencing stage, the proponent brings forward an idea for consideration by the EIA process. At a minimum, it should outline the 'project's physical components, location, and operational activities … [and] include a clear rationale for why the project is needed' (Arnold and Hanna 2017). Depending on the jurisdiction, the development of the proposal may be done in consultation with the governing agency to help produce a more acceptable proposal for consideration (Hanna 2016).

2. Screening

At this juncture, the proposal is reviewed by the identified agency – or in the case of a multijurisdictional context, the necessary organizations – to determine if it meets the necessary requirements or thresholds to proceed without further assessment or if it must be reviewed. Criteria is often varied, and may include legal considerations, the type of project, scale/cost, or a combination thereof (Hanna 2016). If an assessment is required, the project proceeds to scoping.

3. Scoping

Upon determination that an assessment is required, this stage focuses on the boundaries of what will be assessed for agency review and approval. Drawing upon the initial proposal, this exercise is meant to target assessment activities on potentially significant issues and impacts, and may be guided to varying degrees by the relevant jurisdictions (Hanna 2016). To ensure sufficient coverage of potential impacts, the 'spatial and temporal scales for assessing impacts should be set beyond the project site and lifespan, and account for reasonably foreseeable cumulative effects' (Arnold and Hanna 2017). Focus should not be simply bound to the physical coverage of the project, but consider how it interfaces with the surrounding area, over time.

4. *Assessment and report preparation*

With clarity of scope – the rules of engagement between proponent and review agency – this stage is concerned with the heart of EIA: the analysis and evaluation of adverse, and potentially positive, impacts of the project (Arnold and Hanna 2017). Baseline data is collected, multiple stakeholders and experts are engaged for input, and predicted impacts are analyzed and assessed for their significance (Arnold and Hanna 2017). Risk treatment options are also identified through mitigation, which 'involves outlining the measures that can be taken to reduce or eliminate the impacts identified' (Hanna 2016). The collection of baseline conditions, predicted impacts, their respective expected significance, and the list of priority recommendations to mitigate the development's risk are documented and submitted to the review body.

5. *Agency review and decision*

The EIA report, once submitted, is assessed by reviewers (government and/or independent experts), recommendations are proposed, and a decision to approve, approve with conditions, or reject the project is made (Hanna 2016). Depending on the jurisdiction, further public participation may be sought to support transparency and inform recommendations and decision-making (Arnold and Hanna 2017). The approach for a final decision is often driven by the significance of impacts; small impacts may be administratively addressed through the appropriate agency, whereas predicted significant impacts may require more formal settings (e.g., public hearing) (Hanna 2016).

6. *Monitoring, compliance, follow-up*

This stage concerns itself with examining the efficacy of the assessment relative to the project and to track compliance against agreed-to recommendations (Arnold and Hanna 2017). These activities may occur during, and after, the completion of the project.

Potential EIA linkages with disaster risk assessment

Some of these stages are offered more insight from disaster risk assessment than others; however, there may be resources that proponents, agency staff and decision-makers may be able to draw upon to complement the activities outside of direct impact assessment analysis. This is not an exhaustive list of risk considerations and resources, but a starting point for practitioners to be able to draw upon to support EIA practice. For applicability, the stages have been grouped by the two predominant roles within the assessment process, both of which may involve EIA practitioners:

1. **Proponent:** The individual or organization proposing the project, and developing the EIA submission
2. **Review Agency:** The organization responsible for screening and reviewing the EIA report, and making recommendations to decision-makers (internal or external to the agency)

While the public and decision-makers play a critical role within the EIA process, their review of the project is highly contingent on the materials provided by the proponent and review agency. The consideration of disaster risk by the proponent and review agency would ideally result in its inclusion through the EIA consultative process(es), where the public and key stakeholders may comment on the project with an improved understanding of its net impact inclusive of risk information generated by both the EIA and DRA processes.

Disaster risk considerations for the proponent

Applicable stages

- Stage 1. Project Proposal – What Is It We Want to Do?
- Stage 4. Assessment and Report Preparation
- Stage 6. Monitoring, Compliance, Follow-Up

For the proponent (and subsequently the review agency), it is important to consider the project within the context of the disaster risk facing the development location. Some guiding questions across stages may include:

- What natural hazards are present in the area? Is the project itself exposed to them, or potentially exacerbate them?
- If the structures with the development were to fail, would there be adverse impacts to the area (environmental, social, health)?
- Has a climate-impacted future been incorporated into design thresholds to consider climate-driven hazards such as inclement weather?

These aim to solicit insight into both upstream and downstream implications; in essence, does the project have the potential to be the source of harm, increase existing hazards, itself be harmed, or a combination thereof. The proponent should be considering the direct and indirect impacts the initiative will have on existing disaster risk, as well as if the project placement subjects itself to potential damage. Does the project potential exacerbate existing risk by contributing to environmental degradation which will increase the frequency or severity of a hazard (e.g., removal of vegetation causing unstable slopes, impermeable surfaces resulting in pluvial flooding)? Does the project potentially contribute to risk drivers (e.g., increases to social vulnerability, contributions to anthropogenic climate change)? If so, building in mitigative actions within the proposal at the outset of the project can reduce potential impacts both *from* the project and *to* the project.

Recognition of the existing disaster risk profile of the area in which the development is proposed can support the proponent in site selection during initial proposal development stage and inform the sufficiency of mitigative actions during assessment analysis. Drawing upon local and regional risk assessments can provide this context. These are typically available from local governments and may be posted on their respective website. Depending on the jurisdiction, regional or national governments may have a repository of this disaster risk information. Other jurisdictions may have complementary information such as hazard maps, probabilistic models, and historical events. In the absence of a comprehensive disaster risk assessment results, these other tools can still provide valuable information pertaining to potential disaster risk.

Disaster risk considerations for a review agency

Applicable stages

- Stage 2. Screening
- Stage 3. Scoping
- Stage 5. Agency Review and Decision
- Stage 6. Monitoring, Compliance, Follow-Up

While not typically considered beyond the consideration of environmental degradation, there is the potential for a screening agency to consider the project regarding the existing hazard or disaster risk it is potentially exposed to or may exacerbate. This can be viewed similarly to cumulative effects assessment (e.g., Franks, Brereton, and Moran 2013); in aggregate, what does the project do *in concert with* disaster risk? Neither the disaster risk nor the project upon individual review may yield sufficient risk to warrant significant concern, but taken together, the harm associated with each may be compounded requiring further mitigation than would typically be required for the project by itself. Herein lies an opportunity for EIA practitioners to consider important alignment of project activities with disaster risk knowledge, to ensure sustainability is not degraded by development activities. This can be a key consideration during project screening.

In the Stage 2 screening phase, the reviewing organization should examine the relationship between disaster risk and the project for both downstream and upstream impacts; while the focus is naturally on how the project may adversely impact the environment, the exposure of the project to disaster risk is also important. This is of particular concern during the scoping stage, and the spatial extent of hazards can be contrasted with that of the project to not only consider reasonable cumulative effects, but the interplay with disaster risk as well. If a project is approved without this context, the mitigative actions may not be sufficient and established thresholds of the project's impacts relative to baseline may be significantly exceeded should an event occur. For example, a mine's tailing pond may be developed in consideration of the mine's lifespan and incorporate climate change resilience within its construction (i.e., engineered resistance); however, should even a minor earthquake cause the failure of the pond's structure, the addition of the mine has contributed to a situation which may result in disaster should local capacity to address the failure be exceeded.

During agency review, the proponents' efforts to mitigate negative impacts can be reviewed for the crosswalk with disaster risk; determining if adequate mitigative actions have been incorporated is important to ascertain the net residual risk – the collection of impacts in consideration of corrective actions – is acceptable based on the net positive benefits of the project. During the solicitation of expert advice during the review process, the agency may engage with hazard and emergency management practitioners to consider local capacity, assessed risks in the area, and potential vulnerabilities with adjoining communities. These practitioners are also valuable for supporting monitoring and compliance, where changes may influence hazards (e.g., changes to waterways influencing flood mitigation, preparedness and response) and should be informed of changes which may require emergency management plans to be updated (e.g., increased change of industrial accidents, new personnel in the area, changes in evacuation protocols).

Concluding thoughts

EIA and disaster risk management practices, while infrequently or loosely integrated, aim to make societies safer and more sustainable. This chapter has provided an overview of disaster risk concepts and processes and endeavored to link disaster risk assessment and management to EIA practice. Mirroring the varied EIA assessment types, disaster risk assessment also varies in terms of intent and scale; DRA applicability to EIA is substantially dependent on the type of EIA being conducted.

The intent of this chapter is not to suggest that one practice is subordinate to another, but to highlight key logical connections between them to support additional value in EIA by illustrating some of the key concepts, tools, and data available from DRA that may prove useful for EIA practitioners. Likewise, while not elaborated, there is inherent value for disaster risk practitioners to draw upon the various assessment approaches and results from EIA practice; richer disaster risk efforts can result from a better understanding of development projects and their impacts to the natural environment and associated social systems. As these practices mature, there is continued value in learning from the other, particularly in consideration of a climate-impacted future and increasing need to support sustainability. Two key topics which will require further integration opportunities include:

1. *An increased cohesion between EIA and DRA when assessing disaster reduction projects.* Given increased disasters, there is an increased focus on climate change adaptation efforts and disaster risk initiatives; some of these efforts involve gray and green mitigative infrastructures, themselves requiring impact assessments, leading to a fascinating, but somewhat circular, assessment process of EIA and DRA. Increased alignment between practices can support improved, environmentally appropriate disaster risk reduction projects.
2. *Increasing resilience to support sustainability.* Also due to increased disaster risk, the notions of resilient development and disaster resilience gain traction; resilience is valuable in building capacities to address both known and unknown risk. How disaster risk practice informs resilience-building within sustainability assessment will be another area that practitioners will need to contribute to in harmonizing the disciplines.

While the reason we conduct impact assessment differs from why we conduct disaster risk assessment, both practices aim to proactively identify, assess, and reduce the risk associated with detrimental activities. Whether safeguarding the environment and society from development activities, or from disaster risk, both disciplines have the aspirational goal to make our planet safer and more sustainable.

Disaster risk resources for EIA practitioners

This section provides a sample of resources from DRA which may provide valued context for EIA practitioners.

- Local disaster risk assessment results
  Predominantly developed by local government, these resources may be available on their respective website, or are available from the local emergency management office. These risk assessments can include valuable information such as:
  - Hazards and threats within the area of development proposal
  - Composition and demographics of local communities
  - Vulnerability assessments of local area
  - Placement of residences and infrastructure within the area
  - Existing mitigative actions (including potential land use considerations)
  - Identified roles and responsibilities

- Prioritization of disaster risks from assessment and evaluation processes
- Linkages to relevant legislation, guidelines, bylaws
- Resilience strategies (e.g., Resilient Cities Network members[1])

  Resilience strategies outline actions and activities being planned for within a given society to advance its safety; however, these strategies often outline acute and chronic conditions – including hazards and vulnerabilities – that can inform EIA practitioners about the current state of the community or region. These strategies may also be valuable in supporting SEA efforts, aligning resilience initiatives with sustainability goals.
- National disaster risk assessments and risk registers

  While NRA results may be too high-level to be of use for local, project-based EIA, they may be valuable to inform SEA and CEA efforts. EU countries predominantly post these on their respective national websites. Leading examples include those from the Netherlands (NRA) and United Kingdom (Risk Register).
- Hazard maps and data, including probabilistic models

  Countries often have rich hazard information that EIA practitioners can draw upon to augment existing environmental information available. The United States Geological Survey (USGS) and National Oceanic and Atmospheric Administration (NOAA), for example, provides a series of online datasets and tools for public use (USGS n.d.; NOAA n.d.).
- Key future-oriented variables

  These can include climate change and socioeconomic projections, which inform not only the engineering tolerance required for projects, but also can inform underpinning risk exposure (e.g., increased density in risk area through projected growth) or contributors to vulnerability (e.g., increasing wealth disparity, aging populations).
- Historical disaster records

  Available on academic and government websites (depending on jurisdiction), historical records of disaster events can trigger a more comprehensive review of risk facing a project. While important to focus on the collection of evidence to inform decisions, historical context can be valuable as a starting point to identify hazards within spatial boundary of the project review area.

  For example, Canada published the Canadian Disaster Database, a searchable repository that can show the estimated spatial extent of events (Public Safety Canada 2019b).
- International disaster risk information

  Several international organizations, including the United Nations and Organisation for Economic Co-operation and Development (OECD), have disaster risk information at the global and national levels. The United Nations, through its PreventionWeb site, publishes high-level risk profiles for member countries, reporting variables from the Sendai Framework for Disaster Risk Reduction, and historical disaster records (UNDRR n.d.b). While high-level, these can inform SEA-type activities within EIA, and may provide regional context for other EIA-based assessments. The OECD iLibrary allows users to search by country, theme, and key words, such that practitioners can focus efforts to one or more countries and examine OECD research on disaster and development practice, including EIA (OECD 2020).

Note

1 The Resilient Cities Network was created to sustain efforts from the 100 Resilient Cities Program by the Rockefeller Foundation which was sunset in 2019. The listing of member cities can be found at https://resilientcitiesnetwork.org/.

References

Arnold, Lauren, and Kevin Hanna. 2017. 'Best Practices in Environmental Assessment: Cases Studies and Application to Mining'. Canadian International Resources and Development Institute (CIRDI). http://ok-cear.sites.olt.ubc.ca/files/2018/01/Best-Practices-in-Environmental-Assessment.pdf.

Burns, Tom R., and Nora Machado Des Johansson. 2017. 'Disaster Risk Reduction and Climate Change Adaptation – A Sustainable Development Systems Perspective'. *Sustainability (Basel, Switzerland)* 9 (2): 293. https://doi.org/10.3390/su9020293.

Cardona, Omar-Dario, van Aalst, Maarten K. , Birkmann, Jörn , Fordham, Maureen , McGregor, Glenn, Perez, Rosa , Pulwarty, Roger S. , et al. 2012. 'Determinants of Risk: Exposure and Vulnerability'. In *Managing the Risks of Extreme Events and Disasters to Advance Climate Change Adaptation*, edited by Christopher B. Field, Vicente Barros, Thomas F. Stocker, and Qin Dahe, 65–108. Cambridge: Cambridge University Press. https://doi.org/10.1017/CBO9781139177245.005.

Casajus Valles, Ainara, Alfred De Jager, Francesco Dottori, Luca Galbusera, Blanca García Puerta, Georgios Giannopoulos, Serkan Girgin, et al. 2019. *Recommendations for National Risk Assessment for Disaster Risk Management in EU: Approaches for Identifying, Analysing and Evaluating Risks: Version 0*. http://publications .europa.eu/publication/manifestation_identifier/PUB_KJNA29557ENN.

Centre for Research on the Epidemiology of Disasters. 2020. 'The Human Cost of Disasters: An Overview of the Last 20 Years (2000–2019)'. https://www.undrr.org/publication/human-cost-disasters-overview -last-20-years-2000-2019.

Collins, Andrew E. 2018. 'Advancing the Disaster and Development Paradigm'. *International Journal of Disaster Risk Science* 9 (4): 486–495. https://doi.org/10.1007/s13753-018-0206-5.

Coppola, Damon P. 2015. *Introduction to International Disaster Management*. Third. Book, Whole. Amsterdam: Elsevier/Butterworth-Hein.

Cutter, Susan L., Lindsey Barnes, Melissa Berry, Christopher Burton, Elijah Evans, Eric Tate, and Jennifer Webb. 2008. 'A Place-Based Model for Understanding Community Resilience to Natural Disasters'. *Global Environmental Change* 18 (4): 598–606. https://doi.org/10.1016/j.gloenvcha.2008.07.013.

Fekete, Alexander, Gabriele Hufschmidt, and Sylvia Kruse. 2014. 'Benefits and Challenges of Resilience and Vulnerability for Disaster Risk Management'. Edited by Alexander Fekete and Gabriele Hufschmidt. *International Journal of Disaster Risk Science* 5 (1): 3–20. https://doi.org/10.1007/s13753-014-0008-3.

Frame, Bob. 2008. '"Wicked", "Messy", and "Clumsy": Long-Term Frameworks for Sustainability'. *Environment and Planning. C, Government & Policy* 26 (6): 1113–1128. https://doi.org/10.1068/c0790s.

Franks, Daniel M., David Brereton, and Chris J. Moran. 2013. 'The Cumulative Dimensions of Impact in Resource Regions'. *Resources Policy* 38 (4): 640–647. https://doi.org/10.1016/j.resourpol.2013.07.002.

Frazier, Tim G., Courtney M. Thompson, and Raymond J. Dezzani. 2014. 'A Framework for the Development of the SERV Model: A Spatially Explicit Resilience-Vulnerability Model'. *Applied Geography (Sevenoaks)* 51 (Journal Article): 158–172. https://doi.org/10.1016/j.apgeog.2014.04.004.

Government of Canada. 2007. 'Emergency Management Act'. August 3, 2007. https://laws-lois.justice.gc .ca/eng/acts/e-4.56/FullText.html.

Gunderson, Lance. 2010. 'Ecological and Human Community Resilience in Response to Natural Disasters'. *Ecology and Society* 15 (2): 18. https://doi.org/10.5751/es-03381-150218.

Hanna, Kevin S. 2016. *Environmental Impact Assessment: Practice and Participation*. Third. Book, Whole. Don Mills, Ontario, Canada: Oxford University Press.

Jabareen, Yosef. 2013. 'Planning the Resilient City: Concepts and Strategies for Coping with Climate Change and Environmental Risk'. *Cities* 31 (Journal Article): 220–229. https://doi.org/10.1016/j.cities .2012.05.004.

Joakim, Erin P., Linda Mortsch, and Greg Oulahen. 2015. 'Using Vulnerability and Resilience Concepts to Advance Climate Change Adaptation'. *Environmental Hazards* 14 (2): 137–155. https://doi.org/10.1080 /17477891.2014.1003777.

Kim, Donghyun, and Up Lim. 2016. 'Urban Resilience in Climate Change Adaptation: A Conceptual Framework'. *Sustainability (Basel, Switzerland)* 8 (4): 405. https://doi.org/10.3390/su8040405.

Lamorgese, Lydia, and Davide Geneletti. 2013. 'Sustainability Principles in Strategic Environmental Assessment: A Framework for Analysis and Examples from Italian Urban Planning'. *Environmental Impact Assessment Review* 42: 116–126. https://doi.org/10.1016/j.eiar.2012.12.004.

Lavell, Allan, Michael Oppenheimer, Cherif Diop, Jeremy Hess, Robert Lempert, Jianping Li, Robert Muir-Wood, et al. 2012. 'Climate Change: New Dimensions in Disaster Risk, Exposure, Vulnerability, and Resilience'. In *Managing the Risks of Extreme Events and Disasters to Advance Climate Change Adaptation*, edited by Christopher B. Field, Vicente Barros, Thomas F. Stocker, and Qin Dahe, 25–64. Cambridge: Cambridge University Press. https://doi.org/10.1017/CBO9781139177245.004.

National Institute of Building Science. 2019. 'Natural Hazard Mitigation Saves Report'. 2019. https://www.nibs.org/page/mitigationsaves.

NOAA. n.d. 'Natural Hazards Viewer'. Accessed November 16, 2020. https://maps.ngdc.noaa.gov/viewers/hazards/.

Noble, Bram F. 2015. *Introduction to Environmental Impact Assessment: A Guide to Principles and Practice*. Third. Book, Whole. Don Mills, Ontario: Oxford University Press.

OECD. 2020. 'OECD ILibrary'. Text. 2020. https://www.oecd-ilibrary.org/.

Oliver-Smith, A., Alcántara-Ayala, I, Burton, Ian, and Lavell, Allan. n.d. 'Forensic Investigations of Disasters (FORIN)', https://www.preventionweb.net/publication/forensic-investigations-disasters-forin-conceptual-framework-and-guide-research

Pelling, Mark. 2007. 'Learning from Others: The Scope and Challenges for Participatory Disaster Risk Assessment'. *Disasters* 31 (4): 373–385. https://doi.org/10.1111/j.1467-7717.2007.01014.x.

Public Safety Canada. 2019a. 'Emergency Management Strategy for Canada: Toward a Resilient 2030'. March 2, 2019. https://www.publicsafety.gc.ca/cnt/rsrcs/pblctns/mrgncy-mngmnt-strtgy/index-en.aspx.

Public Safety Canada. 2019b. 'The Canadian Disaster Database'. 2019. https://www.publicsafety.gc.ca/cnt/rsrcs/cndn-dsstr-dtbs/index-en.aspx.

Resilient Cities Network. n.d.. https://resilientcitiesnetwork.org.

Smith, Adam B. 2020. 'U.S. Billion-Dollar Weather and Climate Disasters, 1980 – Present (NCEI Accession 0209268)'. NOAA National Centers for Environmental Information. https://doi.org/10.25921/STKW-7W73.

Tajima, Ryo, and Thomas B. Fischer. 2013. 'Should Different Impact Assessment Instruments Be Integrated? Evidence from English Spatial Planning'. *Environmental Impact Assessment Review* 41: 29–37. https://doi.org/10.1016/j.eiar.2013.02.001.

UNSDR. n.d. a 'Terminology'. Accessed December 23, 2021. https://www.undrr.org/terminology.

USGS. n.d. 'Web Tools'. Accessed December 23, 2021. https://www.usgs.gov/products/web-tools

UNDRR. n.d.b. 'Disaster Data & Statistics | PreventionWeb.Net'. Accessed December 23, 2021. https://www.preventionweb.net/understanding-disaster-risk/disaster-losses-and-statistics

UNISDR. 2017. 'National Disaster Risk Assessment'. United Nations Office for Disaster Risk Reduction. https://www.preventionweb.net/files/52828_nationaldisasterriskassessmentwiagu.pdf.

Zhou, Hongjian, Xi Wang, and Jing'ai Wang. 2016. 'A Way to Sustainability: Perspective of Resilience and Adaptation to Disaster'. *Sustainability (Basel, Switzerland)* 8 (8): 737. https://doi.org/10.3390/su8080737.

10

REGIONAL ASSESSMENT

Lauren Arnold, Chris G. Buse, Rob Friberg, and
Bram Noble, Kevin Hanna

Introduction

Decisions about land-use and resource development, and the resulting environmental changes, can cause impacts to ecological and social systems that materialize on multiple scales – from the local to the global. This chapter describes regional assessment (RA) as an approach to impact assessment and explores what it means to apply a regional perspective to impact assessment processes. RA can be conceptualized as a distinct form of impact assessment, or as an approach to other forms of impact assessment, including environmental impact assessment (EIA), cumulative effects assessment (CEA) (see Chapter 3 in this book), and strategic environmental assessment (SEA) (see Chapter 2 in this book) to name a few. This chapter discusses the key principles and goals of RA, whether practiced as a stand-alone assessment model or as an approach to inform other types of impact assessment, and the complexity of defining and differentiating RA from other impact assessment approaches. We also highlight several applications of RA to illustrate its use in practice. Finally, we outline important practical considerations and challenges for implementing RA.

What is a regional assessment?

RA is not new. The need to "think regionally" about impacts, including biophysical, social, and health impacts affecting the environment, has underpinned many large-scale impact assessment and monitoring initiatives since the 1970s (Sadler 2011; Therivel 2012; Gunn and Noble 2015). This has especially been the case in Canada, from the Mackenzie Valley Pipeline Inquiry (1974–1977) and Beaufort Sea Hydrocarbon Review (1982–1984), to the Northern Rivers Ecosystem Initiative (1998–2003), and the forthcoming RA focused on northern Ontario's mineral-rich Ring of Fire region (2020).

The interest in RA stems from the recognized need to extend impact analyses, monitoring, and management efforts beyond the footprint of an individual development project (Gunn and Noble 2011; Elvin and Fraser 2012). The application and efficacy of EIA, for example, can be limited by project-specific spatial and temporal scales. Environmental changes resulting from land-use and development actions can have transboundary implications for ecological and social components that manifest beyond the site of construction or operation of any single project

DOI: 10.4324/9780429282492-11

(Gillingham and Johnson 2016; Noble 2008). In more recent years, the surge of interest in RA is coupled with increasing concerns about cumulative environmental effects (Gunn and Noble 2015), the scaling back of many regulatory EIA systems to facilitate timely development approvals (Fidler and Noble 2012), the role of Indigenous peoples in the planning and management of development on their traditional territories (see Porta and Banks 2011), and the need to ensure that individual decisions are made within the broader regional context of ecological and socio-economic sustainability.

Gunn and Noble (2015) describe RA as more than expanding the spatial scale of impact assessment; rather, RA is increasingly viewed as an integration of planning and assessment instruments in which ecological and social sustainability are foundational to evaluation and decision-making processes. In this regard, RAs typically consider all development or land uses and their resulting impacts within a defined region, with the key goal of understanding regional impact pathways and outcomes. In other words, RA is about understanding the broader regional context of development, and understanding whether or how multiple projects or land-use plans create key threats and vulnerabilities to human and ecological valued components (VCs).[1] This necessarily includes the state of VCs in the past, present, and future, and setting out management solutions to achieve regional sustainability goals or objectives (Beanlands and Duinker 1984; CCME 2009; Gunn and Noble 2015). RA is thus anticipatory, planning-oriented, and can be strategic in its assessment outlook.

A fundamental challenge for RA is drawing clear boundaries around its meaning and application among the multiple forms and variations of assessment that exist. Provisions do sometimes exist for RA in legislation, such as in the Canadian *Impact Assessment Act* (2019), but RA itself is often not explicitly defined. The Impact Assessment Agency of Canada, for example, broadly describes RA as: "studies conducted in areas of existing projects or anticipated development to inform planning and management of cumulative effects and inform project impact assessments" (2019). In essence, these RAs are intended to inform regional baselines, thresholds, and mitigation measures that could be used to assist in subsequent project-based EIA; identify impacts to Indigenous people's rights and interests; and provide guidance for managing cumulative effects (Impact Assessment Agency of Canada 2019). RAs under the Canadian *Impact Assessment Act* (2019) may be specifically designed for a number of different purposes, such as data gathering and trend analysis, setting thresholds and mitigation standards, and/or regional development or planning, but consistent operational guidance is not yet available and practical applications and are forthcoming.

For example, a RA under the Canadian *Impact Assessment Act* is being completed in the Ring of Fire (2020), an area of mineral deposits in northern Ontario that has experienced decades of mineral exploration and development, and where a coordinated effort to understand regional impacts and cumulative effects has been called for by Indigenous nations and public stakeholders. An RA was also recently completed for Offshore Oil and Gas Exploratory Drilling east of the province of Newfoundland and Labrador (2020). This assessment was initiated under the former *Canadian Environmental Assessment Act* (2012), under the Act's provision for "regional studies" and is intended to improve the efficiency and effectiveness of assessment processes for subsequent projects in the study area; though, details in terms of the implementation of the results and the role of the RA in supporting decision-making and other assessments is not yet clear. In the early 2000s, the Canadian Environmental Assessment Agency's Research and Development program prioritized "regional frameworks", noting that "working at the regional scale can provide proponents, government decision makers and affected publics with a better understanding of [...] cumulative effects" (CEAA 2000–2003). Thus, while RA is not a new idea or concept, and its value has long been recognized, there is consistent definition for RA and the process itself. In the international context, RA is perhaps even less clearly defined and present more often as a principle, ideal, or element of other assessments.

A single, clear definition of RA is elusive in part because its implementation varies, and RA is often used synonymously with, or folded into, other forms of impact assessment, such as CEA or SEA, which are covered in other chapters in this volume. The term RA becomes confusing as yet another label and another form of assessment which is sometimes, though not always, explicitly distinguished from other types of assessments that also attempt to include a regional focus. CEA and SEA both ideally adopt a regional perspective, though this is not always the case in practice (Harriman Gunn and Noble 2009; Halseth et al. 2016; Noble et al. 2019). An RA may be initiated with the intent to provide information on the cumulative impacts of proposed developments within a region, or to assist with strategic land-use planning or decision-making by way of exploring future regional land-use scenarios, but an RA does not necessarily imply an explicit cumulative or strategic focus in every application. RAs may be conducted for the purpose of understanding change in regional baseline conditions, understanding the vulnerability of certain values components or ecological systems, or establishing thresholds or informing management actions – with no real intent to model cumulative effects or explore future scenarios of land-use and development pressures. CEA, SEA, EIA, and RA are not always completely distinct and independent, nor are they completely overlapping or synonymous. RA may also be applied in a wide range of other impact assessment processes, including social impact assessment, health impact assessment, economic impact assessment, and even regional planning process. Perhaps what is important in understanding or defining RA is understanding its underlying principles, the questions it is meant to answer, and the decisions or issues it is meant to inform.

Dissecting the concepts included in RA can provide insight into these principles. The term *regional* can be understood in a number of ways. For instance, regional may be conceptualized based on the need to integrate and understand different spatial scales of assessment (Buse et al. 2020). This might mean defining a regional boundary or ecologically defined study areas in addition to local study areas or individual project footprints. Such spatial distinctions are important for understanding impact drivers, pathways, and ecological systems in which VCs operate (Willsteed et al. 2018; Therivel and Ross 2007; Squires and Dubé 2013). For instance, it may be important to spatially define a specific ecosystem of interest, a range for a species that has been disturbed, complete a landscape-level assessment of vulnerability (Pavlickova and Vyskupova 2015), or otherwise track critical VCs that are inherently borderless (e.g., air quality, water flows, human mobility, etc.).

One useful heuristic that could apply across ecological and human social and health considerations may be macro, meso, and micro scales. When applied to ecological components, a macro-level of analysis refers to the ecosystem itself, while meso-level analysis is focused on an ecological unit such as the watershed, and micro-level analysis is focused on a single stream. When applied to human socioeconomic systems, a macro-level analysis may refer to the scale of society itself, meso-level analysis is focused on specific populations and communities, and micro-level analysis is focused on the interactions and actions of individuals (Beaussier et al. 2019). Similarly, macro, meso, and micro could refer to population-level public health considerations, community health, and individual health, respectively. Socioeconomic and health systems operate within and across these human scales as well as across space and time. It is often necessary to consider multiple spatial scales to assess environmental, social, and health impacts, and the scale of analysis will dictate important questions pertaining to management thresholds, triggers and associated interventions, and different ethical imperatives depending on the level of analysis (Buse, Smith, and Silva 2019). For instance, the socioeconomic effects of resource development may be distributed differently at a regional or national scale, where the impacts may be relatively positive when compared with the local scale where negative impacts and strain on communities may be more severe (Loayza and Rigolini 2016; Banks 2013; Fedorova and Pongrácz 2019).

Regional also refers to the management or political scales of analysis such as municipal, state/provincial, national, and international boundaries. Such jurisdictional spatial boundaries may be important for conducting assessments and making decisions in relation to regional management areas, conservation areas, and planning decisions at the province or state level (Jiliberto 2011; Chilima et al. 2017) or, as is increasingly the case in Canada and other post-colonial countries, for assessments in Indigenous land-use or title areas (Larsen et al. 2017; Gillingham et al. 2016).

Essential for understanding the *regional* focus of RA is the application of a systems-based perspective, whereby emphasis is placed on the interconnectedness of VCs and the pathways that lead to environmental and social impacts across multiple spatial scales. This brings clarity to the meaning of *assessment* in the context of RA. RAs are not simply "regional studies", which may focus on describing the current baseline or state of a region – such as an ecosystem or watershed, or socioeconomic setting. Rather, RA must involve some form of analysis of trends and disturbances that have shaped the baseline, and identify important drivers of change, impacts, and pathways. The distinction between a regional "study" versus regional "assessment" is often not very clear in guidance materials or regulations, but such a distinction is important for distinguishing RA from data aggregation exercises, or from simple descriptions of current baselines (such as the state of a VC) within a regional boundary. An assessment implies not only an account of the current environmental state, including key drivers, trends, and causes, but also an interpretation of what these changes mean in the ecological and social context, and, ideally, an evaluation of management options.

Key principles for RA then, are the application of a *regional* and *systems perspective* to the *assessment* of the environment and VCs, their *relationships* and *interactions*, and the *impacts* and *impact pathways* that affect them. RA refers to an assessment process, and a set of principles that may be integrated into a wide range of applications and applied within or alongside other assessments. In a sense, it may not matter if the assessment in question is labeled as a regional CEA, a regional SEA, or a RA that is strategically applied with a goal of understanding cumulative effects, but rather what the assessment actually includes. However, in another sense, terminology, distinction, and clear guidance does matter if we are to understand how to practically carry out assessments. In the following sections, with this conceptual fuzziness in mind, we discuss well-established examples of RA as it has been integrated within, or triggered by, EIA, CEA, and/or SEA, with the intent to illustrate principles of RA in practice.

Applications of regional assessment

Regional environmental impact assessment

The ideals behind RA are not new and, as we have noted, emerged in response to the limitations of early applications of project-based EIA. While EIA is project centered in scope, and often in scale, project impacts can have regional consequences, and it is important to understand the broader biophysical and human systems within which impacts occur. RA may be carried out in response to a particular project, or as part of an EIA process, but the analysis or assessment of impacts must extend beyond the project-specific focus. An early Canadian example is the Mackenzie Valley Pipeline Inquiry. The inquiry was initiated in 1974 to examine the potential impacts of a natural gas pipeline project proposed in northern Canada, and the perspectives of the Indigenous nations and groups who lived along the route (Berger 1977). The final report produced by Justice Thomas Berger would become a key turning point in impact assessment as a case where social impacts were given significant attention, and Indigenous groups received formal funding and opportunities to present their own evidence (Ross 1990; Burdge and Vanclay 1996). The inquiry is also significant as an RA example of a large-scale project triggering the need to understand environmental and social impacts operating across multiple scales, from the project infrastructure footprint to the larger regional implications of a pipeline fragmenting social and ecologically important habitats for caribou and the implications for the way of life across the Canadian north.

It has been argued by many that assessments of regions, and of regional impacts, are best carried out independently from any one EIA process and the constraints in terms of time, resources, and conceptual scope that often accompany them (Noble, Liu, and Hackett 2017; Duinker and Greig 2006; Atlin and Gibson 2017). Many of these arguments are related to the need to consider cumulative impacts operating across large spatial and temporal scales and appropriate ecological scales through which the impact and functioning of ecological components can be understood, such as the need to consider the entire watershed in order to understand impacts to freshwater (Seitz, Westbrook, and Noble 2011; Squires and Dubé 2013; Franks, Brereton, and Moran 2013). These are important initiatives, though the definition of regional and local study areas, and understanding cumulative impacts, also remains an important part of EIA in order to consider the ecological and social scales at which VCs are affected, and how these components are connected through pathways and impact drivers operating at multiple scales (Arnold, Hanna, and Noble 2019; Noble, Liu, and Hackett 2017). RAs may be an important aspect of an EIA in order to obtain a clearer picture of impacts and to inform decision-making about project implementation. In this respect, RAs can be separate from EIAs, and RA can be linked to and inform EIA outcomes and decisions, though adopting a regional study boundary in EIA does not equal an RA. RA may also be used to help to decide appropriate projects and development paths before project-specific EIAs are undertaken, by providing an understanding of baseline trends, limits, thresholds, and what types of developments are "acceptable" in a particular region.

Regional cumulative effects assessment

RA has become a key dimension of CEA, and many scholars have argued that cumulative effects cannot be properly assessed in the absence of a regional approach. Cumulative effects refer to the incremental, additive, and synergistic impacts on people, non-human animals, and the environment across time and space (Franks, Brereton, and Moran 2010). One of the primary tensions for CEA is the scale at which it is, and should be, implemented. Most EIA regulations require at least some consideration of cumulative effects, but a project-level focus is not conducive to the large spatial and temporal scales required for CEA, or for supporting the time, resource, and organizational structures needed to effectively carry it out (Sinclair, Doelle, and Duinker 2017; Jones 2016). CEA remains an important, if not always realized, part of an EIA process and project-based decision-making where understanding the cumulative context of the project and its impacts is vital, but regional CEA (RCEA) has emerged to address the notion that cumulative effects are best assessed on ecologically relevant scales of analysis (Harriman Gunn and Noble 2009; Johnson et al. 2016).

RCEA is preferably completed independently from any one project EIA and would provide information that is usable across a range of development options, industries, and levels of decision-making. The regional dimension of RCEA indicates attention to the full spatial scale over which a VC functions. Numerous frameworks and methods have been put forward to support the regional analysis of VCs (Hodgson and Halpern 2019; Ho, Eger, and Courtenay 2018; Jones et al. 2017), and also for the analysis of cumulative impacts from specific resource development industries (Franks, Brereton, and Moran 2013; Atlin and Gibson 2017; Athayde et al. 2019; Johnson et al. 2016). Importantly, the spatial scale of assessment must be adjusted according to the ecological or the socioeconomic context of the VC. The nature of cumulative impacts also implies attention to broad temporal scales and including both a retrospective and prospective analysis of VCs, ideally with an effective link to decision-making at, and beyond, the project level.

A crucial point about RCEA is that it is ideally not just about scaling up boundaries, but it is about applying a different approach to understanding relationships between VCs, decision-making, and assessment. In practice, RCEA may be applied to a single project, multiple projects, strategic decisions and planning for a particular industry, or to multiple industries (Harriman Gunn and Noble 2009). The strategic applications are typically considered to be more effective in providing opportunities to link meaningfully to decision-making (Noble and Nwanekezie 2017; Bidstrup, Kørnøv, and Partidário 2016; Harriman Gunn and Noble 2009; Harriman and Noble 2008).

Regional strategic environmental assessment

SEAs are usually conceptualized as assessments of specific policies, plans, or programs of a particular industry or sector (Partidário 2015). RSEA has been further defined as "a process designed to systematically assess the potential environmental effects, including cumulative effects, of alternative strategic initiatives, policies, plans, or programs for a particular region" (CCME 2009, 6). The state of knowledge for RSEA is rapidly growing, and a range of applications and methodologies have been developed (Buse et al. 2020).

Central to the effectiveness of RSEA, and important for differentiating it from other forms of assessment, is its strategic nature (Noble et al. 2019). RSEA is necessarily oriented toward directing the design and implementation of development activities, exploring alternatives and options, and aligning with decision-making and planning. The interpretation of *strategic* has evolved beyond appraising policies, plans, or programs and assessing their impacts or compliance to objectives; RSEA is meant to shape or form policies, plans, or programs and explore alternatives in order to achieve desired goals or outcomes (Partidário 2015; Pang, Mörtberg, and Brown 2014; White and Noble 2013; Jiliberto 2011; Slootweg and Jones 2011). The core principles of RSEA, in addition to a strategic focus, are that it is oriented toward thinking about sustainability and resilient futures, it is cumulative effects focused, multi-scaled, multi-tiered, multi-sectoral, participatory, opportunistic, and adaptive (CCME 2009; Harriman Gunn and Noble 2009; Noble and Nwanekezie 2017).

An example of an RSEA is the Great Sand Hills Regional Environmental Study completed in Saskatchewan, Canada (Noble 2008). The Great Sand Hills is a region of native prairie and sand dunes that has been impacted by a range of development activities and human uses and holds significant ecological, economic, and cultural significance. A regional lens was important in this case, and the SEA framework was designed with a cumulative impact focus and used a multi-tiered spatial scale that considered biophysical, socioeconomic, and cultural boundaries. For instance, the landscape scale was an important starting point for ecological components, while socioeconomic boundaries were based on a much larger area inclusive of multiple municipalities, and the cultural boundary was extended beyond the Great Sand Hills area and across provincial boundaries to account for the cultural ties and movements of First Nation groups. The assessment was also fundamentally forward-thinking while being cognizant of existing policies, plans, and land uses as well as potential development scenarios, objectives, and environmental and land-use management targets.

Another RSEA example is the assessment for the Browse Liquefied Natural Gas (LNG) Precinct conducted in the Kimberly coastline of Australia (Western Australian Government 2010). The LNG precinct was proposed with the intention of preventing disjointed project-by-project development and associated cumulative impacts in the region and to facilitate economic synergy within the industry. The SEA process was key for evaluating site alternatives for the precinct and providing a framework to minimize potential impacts and issues for future LNG developments. The effectiveness of the SEA, and particularly the social assessment components and the meaningful inclusion of Indigenous peoples, has been debated (Mills 2019; O'Faircheallaigh 2010, 2009). But aside from its outcomes, the application of this assessment in a region where extensive LNG development is likely is an example of an RSEA providing an opportunity for regional impact assessment and targeted planning for a particular industry of interest.

Practical considerations and challenges for regional assessment

The regional lens

Fundamental to RA is the application of a regional lens or perspective, but the ambiguity of terminology for RA applications and the fuzzy lines between the different (yet related) assessment processes described above is a potential challenge for practitioners. As previously noted, hard distinctions between RA, EIA, RSEA, RCEA may not always be necessary. It is perhaps the content of the assessment and its underlying principles that are most important, rather than the assessment label. However, it is also vital to understand that these processes are not synonymous. That is, EIA, CEA, and SEA are not necessarily always applied regionally, and RA is not necessarily always applied in a project-based, cumulative, or strategic context, but rather may be applied independently or in any number of other assessment processes.

An important note for practitioners is that best practices for SEA, CEA, and RA include effectively integrating elements of strategic, cumulative, and regional assessments. For instance, an ideal CEA is regional and forward-thinking/strategic. A SEA that is intended to address land-use issues and planning for development should be regional in its approach and include a broad perspective of sustainability, and attention to cumulative impacts (Bidstrup, Kørnøv, and Partidário 2016; Blakely et al. 2017; Harriman Gunn and Noble 2009). The flexibility in applications of RA is in a sense a key strength in terms of its ability to support a wide range of assessment processes and decision-making contexts. But this flexibility in applications and methods is also a key challenge for the practitioner seeking tangible information on what RA should look like. We know that there is no one approach – RA may be adaptive in its implementation and make use of the full suite of interdisciplinary methods present in EIA, RCEA, RSEA, and other impact assessment research and practice-based experiences

Tools that can facilitate regional spatial analysis, such as Geographic Information Systems (GIS) and remote sensing, are especially useful. Spatial analysis has been used extensively across EIA, CEA, and RSEA to model and visualize environmental change and to support decision-making (Choi and Lee 2016; Diez-Rodríguez, Zio, and Fischer 2019; Martinez-Grana et al. 2014; Sadler 2011). In the Great Sand Hills RSEA, for instance, GIS was used to project spatial patterns of potential development and disturbance scenarios, which supported the regional level assessment and also allowed the visualization of possible futures that were used in participatory workshops (Noble 2008). Increasingly these tools are also being leveraged for the integration of multiple values, including environment, social, and health impact assessment (Huang and London 2016; Sadd et al. 2011; Tyson, Lantz, and Ban 2018; Pavlickova and Vyskupova 2015; Martinez-Grana et al. 2014). For instance, Cumulative Environmental Vulnerability Assessment and mapping tools have been developed for California to identify communities that are vulnerable to pollution impacts. Huang and London combined an environmental hazard index, a social vulnerability index, and a health index, which were informed by participatory research and consultation, to visualize community vulnerability to cumulative impacts in the San Joaquin Valley and the Coachella Valley (Huang and London 2012, 2016). A key aspect of spatial visualization is also the ability to "zoom in" and "zoom out" at different levels of analysis, enabling in-depth consideration of specific landscape units without losing a sense of the whole system.

Qualitative methods such as interviews, surveys, ethnography, and participatory models such as matrices, network diagrams, and Bayesian Belief Networks have also been advocated as valuable tools for assessing environmental and social impacts (Hodgson and Halpern 2019; Larsen et al. 2017; Christensen, Krogman, and Parlee 2010; Mantyka-Pringle et al. 2017). For example, Mantyka-Pringle et al. utilized Bayesian Belief Networks, a probability-based model to represent key factors and their potential influences, to operationalize a "two-eyed seeing approach" for cumulative impacts using scientific knowledge and Indigenous knowledge in the Slave River Delta, Northwest Territories (Mantyka-Pringle et al. 2017). Interdisciplinary and structured, but adaptive and participatory, approaches are encouraged for RCEA and RSEA (CCME 2009; Sadler and Dalal-Clayton 2012).

As previously explained, the aim for RA is not simply to describe a region or condition, but to complete an assessment of impacts, drivers, outcomes, and system functioning. The tools described above, such as GIS and remote sensing, are commonly used to describe regions or produce maps of existing conditions. Importantly for RA, these tools may be leveraged as part of a dynamic systems model to functionally relate development pressures to response indicates and to help to understand the connections between ecological and social systems and how they respond to stress (Dubé et al. 2013).

Regardless of the specific methods applied, an important practical consideration is that when a "regional", or a RA, dimension is added to EIA, CEA, or SEA, there is a difference in terms of the spatial scale considered, but also in the advantages provided by adopting the regional perspective in how the nature of change for a VC is understood. The RA lens allows a better understanding of the complexity of factors affecting VCs, the relationships between them, and of the system itself, which improves insight into specific impacts and disturbances and possible mitigation solutions. RA as a process, lens, and area of research provides an opportunity to explore a wide range of methodologies and approaches that can be applied in diverse settings and assessment contexts.

Integration and tiering

The term *integration* has been used in a number of ways in impact assessment fields to refer to a several different goals, such as establishing effective relationships between multiple scales of assessment, combining multiple disciplines and approaches, and considering multiple VCs and impact types (Buse et al. 2020). One of the ways that integration is understood is to refer to the extent to which an assessment considers multiple land-use values, including biophysical environmental components and human social components, including human health, and the relationships between them (Buse et al. 2020). Another way integration is used is to refer to "tiering" or the extent to which an assessment process is effectively linked to other forms of assessment, and the decision-making and planning systems that it intends to influence (CCME

2009; Buse et al. 2020). Both these types of integration are important challenges and considerations for RA design and implementation.

Since its inception, impact assessment has been intended to include impacts beyond those to the biophysical environment, such as those to human systems that result from land-use changes, industrial activities, and from the pathways between biophysical impacts and human systems. From their earliest iterations, RSEA and RCEA have attempted to leverage insights from environmental, social and health impact assessment methods in order to understand the impacts of projects on their respective values (Ho, Eger, and Courtenay 2018; Sadler and Dalal-Clayton 2012). There is no single agreed-upon approach to achieve such integration in RSEA/RCREA, and a number of methods are being explored: life cycle costing (Brandão et al. 2010); Bayesian networks, coupled component models, agent-based models, and knowledge-based models (Kelly et al. 2013); directional distance functions to quantify the extent of maximizing desirable outcomes and mitigating negative outcomes (Macpherson, Principe, and Smith 2010); dynamic evaluation procedures to build consensus among diverse stakeholder needs (Naddeo et al. 2013); and the Drivers–Pressures–Conditions–Response framework (Harwell et al. 2019). Such multi-criteria analysis may enable an understanding of weighing diverse environment, community, and health values and trade-offs for proposed policy alternatives, such as integrating the Drivers–Pressures–Conditions–Response framework into SEA to explicitly map the connections between ecosystems services and their influence on human communities and systems (Hardwell et al. 2019).

But, the effective consideration of human social components is considered a weak aspect of RCEA, RSEA, and EIA, each of which tends to remain weighted toward biophysical considerations and components (Hodgson, Halpern, and Essington 2019; Gillingham and Johnson 2016; Mitchell and Parkins 2011). There is a significant bias in the existing research base toward quantitative analysis and indicators, which can inhibit the selection of diverse values for assessment and result in less "tangible" and less easily measured social components and local values being considered more anecdotally (Buse et al. 2020).

The conceptual support for improving the integration of diverse VCs and exploring a wider range of methods is strong, but the practical guidance is lacking; essentially, there is consensus that assessment *should* be integrated, but little instruction on how to achieve it. Knowledge on how to incorporate social and health impacts in RCEA and RSEA is particularly sparse when compared to the body of work focused on biophysical considerations (Buse et al. 2020; Moran, Franks, and Sonter 2013; Willsteed et al. 2018; Mitchell and Parkins 2011). Even less common is the effective merging or understanding of the interactions between biophysical, social, and health components and insight on how decisions are, and can be, made in situations defined by the complex interrelationships between these components and priorities for human and ecological systems (Buse et al. 2020). These are important considerations for any application of RA and requires careful attention to the identification of VCs, assessment design, and methods. RA ideally can be an important forum where integration of diverse VCs might be achieved with a regional and/or system-based perspective.

Effective tiering or integration between assessment types and levels presents a related challenge for RA. A robust assessment that integrates environmental, social, and health systems is still not meaningful unless the inputs have been informed by other relevant processes, initiatives, or studies, and the outputs are effectively used to influence some level of decision-making or process. There are a wide range of processes under the broad umbrella of impact assessment that may be most effective when they are informed and influence each other. For example, CEA is ideally completed in a SEA framework and embedded in regional planning and decision-making processes, and also useful for informing subsequent project EIAs (Harriman Gunn and

Noble 2009; Halseth et al. 2016; Hegmann and Yarranton 2011; Gunn and Noble 2011). Tiering is often used to refer to this goal of effective relationships among types and levels of assessment and to the decision-making and planning processes they are intended to influence.

This is important in the context of RA as a process that is interrelated with and used to frame diverse impact assessment applications. The practical applications that have been described in this chapter are predominantly cases in which the assessment or the RA has been reactive and triggered by a need to think regionally, either to better understand an individual project or to shift to a strategic or cumulative focus after significant existing impacts are realized. However, all of the assessment processes described are intended to be forward-thinking and connected to decision-making at some level. In many ways, RA implies a connection to proactive regional planning, even if it is applied during an EIA, because it is entrenched in the intention to understand the context of the project and the potential impacts over a region, society, or management area. An effective assessment must not only be sound technically, but it must be meaningful. Doing "better" integrated assessments will not result in "better" management unless the relationships between assessment types and levels of decision-making are effective. Regardless of how and in what context it is implemented, RA outputs need to be adopted as inputs to "next-level" assessment and decision processes, whether in project EIAs (i.e., setting terms of reference for EIAs or setting monitoring indicators or mitigations) or in regional management plans (i.e., informing land-use zoning or limits). An important consideration is the extent to which the design of an RA, or an assessment with a core regional dimension such as RSEA and RCEA, facilitates its use in the types of decision-making and planning discussions that it is intended to drive.

Conclusion

Regional assessment is a process in itself, but it also encompasses a set of principles and considerations that may be applied in a number of assessment protocols. RA is about more than scaling up an assessment to a broader spatial scale; it implies a different way of thinking and adopting a systems perspective to impacts and environmental changes, and how VCs that define the assessment take on new meanings at different scales of analysis. It is planning-oriented. An effective RA, connected to decision-making, can support diverse assessment applications and decision-making at the project level, and at the strategic planning level by improving insight into impact pathways and cumulative impacts. RA allows the application of a wide range of methods to integrate diverse values and ecological and social components. Large-scale and strategic initiatives are rarely easy to apply in practice, but a regional assessment lens can more effectively support sustainable planning and regional decision-making frameworks, than approaches that focus on smaller scales and shorter time frames.

Note

1 Valued components (VCs) refer to resources, species, social values, or ecological components that have been identified by stakeholders as critical. The use of VCs is intended to facilitate focused assessment processes.

References

Arnold, Lauren M., Kevin Hanna, and Bram Noble. 2019. "Freshwater Cumulative Effects and Environmental Assessment in the Mackenzie Valley, Northwest Territories: Challenges and Decision-Maker Needs".

Impact Assessment and Project Appraisal 37 (6): 516–525. https://doi.org/10.1080/14615517.2019.1596596.

Athayde, Simone, Carla G. Duarte, Amarilis L. C. F. Gallardo, Evandro M. Moretto, Luisa A. Sangoi, Ana Paula A. Dibo, Juliana Siqueira-Gay, and Luis E. Sánchez. 2019. "Improving Policies and Instruments to Address Cumulative Impacts of Small Hydropower in the Amazon". *Energy Policy* 132 (October 2018): 265–271. https://doi.org/10.1016/j.enpol.2019.05.003.

Atlin, Cole, and Robert Gibson. 2017. "Lasting Regional Gains from Non-Renewable Resource Extraction: The Role of Sustainability-Based Cumulative Effects Assessment and Regional Planning for Mining Development in Canada". *Extractive Industries and Society* 4 (1): 36–52. https://doi.org/10.1016/j.exis.2017.01.005.

Banks, Glenn. 2013. "Little by Little, Inch by Inch: Project Expansion Assessments in the Papua New Guinea Mining Industry". *Resources Policy* 38 (4): 688–695. https://doi.org/10.1016/j.resourpol.2013.03.003.

Beanlands, G.E. & Duinker, P.N. 1984. An ecological framework for environmental impact assessment. *Journal of Environmental Management* 18(3): 267–277.

Beaussier, Thomas, Sylvain Caurla, Véronique Bellon-Maurel, and Eleonore Loiseau. 2019. "Coupling Economic Models and Environmental Assessment Methods to Support Regional Policies: A Critical Review". *Journal of Cleaner Production* 216: 408–421. https://doi.org/10.1016/j.jclepro.2019.01.020.

Berger, Thomas R. 1977. *Northern Frontier, Northern Homeland*. Toronto: J. Lorimer & Company.

Bidstrup, Morten, Lone Kørnøv, and Maria Rosário Partidário. 2016. "Cumulative Effects in Strategic Environmental Assessment: The Influence of Plan Boundaries". *Environmental Impact Assessment Review* 57: 151–158. https://doi.org/10.1016/j.eiar.2015.12.003.

Blakely, Jill A. E., Peter Duinker, Lorne Grieg, George Hegmann, and Bram Noble. 2017. "Cumulative Effects Assessment: FASTIPS No. 16".

Brandão, M., R. Clift, L. Milài Canals, and L. Basson. 2010. "A Life-Cycle Approach to Characterising Environmental and Economic Impacts of Multifunctional Land-Use Systems: An Integrated Assessment in the UK". *Sustainability*, 2(12):3747–3776.

Burdge, Rabel J., and Frank Vanclay. 1996. "Social Impact Assessment: A Contribution To the State of the Art Series". *Impact Assessment* 14 (1): 59–86. https://doi.org/10.1080/07349165.1996.9725886.

Buse, C. G., R. Friberg, L. Arnold, and K. Hanna. 2020. "Unlocking the Promise of 'Integrated' Regional and Strategic Environmental Assessments Based on a Realist Review of the Scholarly Literature". Kelowna, BC: Centre for Environmental Assessment Research, University of British Columbia.

Buse, C. G., M. Smith, and D. S. Silva. 2019. "Attending to Scalar Ethical Issues in Emerging Approaches to Environmental Health Research and Practice". *Monash Bioethics Review* 37 (1–2): 4–21.

Canadian Environmental Assessment Act. (S.C. c. 19 s. 52, 2012). Government of Canada.

CCME (Canadian Council of Ministers of the Environment). 2009. *Regional Strategic Environmental Assessment in Canada: Principles and Guidance*. Winnipeg, MB.

CEAA (Canadian Environmental Assessment Agency). 2000. *Research and Development Program: Research Priority Areas 2000–2003*. Hull, QC: CEAA.

Chilima, Jania S., Jill A. E. Blakely, Bram F. Noble, and Robert J. Patrick. 2017. "Institutional Arrangements for Assessing and Managing Cumulative Effects on Watersheds: Lessons from the Grand River Watershed, Ontario, Canada". *Canadian Water Resources Journal* 42 (3): 223–236. https://doi.org/10.1080/07011784.2017.1292151.

Choi, H. S., and G. S. Lee. 2016. "Planning Support Systems (PSS)-Based Spatial Plan Alternatives and Environmental Assessments". *Sustainability* 8 (3): 286

Christensen, Lisa, Naomi Krogman, and Brenda Parlee. 2010. "A Culturally Appropriate Approach to Civic Engagement: Addressing Forestry and Cumulative Social Impacts in Southwest Yukon". *Forestry Chronicle* 86 (6): 723–729. https://doi.org/10.5558/tfc86723-6.

Diez-Rodríguez, J. J., Simone Di Zio, and T. B. Fischer. 2019. "Introducing a Group Spatial Decision Support System for Use in Strategic Environmental Assessment of Onshore Wind Farm Development in Mexico". *Journal of Cleaner Production* 222: 1239–1254.

Dubé, M. G., P. Duinker, L. Greig, M. Carver, M. Servos, M. McMaster, ... K. R. Munkittrick. 2013. "A Framework for Assessing Cumulative Effects in Watersheds: An Introduction to Canadian Case Studies". *Integrated Environmental Assessment and Management* 9(3), 363–369. https://doi.org/10.1002/ieam.1418

Duinker, Peter N., and Lorne A. Greig. 2006. "The Impotence of Cumulative Effects Assessment in Canada: Ailments and Ideas for Redeployment". *Environmental Management* 37 (2): 153–161. https://doi.org/10.1007/s00267-004-0240-5.

Elvin, S. and G. Fraser. 2012. "Advancing a National Strategic Environmental Assessment for the Canadian Offshore Oil and Gas Industry with Special Emphasis on Cumulative Effects". *Journal of Environmental Assessment Policy and Management* 14 (4). 1250015

Fedorova, Elena, and Eva Pongrácz. 2019. "Cumulative Social Effect Assessment Framework to Evaluate the Accumulation of Social Sustainability Benefits of Regional Bioenergy Value Chains". *Renewable Energy* 131: 1073–1088. https://doi.org/10.1016/j.renene.2018.07.070.

Fidler, C., and B. Noble. 2012. "Advancing strategic environmental assessment in the offshore oil and gas sector: Lessons from Norway, Canada, and the United Kingdom.". *Environmental Impact Assessment Review,*, 34: 12–21.

Franks, Daniel M., David Brereton, and Chris J. Moran. 2010. "Managing the Cumulative Impacts of Coal Mining on Regional Communities and Environments in Australia". *Impact Assessment and Project Appraisal* 28 (4): 299–312. https://doi.org/10.3152/146155110X12838715793129.

———. 2013. "The Cumulative Dimensions of Impact in Resource Regions". *Resources Policy* 38 (4): 640–647. https://doi.org/10.1016/j.resourpol.2013.07.002.

Gillingham, Michael P., Greg R. Halseth, Chris J. Johnson, and Margot W. Parkes. 2016. "Exploring Cumulative Effects and Impacts Through Examples". In *Integration Imperative: Cumulative Environmental, Community and Health Effects of Multiple Natural Resource Developments*, edited by Michael P. Gillingham, Greg R. Halseth, Chris J. Johnson, and Margot W. Parkes, 153–193. Switzerland: Springer International Publishing.

Gillingham, Michael P., and Chris J. Johnson. 2016. "Cumulative Impacts and Environmental Values". In *Integration Imperative: Cumulative Environmental, Community and Health Effects of Multiple Natural Resource Developments*, edited by Michael P. Gillingham, Greg R. Halseth, Chris J. Johnson, and Margot W. Parkes, 49–83. Switzerland: Springer International Publishing.

Government of Western Australia. 2010. "Browse LNG - Environment: Strategic Assessment Report". Retrieved from https://www.jtsi.wa.gov.au/what-we-do/offer-project-support/lng-precincts/browse-kimberley/browse-lng---environment.

Gunn, Jill, and Bram F. Noble. 2011. "Conceptual and Methodological Challenges to Integrating SEA and Cumulative Effects Assessment". *Environmental Impact Assessment Review* 31 (2): 154–160. https://doi.org/10.1016/j.eiar.2009.12.003.

Gunn, J., and B. F. Noble. 2015. "Sustainability considerations in regional environmental assessment". In *Sustainability Assessment Handbook*, edited by A. Morrison-Saunders, J. Pope, and A. Bond, 79–102. Cheltenham, UK: Edward Elgar.

Halseth, Greg R., Michael P. Gillingham, Chris J. Johnson, and Margot W. Parkes. 2016. "Cumulative Effects and Impacts: The Need for a More Inclusive, Integrative, Regional Approach". In *Integration Imperative: Cumulative Environmental, Community and Health Effects of Multiple Natural Resource Developments*, edited by Michael P. Gillingham, Greg R. Halseth, Chris J. Johnson, and Margot W. Parkes, 3–21. Switzerland: Springer International Publishing.

Harriman Gunn, Jill, and Bram F. Noble. 2009. "Integrating Cumulative Effects in Regional Strategic Environmental Assessment Frameworks: Lessons From Practice". *Journal of Environmental Assessment Policy and Management* 11 (03): 267–290. https://doi.org/10.1142/S1464333209003361.

Harriman, Jill a. E., and Bram F. Noble. 2008. "Characterizing Project and Strategic Approaches to Regional Cumulative Effects Assessment in Canada". *Journal of Environmental Assessment Policy and Management* 10 (01): 25–50. https://doi.org/10.1142/S1464333208002944.

Harwell, M. A., J. H. Gentile, L. D. McKinney, J. W. Tunnell, W. C. Dennison, R Heath Kelsey, K. M. Stanzel, G. W. Stunz, K. Withers, and J. Tunnell. 2019. "Conceptual Framework for Assessing Ecosystem Health". *Integrated Environmental Assessment and Management* 15 (4): 544–564.

Hegmann, G. and G. A. Yarranton. 2011. "Alchemy to Reason: Effective use of Cumulative Effects Assessment in Resource Management". *Environmental Impact Assessment Review* 31(5): 484–490. https://doi.org/10.1016/j.eiar.2011.01.011

Ho, Elaine, Sondra Eger, and Simon C. Courtenay. 2018. "Assessing Current Monitoring Indicators and Reporting for Cumulative Effects Integration: A Case Study in Muskoka, Ontario, Canada". *Ecological Indicators* 95 (November 2017): 862–876. https://doi.org/10.1016/j.ecolind.2018.08.017.

Hodgson, Emma E., and Benjamin S. Halpern. 2019. "Investigating Cumulative Effects across Ecological Scales". *Conservation Biology* 33 (1): 22–32. https://doi.org/10.1111/cobi.13125.

Hodgson, Emma E., Benjamin S. Halpern, and Timothy E. Essington. 2019. "Moving beyond Silos in Cumulative Effects Assessment". *Frontiers in Ecology and Evolution* 7 (JUN): 1–8. https://doi.org/10.3389/fevo.2019.00211.

Huang, Ganlin, and Jonathan K. London. 2012. "Cumulative Environmental Vulnerability and Environmental Justice in California's San Joaquin Valley". *International Journal of Environmental Research and Public Health* 9(5): 1593–1608. https://doi.org/10.3390/ijerph9051593

———. 2016. "Mapping in and out of 'Messes': An Adaptive, Participatory, and Transdisciplinary Approach to Assessing Cumulative Environmental Justice Impacts". *Landscape and Urban Planning* 154: 57–67. https://doi.org/10.1016/j.landurbplan.2016.02.014.

Impact Assessment Act. (S.C. c. 28 s. 1, 2019). Government of Canada.

Impact Assessment Agency of Canada. 2019 "Regional Assessment under the Impact Assessment Act". Retrieved from https://www.canada.ca/en/impact-assessment-agency/services/policy-guidance/regional-assessment-impact-assessment-act.html.

Impact Assessment Agency of Canada. 2020. "Regional Assessment in the Ring of Fire Update". Retrieved from https://iaac-aeic.gc.ca/050/evaluations/document/136403

Impact Assessment Agency of Canada. 2020. "Reginal Assessment of Offshore Oil and Gas Exploratory Drilling East of Newfoundland and Labrador: Final Report". Retrieved from https://iaac-aeic.gc.ca/050/evaluations/document/134141

Jiliberto, Rodrigo. 2011. "Recognizing the Institutional Dimension of Strategic Environmental Assessment". *Impact Assessment and Project Appraisal* 29 (2): 133–140. https://doi.org/10.3152/146155111X12959673795921.

Johnson, Chris J., Michael P. Gillingham, Greg R. Halseth, and Margot W. Parkes. 2016. "A Revolution in Strategy, Not Evolution of Practice: Towards an Integrative Regional Cumulative Impacts Framework". In *Integration Imperative: Cumulative Environmental, Community and Health Effects of Multiple Natural Resource Developments*, edited by Michael P. Gillingham, Greg R. Halseth, Chris J. Johnson, and Margot W. Parkes, 217–243. Switzerland: Springer International Publishing.

Jones, F. Chris. 2016. "Cumulative Effects Assessment: Theoretical Underpinnings and Big Problems". *Environmental Reviews* 24 (2): 187–204. https://doi.org/10.1139/er-2015-0073.

Jones, F. Chris, Rachel Plewes, Lorna Murison, Mark J. MacDougall, Sarah Sinclair, Christie Davies, John L. Bailey, Murray Richardson, and John Gunn. 2017. "Random Forests as Cumulative Effects Models: A Case Study of Lakes and Rivers in Muskoka, Canada". *Journal of Environmental Management* 201: 407–424. https://doi.org/10.1016/j.jenvman.2017.06.011.

Kelly, R. A., A. J. Jakeman, O. Barreteau, M. E. Borsuk, S. ElSawah, S. H. Hamilton, H. J. Henriksen, et al. 2013. "Selecting among Five Common Modelling Approaches for Integrated Environmental Assessment and Management". *Environmental Modelling and Software* 47: 159–181.

Larsen, Rasmus Kløcker, Kaisa Raitio, Marita Stinnerbom, and Jenny Wik-Karlsson. 2017. "Sami-State Collaboration in the Governance of Cumulative Effects Assessment: A Critical Action Research Approach". *Environmental Impact Assessment Review* 64: 67–76. https://doi.org/10.1016/j.eiar.2017.03.003.

Loayza, Norman, and Jamele Rigolini. 2016. "The Local Impact of Mining on Poverty and Inequality: Evidence from the Commodity Boom in Peru". *World Development* 84: 219–234. https://doi.org/10.1016/j.worlddev.2016.03.005.

Macpherson, A. J., P. P. Principe, and E. R. Smith. 2010. "A Directional Distance Function Approach to Regional Environmental-Economic Assessments". *Ecological Economics* 69 (10): 1918–1925.

Mantyka-Pringle, Chrystal S., Timothy D. Jardine, Lori Bradford, Lalita Bharadwaj, Andrew P. Kythreotis, Jennifer Fresque-Baxter, Erin Kelly, et al. 2017. "Bridging Science and Traditional Knowledge to Assess Cumulative Impacts of Stressors on Ecosystem Health". *Environment International* 102: 125–137.

Martinez-Grana, A. M., J. Goy y Goy, I. Bustamante Gutierrez, and Cz. Cardena. 2014. "Characterization of Environmental Impact on Resources, Using Strategic Assessment of Environmental Impact and Management of Natural Spaces of 'Las Batuecas-Sierra de Francia' and 'Quilamas' (Salamanca, Spain)". *Environmental Earth Sciences* 71 (1): 39–51.

Mills, L. N. 2019. "The Conflict over the Proposed LNG Hub in Western Australia's Kimberley Region and the Politics of Time". *The Extractive Industries and Society* 6 (1): 67–76.

Mitchell, Ross E., and John R. Parkins. 2011. "The Challenge of Developing Social Indicators for Cumulative Effects Assessment and Land Use Planning". *Ecology and Society* 16 (2). https://doi.org/10.5751/ES-04148-160229.

Moran, C. J., D. M. Franks, and L. J. Sonter. 2013. "Using the Multiple Capitals Framework to Connect Indicators of Regional Cumulative Impacts of Mining and Pastoralism in the Murray Darling Basin, Australia". *Resources Policy* 38 (4): 733–744. https://doi.org/10.1016/j.resourpol.2013.01.002.

Naddeo, V., V. Belgiorno, T. Zarra, and D. Scannapieco. 2013. "Dynamic and Embedded Evaluation Procedure for Strategic Environmental Assessment". *Land Use Policy* 31: 605–612.

Noble, B., R. Gibson, L. White, J. Blakley, P. Croal, K. Nwanekezie, and M Doelle. 2019. "Effectiveness of Strategic Environmental Assessment in Canada under Directive-Based and Informal Practice". *Impact Assessment and Project Appraisal* 37 (3–4): 344–355.

Noble, Bram. 2008. "Strategic Approaches to Regional Cumulative Effects Assessment: A Case Study of the Great Sand Hills, Canada". *Impact Assessment and Project Appraisal* 26 (2): 78–90. https://doi.org/10.3152/146155108X316405.

Noble, Bram, Jialang Liu, and Paul Hackett. 2017. "The Contribution of Project Environmental Assessment to Assessing and Managing Cumulative Effects: Individually and Collectively Insignificant?" *Environmental Management* 59 (4): 531–545. https://doi.org/10.1007/s00267-016-0799-7.

Noble, Bram, and Kelechi Nwanekezie. 2017. "Conceptualizing Strategic Environmental Assessment: Principles, Approaches and Research Directions". *Environmental Impact Assessment Review* 62: 165–173. https://doi.org/10.1016/j.eiar.2016.03.005.

O'Faircheallaigh, Ciaran. 2009. "Effectiveness in Social Impact Assessment: Aboriginal Peoples and Resource Development in Australia". *Impact Assessment and Project Appraisal* 27 (2): 95–110. https://doi.org/10.3152/146155109X438715.

———. 2010. "Public Participation and Environmental Impact Assessment: Purposes, Implications, and Lessons for Public Policy Making". *Environmental Impact Assessment Review* 30 (1): 19–27. https://doi.org/10.1016/j.eiar.2009.05.001.

Pang, Xi, Ulla Mörtberg, and Nils Brown. 2014. "Energy Models from a Strategic Environmental Assessment Perspective in an EU Context – What Is Missing Concerning Renewables?" *Renewable and Sustainable Energy Reviews* 33: 353–362. https://doi.org/10.1016/j.rser.2014.02.005.

Partidario, Maria Rosario. 2015. "A Strategic Advocacy Role in Sea for Sustainability". *Journal of Environmental Assessment Policy and Management* 17 (1): 1–8. https://doi.org/10.1142/S1464333215500155.

Pavlickova, Katarina, and Monika Vyskupova. 2015. "A Method Proposal for Cumulative Environmental Impact Assessment Based on the Landscape Vulnerability Evaluation". *Environmental Impact Assessment Review* 50: 74–84. https://doi.org/10.1016/j.eiar.2014.08.011.

Porta, L. and N. Banks. 2011. *Becoming Arctic Ready: Policy Recommendations for Reforming Canada's Approach to Regulating Offshore Oil and Gas in the Arctic.* Washington, DC: PEW Group.

Ross, Helen. 1990. "Progress and Prospects in Aboriginal Social Impact Assessment". *Australian Aboriginal Studies* (1): 11–17 http://search.informit.com.au/documentSummary;dn=154898446084292;res=IELIND.

Sadd, James L., Manuel Pastor, Rachel Morello-Frosch, Justin Scoggins, and Bill Jesdale. 2011. "Playing It Safe: Assessing Cumulative Impact and Social Vulnerability through an Environmental Justice Screening Method in the South Coast Air Basin, California". *International Journal of Environmental Research and Public Health* 8 (5): 1441–1459. https://doi.org/10.3390/ijerph8051441.

Sadler, B., , Durisk, J. Fischer, T., Partidario, M., Verham, R. Aschemann, R. ed. 2011. *Handbook of Strategic Environmental Assessment.* Routledge: Earthscan,

Sadler, B., and D. B. Dalal-Clayton. 2012. *Strategic Environmental Assessment: A Sourcebook and Reference Guide to International Experience.* Routledge: Earthscan.

Seitz, Nicole E., Cherie J. Westbrook, and Bram F. Noble. 2011. "Bringing Science into River Systems Cumulative Effects Assessment Practice". *Environmental Impact Assessment Review* 31 (3): 172–179. https://doi.org/10.1016/j.eiar.2010.08.001.

Sinclair, A. John, Meinhard Doelle, and Peter N. Duinker. 2017. "Looking up, down, and Sideways: Reconceiving Cumulative Effects Assessment as a Mindset". *Environmental Impact Assessment Review* 62: 183–194. https://doi.org/10.1016/j.eiar.2016.04.007.

Slootweg, Roel, and Mike Jones. 2011. "Resilience Thinking Improves SEA: A Discussion Paper". *Impact Assessment and Project Appraisal* 29 (4): 263–276. https://doi.org/10.3152/146155111X12959673795886.

Squires, Allison J., and Monique G. Dubé. 2013. "Development of an Effects-Based Approach for Watershed Scale Aquatic Cumulative Effects Assessment". *Integrated Environmental Assessment and Management* 9 (3): 380–391. https://doi.org/10.1002/ieam.1352.

Therivel, R. 2010. *Strategic Environmental Assessment in Action.* 2nd ed. Routledge: Earthscan.

Therivel, Riki, and Bill Ross. 2007. "Cumulative Effects Assessment: Does Scale Matter?" *Environmental Impact Assessment Review* 27 (5): 365–385. https://doi.org/10.1016/j.eiar.2007.02.001.

Tyson, William, Trevor C. Lantz, and Natalie C. Ban. 2018. "Cumulative Effects of Environmental Change on Culturally Significant Ecosystems in the Inuvialuit Settlement Region". *Arctic* 69 (4): 391–405.

White, Lisa, and Bram F. Noble. 2013. "Strategic Environmental Assessment for Sustainability: A Review of a Decade of Academic Research". *Environmental Impact Assessment Review* 42: 60–66. https://doi.org/10.1016/j.eiar.2012.10.003.

Willsteed, Edward A., Silvana N. R. Birchenough, Andrew B. Gill, and Simon Jude. 2018. "Structuring Cumulative Effects Assessments to Support Regional and Local Marine Management and Planning Obligations". *Marine Policy* 98 (September): 23–32. https://doi.org/10.1016/j.marpol.2018.09.006.

11

GENDER ANALYSIS AND ENVIRONMENTAL IMPACT ASSESSMENT

Challenges and opportunities for transformative approaches

Priya Bala-Miller, Nicole Peletz and Kevin Hanna[1]

Introduction

2020 not only marked the 25th anniversary of the Fourth World Conference on Women and adoption of the Beijing Declaration and Platform for Action (UN 1995), but it also offered a five-year benchmark of progress toward the 2030 Agenda for Sustainable Development and Related Goals (SDGs) that include global commitments to gender equality and the empowerment of women and girls (Goal 5) (UN 2015).

These milestones focused the attention of policymakers, the private sector, and civil society on progress toward gender equality. Underlying principles of fairness and equity within these initiatives have prompted an array of measures that tackle social, economic, and historical disadvantages that prevent equality of opportunity for all, regardless of their gender (GAC 2017a).

Global campaigns for wage equity and board diversity among Fortune 500 companies, gender mainstreaming[2] in global governance initiatives, and a plethora of technical capacity-building initiatives focused exclusively on uplifting women[3], offer strong signals of the uptake in the business and policy case for gender equity, as well as a burgeoning market for gender-specific data and disclosures by all stakeholders. Organizations such as the World Bank Group and the Organisation for Economic Co-operation and Development (OECD) provide qualitative and quantitative data related to human development, gender equality, education, health, and other factors (Sauer 2013). Gender-specific statistical databases are available and accessible through portals such as the gender development research service – BRIDGE – as well as the United Nations Statistics Division (UNSD).

Environmental impact assessment (EIA) has also encountered these developments (Bice 2020, Parsons 2020; Bond and Dusik 2020). As a tool of analysis within EIA, *gender analysis* or *gender-based analysis* sheds light on the context, impacts, and potential costs and benefits of an initiative or project for gender-diverse groups. Gender analysis (GA) encompasses a variety of methods used to understand the relationships between gender identity groups, their access to

DOI: 10.4324/9780429282492-12

resources, activities, and the constraints they face relative to each other (GAC 2017b, Srinivasan and Mehta 2003, Hill and Newell 2009).

Sauer's recent analysis of the state of the art on gender and impact assessment contends that

> studies finding that women feel betrayed by EIAs and their negative consequences for women (e.g., in tax reforms, labour market or social benefit reforms) still outnumber the optimistic views. Science and evidence-based policy advice have largely ignored questions of gender justice. Contrary to other fields, such as environmental research, where scholarly research and EIA appraisals usually inform each other, gender research and EIA practices have not yet interacted deeply enough.
>
> *(421)*

Other researchers have highlighted the gap in implementation of gender mainstreaming[4] and gender analysis tools, as well as their lack of effectiveness (Scala and Paterson 2017; Lukatela 2014; Paterson 2010; Lang 2009; Squires 2005; Daly 2005; Hankivsky 2005). For instance, Dalseg et al. (2018) undertook a thematic analysis of environmental assessment cases for three resource-extraction projects in the Canadian North. It was noted that the assessment processes scarcely included gender considerations and lacked meaningful participation of Indigenous women (Dalseg et al. 2018). This observation supports the analysis undertaken by the Feminist Northern Network (FemNorthNet), who determined that there is little integration of EIA and GBA in Canadian jurisdictions (FemNorthNet 2016).

This chapter an overview of the conceptual setting, contemporary tools, and frameworks in use by policymakers and development practitioners, and both the gaps and opportunities for EIA processes to contribute to gender-transformative outcomes in policy and practice.

Conceptual terrain and rationale

As a planning and decision-support tool, EIA identifies needs, risks, positive and negative impacts, and potential mitigants for a proposed development initiative (for example, IAIA 2009).[5] In the EIA context, *gender* can be understood as a social construct that refers to the cultural characteristics, social behaviors, expressions, and identities of all individuals (SWC 2017a). Gender is distinct from *sex*, which refers to the biological attributes one is assigned at birth (male, female, both, or neither). Contemporary understandings of gender identity and gender expression have expanded beyond the male-female binary to reflect a broader range of lived experience (SWC 2017a).

For newcomers to GA, there are two helpful glossaries available online. The first is the UN Women Training Centre's Glossary (2011–2017), a tool that provides concepts and definitions with gender perspective structured according to the thematic areas of UN Women available in English, Spanish, and French. It includes gender concepts as well as international conferences, agendas, initiatives, and partnerships related to gender equality. The second is the EIGE's (2020) *Gender Equality Glossary & Thesaurus*, a specialized terminology tool focusing on gender equality. It fosters a common understanding of gender equality terms across the EU and promotes gender-fair and inclusive language. Similar compendiums based on intersectional identities and Indigenous worldviews would offer a much-needed alternative to Eurocentrism in bridging gender analysis with EIA.[6]

As mentioned above, *gender analysis* or *gender-based analysis* is not only about disaggregated data on context, impacts, costs, and benefits on gender-diverse groups; it is also a way to shed light on how these groups interact, differ in their access to power and material resources, and their ability to affect change. Section K of the Beijing Declaration and Platform for Action (UN, 1995) tackles the links between development, gender, and the environment:

> The strategic actions needed for sound environmental management require a holistic, multidisciplinary and intersectoral approach. Women's participation and leadership are essential to every aspect of that approach. The recent United Nations global conferences on development … have all acknowledged that sustainable development policies that do not involve women and men alike will not succeed in the long run… Women's experiences and contributions to an ecologically sound environment must therefore be central to the agenda for the twenty-first century.
>
> *(251–252)*

For EIA practice, GA can provide opportunities for historically marginalized and groups to have greater access and control over resources, and be in a stronger rights and benefits position stemming from development projects (Vernooy 2006). Gender analysis can also help to identify impacts and their effects on all individuals or groups. This knowledge helps mitigate the negative impacts and promote opportunities and benefits for men, women, and gender-diverse people, as well as promote best practices for future developments (Peletz and Hanna 2019; see also Bacchi and Eveline 2010; Colfer et al. 2018; Hill et al. 2017). A systematic review of academic and industry sources conducted by Walker, Reed, and Thiessen (2019) further confirms that other reasons for integrating gender and diversity into EIA practices include "fostering better public engagement and more effective mitigation strategies; contributing to health and well-being; avoiding costly conflict, providing local employment opportunities; and enhancing growth, profitability, and the social license to operate" (4) (see Table 11.1).

Approaches

Although GA has been well-established in development sector initiatives,[7] its application within assessment practice is evolving. The application of gender analysis is further complicated because

Table 11.1 Mini-case example (adapted from Baikie and Dean 2015)

Project	Muskrat Falls Hydroelectric Development on the Churchill River.
Location	Happy Valley-Goose Bay, Central Labrador, Canada.
Tool	Gender-specific consultation.
Methodology	Add-on analysis conducted post environmental assessment (EA) via community-based participatory action research with 100 local women.
Outcomes	The consultation confirmed that local women have a significant connection to the Churchill River through hunting, trapping and foraging activities. The River also served as an important meeting place for families and friends.
Potential benefits	By making these concerns visible, the proponent may be better equipped to develop relevant strategies that generate more inclusive benefits. These could include targeted employment opportunities for local women who use the river for food security and livelihoods, as well as identification and protection of cultural capital by protecting certain areas of the river for safe, continued use by the community.

the issues to be analyzed are typically not conducive to a comprehensive "one-size-fits-all" methodology for practical application in EIA (Peletz and Hanna 2019), and although evaluative frameworks with gender considerations exist, empirical cross-jurisdictional comparative analysis of their application in practice remains rare.

Among diverse approaches, gender may be mainstreamed across all dimensions of planning or assessment frameworks, or in other cases GA may be a stand-alone component of the *ex-ante* assessment process (Peletz and Hanna 2019). Gender analysis can also be used as an analytical framework for monitoring and evaluation of project outcomes throughout the project life cycle, and for monitoring and follow-up. The latter form of gender-related impact evaluation has been taken up by development organizations who can view monitoring and evaluation throughout a project or program's life cycle as an important element for impact measurement and knowledge mobilization and for accountability. These global and national commitments are a key push factor for the increasing sophistication of gender impact assessment frameworks (see Table 11.2).

Some scholars and practitioners advocate that the most effective approach is through the incorporation of a gender lens within holistic social impact assessment (SIA) processes. The argument in favor of this route is that SIAs have already established and field-tested methodologies, standard tools, and approaches geared toward unpacking social, political, and relational variables in the context of EIA. This approach has the potential to ensure that the issues and qualities that gender analysis works to reveal are integrated into assessments of projects, policies, and plans (Peletz and Hanna 2019; Lahiri-Dutt and Ahmad 2012).

Identifying and analyzing indicators such as population change and relocations, changes to the community and institutional organizations, as well as community infrastructure needs, is central to SIA (Parkins and Mitchell 2016). SIA can account for historical and cultural context, enable inclusive participation and decision-making, and consider concepts of vulnerability and resilience (Burdge 2004; Mass and Liket 2011; Parkins and Mitchell 2016; DFID 1999; Murray and Ferguson 2001; Parkins and Mitchell 2016). As such, there is a synergy between methodologies and approaches

Table 11.2 Illustrative development assistance policy frameworks for gender analysis

National	United Kingdom's Department for International Development	*Strategic Vision for Gender Equality* (2018)
	Global Affairs Canada	*Canada's Feminist International Assistance Policy* (2017a)
Inter-governmental	Organisation for Economic Cooperation and Development (OECD) Development Assistance Committee (DAC)	*Gender Equality Policy Marker* (2016)
	European Union	*Framework for Gender Equality and Women's Empowerment: Transforming the Lives of Girls and Women through EU External Relations* (O'Connell and Gavas 2015)
	African Union	Gender equality reporting framework: *The Solemn Declaration on Gender Equality in Africa (SDGEA)* (2004)
International Financial Institutions	World Bank	*World Bank Group Gender Strategy (2015)*
	Asian Development Bank	*Policy on Gender and Development* (2003)

of broader SIA's and the equity-seeking aims of gender analysis (Peletz and Hanna 2019). Overall, there is some consensus across EIA practitioners and scholars that SIA approaches offer the best space for adopting a gender lens, regardless of the stages in the EIA process at which GA is used.

Conceptual clarity and policy coherence are needed on whether individual assessments are focused on one, some, or all of the following functions: assessing needs, assessing actual impacts, or making credible forecasts about anticipated impact within the context of development projects. Each function may entail similar or distinct conceptual and analytical choices. These choices in turn can significantly shape chosen indicators, as well as how these indicators are collated, analyzed and weighted in a particular assessment process. Cumulatively, these choices present their own ethical, political and epistemological trade-offs. Although some in the field acknowledge that a split between the "hard science" of environmental/bio-physical data and often qualitative social relations data is not helpful, the adoption of mixed-method and holistic analytic approaches continues to be a practice challenge (Becker and Vanclay 2003; Vanclay and Esteves 2011).[8]

Sample frameworks and tools (2000–present)

Gender analysis tools and frameworks have been widely used in development practice for almost 50 years (Warren 2007). Focusing on the temporal period from 2000 to the present, this section provides an illustrative compendium of five distinct tools and frameworks that are likely to resonate with practitioners and policymakers alike. Each initiative is briefly described below:

Gender-based analysis plus (GBA+)

In Canada, the *GBA+ framework* is a seven-step process outlined by SWC the name for Status of Women Canada, SWC, has been changed to Women and Gender Equality Canada (2017b):

1. Identification of the issue and surrounding social, cultural, and economic contexts
2. Challenge assumptions related to gender norms
3. Gather facts through research and consultation
4. Develop options and make recommendations based on research and consultations
5. Monitor and evaluate the initiative
6. Communication of findings and recommendations
7. Documentation of the analysis

Steps 6 and 7 can be taken throughout the GBA+ process, alongside steps 1–5. SWC provides a set of guiding key questions to inform the analysis for each process step (see SWC 2017b for guiding questions and information). An example of how the GBA+ framework might be used is provided in module three of the SWC GBA+ online training course (see SWC 2017a). This illustration uses a hypothetical scenario based on the forestry sector in Canada (Table 11.3).

The European Institute for Gender Equality's (EIGE) *Gender Impact Assessment: Gender Mainstreaming Toolkit*

EIGE's *Gender Impact Assessment: Gender Mainstreaming Toolkit* (2016) outlines a five-step process on how to apply and implement the framework. Similar to the SWC GBA+ framework, the GIA framework includes guiding questions for each step of the process. The guide further outlines GIA policies in several EU nation-states, including Austria, Belgium, Denmark, Germany, Finland, Sweden, and Spain. There are five parts to the framework:

Table 11.3 Case scenario (adapted from SWC 2017a)

Scenario	Federal government analysis to support the environmental and economic sustainability of forest-based communities through strategic funding programs
Location	Canada
Tool	GBA+
Methodology	Qualitative and quantitative data gathering and analysis, stakeholder consultations.

Applying GBA+ to the scenario

Step 1: **Identify issues and context**	The practitioner identifies that Canada's forest industry has experienced a significant change in recent decades. Consumers are seeking environmentally sustainable practices and products. Communities dependent on the forestry industry are facing challenges such as declining employment, decreasing timber supplies, trade uncertainties, and greater international competition. New developments in the industry, such as the use of waste for value-added products, greater eco-tourism opportunities, and the development of bio-energy systems, could potentially lead to a more sustainable industry while diversifying and supporting the economies of forest-based communities.
Step 2: **Challenge assumptions**	The practitioner challenges assumptions about the forestry sector as well as forest-based communities. Examples of assumptions include the ideas that all forest-dependent communities can be treated the same way, that everyone based in those communities wants to be involved in forestry, and that those involved in the sector all have the same needs and wants.
Step 3: **Research and consult**	The practitioner focuses on the collection of demographically disaggregated data and gathering facts. The perspectives of stakeholders are also necessary to develop a thorough understanding of the issues at play. Because forestry workers who occupy supervisory and managerial roles are often consulted more regularly than under-represented groups, an effort to engage with groups such as women and Indigenous workers is essential to an inclusive participation process. Consulting individuals with knowledge of local forests and ecosystems is also essential; this could include community elders, forest-sector organizations, and/or technical experts. Examples of facts and disaggregated data that could emerge: • 12% of forestry workers are women, while only 6% of upper-level positions are filled by women; • Indigenous workers make up 3.5% of full-time forestry workers compared to 1.8% of full-time workers in all other industries; • Women account for 14% of the Indigenous labor force and have the lowest average income of all forestry workers.
Step 4: **Develop options and recommendations**	Practitioners refine issues based on facts and data. For example, it is now known that communities are experiencing uneven changes in the industry; that Indigenous women face disproportionate challenges compared to other forestry workers; and that in many Indigenous communities, forests hold a significant cultural and traditional value that cannot be overlooked. Based on the findings, three recommendations are proposed: **Option 1: The initiative to promote innovation and economic diversification will be integrated uniformly across all communities in the sector.**

(Continued)

Table 11.3 (Continued)

	Although the GBA+ demonstrates that local settings and contexts are variable among communities and that a uniform approach may lead to unequal impacts and benefits, it is often the case that uniform approaches are chosen due to feasibility and scope. When this is the case, it is imperative to highlight limitations and outline how data collection and analysis will be done.
	Option 2: The initiative will focus on integrating and engaging under-represented and marginalized groups in the forestry sector, including women and Indigenous people.
	The implementation of this aspect would include the promotion of family-friendly practices and policies, as well as unique training opportunities targeted at specific groups.
	Option 3: The initiative will include partnerships with communities and local governments to promote bottom-up decision-making and develop the best options to diversify the local economy.
	This could include weaving traditional knowledge into decision-making alongside current practices and diversification efforts in order to best determine how to support innovation for all stakeholders.
Step 5: Monitor and evaluate	Once an initiative is developed and chosen, the final step of the process involves monitoring and evaluating the implemented recommendations in order to ensure that the objectives are being met and that any ongoing issues are identified and addressed.
	Developing communication strategies is important to further engage the local community and to promote the initiative. Examples of products include those aimed at community groups by identifying traditional cultural practices as imperative to the development of the industry; campaigns aimed at younger populations to promote employment through a representation of an inclusive and diverse workforce; and communication through local newspapers, radio, and community events.

1. Definition of the purpose of the policy, program, or project
2. Checking gender relevance by considering target groups, and both direct and indirect impacts on diverse groups
3. Gender-sensitive analysis through information and data collection
4. Weighing the gender impact by accounting for criteria, such as participation and access to and control of resources
5. Document the findings and provide conclusions along with formulated proposals to promote gender equality

For EIGE, GIA is centered on the question – "Does a law, policy or program reduce, maintain or increase the gender inequalities between women and men?" (EIGE 2016, para. 1). It is further referred to as a "learning process", as GIA is continuing to be developed and improved through the identification of data gaps, increased capacity through gender training, and follow-up based on previous experiences (EIGE 2016).

Oxfam guidance for hydropower and extractives projects

Oxfam's *Balancing the Scales: Using Gender Impact Assessment in Hydropower Development* outlines a framework for specific use in hydropower development projects (Simon 2013). The framework includes six steps drawing on various tools developed in earlier gender analysis frameworks. The steps are as follows:

1. Data collection
2. Understand the context
3. Identify risks, issues, and impacts brought on by the project
4. Understand women's needs
5. Develop recommendations
6. Review and audit regularly

Hill et al. (2017) use the Oxfam GIA approach to determine the gender impacts of two hydro-power projects in Laos and Vietnam. Their study determined that women were not adequately consulted prior to development, and as a result unequal gender impacts emerged. For example, both men and women experience a loss of livelihood from fishing practices as a result of the hydro-developments; however, through the GIA it was determined that the new economic opportunities emerging from the development, such as work at the dam and tourism positions, strongly favor men. The study concludes by emphasizing the use of a gender analysis prior to hydropower developments.

Oxfam also published *A Guide to Gender Impact Assessment for the Extractive Industries* (Hill, Madden, and Collins 2017) with a specific focus on mining and oil and gas projects. The guide offers a four-step GIA framework specific to mining and oil and gas developments. The framework has five steps:

1. Identify and collect baseline data and information regarding the impacted community: Baseline data should include information related to the gender division of labor, as well as access and control of resources. The guide includes two detailed matrices to assist with the collection of data. However, the guide emphasizes that the tools should be used as templates and adapted based on the relevant context. For example, the matrix for access and control of resources identifies land, water, infrastructure, and labor among the resources; these could be removed or adjusted according to the particular context at hand.
2. Analyze the collected information through community consultations and public partici-pation: This is achieved through the analysis of four key issues, including structural and institutional causes of inequality, barriers of participation in decision-making, practical and strategic gender needs, and identifying and mitigating negative project impacts.
3. Develop a gender action plan: The plan should enable gender-responsive engagement and be publicly available. Furthermore, the plan should be developed in close consultation with the impacted community as well as discussed openly to ensure that all impacted members understand the developer's commitments.
4. Review the plan and follow-up: Monitoring and evaluation should be undertaken by someone outside of the development company to ensure transparency and objectivity, as well as engaging with community members regarding their perceptions of effectiveness and ideas for improvements.

World Bank Group's rapid assessment toolkit for artisanal and small-scale mining (ASM)

The World Bank Group published a *Rapid Assessment Toolkit* for the ASM sector to help identify gender issues and promote gender equality (Eftimie et al. 2012). The toolkit includes six modules outlining an introduction to gender in ASM, a gender and ASM framework, specific gender and ASM tools, as well as three detailed cases of the toolkit in practice. The framework breaks down the ASM value chain into five primary components (prospecting and exploration, mining, processing, goods and services, marketing of minerals) and seeks to identify and address gender issues at each component based on issues concerning roles and responsibilities, access and control, as well as impact and benefits. Detailed sample questions are identified for each issue at all value-chain components (see Table 2.1 in Eftimie et al. 2012, 23).

The gender and ASM framework has three primary sections: design and planning; data collection; interpretation, validation, and write-up. Each section consists of three to six sub-steps, including an estimated number of days needed to complete the step as well as identified ASM tools to be used at each step. The analytical tools described in the toolkit include collecting background information, key informant interviews, ASM site visits, participatory focus groups, and surveys (Eftimie et al. 2012). A Tanzanian case study illustrates the use of the Gender and ASM Framework. The case identified disproportionate impacts on men and women. For instance, women are often responsible for mineral and gemstone processing which is associated with significant health risks. The recommendations included the need to improve working conditions for women by getting the mining industry to review the adequacy of and adherence to safety regulations and best practices (Hinton and Wagner 2010; Eftimie et al. 2012; Peletz and Hanna 2019).

UN Women's *Inclusive Systemic Evaluation for Gender Equality, Environments, and Marginalized Voices* (ISE4GEMs)

The UN Women Independent Evaluation Office, along with Australian and American researchers, have written and piloted an evaluation guide entitled: *Inclusive Systemic Evaluation for Gender Equality, Environments and Marginalized Voices (ISE4GEMs): A New Approach for the SDG Era* (2018). Referred to as the "ISE4GEMs", this guide brings together transdisciplinary evaluation methods, re-thinks systemic evaluation methodology, and introduces the gender equality, environments, and marginalized voices (GEMs) framework. The guide is written in two parts. Part A contains key concepts from systems thinking, including boundary analysis, emergence, and the difference between systemic and systematic thinking (Stephens, Lewis, and Reddy 2018). The GEMs framework for complex intersectional analysis is explained and each GEMs dimension is described and key elements for practice. Part B provides practical steps to walk through the planning, conduct and analysis phases of an evaluation. The process is participatory and contributes to capacity development, learning and empowerment of participants and stakeholders. An accompanying set of tools and extensive resources are provided in the guide's annexe section.

Looking ahead: gender-transformative EIAs

By seeing EIA as a distinct field of policy and practice, the degree to which gender is considered and addressed can be mapped on a continuum from exploitive to accommodating and finally to transformative approaches (adapted from UNFPA 2020):

Exploitive

- Gender-Unequal: The EIA process and outcomes perpetuate gender inequalities
- Gender-Blind: EIA ignores gender norms, discrimination, and inequalities

Accommodating

- Gender-Aware: EIA acknowledges but does not address gender inequalities
- Gender-Responsive: EIA acknowledges and considers gender-differentiated needs, impacts, and opportunities

Transformative

- Gender-Transformative: EIA addresses the causes of gender-based inequalities and works to transform harmful gender roles, norms, and power relations.

Over time there has been a noticeable (but not widely diffused) evolution in EIA research and practice along a continuum from accommodating to more recently toward gender-transformative approaches. While it would be helpful to systematically assess these shifts over time, this is difficult due to the lack of documented examples of the implementation of early frameworks in specific sectors. These data gaps limit the ability to benchmark frameworks along the continuum, considering factors such as usability, accessibility, and the role of stakeholder participation and involvement.

Looking ahead, aligning gender analysis within EIA toward transformative approaches is an emerging frontier for policy and practice. There are a few considerations that could facilitate this alignment. Drawing on similarities from frameworks mentioned above, key themes emerge: understanding and adapting to context-specificity; considering intersectionality and demographically diverse data; facilitating meaningful, inclusive and diverse participation; and attention to social power relations (including consideration of gender roles and gender relations). Each of these themes are discussed below, along with examples of how these ideas can be supported in EIA.

Context-specificity

Understanding context is essential for establishing an appropriate and practical approach when adopting a gender lens in EIA, as cultures, geographies, histories, languages, politics, and values can all influence the indicators/determinants required for the analysis. The need for context-specificity is highlighted by the culturally relevant gender-based analysis (GBA) framework developed by the Native Women's Association of Canada (NWAC). The framework's application protocol (2010) acknowledges that although there have been improvements in the status of women since the 1970s,

> Canada's continued noncompliance with international protocols calling for the use of GBA tools allows gender disparities in the health, social, and economic sectors to persist, and that these disparities are magnified within the Indigenous population.[9] Indigenous women's identity – their gender and their culture – are especially marginalized by the intertwined effects [of] sexism and racism.
>
> *(NWAC 2010, 1)*

The protocol states, mainstream GBA frameworks are not adequately equipped to account for the cultural identity of Indigenous women. The tool also considers distinctive constitutional safeguards for the rights of Indigenous women, and culturally specific guiding principles that seek to reverse colonial legacies by revitalizing their traditional roles and status.

The Pauktuutit Inuit Women of Canada's (2016) culturally specific GBA model encourages consideration of resource-extraction issues and how they relate to (1) The Inuit Way (Elders, culture, language, family, community, and spirituality); (2) Traditional Influences on the Inuit Way (land, weather, animals, and country food); (3) Contemporary Influences on the Inuit Way (institutions, policies, laws, climate change, globalized, and capitalist economies); and (4) Assessing Gender Impacts in an Inuit Cultural Context (pulling it all together). For instance, the use of country (traditional) food consumption as an indicator in the Inuit GBA framework would demonstrate large discrepancies among Inuit men, women, girls, and boys, where men generally consume greater amounts of country food compared to women, until women reach the age of 60 at which point they begin to consume more than men, while Inuit girls consume significantly more than boys (Pauktuutit Inuit Women of Canada 2008). Rasmussen and Guillou's (2012) allusion to a feeling of disconnect between Inuit community members and outside researchers or practitioners due to differing values and worldviews has also been expressed by other researchers and practitioners working outside of their own cultural context. The use of Inuit specific indicators, developed by communities, helps to ensure that the needs of the population are relevant and respectfully addressed.

Intersectionality and demographically diverse data

The concept of intersectionality is attributed to Kimberlé Crenshaw's 1989 paper titled "Demarginalizing the Intersection of Race and Sex: A Black Feminist Critique of Antidiscrimination Doctrine, Feminist Theory and Antiracist Politics" (cited in Peletz and Hanna 2019). In the context of gender-based analysis as a tool for policymaking, the concept was also suggested in the wording of the 1995 Beijing Declaration and Platform for Action that noted the need to

> (i)ntensify efforts to ensure equal enjoyment of all human rights and fundamental freedoms for all women and girls who face multiple barriers to their empowerment and advancement because of such factors as their race, age, language, ethnicity, culture, religion, or disability, or because they are Indigenous people.
>
> *(UN 1995, para. 32)*

Intersectionality is the consideration of various characteristics and identity factors, including gender, that are interwoven and acting together rather than unconnected and exist in isolation from each other (Peletz and Hanna 2019). The concept emphasizes that people are often disadvantaged by multiple factors. For planning, assessment, and policy, intersectionality offers a frame for viewing a person, a group, or a socioeconomic issue as being shaped by an array of biases or discriminations (Peletz and Hanna 2019). From an intersectionality perspective, impact assessment requires "a fundamental shift away from emphasizing only broad-scale trends among men and women as groups, and assuming that what members of these groups hold in common is most analytically salient" (O'Shaughnessy and Krogman 2011, 136, in Walker, Reid, and Thiessen 2019).

The importance of theory is important if the field is to avoid what Lahiri-Dutt and Ahmad (2012) describe as a falling into a "theory vacuum" where women are incorporated in the existing analyses as an "add-on" measure to satisfy trendy approaches, rather than truly questioning why a gender lens should be adopted in EIA. Explanation is a necessary precondition for social transformation.

Considerations of intersectionality also extend to the data sets that EIA practitioners create and use in order to craft their assessments. In this regard, leading practices suggest that, where possible, data should be disaggregated based on gender and other relevant demographic attributes. While it is difficult to determine indicators in the absence of available and reliable baseline data, these data gaps should be addressed and made explicit in EIA practice (EIGE 2016).

Efforts by Canada's federal government are instructive on the role that policymakers can play in enabling this form of analysis. Canada's federal Impact Assessment Act has provisions for the consideration of sex, gender, and diversity in EIA processes. These requirements are made in two sections. First, the preamble of the Act indicates that the

> Government of Canada is committed to assessing how groups of women, men and gender diverse people may experience policies, programs and projects and to taking actions that contribute to an inclusive and democratic society and allow all Canadians to participate fully in all spheres of their lives.

Second, and the key active section related to gender and diversity, the Impact Assessment Act specifies that an assessment of a designated project must consider "the intersection of sex and gender with other identity factors" (S. 22 [1.s] in Walker et al. 2019).

Walker et al. (2019) provide examples and case studies of the use of intersectional data in impact assessment. These include Archibald and Crnkovich's (1999) analysis of gendered exclusions from an environmental assessment of a proposed nickel mine at Voisey's Bay in Labrador, Canada, and Kojola's (2018) examination of the ways in which race, class, gender, and Indigeneity intersected in formal state institutions to influence which types of knowledge were considered and how they informed the engagement processes associated with environmental impact assessments of proposed copper-nickel mines in Minnesota.

Facilitating meaningful, inclusive, and diverse participation

Along with the scholarship noted above, Reed and Davidson's work (2011) found that gender, class, and race play significant roles in how (and which) individuals are selected to participate, what they bring to the table, and how they behave and share knowledge on forest advisory committees in Canada. These findings underscore the importance of EIA mechanisms to foster inclusive and diverse participation in gender analysis and related decision-making, particularly for marginalized groups. In this regard, attention to potential participation barriers should be recognized through certain key leading practices identified in the literature that ensure: participation opportunities are open and accessible; all groups are comfortable speaking in each other's presence; materials are translated appropriately and accessible to everyone; gender-diverse facilitators are deployed to diminish bias; and feedback sessions are facilitated for all groups of people (Hill, Madden, and Collins 2017).

Inclusive participation has been demonstrated to improve efficiency, effectiveness, equity, and sustainability of resource developments and resource management projects (Johnson et al., 2004). An inclusive representation of all stakeholders ensures that the analysis reflects all identified needs, concerns, and priorities, notably those that could otherwise be overlooked. It is important to stress that the adoption of a gender lens should not be conflated with inclusive participation as the former does not necessarily guarantee the latter. In the Canadian context,

GBA has been critiqued for exacerbating gender differences through the exclusion of other oppressed groups and neglecting intersectionality (Hankivsky 2005), and for favoring expert opinions over those of women's groups and others who are not often heard (Hankivsky 2009; Rankin and Wilcox 2004).

The requirement that gender-based analysis requires a "gender expert can inadvertently favor particular (colonial) forms of knowledge and knowledge production" (Paterson 2010). This outcome results in part because the benchmarks of who counts as a "gender expert" within development practice tend to prioritize Western education credentials and technical training in related domains. If deployed with inclusion in mind, however, a gender analysis can be used in the assessment of projects to unearth a range of needs, determine opportunities for different groups, and then outline provisions needed to support a diverse set of stakeholders. When paired with effective participatory approaches, applying a gender lens can mitigate the limitations of an expert-driven approach (Bacchi 2003; Kabeer 1994). Such equality considerations should be explicit in program, project, and policy objectives to ensure the effectiveness and transformative potential of the analysis (Verloo 2002; Warren 2007).

For example, a mining company in Pakistan identified capacity constraints in the health sector and decided to undertake an initiative to train female health workers to later work on the mine and offer further health and hygiene services to fellow workers. The initiative created capacity and promoted empowerment (Yasmeen 2015; Hill, Madden, and Collins 2017). After the mine was decommissioned, the women were able to use their training and gain employment elsewhere in the economy. This contributes to advancing gender-responsive outcomes and supports broader sustainability objectives by supporting skill and capacity development that outlasts the project's lifetime (Hill, Madden, and Collins 2017).

Attention to social power relations and norms

An understanding of the difference between access to resources and control of resources is central to using EIA as a tool for changing unjust social hierarchies, including on the basis of gender. The United Nations Development Programme (2001) explains that access to resources refers to the availability or opportunity for use, while control of the resource implies decision-making authority. They further distinguish between three categories of resources: *economic/productive resources* including land, credit, income, employment; *political resources* including education, representation in policy, leadership; and *time* as a general resource (UNDP 2001). Furthermore, identification and understanding of the benefits of resource control and use, including satisfaction of both practical and strategic needs, is essential in any analysis (SIDA 2015).

A gender analysis that has transformative potential as an ascribed value should, for example, not only consider who is using agricultural land, but also consider who is deciding how it is being used and managed. For example, in Eastern India, it has been documented that although women make up a large portion of farm laborers, they only own approximately 10% of the land (Lahiri-Dutt and Ahmad 2012). As mining encroaches on agricultural land, men gain benefits from ownership of land and new jobs in mining, while women lose access to the land and resources that traditionally sustain livelihoods (Lahiri-Dutt and Ahmad 2012).

Some point out that gender analysis assumes that problems and inequalities can be readily mitigated through more information, rather than by addressing the underlying structures, processes, and causes under which gender and diversity inequalities are formed (Paterson 2010, Bacchi and Eveline 2003). It is suggested that through reflexive framing – placing oneself (in this case, the gender expert) inside the analysis, and attempting to understand how problems are created and reproduced – the use of GBA could better influence social relations (Paterson 2010).

Consideration of gender norms and values such as division of labor, cultural norms, and organization of private life should be recognized and evaluated to develop information and knowledge that may help promote the equal social value of women, men, and gender-diverse people.

Regarding work and labor, understanding the distinction between productive work, reproductive work, and community work is also imperative. Productive work refers to the production of goods or services, often generating an income, whereas reproductive refers to work in the household such as the care and maintenance of the home and children, and community work refers to work related to political, religious, or social work in the community (SIDA 2015). For example, as reproductive work can often be undervalued or overlooked, an understanding of value associated with different types of labor can ensure that benefits are relevant and meaningful to diverse groups of workers.

Dool-Soon and Kang's (2016) work provides illustrative evidence on the importance of considering how prevailing norms and institutions perpetuate inequality. They observe that despite the increased development and implementation of gender impact assessment frameworks in Korea, the results from gender analysis are not being applied in practice due to underlying patriarchal values (Dool-Soon and Kang 2016). Likewise, in Taiwan, the implementation of gender analysis through gender impact assessment (GIA) has encountered problems such as resistance and passivity from bureaucrats, lack of baseline gender-related data, lack of specific gender training, and inadequate auditing and monitoring programs (Peng 2015).

Conclusion

In EIA, the transformative potential of gender analysis can best be achieved through an understanding of the social and environmental context in which development occurs; collecting disaggregated data through diverse sources such as databases as well as consultation processes; undertaking inclusive and diverse participation as well as a collaboration with diverse stakeholders; and identifying and analyzing distribution, access, and control of resources, as well as gender norms and values, through consideration of both gender roles and gender relations.

Advancing best practices requires ongoing work. First, if it is to be implemented and used, gender analysis (or gender assessment) requires practical and specific methods and tools that can be applied by practitioners at project-level assessment. In part, these can be adapted from social/economic impact assessment. Second, the role of GA in contexts where social structures are not necessarily amenable to changing gender roles or some notions of inclusion will require careful approaches that will involve more time and resources. Research that can outline best practices and provide examples will be of value to development organizations and businesses. Third, the effectiveness of GA approaches should be assessed to understand the actual impacts and outcomes and to provide data and knowledge that can be used to advance best practices and refine approaches and tools. Finally, the language of gender analysis can be difficult to navigate. For EIA, guidance must provide clear language and methods, and be oriented toward applied outcomes and practical use.

Hanna and Peletz (2019) identify seven key lessons for EIA practice:

1. The intersection of gender with other identity factors will be present to some extent and in some form for all projects.
2. Thinking about gender and other identity factors should occur early in the project process, for example, at the point of the project concept. This will help anticipate potential impacts.

3. Gender analysis begins with data collection in order to create an understanding (information and knowledge) of the social and economic setting, and to begin to be able to understand the impacts of the project on different groups of people.

4. Gender analysis will be more effective and beneficial if it is integrated into all stages of a project's life cycle (design, inception, operation, closure, and continuous, long-term monitoring and evaluation).

5. Gender analysis can be implemented through the lens of social/economic impact assessment, where established tools and approaches can be adapted to include gendered perspectives and issues. This can help ensure that the analysis is not an add-on, but instead becomes part of the routine consideration of project effects – both positive and adverse.

6. Using gender analysis can provide an opportunity to advance project benefits and develop innovative approaches to employment, community engagement, and timely project implementation and operation. There is a tendency to focus on the negative outcomes of projects. These should not be ignored; however, the positive benefits of gender assessment should also be recognized and indeed emphasized within a frame of opportunity and innovation.

7. A gender analysis can contribute to risk management by providing additional information and knowledge about project impacts and opportunities for mitigation and benefit enhancement.

The field of gender analysis is conceptually well-developed. There are multiple examples from a range of jurisdictions and development settings, and there are general templates that can provide basic guidance for practitioners. In EIA, while the application of gender analysis is progressing, there is a need for more case examples, more rigorous theorizing, better integration into existing impact assessment tools, and further guidance and capacity building for proponents, regulators, and communities in order for GA to be applied in practice (Bice 2020).

There is a need for further research that highlights the development of real-world applied methods and comprehensive approaches that can be used by proponents and regulators. This will require drawing on existing EIA methodologies and adapting other gender analysis frameworks and methods to the specific needs of EIA practice.

Notes

1 This chapter substantially draws on an earlier report prepared by Nicole Peletz and Kevin Hanna. 2019. *Gender Analysis and Impact Assessment: Canadian and International Experiences.* Canadian International Resources and Development Institute (CIRDI), Vancouver. The paper is adapted here with permission.

2 Gender mainstreaming refers to the integration of a gender perspective in policies and programmes. The Beijing Declaration and Platform for Action (1995) was a critical juncture in mandating the analysis of particular development interventions based on gender-disaggregated effects.

3 Illustrative examples include the UN's #stoptherobbery campaign (UN Women 2017), the Equal Pay International Coalition (2017), the UN Women and the UN Global Compact's Women's Empowerment Principles (WEP n.d) and CitiGroup's (2019) "The Moment" campaign. See also: on wage equity (Christiansen et al. 2016) and on board diversity (Ellis and Eastman 2018).

4 Gender mainstreaming refers to the integration of a gender perspective in policies and programmes. The Beijing Declaration and Platform for Action (UN 1995) was a critical juncture in mandating the analysis of particular development interventions based on gender-disaggregated effects.

5 Other chapters in this volume outline the various types of assessments that comprise the field, and for the sake of brevity are not revisited here.

6 On this note, the 2019 special issue published by leading Brazilian international relations journal *Contexto Internacional* is informative. The peer-reviewed articles outline the complexities and contradictions involved in gender analysis, and "underscore(s) the need to remain vigilant about any simplistic

inclusion of the category of 'women' or 'gender' into the already settled grammars" (Natália Maria Félix de Souza 2019, 249).

7 Note there is a considerable body of critical feminist scholarship that takes issue with what is described as the "women in development" or "ecofeminist" approach to gender mainstreaming. A key fault line across these scholarly traditions is that the latter tend to homogenize women as an analytical category, thereby perpetuating existing class-based, economic, political, and power hierarchies. See, for example, (Lahiri-Dutt and Ahmad 2012; Manning, 2014; Koutouki et al. 2018; Nightingale et al. 2017; Parmenter 2011, in Walker, Reed, and Thiessen 2019).

8 See, for example, Kobayashi, Peters, and Khan's (2015) discussion on the hybridization of life cycle assessment and quantitative risk assessment methodologies, as well as other works such as, Montini (2013), and Riha, Levitan, and Hutson (1996).

9 The term *Aboriginal* was used in the original report. Marie-Céline Charron (2019), an expert in Indigenous Affairs and a member of the Naskapi Nation of Kawawachikamach, notes that, in Canada, the term *Indigenous* is increasingly replacing the term *Aboriginal*. This is because the former is recognized internationally, for instance, by the United Nations' Declaration on the Rights of Indigenous Peoples. The term *Indigenous* may be used to refer to Métis, Inuit, and First Nations communities. Indigenous Nations, communities, organizations, and people are distinct, requiring sensitivity to their respective name, spelling, and identity.

References

African Union. 2004. Gender Equality Reporting Framework: The Solemn Declaration on Gender Equality in Africa (SDGEA). Accessed September 23, 2020. https://au.int/en/documents/20200708/solemn-declaration-gender-equality-africa.

Archibald, L., and M. Crnkovich. 1999. *If Gender Mattered: A Case Study of Inuit Women, Land Claims and the Voisey's Bay Nickel Project*. Ottawa: Status of Women Canada. http://epe.lac-bac.gc.ca/100/200/301/swc-cfc/if_gender_mattered-e/archibald-e.pdf.

Asian Development Bank. 2003. Policy on Gender and Development. Accessed September 23, 2020. https://www.adb.org/documents/policy-gender-and-development.

Bacchi, Carol Lee. 2010. "Gender/Ing Impact Assessment: Can It Be Made to Work?" In: *Mainstreaming Politics: Gendering Practices and Feminist Theory*, Edited by Carol Lee Bacchi and Joan Eveline, 17–38. Adelaide: University of Adelaide Press.

Bacchi, Carol Lee, and Joan Eveline. 2003. Mainstreaming and Neoliberalism: A Contested Relationship. *Policy and Society* 22(2): 98–118.

Baikie, G., and L. Dean. 2015. *Claiming Our Place: Local Women Matter in Natural Resource Development*. Ottawa: Canadian Research Institute for the Advancement of Women. Accessed February 2018. http://fnn.criaw-icref.ca/images/userfiles/files/ClaimingOurPlace.pdf.

Becker, H.A., and Frank Vanclay. 2003. *The International Handbook of Social Impact Assessment: Conceptual and Methodological Advances*. Cheltenham: Edward Elgar.

Bernauer, Warren. 2011. *Mining and the Social Economy in Baker Lake, Nunavut*. Saskatoon: Universit of Saskatchewan. Prepared for the Northern Ontario, Manitoba, and Saskatchewan Regional Node of the Social Economy Suite.

Bice, Sara. 2020. The Future of Impact Assessment: Problems, Solutions and Recommendations. *Impact Assessment and Project Appraisal*, 38(2): 104–108.

Bond, A., and Jiří Dusík. 2020. Impact Assessment for the Twenty-First Century – Rising to The Challenge. *Impact Assessment and Project Appraisal*, 38(2): 94–99.

Burdge, R.J. 2004. *The Concepts, Process, and Methods of Social Impact Assessment*. Middleton, Wis.: Social Ecology Press.

Charron, Marie-Céline. 2019. "No Perfect Answer: Is It First Nations, Aboriginal or Indigenous?" Perspectives: Indigenous Affairs. March 6, 2019. Accessed September 23, 2020. https://www.national.ca/en/perspectives/detail/no-perfect-answer-first-nations-aboriginal-indigenous/.

Christiansen, Lone, Huidan Lin, Joana Pereira, Petia B. Topalova, and Rima Turk-Ariss. 2016. Gender Diversity in Senior Positions and Firm Performance: Evidence from Europe. IMF Working Paper No. 16/50. https://ssrn.com/abstract=2759759.

CitiGroup. 2019. "Citi Furthers Commitment to Global Pay Equity with New Campaign 'The Moment'". Accessed September 23, 2020. https://www.citigroup.com/citi/news/2019/191010a.htm.

Colfer, C.J.P., B.S. Basnett, and M. Ihalainen. 2018. Making Sense of "Intersectionality": A Manual for Lovers of People and Forests. Occasional Paper 184. Bogor, Indonesia: CIFOR.

Crenshaw, Kimberlé. 1989. Demarginalizing the Intersection of Race and Sex: A Black Feminist Critique of Antidiscrimination Doctrine, Feminist Theory and Antiracist Politics. *U. Chi. Legal F.*, 139.

Daly, Mary E. 2005. Gender Mainstreaming in Theory and Practice. *Social Politics: International Studies in Gender, State and Society* 12(3): 433–450.

de Souza, Natália Maria Félix. 2019. Introduction: Gender in the Global South: A Complex and Contradictory Agenda. *Contexto Internacional* 41(2): 249–253.

Department for International Development. (DFID – UK). 1999. Sustainable Livelihoods Guidance Sheet: Introduction. PDF. Accessed March 2018. https://www.ennonline.net/attachments/871/dfid-sustainable-livelihoods-guidance-sheet-section1.pdf.

DFID. 2017. *Strategic Vision for Gender Equality (2018)*. Accessed September 23, 2020. https://assets.publishing.service.gov.uk/government/uploads/system/uploads/attachment_data/file/708116/Strategic-vision-gender-equality1.pdf.

Dool-Soon, Kim, and Minah Kang. 2016. Rapid Growth-What's Next for Gender Mainstreaming? Analyzing the Gender Impact Assessment System in Korea. *Journal of Women, Politics and Policy* 37(2): 168.

Eftimie, A., K. Heller, J. Strongman, J. Hinton, K. Lahiri-Dutt, and N. Mutemeri. 2012. *Gender Dimensions of Artisanal and Small-Scale Mining: A Rapid Assessment Toolkit*. Washington, DC: World Bank.

EIGE (European Institute for Gender Equality). 2016. *Gender Impact Assessment: Gender Mainstreaming Toolkit*. Luxembourg: Publications Office of the European Union. Accessed January 2018. https://eige.europa.eu/rdc/eige-publications/gender-impact-assessment-gender-mainstreaming-toolkit.

EIGE. 2020. Glossary and Thesaurus. Accessed September 23, 2020. https://eige.europa.eu/thesaurus/overview.

Ellis, Morgan, and Meggin Thwing Eastman. 2018. Women on Boards: Progress Report 2018. MSCI Inc. ESG Research Products and Services. Accessed September 23, 2020. https://www.msci.com/documents/10199/36ef83ab-ed68-c1c1-58fe-86a3eab673b8.

Environment Assessment Process Expert Review Panel. 2017. *Building Common Ground: A New Vision for Impact Assessment in Canada* (April 5, 2017). PDF. Accessed February 2018. https://www.canada.ca/content/dam/themes/environment/conservation/environmental-reviews/building-common-ground/building-common-ground.pdf.

FemNorthNet. 2016. *Gender Based Analysis Meets Environmental Assessment*. PDF. Accessed December 2017. http://fnn.criaw-icref.ca/images/userfiles/files/GBAMeetsEnviroAssessPP.pdf.

GAC (Global Affairs Canada). 2017a. Policy on Gender Equality. Accessed June 2017. http://international.gc.ca/world-monde/funding-financement/policy-politique.aspx?lang=eng#a3.

GAC (Global Affairs Canada). 2017b. Gender Analysis. Accessed June 2017. http://international.gc.ca/world-monde/funding-financement/gender_analysis-analyse_comparative.aspx?lang=eng.

GAC. 2019. *Canada's Feminist International Assistance Policy*. Modified January 14, 2020. https://www.international.gc.ca/world-monde/issues_development-enjeux_developpement/priorities-priorites/policy-politique.aspx?lang=eng

Hankivsky, Olena. 2005. Gender vs. Diversity Mainstreaming: A Preliminary Examination of the Role and Transformative Potential of Feminist Theory. *Canadian Journal of Political Science* 38(4) December: 977–1001.

Hankivsky, Olena. 2009. "Gender Mainstreaming in Neoliberal Times: The Potential of 'Deep Evaluation'". In: *Public Policy for Women: The State, Income Security and Labor Market Issues*, Edited by Cohen and Pulkingham. Toronto: University of Toronto Press.

Hill, C and K. Newell. 2009. *Women, communities and mining: The gender impacts of mining and the role of gender impact assessment*. Melbourne: Oxfam Australia.

Hill, Christina, Chris Madden, and Nina Collins. 2017. *A Guide to Gender Impact Assessment for the Extractive Industries*. Melbourne: Oxfam.

Hill, Christina, Phan Thi Ngoc Thuy, Jacqueline Storey, and Silavanh Vongphosy. 2017. Lessons Learnt from Gender Impact Assessments of Hydropower Projects in Laos and Vietnam. *Gender and Development* 25(3): 455.

Hinton, J., and S. Wagner. 2010. *Gender and Artisanal & Small Scale Mining (ASM): A Case Study in Merirani, Tanzania*. Tanzania Draft Pilot Study Report. World Bank.

IAIA (International Association for Impact Assessment). 2009. What Is Impact Assessment? PDF. Accessed January 2018. https://www.iaia.org/uploads/pdf/What_is_IA_web.pdf.

Johnson, N. et al. 2004. The Practice of Participatory Research and Gender Analysis in Natural Resource Management. *Natural Resources Forum* 28(3): 189–200.

Kabeer, Naila. 1994. *Reversed Realities: Gender Hierarchies in Development Thought*. London; New York: Verso.

Dalseg, K.S., R.Kuokkanen, S.Mills and D. Simmons. 2018. Gendered Environmental Assessments in the Canadian North: Marginalization of Indigenous Women and Traditional Economies. The Northern Review 47 (47): 135–166.

Kobayashi, Y., G.M. Peters, and S.J. Khan. 2015. Towards More Holistic Environmental Impact Assessment: Hybridisation of Life Cycle Assessment and Quantitative Risk Assessment. *Environmental Science and Technology* 49(15): 8924–8931.

Kojola, E. 2018. Indigeneity, Gender and Class in Decision-Making about Risks from Resource Extraction. *Environmental Sociology*. Accessed September 23, 2020. https/doi.org/10.1080/23251042 .2018.1426090

Koutouki, K., Lofts, K., and Davidian, G. 2018. A rights-based approach to indigenous women and gender inequities in resource development in northern Canada. Review of European, Comparative & International Environmental Law, 27(1): 63–74. https://doi.org/10.1111/reel.12240

Lahiri-Dutt, K., and N. Ahmad. 2012. "Considering Gender in Social Impact Assessments". In: *New Directions in Social Impact Assessments: Conceptual and Methodological Advances*, Edited by F. Vanclay and A.M. Esteves, 117–137. Cheltenham: Edward Elgar Publishing.

Lang, Sabine. 2009. Assessing Advocacy: European Transnational Women's Networks and Gender Mainstreaming. *Social Politics: International Studies in Gender, State and Society* 16(3): 327–357.

Lukatela, Ana Stephanie. 2014. Gender Mainstreaming Strategies in the International Development Context: Why Practice Has Not Made Perfect. Dissertation. University of British Columbia.

Maas, K., and K. Liket. 2011. Social Impact Measurement: Classification of Methods. In: *Environmental Management Accounting and Supply Chain Management*. Eco-Efficiency in Industry and Science, vol. 27, Edited by R. Burritt, S. Schaltegger, M. Bennett, T. Pohjola, and M. Csutora. Dordrecht: Springer. https://doi.org/10.1007/978-94-007-1390-1_8

Manning, Susan. 2014. Feminist Intersectional Policy Analysis: Resource Development and Extraction Framework. FemNorthNet and the Canadian Research Institute for the Advancement of Women. Accessed February 2018. http://fnn.criaw-icref.ca/images/userfiles/files/FIPAFramework.pdf.

Montini, Massimiliano. 2013. Towards a New Instrument for Promoting Sustainability Beyond the EIA and the SEA: The Holistic Impact Assessment (HIA) (2013). VIU Working Papers 10.2013, https://ssrn .com/abstract=2797228

Murray, J., and Ferguson, Mary. 2001. *Women in Transition out of Poverty*. Toronto: Women and Economic Development Consortium. Accessed January 2018. https://ccednet-rdec.ca/en/toolbox/women -transition-out-poverty-asset-based-approach-building.

Nightingale, Elana, Karina Czyzewski, Frank Tester, and Nadia Aaruaq. 2017. The Effects of Resource Extraction on Inuit Women and Their Families: Evidence from Canada. *Gender and Development* 25(3): 367.

NWAC (Native Women's Association of Canada). 2010. Culturally Relevant Gender Application Protocol: A Workbook. Accessed September 23, 2020. https://www.nwac.ca/wp-content/uploads/2015/05 /2010-NWAC-Culturally-Relevant-Gender-Application-Protocol-A-Workbook.pdf.

O'Connell, Helen, and Mikaela Gavas. 2015. *The European Union's New Gender Action Plan 2016–2020: Gender Equality and Women's Empowerment in External Relations*. London: Overseas Development Network. Accessed September 23, 2020. https://www.odi.org/publications/10021-gender-equality -empowerment-eu-external-relations-gap-2016_2020.

O'Shaughnessy, S., and Krogman, N.T. 2011. Gender as contradiction: From dichotomies to diversity in natural resource extraction. Journal of Rural Studies, 27: 134–143, https://doi.org/10.1016/j.jrurstud .2011.01.001

OECD (Organisation for Economic Cooperation and Development). 2016. *Handbook on the OECD Gender Equality Policy Marker*. Development Assistance Committee (DAC). Accessed September 23, 2020. https://www.oecd.org/dac/gender-development/Handbook-OECD-DAC-Gender-Equality -Policy-Marker.pdf.

Parkins, John R., and Ross E. Mitchell. 2016. "Social Impact Assessment: A Review of Academic and Practitioner Perspectives and Emerging Approaches". In: *Environmental Impact Assessment: Practice and Participation*, Edited by Kevin Hanna, 122–140. Oxford: University Press.

Parsons, Richard. 2020. Forces for Change in Social Impact Assessment. *Impact Assessment and* Project Appraisal 38(4): 278–286.

Paterson, Stephanie. 2010. What's the Problem with Gender-Based Analysis? Gender Mainstreaming Policy and Practice in Canada. *Canadian Public Administration* 53(3): 395.

Pauktuutit Inuit Women of Canada. 2008. Pauktuutit Inuit Women of Canada. *Canadian Woman Studies* 26(3/4): 135. Accessed February 2018. https://cws.journals.yorku.ca/index.php/cws/article/viewFile/22122/20776.

Pauktuutit Inuit Women of Canada and School of Social Work. University of British Columbia. 2016. *The Impact of Resource Extraction on Inuit Women and Families in Qamani'tuaq, Nunavut Territory: A Quantitative Assessment.* Accessed September 23, 2020. https://www.pauktuutit.ca/wp-content/uploads/Quantitative-Report-Final.pdf.

Peletz, N., and K. Hanna. 2019. *Gender Analysis and Impact Assessment: Canadian and International Experiences.* Canadian International Resources and Development Institute (CIRDI), Vancouver: University of British Columbia.

Peng, Yen-Wen. 2015. "Gendering Policy Analysis? the Problems and Pitfalls of Participatory 'Gender Impact Assessment'". In: *Policy Analysis in Taiwan*, Edited by Kun Yu-Ying, 1st ed. Vol. 5;5. Bristol, UK: Policy Press.

Rankin, L. Pauline, and Krista D. Wilcox. 2004. De-gendering Engagement?: Gender Mainstreaming, Women's Movements and the Canadian Federal State. *Atlantis* 29(1): 52.

Rasmussen, Derek, and Jessica Guillou. 2012. Developing an Inuit-Specific Framework for Culturally Relevant Health Indicators Incorporating Gender-Based Analysis. *Journal of Aboriginal Health* 8(2): 24.

Reed, M.G., and D. Davidson. 2011. Terms of Engagement: The Involvement of Canadian Rural Communities in Sustainable Forest Management. In: *Reshaping Gender and Class in Rural Spaces*, Edited by B. Pini and B. Leach, 199–220. Aldershot, England: Ashgate Publishing.

Riha, S., L. Levitan, and J. Hutson. 1997 *Environmental Impact Assessment: The Quest for a Holistic Picture.* Part 3 Chapter 2. U.S. Dept. of Agriculture, ERS. Paper presented at the symposium/workshop "Broadening support for 21st-century IPM" held February 27–March 1, 1996, Washington, D.C. Accessed September 20, 2020. https://agris.fao.org/agris-search/search.do?recordID=US9742899

Sauer, Arn. T. 2013. Gender Impact Assessment. *LIAISE Toolbox.* Accessed March 2014. http://www.liaise-kit.eu/ia-method/gender-impact-assessment-0.

Sauer, Arn T. 2018. Equality Governance *via* Policy Analysis? The Implementation of Gender Impact Assessment in the European Union and Gender-Based Analysis in Canada. *Transcript Publishing. Political Science* 68.

Scala, Francesca, and Stephanie Paterson. 2017. Gendering Public Policy or Rationalizing Gender? Strategic Interventions and GBA+ Practice in Canada. *Canadian Journal of Political Science* 50(2): 427.

SIDA (Swedish International Development Cooperation Agency). 2015. Gender Analysis – Principles & Elements. Accessed February 2018. https://www.sida.se/contentassets/a3f08692e731475db106fdf84f2fb9bd/gender-tool-analysis.pdf.

Simon, M. 2013. *Balancing the Scales: Using Gender Impact Assessment in Hydropower Development.* Melbourne: Oxfam Australia.

Stephens, Anne, Ellen D. Lewis, and Shravanti Reddy. 2018. *Inclusive Systemic Evaluation for Gender Equality, Environments and Marginalized Voices (ISE4GEMs): A New Approach for the SDG Era.* New York: UN Women.

Squires, Judith. 2005. Is Mainstreaming Transformative? Theorizing Mainstreaming in the Context of Diversity and Deliberation. *Social Politics: International Studies in Gender, State and Society* 12(3): 366–388.

Srinivasan, B., and L. Mehta. 2003. Assessing Gender Impacts. Chapter 11, pp 161–178, In H.A. Becker and F. Vanclay (eds). *The International Handbook of Social Impact Assessment: Conceptual and Methodological Advances.* Cheltenham: Edward Elgar.

SWC (Status of Women Canada). 2016. Gender-Based Analysis Plus. Accessed May 2017. http://www.swc-cfc.gc.ca/gba-acs/index-en.html.

SWC (Status of Women Canada). 2017a. Online Course: Introduction to GBA+. Accessed September 2018. http://www.swc-cfc.gc.ca/gba-acs/course-cours-2017/eng/mod00/mod00_01_01.html.

SWC (Status of Women Canada). 2017b. Demystifying GBA+ Job Aid. PDF. Accessed March 2018. http://www.swc-cfc.gc.ca/gba-acs/course-cours-2017/assets/modules/Demystifying_GBA_job_aid_EN.pdf.

UN (United Nations). 1995. *The Beijing Declaration and the Platform for Action.* New York: UN Department of Public Information.

UN. 1997. *Report of the Economic and Social Council for 1997.* A/52/3. September 18, 1997.

UN. 2015. *Transforming Our World, the 2030 Agenda for Sustainable Development.* General Assembly Resolution A/RES/70/1. October 21, 2015.

UN Women Training Centre. 2011–2017. Gender Equality Glossary. Resource Centre. Accessed September 23, 2020. https://trainingcentre.unwomen.org/mod/glossary/view.php?id=36&lang=en.

UNDP (United Nations Development Programme). 2001. *Learning and Information Pack: Gender Analysis*. Gender and Development Programme. PDF. Accessed March 2018. http://www.undp.org/content/dam/undp/library/gender/Institutional%20Development/TLGEN1.6%20UNDP%20GenderAnalysis%20toolkit.pdf.

Vanclay, F., and Ana Maria Estevez. 2011. *New Directions in Social Impact Assessment: Conceptual and Methodological Advances*. Cheltenham: Edward Elgar.

Verloo, M. 2002. The Development of Gender Mainstreaming as a Political Concept for Europe. Paper presented at the Conference on Gender Learning, Leipzig, September 6–8.

Vernooy, R. 2006. *Social and Gender Analysis in Natural Resource Management: Learning Studies and Lessons from Asia*. Thousand Oaks; New Delhi: SAGE Publications.

Walker, Heidi A., Maureen G. Reed, and Bethany Thiessen. 2019. Gender and Diversity Analysis in Impact Assessment. The Canadian Environmental Impact Assessment Agency. Accessed September 23, 2020. https://research-groups.usask.ca/reed/documents/CEAA%20Report.FINAL.%20Walker%20Reed%20Thiessen.%20Gender%20Diversity%20in%20IA.Feb%208%202019.pdf.

Warren, H. 2007. Using Gender-Analysis Frameworks: Theoretical and Practical Reflections. *Gender and Development* 15(2): 187–198.

Women's Empowerment Principles. n.d. About WEP. Accessed September 23, 2020 https://www.weps.org/about.

UN Population Fund (UNFPA). 2020. *Technical Note on Gender Transformative Approaches in the Global Programme to End Child Marriage Phase 2: A Summary for Practitioners*. New York: UNFPA. Accessed 20 December 2021 at: https://www.unfpa.org/sites/default/files/resource-pdf/Technical_Note_on_Gender-Transformative_Approaches_in_the_GPECM_Phase_II_A_Summary_for_Practitioners-January-2020.pdf

UN Women 2017. "InFocus: Equal Pay for Work of Equal Value". Commission on the Status of Women (CSW) 61. Women's Economic Empowerment in the Changing World of Work. Accessed September 23, 2020. https://www.unwomen.org/en/news/in-focus/csw61/equal-pay.

World Bank Group. 2015. Gender Strategy (FY16-23): Gender Equality, Poverty Reduction and Inclusive Growth. Accessed November 10, 2020. http://hdl.handle.net/10986/23425.

Yasmeen, S. 2015. *Social Impact of Mining on Women: Balochistan and Sangatta Compared*. Brisbane: International Mining for Development Centre.

12

APPLICATIONS OF GEOGRAPHIC INFORMATION SYSTEMS, SPATIAL ANALYSIS, AND REMOTE SENSING IN ENVIRONMENTAL IMPACT ASSESSMENT

Mathieu Bourbonnais

Introduction

An environmental impact assessment (EIA) is inherently geographic. Each impact assessment requires detailed and specific information on environmental conditions, including both physical and socioeconomic, that are likely to be affected by the proposed project over space and time (Glasson and Therivel 2019). Proponents are required to collect and manage the geographic data needed to understand baseline environmental conditions, analyze potential effects and mitigation options, and communicate assessment results for the proposed project.

Geographic information systems (GIS) are software used to manage, analyze, and visualize geographic data (Goodchild 2000). Widely recognized as the first GIS, the Canadian Geographic Information System developed by Dr. Roger Tomlinson in the 1960s was designed to digitize maps of natural resources and land use and store results in a spatially referenced database (Goodchild 2018). In most cases, early GIS databases were used for the inventory of spatially referenced natural resource (e.g., soils, forests, crops, minerals) or socioeconomic (e.g., census) data that could be queried and mapped to support land-use planning and resource management by governments and industry (Martin 2009). Access to reliable and fast desktop computing and web-based GIS combined with the increasing availability of geographic data, including open data and remote-sensing imagery, resulted in rapid growth and uptake of GIS and related geospatial technologies. Today, GIS and geospatial technologies are a multi-billion-dollar industry, with numerous mature software available (e.g., Environmental Systems Research Institute (ESRI) ArcPro/ArcGIS) and a growing number of open-source alternatives (e.g., QGIS; Coetzee et al. 2020), used across diverse fields including urban planning, resource and landscape management, and environmental and social sciences. While GIS are most widely known for cartographic applications, statistical approaches for analyzing spatial data (i.e., spatial analysis, see Goodchild 2000) and examining relationships among spatial patterns and processes are now common.

GIS and geographic data fall more broadly in the field of geomatics, which includes related disciplines of surveying and remote sensing, as well as techniques in global positioning systems (GPS), digital terrain analysis, and the development and use of decision support systems (Gomarasca 2010). Remote sensing is the use of aerial or satellite imagery (i.e., Earth observation data) for collecting information about the Earth's surface and represents a promising if somewhat underused source of information on baseline conditions and as a tool for long-term monitoring in EIA (Harker et al. 2021). Many governments now maintain databases of air photos, and several Earth observation missions, including Landsat and Sentinel, offer freely available imagery with global coverage used in environmental monitoring and management (Wulder and Coops 2014).

Box 12.1 Terminology and definitions

Geographic information system	A computer-based system used to manage, manipulate, analyze, and visualize geographic data.
Geographic data	Georeferenced data that has a xy coordinate allowing features to be mapped.
Spatial analysis	Techniques and methods are used to manipulate and analyze geographic data to understand spatial patterns and relationships.
Geomatics	Field focused on approaches, tools, and technologies for collecting, storing, manipulating, modeling, and visualizing geographic data.
Vector data	Object-based data model representing geographic data as points, lines, or polygons (areas).
Raster data	Field-based data model representing geographic data as a regular grid with a defined spatial resolution.
Multi-criteria analysis	Decision support tools are used to summarize and map possible outcomes based on defined criteria and geographic data. In EIA, multi-criteria analysis is often used for screening and planning.
Geovisualization	Tools and techniques, which are increasingly interactive, allowing the user to explore and analyze geographic data. Geovisualization is primarily used for knowledge discovery and can include maps, images, 3D visualization, and other interactive media.
Spatial interpolation	Methods used to predict values at unsampled locations resulting in a spatially continuous surface.
Earth observation	Collecting information about Earth's biological, chemical, and physical properties using remote-sensing techniques.
Remote sensing	Techniques used to detect and monitor physical characteristics of an object or area at a distance using reflected or emitted radiation. Common platforms used for Earth observation and remote sensing include satellites, manned aircraft, and unmanned aerial vehicles (i.e., drones).

The use of GIS and geographic data in the various stages of an EIA is not a recent development. The ability to manage and analyze data from multiple sources using a GIS and communicate results using maps and statistics represented an advancement over geographic methods previously used in EIA (Eedy 1995). There have been several reviews summarizing the use and potential applications of GIS in EIA (e.g., Eedy 1995; João and Fonseca 1996; Rodriguez-Bachiller and Glasson 2004; Atkinson and Canter 2011; Del Campo 2012; Gharehbaghi and Scott-Young 2018). A common theme is that GIS, geospatial methods, and geographic data can contribute substantially to many phases of the EIA process; however, reviews generally suggest methods remain underused or their application in various stages of an EIA remains unclear. This is common when GIS are used primarily as a tool for data management and manipulation, and visualization. For example, many commercial GIS software include a spatial database that allows users to store and manage spatial and non-spatial data streamlining processes of data collection and management in various stages of the EIA process. Consequently, spatial methods and various geographic data are often an important part of "desk-based studies" during an EIA for data management and visualization (i.e., maps), which may limit the scope of their application in different stages of an EIA.

Despite this, research applications of spatial methods and Earth observation data for environmental monitoring are extensive. The goal of this chapter is to highlight applications of spatial analysis, geovisualization, Earth observation data, and remote-sensing techniques in EIA with a discussion of the strengths and weaknesses of commonly used methods and data (Box 12.1). Two case studies demonstrate how spatial analysis and Earth observation data can be integrated into participatory processes, understanding baseline environmental conditions, and contribute to long-term monitoring of potential impacts.

Map overlay and multi-criteria analysis

The digital representation of geographic information in a GIS relies primarily on two data models. The vector data model is *object*-based and represents geographic features as points, lines, or polygons with associated attribute data describing each feature stored in a linked table. The raster data model is *field*-based and represents geographic features as an array of values or classes using cells (i.e., pixels) with a defined resolution (e.g., 30 m × 30 m) and extent. Earth observation satellite data are commonly provided as raster data.

Geographic data in a GIS are often handled and conceptualized as layers, each representing a specific theme (e.g., soil or forest type, land cover, surficial geology, social values, etc.) that can be visualized and analyzed based on spatial relationships. Spatial methods that are widely used in EIA include overlay and distance-based analyses that allow users to identify spatial relationships among geographic features and layers (Rodriguez-Bachiller and Glasson 2004). These relationships might include areas and locations of intersection between a project and sensitive habitat, watersheds, or cultural values, allowing assessment of potential impacts (João and Fonseca 1996). Simple map overlays represent spatial relationships as Boolean (e.g., the project area intersects sensitive habitat, or it does not), which may not capture potential impacts. Spatial relationships may be better characterized using distance-based methods. For example, Euclidean distance to hydrological features and potential for downstream impacts, as well as the distance to other developments and communities, are critical considerations for project and mitigation planning and monitoring (Suh et al. 2017).

Map overlays are commonly applied in a spatial multi-criteria decision analysis (SMCDA) for site suitability and risk analysis (Malczewski 2006). SMCDAs can be used as screening and planning tools in EIA, as they summarize and map possible outcomes (e.g., potential locations for development) based on defined criteria and geographic data (Huang, Keisler, and

Linkov 2011). Some examples include assessing site suitability for renewable energy projects (Cradden et al. 2016) agriculture (Feizizadeh and Blaschke 2013), transportation corridors (López and Monzón 2010), and landfills (Ersoy and Bulut 2009), evaluating environmental hazards (Feizizadeh et al. 2014), multi-stakeholder land-use planning (Zhang, Li, and Fung 2012), and potential habitat suitability and conservation planning (Adem Esmail and Geneletti 2018). A GIS is used to standardize or classify input layers in a SMCDA using either statistical approaches or defined criteria, and each layer is weighted based on its perceived importance in the model. Classes and weights are often defined based on expert opinion and can incorporate information from stakeholders and the general public (Antunes, Santos, and Jordão 2001). Geographic data can be incorporated from multiple sources (Box 12.2) based on the evaluation criteria and converted to a common data type (i.e., vector or raster) and spatial resolution. Once layers have been standardized and weighted, they are summed using a Weighted Linear Combination (although other approaches exist for combining layers, including Ordered Weighted Averaging; Drobne and Lisec 2009) and mapped, providing a scaled characterization of suitability or risk based on the defined criteria.

An early example of a simple multi-criteria map overlay for assessing the environmental vulnerability of a road development included geographic data characterizing land cover and topography, social factors (e.g., distance to the community), and environmental conditions (e.g., hydrology, air pollution, and noise; Li et al. 1999). Each polygon layer was classified based on potential environmental impacts, and a GIS overlay was used to identify the route and potential alternatives with the lowest cumulative impact. This example highlights limitations inherent in spatial modeling of geographic data, which represent an aggregate model of real-world features. For example, a polygon or cell classified as "forest" may contain multiple tree species or sensitive habitats that are not represented in the spatial layer due to the resolution of the data and how they were defined and sampled (Gonzalez and Enríquez-De-Salamanca 2018). As a result, data may not reflect the spatial scale needed for analysis of potential impacts and may introduce uncertainty in the analysis and propagate error in the model (Crosetto and Tarantola 2001). Understanding the spatial and temporal limitations of geographic data is important for interpreting model outputs and how they can be used effectively in decision-making processes.

Assigning layer weights and classes can be subjective, which influences model outputs and interpretation. The analytic hierarchy process (AHP; Saaty 1980) uses pairwise comparisons of all input layers based on preference factors or a standardized scale of importance and returns principal weights for each layer which can be aggregated. The AHP approach is the most widely used SMCDA method in environmental research (Cegan et al. 2017; Huang, Keisler, and Linkov 2011), and AHP plugins are available for many desktop GIS software. Alternatives to AHP include the analytic network process, a nonlinear form of AHP, and fuzzy AHP, which accounts for imprecision in the exact numeric pairwise ranking of the input layers required by AHP (Mosadeghi et al. 2015). Modifying layer weights or classes can strongly influence mapped suitability outputs, thereby creating uncertainty in the decision and planning process. It is important to define the criteria required to make an informed decision at the outset to ensure suitable geographic data are available, which can be difficult in EIA. Performing a sensitivity analysis or error propagation analysis to determine the influence of data imprecision or inaccuracy, as well as different layer weights and classes on model outputs is recommended (Chen, Yu, and Khan 2010), although, in practice, model sensitivity is not always assessed (Huang, Keisler, and Linkov 2011).

Box 12.2 Spatial multi-criteria decision analysis and engagement to support Indigenous governance

The *Impact Assessment Act* passed in 2019 by Canada's federal government requires proponents to consult with and mitigate potential impacts to Indigenous peoples and to include and explain the use of Indigenous knowledge in impact assessment decision-making. While the Act represents an advancement over previous legislation which did not recognize Indigenous title (Paci, Tobin, and Robb 2002), many obstacles for Indigenous participation in EIA remain, including a lack of opportunities for cross-cultural and technical training (Eckert et al. 2020). Legal requirements to include Indigenous knowledge (or Traditional ecological knowledge) of environmental (e.g., water, habitat, wildlife) and cultural (e.g., traditional lands) values in an assessment are also complex because of challenges inherent in summarizing different forms of knowledge for management and decision-making (i.e., Indigenous knowledge and western science; see Raymond et al. 2010; Cooke et al. 2016; Bartlett, Marshall, and Marshall 2012).

Culturally modified trees are living or dead trees with evidence of traditional or cultural use, including bark removal, removal of fiber for construction of buildings, canoes, and carvings, or pitch collection, by First Nation Peoples (Benner et al. 2019). These culturally modified trees have significant biocultural value to First Nation Peoples in the Great Bear Rainforest of British Columbia, Canada and are threatened by commercial logging and other industrial activities. Working with the Kitasoo/Xai'xais Nation, DeRoy et al. (2021) developed a spatially explicit multi-criteria evaluation (Eastman 1999) for predicting the distribution of culturally modified trees. Their multi-criteria model incorporated Indigenous knowledge of the distance to known habitation sites and canoe landing sites with biophysical indicators of topography and forest height derived from light detection and ranging (LiDAR) and canopy cover attributed as either western red-cedar or yellow-cedar. Variables were weighted using AHP (Saaty 1980) and combined in a GIS to evaluate the relative importance of each for predicting site suitability and mapping the distribution of culturally modified trees throughout Kitasoo/Xai'xais Territory. In the context of an EIA, culturally modified trees may be defined as a valued ecosystem component (VEC), which is an environmental feature of concern to Indigenous peoples, government, or the public that may be affected by a project (Broderick, Durning, and Sánchez 2018). The modeling approach described provides a framework for including Indigenous knowledge with remote-sensing data and other geographic information for supporting First Nations governance and providing valuable baseline information for mitigating impacts to culturally important resources that may be affected by future projects or development (DeRoy et al. 2021).

The layer-based approach inherent in GIS is conceptually linked to cumulative effects assessment (CEA; see Chapter 3 in this book). Individual layers can represent past projects, unique biophysical and social values (Box 12.2), and temporal snapshots (Box 12.3) of environmental conditions (Atkinson and Canter 2011). A GIS is often used to determine if the locations of VEC's overlap spatially or temporally with stressors (e.g., previous development, projects, and disturbance) that are ranked based on their potential impacts (Duinker et al. 2013). Spatial approaches can also help determine the influence of geographic extent and scale on the analysis in a CEA (Therivel and Ross 2007). While the geographic extent of a CEA or EIA is often defined primarily based on data availability, spatial approaches for defining thresholds of change in environmental variables and mapping uncertainty are increasingly available (Peeters et al. 2018).

Exploring spatial patterns

Information used for environmental monitoring and management is frequently collected at point locations, often for logistical reasons. Examples include data on air and water quality, soil properties, pollution and contamination concentrations, climate variables, and boreholes for mineral exploration. Often these data represent samples of spatially continuous phenomena, and interpreting and visualizing them as continuous surfaces can support environmental management and decision-making (Li and Heap 2011). Methods for spatial interpolation are used to predict values at unsampled locations resulting in a spatially continuous surface (Figure 12.1). There are numerous methods for spatial interpolation available, as predicting unsampled locations is a fundamental statistical question across several fields. A useful summary of methods for spatial interpolation in environmental sciences is provided by Li and Heap (2014), who categorize methods as geostatistical, non-geostatistical, and other multivariate predictive methods, including machine learning and geographically weighted regression.

Geostatistics are methods used to understand spatial and temporal variation and correlation (i.e., spatial autocorrelation) of sampled values based on distance (Burrough 2001). Kriging, which includes several generalized least-squares regression algorithms, is the most commonly used geostatistical approach for spatial interpolation in geosciences, mining, hydrology, and soil and environmental sciences (Li and Heap 2014). Compared to non-geostatistical approaches for spatial interpolation, which includes methods such as inverse distance weighting and nearest neighbor, kriging (i.e., ordinary kriging) provides unbiased linear estimates (with several assumptions) with known variance and can incorporate related variables (i.e., co-kriging) to improve predictions (Webster and Oliver 2007). In mineral exploration, estimates of the location, grade, and quantity of deposits can help with mine planning and extraction. Kriging has been used as an exploratory method to predict and map the spatial distribution of mineral resources and

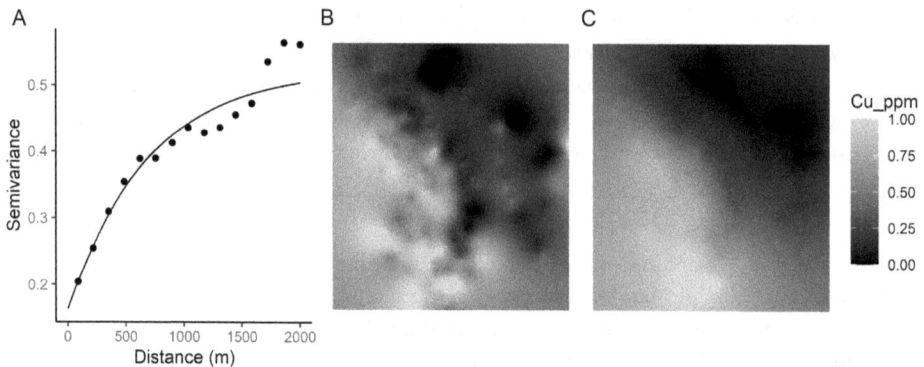

Figure 12.1   Methods for spatial interpolation allow samples to be visualized as a spatially continuous field which may provide insight into underlying processes. Here, over 2,000 soil assays for copper (parts per million; Cu_ppm) are modeled using a semivariogram (A) and interpolated using ordinary kriging (B). The kriging model accounts for localized patterns of spatial autocorrelation when compared to the interpolation using inverse distance weighting (C).

reserve size in the planning phase (Haldar 2018) and for risk assessment of groundwater contamination and remediation planning (Wycisk et al. 2009).

Regardless of the method, data exploration is a critical step in spatial interpolation as methods and results are sensitive to sample size and design, distribution and variance, specification of parameters, and directional trends in the values (Li and Heap 2014). The accuracy and sensitivity of interpolated surfaces can be assessed using cross-validation (Roberts et al. 2017), which can also help determine if the sample size and design is sufficient for understanding a spatially continuous process. Considering the cost and complexity in monitoring environmental factors (e.g., hydrology, soils, contaminants) in EIA, interpolated surfaces can provide an effective approach for integrating sampled data with other spatially continuous data (e.g., topography, land use) in a GIS for decision support and planning.

Quantifying spatial patterns is also a fundamental question in spatial analysis. Here, spatial pattern refers to a characteristic arrangement of geographic features (i.e., clustered, dispersed, or random) based on distance. Most GIS software includes several approaches to quantify and classify spatial patterns using spatial statistics, including global and local measures of spatial association such as Moran's I and Getis-Ord Gi$\star$ (Bivand and Wong 2018). Global measures provide a single estimate characterizing spatial patterns for a study area, while local measures provide estimates for each geographic feature. Local measures are often used to identify "hotspots", where high values are spatially clustered, providing insight on underlying processes that can be useful for environmental monitoring. For example, local Moran's I has been used to identify hotspots of high soil pollution concentration in urban areas related to various human activities, including industrial development and vehicle emissions (Yuan, Cave, and Zhang 2018; Zhang et al. 2008; Li et al. 2014). While these methods are largely exploratory, characterizing and mapping spatial patterns of environmental variables using spatial statistics can help understand underlying ecological or social processes.

Participatory GIS and geovisualization

Decision-making and resource management through EIA is intended to be a participatory process (see Chapter 14) and GIS and geographic data can play a central role in public engagement. Maps of project sites and baseline environmental conditions are commonly used to communicate proposals and areas with potential impacts to the public. However, transparency on data collection, methodology, prediction, and uncertainty can be a major issue in an impact assessment (Tennøy, Kværner, and Gjerstad 2006) and can influence engagement and trust in EIA processes (Hanna 2016). Data availability, complex workflows, and difficulties translating results and methodologies across space and time also create challenges for reproducible and replicable geospatial research, although this is improving as data and code are increasingly made available (Kendron et al. 2021)

Access to data used in an EIA analysis can help ensure transparency and reproducibility as well as trust in the scientific process. Increasingly, there is a legal requirement to make data incorporated in an EIA accessible. For example, the tailored impact statement guidelines for the Canadian *Impact Assessment Act* (S.C. 2019, C. 28, s. 1) requires proponents to submit maps of the project location or footprint and any other relevant baseline or geographic data incorporated in the assessment using specific geospatial data file and metadata standards (i.e., ISO 19115), which can then be made publicly available. These data can be used by external organizations (e.g., Indigenous nations, NGO's, government) to assess potential impacts to aid with consultation and decision-making.

Web-based GIS that incorporates simple geospatial functionality may also help increase participation in EIA. As an example, the Impact Assessment Agency of Canada recently added GIS functionality to the *Canadian Impact Assessment Registry* (https://iaac-aeic.gc.ca/050/evaluations /050?culture=en-CA). The new tool allows users to search for projects using spatial queries and by distance from a location. The tool also allows users to upload other geographic data to visualize potential spatial relationships with projects and to comment on proposals. However, the expertise needed to use many desktop or web-based GIS can limit public participation even when data or public platforms are available.

Other visualization techniques widely employed in impact assessment include terrain analysis using digital elevation models (DEM), which allows visual exposure of projects to be mapped based on project locations, observation points and topographic variability (Wilson 2018). Viewsheds, which represent the area visible from a given geographic location as a polygon(s), are broadly used to quantify and manage the potential visual impacts of a proposed project. Viewshed results can be integrated with other geospatial layers in a GIS to help inform site selection for projects that often meet resistance, including landfill site selection (Geneletti 2010) and the location and density of wind turbines (Gibbons 2015). Limitations of viewsheds for quantifying visual impacts are primarily due to DEMs which do not include information on the natural (e.g., trees) or built (e.g., buildings) environment that might obscure visible areas, and all visible areas are equivalent regardless of the visual prominence of features (Nutsford et al. 2015). The use of higher spatial resolution DEMs and digital surface models (DSM) derived from LiDAR can help overcome some of these limitations (Klouček, Lagner, and Šímová 2015).

Techniques for geovisualization in GIS and their integration in spatial decision support systems are becoming increasingly sophisticated. The ability to develop immersive and interactive digital 3D visualizations with multiple viewpoints using GIS, computer animation, and computer-aided design provide new mediums for knowledge discovery when planning complex projects. For example, 3D visualizations of proposed buildings and projects using geographic data (e.g., aerial photography, building footprints, LiDAR) allow developers to conceptualize space use and add an element of realism in urban planning (Herbert and Chen 2015). More generally, geovisualization and 3D models of landscape development are being used to compare scenarios for stakeholder engagement to inform environmental planning and management (e.g., Newell, Canessa, and Sharma 2017; Xu and Coors 2012). While specialized software for 3D geovisualization is available, geovisualization development remains complex and computationally intensive. Effective use of these tools in planning processes and stakeholder engagement also depends on a realistic representation of presence and place (Jaalama et al. 2021). Despite these challenges, the use of geovisualization to present project plans and scenarios represents an emerging area of stakeholder engagement and participatory planning that holds considerable promise in impact assessments.

Earth observation data

Earth observation data (i.e., remote-sensing imagery) collected using satellites, aircraft, and increasingly unmanned aerial vehicles (UAVs or drones) are an important data source in GIS (Figure 12.2). There are several advantages to using Earth observation data in an EIA for preliminary surveys, collecting baseline information, and long-term monitoring. Remote sensing provides repeat observations of the Earth's surface using consistent methods that allow continuous monitoring of environmental conditions and processes. Many jurisdictions maintain databases of orthophotos (geometrically corrected aerial photographs) that span decades allowing users to assess landscape change and baseline environmental conditions for decision-making

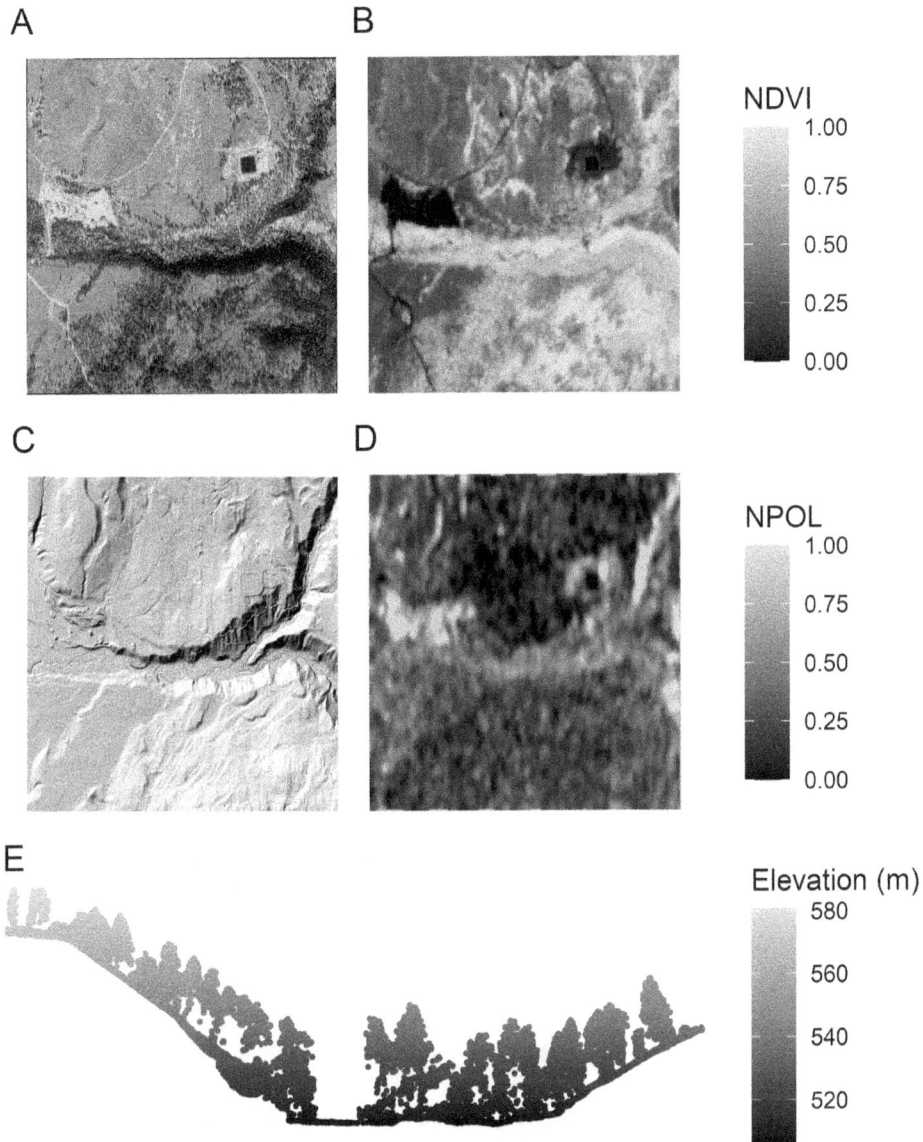

Figure 12.2 There are numerous types of Earth observation imagery and derived products that vary in spatial resolution and can be used in EIA, including orthophotos (A; 0.1 m); the Normalized Difference Vegetation Index (NDVI) derived from multispectral Sentinel-2 imagery (B; 10 m); digital terrain models (DTM) derived from airborne LiDAR data presented as a hillshade (C; 1 m); the normalized polarization derived from synthetic aperture radar (SAR) Sentinel-1 backscatter products (D; 10 m); and point clouds from airborne LiDAR data (E). Panels A–D are 2D raster data for the same 4 km² area, and panel E is a 300 m transect of a 3D LiDAR point cloud showing z (elevation) values for each point.

(Figure 12.2A; Morgan, Gergel, and Coops 2010). Similarly, many Earth observation satellite missions provide decades of freely available global imagery with a consistent spatial (i.e., the pixel size) and temporal resolution (i.e., the time required to revisit the same location). For example, the Landsat program has provided continuous terrestrial monitoring at a spatial resolution of 15–100 m and temporal resolution of 16 days since 1972, which will be extended with the scheduled launch of Landsat 9 in 2021 (Masek et al. 2020).

Multispectral imagery

Open access to Earth observation satellite data has resulted in a marked increase in applied research and methodological development for environmental monitoring (Wulder and Coops 2014; Zhu et al. 2019). Multispectral imagery used for terrestrial and marine monitoring are freely available from the Landsat program (the United States Geological Survey (USGS) & National Aerospace Science Administration (NASA)), the Moderate Resolution Imaging Spectroradiometer (MODIS) on the Terra and Aqua satellites (NASA), the Advanced Very High Resolution Radiometer (AVHRR; U.S. National Ocean and Atmospheric Administration (NOAA)) and the Sentinel-2 and Sentinel-3 satellites (European Space Agency (ESA)). In terrestrial applications, multispectral satellite imagery are commonly used to identify, map, and monitor land cover (Box 12.3; Wulder et al. 2018), disturbance (Hansen et al. 2013; Curtis et al. 2018), ecosystem function (Pettorelli et al. 2018), biodiversity (Randin et al. 2020; Wang and Gamon 2019), and land use (Liou, Nguyen, and Li 2017; Chaves, Picoli, and Sanches 2020).

Applications of land cover monitoring using Earth observation data

Mapping and monitoring land use and land cover is important for sustainable resource and land management (Foley et al. 2005) and has been a primary focus of remote-sensing research (Pandey et al. 2021). In Canada, forested ecosystems cover nearly half the country and timber and natural resources from forests contribute billions of dollars to the national economy. Given the size of the country (9.985 million km²), Earth observation data are used extensively to monitor the influence of human activities (e.g., forest harvest, mining, oil and gas exploration, agriculture, road development) and natural disturbance (e.g., forest fires, insect infestation, drought, and disease) on forest ecosystems to support planning and sustainable management (Banskota et al. 2014; Wulder et al. 2008).

In the early 2000s, the Earth Observation for Sustainable Development of Forests program developed a nationwide land cover map for 800 Mha of forested ecosystems in Canada (Wulder et al. 2003). The land cover map was created using an unsupervised approach combining Landsat multispectral imagery with air photos and plot data to hierarchically classify pixels into increasing levels of detail (e.g., land cover, vegetation type, and density; Wulder et al. 2008). Similar land cover data products are for many countries and regions, including the National Land Cover Data in the United States of America (USGS) and coordination of information on the environment (CORINE) land cover in Europe (European Environmental Agency), as well as globally (Tsendbazar, de Bruin, and Herold 2015), including the MODIS global land cover data product (MCD12Q1) and the ESRI 2020 land cover (https://livingatlas.arcgis.com/landcover/).

These land cover products can provide valuable baseline information on physical and environmental conditions used for land-use planning and resource management. Land cover data products are generally provided as a raster, which can be used as inputs in weighted multi-criteria assessments for quantifying and mapping ecosystem services (Koschke et al. 2012; Fontana et al. 2013), land-use suitability (Pérez-Hoyos, Udías, and Rembold 2020; Feizizadeh and Blaschke 2013), site selection and route planning for projects (Latinopoulos and Kechagia 2015; Watson and Hudson 2015) and resource management (Yemshanov et al. 2015). Land cover data are also widely used for quantifying landscape patterns, habitat, and ecological processes influenced by land use and disturbance (Long, Nelson, and Wulder 2010; Haddad et al. 2015).

Land cover, however, is not static and changes as a result of human activities and natural disturbance. Recently, the availability of open access Earth observation satellite data has led to the development of dynamic land cover products that reflect temporal change (i.e., "Land cover 2.0" *sensu* Wulder et al. 2018). Forest harvest and wildfire represent the two major disturbances that alter forest land cover in Canada. To detect and attribute forest disturbances in Landsat imagery as part of the National Terrestrial Ecosystem Monitoring System, researchers developed annual spectral image composites from 1984–2012 using a best-available pixel approach where spectral values from Landsat scenes are scored based on data quality (e.g., cloud or shadow), sensor type (e.g., Landsat 5, 7, or 8), day of the year, and opacity (White et al. 2014). The annual image composites were used in a time series change detection to identify breakpoints (change events by year) in spectral indices (Hermosilla et al. 2015b) that were attributed by disturbance type (e.g., forest harvest or wildfire) based on geometric properties, trend analysis of spectral properties, and pre- and post-change spectral metrics of the pixel-based change events (Hermosilla et al. 2015a).

The image composites and annual forest change maps were used as inputs to a "Virtual Land Cover Engine" which used a Hidden Markov model to predict land cover class transitions post-disturbance over time (Hermosilla et al. 2018). The resulting annual land cover maps show state transitions among various classes (e.g., coniferous, broadleaf, mixedwood, shrub, wetland, herb) informed by disturbance and recovery. While there are limitations to the annual land cover maps (e.g., not all disturbance is detected and attributed as a result of the minimum mapping unit of Landsat imagery), annual monitoring of landscape change and recovery using Earth observation imagery can help overcome limitations with out-of-date geographic data and contribute to cumulative effects assessments and long-term monitoring in EIA. For example, vegetation and ecosystem condition can be monitored almost continuously over time using vegetation indices like the Normalized Difference Vegetation Index (NDVI, Figure 12.2B; Quiñonez-Piñón, Mendoza-Durán, and Valeo 2007; Pettorelli 2013).

Synthetic aperture radar

Active sensors, which emit pulses of energy and record returns, including synthetic aperture radar and light detection and ranging, are also widely used in Earth observation. There are a number of active satellites available for synthetic aperture radar (SAR), including Sentinel-1 (Torres et al. 2012) and the RADARSAT Constellation Mission (Dabboor, Iris, and Singhroy 2018), as well as several legacy missions (e.g., the Shuttle Radar Topography Mission, TanDEM-X, Advanced Land Observation Satellite). SAR uses microwave energy, generally organized in the X-band (3.8–2.4 cm), C-band (7.5–3.8 cm), and L-band (30–15 cm). Unlike optical satellites (e.g., Landsat), SAR data acquisition is not impacted by weather, atmospheric interference, or time-of-day, and longer bands (i.e., L-band) can penetrate forest canopies and other types of land cover (see Moreira et al. 2013 for an overview of SAR technologies).

Various approaches are available for analyzing SAR data, including analysis of backscatter and SAR interferometry (InSAR) which provides precise measurements of phase difference in values of SAR imagery acquired at different times or positions (Pepe and Calò 2017). InSAR can be used to measure Earth surface displacements and deformation in land cover (El Kamali et al. 2020), and has been used extensively to examine land subsidence and uplift (Zhou, Chang, and Li 2009), slope instability and landslide detection (Colesanti and Wasowski 2006), and can help inform risk management (Raspini et al. 2017). SAR data are also used extensively in hydrological applications, including mapping soil moisture (Paloscia et al. 2013), flooding and disaster monitoring (Clement, Kilsby, and Moore 2018) and sensitive ecosystems. For example, backscatter properties related to water and vegetation penetration using SAR is particularly useful for mapping and monitoring of wetlands (Figure 12.2D; White et al. 2015). Wetlands are biodiversity hotspots, provide numerous ecosystem services, and globally are among the most threatened ecosystems from human activities and climate change (Kingsford, Basset, and Jackson 2016). However, wetlands are generally poorly accounted for in impact assessment despite their importance (Noble, Hill, and Nielsen 2011) and wetland maps produced using SAR and other Earth observation data can contribute substantially to their monitoring and management at landscape scales.

Light detection and ranging

LiDAR are active sensors that emit high-frequency pulses of energy (generally in the near-infrared, although blue-green lasers are used in hydrological applications) that are reflected by a surface. The time elapsed between emittance and return is recorded and converted to a distance allowing the return to be geolocated. The resulting point cloud provides 3D structural information, and each point includes attributes based on the properties of the return, including the location, intensity, and classification (e.g., ground or non-ground) of the point (Figure 12.2E). LiDAR data are normally collected using either terrestrial or airborne (i.e., manned aircraft or UAVs) laser scanners (Beland et al. 2019), although data are available from spaceborne sensors, including the global ecosystem dynamics investigation (GEDI; Dubayah et al. 2020) and the Ice, Cloud and Land Elevation Satellite (ICESat-2; Neumann et al. 2019).

LiDAR data is used widely across numerous disciplines, including forestry, ecology, biology, archaeology, mining, oil and gas exploration, and urban planning. Derived products, which summarize attributes of the point cloud in a 2D grid (i.e., raster), are commonly used in spatial analysis. For example, point cloud attributes allow biophysical forest structure metrics related to stand height and variation, canopy volume, vegetation biomass, stem density, and basal area to be quantified (van Leeuwen and Nieuwenhuis 2010). These metrics provide structural information on vegetation that can be integrated with multispectral imagery and other Earth observation data to provide baseline information on forest and land cover type and condition pre- and post-disturbance (Wulder et al. 2008; Bolton, Coops, and Wulder 2015; Goodbody et al. 2021).

A digital terrain model (DTM or digital elevation model) is a bare-earth topographic model interpolated using spatial methods from LiDAR ground returns (Figure 12.2C; Chen, Gao, and Devereux 2017). Digital elevation models are available from other sources, for example, the 30m resolution Advanced Spaceborne Thermal Emission and Reflection Radiometer (ASTER) Global Digital Elevation Model; however, DTMs derived from LiDAR point clouds have a much higher spatial resolution ($\leq$ 1m). Topographic information from high-resolution DTMs are frequently used to model landslide risk and surface instability in a GIS to inform project planning (e.g., road placement), as well as for modeling and monitoring hydrological systems (Tarolli 2014; Jaboyedoff et al. 2012). Probabilistic flood inundation models are commonly developed using DTMs incorporated in hydraulic models and a GIS to map floodplains for risk management (Alfonso, Mukolwe, and Di Baldassarre 2016). Considering the continued development of floodplains and increased flood risk due to climate change, airborne LiDAR surveys have been flown specifically to acquire data to map floodplains to support urban planning and environmental management (e.g., the "Okanagan Flood Story" https://okanagan-basin-flood -portal-rdco.hub.arcgis.com/).

Challenges and opportunities applying Earth observation data and spatial analysis

Despite the potential for Earth observation data to contribute to EIA, imagery and derived products are underutilized in impact assessment (Harker et al. 2021; Perminova et al. 2016). However, there are limiting factors for applications of Earth observation data in EIA. Analysis ready data are multispectral imagery that are geometrically and radiometrically consistent, which helps reduce pre-processing requirements, and include pixel quality flags (Dwyer et al. 2018). However, imagery can still be impacted by unusable data resulting from clouds, smoke, or other atmospheric noise that must be accounted for and may limit applications if the analysis is time-sensitive (e.g., imagery is required for a specific season or year). SAR data are not limited because of weather or time of day, but pre-processing requirements are extensive and can require specialized software for analysis. Translating Earth observation imagery into geographic information, including land cover or land-use classifications (Box 12.3), increasingly relies on advanced statistical approaches (e.g., machine learning, Maxwell, Warner, and Fang 2018; Ma et al. 2019), which are not available in most desktop GIS software and are computationally intensive. The availability of free and open-source software and cloud-based processing platforms (e.g., Google Earth Engine; Gorelick et al. 2017) can help overcome some of these limitations, but statistical analysis and modeling generally require the use of specialized statistical software (e.g., R and Python).

Earth observation data availability and acquisition can also be a limiting factor for environmental monitoring. LiDAR data are commonly collected using airborne platforms by government and private companies for specific projects (e.g., forest inventories, floodplain mapping etc.) due to high costs (Goodbody et al. 2021). As a result, few jurisdictions have wall-to-wall LiDAR data (although see Coops et al. 2016), which can limit applications of the technology for environmental monitoring in EIA. However, as the cost of LiDAR sensors decreases, combined with increased use of UAVs and alternative approaches for acquiring point clouds (e.g., structure-from-motion), environmental monitoring and digital terrain analysis using 3D geographic data will become more prominent. Additionally, open data portals that systematically share LiDAR data are becoming more common (Beland et al. 2019). Some examples include the Government of British Columbia LiDAR Portal (https://governmentofbc.maps.arcgis.com/) and the Washington State Department of Natural Resources LiDAR Portal (https://lidarportal .dnr.wa.gov/).

It is important to carefully consider the influence of the spatial and temporal resolution of the data on results in any spatial analysis as the availability of existing geographic data may not match the scale of the environmental processes that require information in the impact assessment. For example, the spatial resolution of a land cover map may not reflect the scale of sensitive habitat or cultural values that might be impacted by a proposed project or the current conditions, depending on when the data were collected. Metadata standards established by the government and regulatory agencies for geographic data can help ensure that proponents understand GIS processes and workflows used to sample or develop geographic data, as well as the inherent limitations. The influence of scale on spatial patterns and statistical results has been examined extensively in the geographical and environmental sciences (Levin 1992; Wu 2004; Goodchild 2011). While the importance of scale and challenges inherent in defining the geographic boundaries of an impact assessment has been recognized in EIA and CEA (Therivel and Ross 2007), there remains little consensus on how to scale should be defined for impact assessments (Harker et al. 2021). Geographic data are sampled or aggregated at specific scales, and manipulating the data in a GIS or modifying the boundaries of a study area can alter spatial patterns and results (i.e., the modifiable areal unit problem). For example, a multi-scale overlay analysis focused on a proposed project site might alter the number and type of existing projects and future developments, which are important considerations for understanding cumulative impacts (Harker et al. 2021). However, interpreting results from multi-scale analyses can be challenging as relationships between observed patterns and ecological processes across scales can be complex and nonlinear. Instead, examining the sensitivity of model results (e.g., spatial multi-criteria analysis outputs) to aggregation using statistical tests (Duque et al. 2018) or using geostatistics and semivariograms to identify characteristic scales of spatial autocorrelation in baseline environmental conditions could provide valuable tools for understanding the influence of scale in EIA.

Conclusion

Geographic information systems and geographic data play an important role in EIA. However, despite the use of GIS at various stages of project planning there are still opportunities to increase the application of fundamental spatial analysis tools and geographic data streams in EIA. The ease with which various types of geographic data can be manipulated and integrated with input from experts and stakeholders in a GIS has led to widespread adoption of spatial multi-criteria decision analysis in EIA. Outputs from these models can help support local decision-making processes, and the development of web-based GIS and tools for 3D visualization hold considerable promise for increasing public participation and input in project planning. Despite focusing on biophysical impact assessment here, it is important to note that many of the spatial analysis methods and approaches described can also incorporate the human, social, and cultural value of biophysical VECs. The availability of open access Earth observation data from several different sensors also holds promise for understanding historic baseline environmental baseline conditions, near real-time monitoring of environmental impacts, and modeling planning scenarios. While this chapter focused primarily on terrestrial examples, Earth observation data are widely available for freshwater and marine ecosystems. Barriers to applications of Earth observation data in impact assessments include computational challenges due to data volumes and the need to apply advanced statistical approaches to derive information (e.g., land cover) from imagery. The availability of open-source GIS and cloud-based platforms, as well as the use of geostatistical approaches for exploring relationships between patterns and processes across spatial and temporal scales, should be valuable tools for practitioners moving forward.

References

Adem Esmail, Blal, and Davide Geneletti. 2018. "Multi-Criteria Decision Analysis for Nature Conservation: A Review of 20 Years of Applications". *Methods in Ecology and Evolution* 9(1): 42–53.

Alfonso, L., M.M. Mukolwe, and G. Di Baldassarre. 2016. "Probabilistic Flood Maps to Support Decision-Making: Mapping the Value of Information". *Water Resources Research* 52(2): 1026–1043.

Antunes, Paula, Rui Santos, and Luís Jordão. 2001. "The Application of Geographical Information Systems to Determine Environmental Impact Significance". *Environmental Impact Assessment Review* 21(6): 511–535.

Atkinson, Samuel F., and Larry W. Canter. 2011. "Assessing the Cumulative Effects of Projects Using Geographic Information Systems". *Environmental Impact Assessment Review* 31(5): 457–464.

Banskota, Asim, Nilam Kayastha, Michael J. Falkowski, Michael A. Wulder, Robert E. Froese, and Joanne C. White. 2014. "Forest Monitoring Using Landsat Time Series Data: A Review". *Canadian Journal of Remote Sensing* 40(5): 362–384.

Bartlett, Cheryl, Murdena Marshall, and Albert Marshall. 2012. "Two-Eyed Seeing and Other Lessons Learned within a Co-Learning Journey of Bringing Together Indigenous and Mainstream Knowledges and Ways of Knowing". *Journal of Environmental Studies and Sciences* 2(4): 331–340.

Beland, Martin, Geoffrey Parker, Ben Sparrow, David Harding, Laura Chasmer, Stuart Phinn, Alexander Antonarakis, and Alan Strahler. 2019. "On Promoting the Use of Lidar Systems in Forest Ecosystem Research". *Forest Ecology and Management* 450: 117484.

Benner, Jordan, Anders Knudby, Julie Nielsen, Meg Krawchuk, and Ken Lertzman. 2019. "Combining Data from Field Surveys and Archaeological Records to Predict the Distribution of Culturally Important Trees". *Diversity and Distributions* 25(9): 1375–1387.

Bivand, Roger S., and David W.S. Wong. 2018. "Comparing Implementations of Global and Local Indicators of Spatial Association". *Test* 27(3): 716–748.

Bolton, Douglas K., Nicholas C. Coops, and Michael A. Wulder. 2015. "Characterizing Residual Structure and Forest Recovery Following High-Severity Fire in the Western Boreal of Canada Using Landsat Time-Series and Airborne Lidar Data". *Remote Sensing of Environment* 163: 48–60.

Broderick, Martin, Bridget Durning, and Luis E. Sánchez 2018. "Cumulative Effects". In Therivel, Riki, and Graham Wood (Eds.). (2017). *Methods of Environmental and Social Impact Assessment* (4th edition). New York: Routledge.

Burrough, P. A. 2001. "GIS and Geostatistics: Essential Partners for Spatial Analysis". *Environmental and Ecological Statistics* 8 (4): 361–377.

Campo, Ainhoa González Del. 2016. "GIS in Environmental Assessment: A Review of Current Issues and Future Needs". *Journal of Environmental Assessment Policy and Management* 14(1): 121–143.

Cegan, Jeffrey C., Ashley M. Filion, Jeffrey M. Keisler, and Igor Linkov. 2017. "Trends and Applications of Multi-Criteria Decision Analysis in Environmental Sciences: Literature Review". *Environment Systems and Decisions* 37(2): 123–133.

Chaves, Michel E.D., Michelle C.A. Picoli, and Ieda D. Sanches. 2020. "Recent Applications of Landsat 8/OLI and Sentinel-2/MSI for Land Use and Land Cover Mapping: A Systematic Review". *Remote Sensing* 12(18): 3062.

Chen, Yun, Jia Yu, and Shahbaz Khan. 2010. "Spatial Sensitivity Analysis of Multi-Criteria Weights in GIS-Based Land Suitability Evaluation". *Environmental Modelling and Software* 25(12): 1582–1591.

Chen, Ziyue, Bingbo Gao, and Bernard Devereux. 2017. "State-of-the-Art: DTM Generation Using Airborne LIDAR Data". *Sensors* 17(1): 150.

Clement, Miles A., C.G. Kilsby, and P. Moore. 2018. "Multi-Temporal Synthetic Aperture Radar Flood Mapping Using Change Detection". *Journal of Flood Risk Management* 11(2): 152–168.

Coetzee, Serena, Ivana Ivánová, Helena Mitasova, and Maria Antonia Brovelli. 2020. "Open Geospatial Software and Data: A Review of the Current State and a Perspective into the Future". *ISPRS International Journal of Geo-Information* 9(2): 1–30.

Colesanti, Carlo, and Janusz Wasowski. 2006. "Investigating Landslides with Space-Borne Synthetic Aperture Radar (SAR) Interferometry". *Engineering Geology* 88(3–4): 173–199.

Cooke, Steven J., Jake C. Rice, Kent A. Prior, Robin Bloom, Olaf Jensen, David R. Browne, Lisa A. Donaldson, Joseph R. Bennett, Jesse C. Vermaire, and Graeme Auld. 2016. "The Canadian Context for Evidence-Based Conservation and Environmental Management". *Environmental Evidence* 5(1): 1–9.

Coops, Nicholas C., Piotr Tompaski, Wiebe Nijland, Gregory J. M. Rickbeil, Scott E. Nielsen, Christopher W. Bater, and J. John Stadt. 2016. "A Forest Structure Habitat Index Based on Airborne Laser Scanning Data". *Ecological Indicators* 67: 346–357.

Cradden, L., Christina Kalogeri, I. Martinez Barrios, George Galanis, David Ingram, and George Kallos. 2016. "Multi-Criteria Site Selection for Offshore Renewable Energy Platforms". *Renewable Energy* 87: 791–806.

Crosetto, Michelle, and Stefano Tarantola. 2001. "Uncertainty and Sensitivity Analysis: Tools for GIS-Based Model Implementation". *International Journal of Geographical Information Science* 15(5): 415–437.

Curtis, Philip G., Christy M. Slay, Nancy L. Harris, Alexandra Tyukavina, and Matthew C. Hansen. 2018. "Classifying Drivers of Global Forest Loss". *Science* 361(6407): 1108–1111.

Dabboor, Mohammed, Steve Iris, and Vern Singhroy. 2018. "The RADARSAT Constellation Mission in Support of Environmental Applications". *Multidisciplinary Digital Publishing Institute Proceedings* 2(7): 323.

DeRoy, Bryant C., Vernon Brown, Christina N. Service, Martin Leclerc, Christopher Bone, Iain McKechnie, and Chris T. Darimont. 2021. "Combining High-Resolution Remotely Sensed Data with Local and Indigenous Knowledge to Model the Landscape Suitability of Culturally Modified Trees: Biocultural Stewardship in Kitasoo/Xai'xais Territory". *Facets* 6(1): 465–489.

Drobne, Samo, and Anka Lisec. 2009. "Multi-Attribute Decision Analysis in GIS: Weighted Linear Combination and Ordered Weighted Averaging". *Informatica* 33(4): 459–474.

Dubayah, Ralph, James Bryan Blair, Scott Goetz, Lola Fatoyinbo, Matthew Hansen, Sean Healey, Michelle Hofton, et al. 2020. "The Global Ecosystem Dynamics Investigation: High-Resolution Laser Ranging of the Earth's Forests and Topography". *Science of Remote Sensing* 1: 100002.

Duinker, Peter N., Erin L. Burbidge, Samantha R. Boardley, and Lorne A. Greig. 2013. "Scientific Dimensions of Cumulative Effects Assessment: Toward Improvements in Guidance for Practice". *Environmental Reviews* 21(1): 40–52.

Duque, Juan C., Henry Laniado, and Adriano Polo. 2018. "S-maup: Statistical Test to Measure the Sensitivity to the Modifiable Areal Unit Problem". *PloS One* 13(11): e0207377

Dwyer, John L., David P. Roy, Brian Sauer, Calli B. Jenkerson, Hankui K. Zhang, and Leo Lymburner. 2018. "Analysis Ready Data: Enabling Analysis of the Landsat Archive". *Remote Sensing* 10(9): 1–19.

Eastman, J. Ronald. 1999. "Multi-Criteria Evaluation and GIS". *Geographical Information Systems* 1(1): 493–502.

Eckert, Lauren E., Nick XEMŦOLTW_ Claxton, Cameron Owens, Anna Johnston, Natalie C. Ban, Faisal Moola, and Chris T. Darimont. 2020. "Indigenous Knowledge and Federal Environmental Assessments in Canada: Applying Past Lessons to the 2019 Impact Assessment Act". *Facets* 5(1): 67–90.

Eedy, Wilson. 1995. "The Use of GIS in Environmental Assessment". *Impact Assessment* 13(2): 199–206.

Ersoy, Hakan, and Fikri Bulut. 2009. "Spatial and Multi-Criteria Decision Analysis-Based Methodology for Landfill Site Selection in Growing Urban Regions". *Waste Management and Research* 27(5): 489–500.

Feizizadeh, Bakhtiar, and Thomas Blaschke. 2013. "Land Suitability Analysis for Tabriz County, Iran: A Multi-Criteria Evaluation Approach Using GIS". *Journal of Environmental Planning and Management* 56(1): 1–23.

Feizizadeh, Bakhtiar, Majid Shadman Roodposhti, Piotr Jankowski, and Thomas Blaschke. 2014. "A GIS-Based Extended Fuzzy Multi-Criteria Evaluation for Landslide Susceptibility Mapping". *Computers and Geosciences* 73: 208–221.

Foley, Jonathan A., Ruth Defries, Gregory P. Asner, Carol Barford, Gordon Bonan, Stephen R. Carpenter, F. Stuart Chapin, et al. 2005. "Global Consequences of Land Use". *Science* 309(5734): 570–574.

Fontana, Veronika, Anna Radtke, Valérie Bossi Fedrigotti, Ulrike Tappeiner, Erich Tasser, Stefan Zerbe, and Thomas Buchholz. 2013. "Comparing Land-Use Alternatives: Using the Ecosystem Services Concept to Define a Multi-Criteria Decision Analysis". *Ecological Economics* 93: 128–136.

Geneletti, Davide. 2010. "Combining Stakeholder Analysis and Spatial Multicriteria Evaluation to Select and Rank Inert Landfill Sites". *Waste Management* 30(2): 328–337.

Gharehbaghi, Koorosh, and Christina Scott-Young. 2018. "GIS as a Vital Tool for Environmental Impact Assessment and Mitigation". *IOP Conference Series: Earth and Environmental Science* 127(1), 012009.

Gibbons, Stephen. 2015. "Gone with the Wind: Valuing the Visual Impacts of Wind Turbines through House Prices". *Journal of Environmental Economics and Management* 72: 177–196.

Glasson, John, and Riki Therivel. 2019. *Introduction to Environmental Impact Assessment*. Routledge.

Gomarasca, Mario A. 2010. "Basics of Geomatics". *Applied Geomatics* 2(3): 137–146.

Gonzalez, Ainhoa, and Álvaro Enríquez-De-Salamanca. 2018. "Spatial Multi-Criteria Analysis in Environmental Assessment: A Review and Reflection on Benefits and Limitations". *Journal of Environmental Assessment Policy and Management* 20(3): 1–24.

Goodbody, Tristan R.H., Nicholas C. Coops, Joan E. Luther, Piotr Tompalski, Christopher Mulverhill, Catherine Frizzle, Richard Fournier, Shane Furze, and Sam Herniman. 2021. "Airborne Laser Scanning for Quantifying Criteria and Indicators of Sustainable Forest Management in Canada". *Canadian Journal of Forest Research* 51(7): 972–985.

Goodchild, Michael F. 2018. "Reimagining the History of GIS". *Annals of GIS* 24(1): 1–8.

Goodchild, Michael F. 2011. "Scale in GIS: An overview". *Geomorphology* 130(1–2): 5–9.

Goodchild, Michael F. 2000. "Part 1. Spatial Analysts and GIS Practitioners: The Current Status of GIS and Spatial Analysis". *Journal of Geographical Systems* 2(1): 5–10.

Gorelick, Noel, Matt Hancher, Mike Dixon, Simon Ilyushchenko, David Thau, and Rebecca Moore. 2017. "Google Earth Engine: Planetary-Scale Geospatial Analysis for Everyone". *Remote Sensing of Environment* 202: 18–27.

Haddad, Nick M., Lars A. Brudvig, Jean Clobert, Kendi F. Davies, Andrew Gonzalez, Robert D. Holt, Thomas E. Lovejoy, et al. 2015. "Habitat Fragmentation and Its Lasting Impact on Earth's Ecosystems". *Science Advances* 1(2): 1–10.

Haldar, Swapan K. 2018. *Mineral Exploration: Principles and Applications.* Amsterdam: Elsevier.

Hanna, Kevin S. 2016. "Environmental Impact Assessment: Process, Setting and Efficacy". In *Environmental Impact Assessment Practice and Participation* (3rd ed.). Oxford University Press.

Hansen, M. C. C., P.V. Potapov, R. Moore, M. Hancher, S. A. a Turubanova, A. Tyukavina, D. Thau, et al. 2013. "High-Resolution Global Maps of 21st-Century Forest Cover Change". *Science* 342(6160): 850–854.

Harker, Karly J., Lauren Arnold, Ira J. Sutherland, and Sarah E. Gergel. 2021. "Perspectives from Landscape Ecology Can Improve Environmental Impact Assessment". *Facets* 6(1): 358–878.

Herbert, Grant, and Xuwei Chen. 2015. "A Comparison of Usefulness of 2D and 3D Representations of Urban Planning". *Cartography and Geographic Information Science* 42(1): 22–32.

Hermosilla, Txomin, Michael A. Wulder, Joanne C. White, Nicholas C. Coops, and Geordie W. Hobart. 2015a. "Regional Detection, Characterization, and Attribution of Annual Forest Change from 1984 to 2012 Using Landsat-Derived Time-Series Metrics". *Remote Sensing of Environment* 170: 121–132.

Hermosilla, Txomin, Michael A. Wulder, Joanne C. White, Nicholas C. Coops, and Geordie W. Hobart. 2015b. "An Integrated Landsat Time Series Protocol for Change Detection and Generation of Annual Gap-Free Surface Reflectance Composites". *Remote Sensing of Environment* 158: 220–234.

Hermosilla, Txomin, Michael A. Wulder, Joanne C. White, Nicholas C. Coops, and Geordie W. Hobart. 2018. "Disturbance-Informed Annual Land Cover Classification Maps of Canada's Forested Ecosystems for a 29-Year Landsat Time Series". *Canadian Journal of Remote Sensing* 44(1): 1–21.

Huang, Ivy B., Jeffrey Keisler, and Igor Linkov. 2011. "Multi-Criteria Decision Analysis in Environmental Sciences: Ten Years of Applications and Trends". *Science of the Total Environment* 409(19): 3578–3594.

Jaalama, Kaisa, Nora Fagerholm, Arttu Julin, Juho-Pekka Virtanen, Mikko Maksimainen, and Hannu Hyyppä. 2021. "Sense of Presence and Sense of Place in Perceiving a 3D Geovisualization for Communication in Urban Planning – Differences Introduced by Prior Familiarity with the Place". *Landscape and Urban Planning* 207: 103996.

Jaboyedoff, Michel, Thierry Oppikofer, Antonio Abellán, Marc Henri Derron, Alex Loye, Richard Metzger, and Andrea Pedrazzini. 2012. "Use of LIDAR in Landslide Investigations: A Review". *Natural Hazards* 61(1): 5–28.

João, Elsa, and Alexandra Fonseca. 1996. "The Role of GIS in Improving Environmental Assessment Effectiveness: Theory vs. Practice". *Impact Assessment* 14(4): 371–387.

Kamali, Muhagir El, Abdelgadir Abuelgasim, Ioannis Papoutsis, Constantinos Loupasakis, and Charalampos Kontoes. 2020. "A Reasoned Bibliography on SAR Interferometry Applications and Outlook on Big Interferometric Data Processing". *Remote Sensing Applications: Society and Environment* 19: 100358.

Kedron, Peter, Wenwen Li, Stewart Fotheringham, and Michael Goodchild. 2021. "Reproducibility and Replicability: Opportunities and Challenges for Geospatial Research". *International Journal of Geographical Information Science* 35(3): 427–445.

Kingsford, Richard T., Alberto Basset, and Leland Jackson. 2016. "Wetlands: Conservation's Poor Cousins". *Aquatic Conservation: Marine and Freshwater Ecosystems* 26(5): 892–916.

Klouček, Tomáš, Ondřej Lagner, and Petra Šímová. 2015. "How Does Data Accuracy Influence the Reliability of Digital Viewshed Models? A Case Study with Wind Turbines". *Applied Geography* 64: 46–54.

Koschke, Lars, Christine Fürst, Susanne Frank, and Franz Makeschin. 2012. "A Multi-Criteria Approach for an Integrated Land-Cover-Based Assessment of Ecosystem Services Provision to Support Landscape Planning". *Ecological Indicators* 21: 54–66.

Latinopoulos, Dionsys, and Kiriaki Kechagia. 2015. "A GIS-Based Multi-Criteria Evaluation for Wind Farm Site Selection. A Regional Scale Application in Greece". *Renewable Energy* 78: 550–560.

Leeuwen, Martin van, and Maarten Nieuwenhuis. 2010. "Retrieval of Forest Structural Parameters Using LiDAR Remote Sensing". *European Journal of Forest Research* 129(4): 749–770.

Levin, Simon A. 1992. "The Problem of Pattern and Scale in Ecology: The Robert H. MacArthur Award Lecture". *Ecology* 73(6): 1943–1967.

Li, Jin, and Andrew D. Heap. 2014. "Spatial Interpolation Methods Applied in the Environmental Sciences: A Review". *Environmental Modelling and Software* 53: 173–189.

Li, Jin, and Andrew D. Heap. 2011. "A Review of Comparative Studies of Spatial Interpolation Methods in Environmental Sciences: Performance and Impact Factors". *Ecological Informatics* 6 (3–4): 228–241.

Li, Wanlu, Binbin Xu, Qiujin Song, Xingmei Liu, Jianming Xu, and Philip C. Brookes. 2014. "The Identification of 'hotspots' of Heavy Metal Pollution in Soil-Rice Systems at a Regional Scale in Eastern China". *Science of the Total Environment* 472: 407–420.

Li, Xiugang, Wei Wang, Fang Li, and Xuejun Deng. 1999. "GIS Based Map Overlay Method for Comprehensive Assessment of Road Environmental Impact". *Transportation Research Part D: Transport and Environment* 4(3): 147–158.

Liou, Yuei An, Anh Kim Nguyen, and Ming Hsu Li. 2017. "Assessing Spatiotemporal Eco-Environmental Vulnerability by Landsat Data". *Ecological Indicators* 80: 52–65.

Long, Jed, Trisalyn Nelson, and Michael Wulder. 2010. "Regionalization of Landscape Pattern Indices Using Multivariate Cluster Analysis". *Environmental Management* 46(1): 134–142.

López, Elena, and Andrés Monzón. 2010. "Integration of Sustainability Issues in Strategic Transportation Planning: A Multi-Criteria Model for the Assessment of Transport Infrastructure Plans". *Computer-Aided Civil and Infrastructure Engineering* 25(6): 440–451.

Ma, Lei, Yu Liu, Xueliang Zhang, Yuanxin Ye, Gaofei Yin, and Brian A. Johnson. 2019. "Deep Learning in Remote Sensing Applications: A Meta-Analysis and Review". *ISPRS Journal of Photogrammetry and Remote Sensing* 152: 166–177.

Malczewski, Jacek. 2006. "GIS-Based Multicriteria Decision Analysis: A Survey of the Literature". *International Journal of Geographical Information Science* 20(7): 703–726.

Martin, David. 2009. "The Role of GIS". In Stewart A. Fotheringham and Peter A. Rogerson (Eds.). *The SAGE Handbook of Spatial Analysis*. Los Angeles: Sage Publications .

Masek, Jeffrey G., Michael A. Wulder, Brian Markham, Joel McCorkel, Christopher J. Crawford, James Storey, and Del T. Jenstrom. 2020. "Landsat 9: Empowering Open Science and Applications through Continuity". *Remote Sensing of Environment* 248: 111968.

Maxwell, Aaron E., Timothy A. Warner, and Fang 2018. "Implementation of Machine-Learning Classification in Remote Sensing: An Applied Review". *International Journal of Remote Sensing* 39(9): 2784–2817.

Moreira, Alberto, Pau Prats-Iraola, Marwan Younis, Gerhard Krieger, Irena Hajnsek, and Konstantinos P. Papathanassiou. 2013. "A Tutorial on Synthetic Aperture Radar". *IEEE Geoscience and Remote Sensing Magazine* 1(1): 6–43.

Morgan, Jessica L., Sarah E. Gergel, and Nicholas C. Coops. 2010. "Aerial Photography: A Rapidly Evolving Tool for Ecological Management". *BioScience* 60(1): 47–59.

Mosadeghi, Razieh, Jan Warnken, Rodger Tomlinson, and Hamid Mirfenderesk. 2015. "Comparison of Fuzzy-AHP and AHP in a Spatial Multi-Criteria Decision Making Model for Urban Land-Use Planning". *Computers, Environment and Urban Systems* 49: 54–65.

Neumann, Thomas A., Anthony J. Martino, Thorsten Markus, Sungkoo Bae, Megan R. Bock, Anita C. Brenner, Kelly M. Brunt, et al. 2019. "The Ice, Cloud, and Land Elevation Satellite − 2 Mission: A Global Geolocated Photon Product Derived from the Advanced Topographic Laser Altimeter System". *Remote Sensing of Environment* 233: 111325.

Newell, Robert, Rosaline Canessa, and Tara Sharma. 2017. "Visualizing Our Options for Coastal Places: Exploring Realistic Immersive Geovisualizations as Tools for Inclusive Approaches to Coastal Planning and Management". *Frontiers in Marine Science* 4: 1–19.

Noble, Bram, Michael Hill, and Jesse Nielsen. 2011. "Environmental Assessment Framework for Identifying and Mitigating the Effects of Linear Development to Wetlands". *Landscape and Urban Planning* 99(2): 133–140.

Nutsford, Daniel, Femke Reitsma, Amber L. Pearson, and Simon Kingham. 2015. "Personalising the Viewshed: Visibility Analysis from the Human Perspective." *Applied Geography* 62: 1–7.

Paci, Chris, Ann Tobin, and Peter Robb. 2002. "Reconsidering the Canadian Environmental Impact Assessment Act a Place for Traditional Environmental Knowledge". *Environmental Impact Assessment Review* 22(2): 111–127.

Paloscia, S., S. Pettinato, E. Santi, C. Notarnicola, L. Pasolli, and A. Reppucci. 2013. "Soil Moisture Mapping Using Sentinel-1 Images: Algorithm and Preliminary Validation". *Remote Sensing of Environment* 134: 234–248.

Pandey, Prem Chandra, Nikos Koutsias, George P. Petropoulos, Prashant K. Srivastava, and Eyal Ben Dor. 2021. "Land Use/Land Cover in View of Earth Observation: Data Sources, Input Dimensions, and Classifiers – a Review of the State of the Art". *Geocarto International* 36(9): 957–988.

Peeters, Luk J.M., Daniel E. Pagendam, Russell S. Crosbie, Praveen K. Rachakonda, Warrick R. Dawes, Lei Gao, Steve P. Marvanek, Yong Qiang Zhang, and Tim R. McVicar. 2018. "Determining the Initial Spatial Extent of an Environmental Impact Assessment with a Probabilistic Screening Methodology". *Environmental Modelling and Software* 109: 353–367.

Pepe, Antonio, and Fabiana Calò. 2017. "A Review of Interferometric Synthetic Aperture RADAR (InSAR) Multi-Track Approaches for the Retrieval of Earth's Surface Displacements". *Applied Sciences* 7(12): 1264.

Pérez-Hoyos, A., A. Udías, and F. Rembold. 2020. "Integrating Multiple Land Cover Maps through a Multi-Criteria Analysis to Improve Agricultural Monitoring in Africa". *International Journal of Applied Earth Observation and Geoinformation* 88: 102064.

Perminova, Tataina, Natalia Sirina, Bertrand Laratte, Natalia Baranovskaya, and Leonid Rikhvanov. 2016. "Methods for Land Use Impact Assessment: A Review". *Environmental Impact Assessment Review* 60: 64–74.

Pettorelli, Nathalie. 2013. *The Normalized Difference Vegetation Index*. Oxford: Oxford University Press.

Pettorelli, Nathalie, Henrike Schulte to Bühne, Ayesha Tulloch, Grégoire Dubois, Macinnis-Ng, Cate, Ana M. Queirós, David A. Keith, et al. 2018. "Satellite Remote Sensing of Ecosystem Functions: Opportunities, Challenges and Way Forward". *Remote Sensing in Ecology and Conservation* 4(2): 71–93.

Quiñonez-Piñón, R., A. Mendoza-Durán, and C. Valeo. 2007. "Design of an Environmental Monitoring Program Using NDVI and Cumulative Effects Assessment". *International Journal of Remote Sensing* 28 (7): 1643–1664.

Randin, Christophe F., Michael B. Ashcroft, Janine Bolliger, Jeannine Cavender-Bares, Nicholas C. Coops, Stefan Dullinger, Thomas Dirnböck, et al. 2020. "Monitoring Biodiversity in the Anthropocene Using Remote Sensing in Species Distribution Models". *Remote Sensing of Environment* 239: 111626.

Raspini, Federico, Federica Bardi, Silvia Bianchini, Andrea Ciampalini, Chiara Del Ventisette, Paolo Farina, Federica Ferrigno, Lorenzo Solari, and Nicola Casagli. 2017. "The Contribution of Satellite SAR-Derived Displacement Measurements in Landslide Risk Management Practices". *Natural Hazards* 86(1): 327–351.

Raymond, Christopher M., Ioan Fazey, Mark S. Reed, Lindsay C. Stringer, Guy M. Robinson, and Anna C. Evely. 2010. "Integrating Local and Scientific Knowledge for Environmental Management". Journal of Environmental Management 91(8): 1766–1777.

Roberts, David R., Volker Bahn, Simone Ciuti, Mark S. Boyce, Jane Elith, Gurutzeta Guillera-Arroita, Severin Hauenstein, et al. 2017. "Cross-Validation Strategies for Data with Temporal, Spatial, Hierarchical, or Phylogenetic Structure". *Ecography* 40(8): 913–929.

Rodriguez-Bachiller, Agustin, and John Glasson. 2004. *Expert Systems and Geographic Information Systems for Impact Assessment*. Boca Raton: CRC Press.

Saaty, T.L. 1980. *The Analytic Hierarchy Process*. New York: McGraw-Hill.

Suh, Jangwon, Sung Min Kim, Huiuk Yi, and Yosoon Choi. 2017. "An Overview of GIS-Based Modeling and Assessment of Mining-Induced Hazards: Soil, Water, and Forest". *International Journal of Environmental Research and Public Health* 14(12): 1463.

Tarolli, Paolo. 2014. "High-Resolution Topography for Understanding Earth Surface Processes: Opportunities and Challenges". *Geomorphology* 216: 295–312.

Tennøy, Aud, Jens Kværner, and Karl Idar Gjerstad. 2006. "Uncertainty in Environmental Impact Assessment Predictions: The Need for Better Communication and More Transparency". *Impact Assessment and Project Appraisal* 24(1): 45–56.

Therivel, Riki, and Bill Ross. 2007. "Cumulative Effects Assessment: Does Scale Matter?" *Environmental Impact Assessment Review* 27(5): 365–385.

Torres, Ramon, Paul Snoeij, Dirk Geudtner, David Bibby, Malcolm Davidson, Evert Attema, Pierre Potin, et al. 2012. "GMES Sentinel-1 Mission". *Remote Sensing of Environment* 120: 9–24.

Tsendbazar, N.E., S. De Bruin, and M. Herold. 2015. "Assessing Global Land Cover Reference Datasets for Different User Communities". *ISPRS Journal of Photogrammetry and Remote Sensing* 103: 93–114.

Wang, Ran, and John A. Gamon. 2019. "Remote Sensing of Terrestrial Plant Biodiversity". *Remote Sensing of Environment* 231: 111218.

Watson, Joss J.W., and Malcolm D. Hudson. 2015. "Regional Scale Wind Farm and Solar Farm Suitability Assessment Using GIS-Assisted Multi-Criteria Evaluation". *Landscape and Urban Planning* 138: 20–31.

Webster, Richard, and Margaret A. Oliver. 2007. *Geostatistics for Environmental Scientists*. Chichester: John Wiley & Sons.

White, J.C., M.A. Wulder, G.W. Hobart, J.E. Luther, T. Hermosilla, P. Griffiths, N.C. Coops, et al. 2014. "Pixel-Based Image Compositing for Large-Area Dense Time Series Applications and Science". *Canadian Journal of Remote Sensing* 40(3): 192–212.

White, Lori, Brian Brisco, Mohammed Dabboor, Andreas Schmitt, and Andrew Pratt. 2015. A Collection of SAR Methodologies for Monitoring Wetlands. *Remote Sensing* 7(6): 7615–7645.

Wilson, John P. 2018. *Environmental Applications of Digital Terrain Modeling*. John Wiley & Sons.

Wu, Jianguo. 2004. "Effects of Changing Scale on Landscape Pattern Analysis: Scaling Relations". *Landscape ecology* 19(2): 125–138.

Wulder, M.A., J.A. Dechka, M.A. Gillis, J.E. Luther, R.J. Hall, A. Beaudoin, and S.E. Franklin. 2003. "Operational Mapping of the Land Cover of the Forested Area of Canada with Landsat Data: EOSD Land Cover Program". *Forestry Chronicle* 79(6): 1075–1083.

Wulder, Michael A., Christopher W. Bater, Nicholas C. Coops, Thomas Hilker, and Joanne C. White. 2008. "The Role of LiDAR in Sustainable Forest Management". *Forestry Chronicle* 84(6): 807–826.

Wulder, Michael A., and Nicholas C. Coops. 2014. "Make Earth Observations Open Access". *Nature* 513: 30–31.

Wulder, Michael A., Nicholas C. Coops, David P. Roy, Joanne C. White, and Txomin Hermosilla. 2018. "Land Cover 2.0". *International Journal of Remote Sensing* 39(12): 4254–4284.

Wulder, Michael A, Joanne C White, Morgan Cranny, Ronald J Hall, Joan E Luther, André Beaudoin, David G Goodenough, and Jeff A Dechka. 2008. "Monitoring Canada's Forests. Part 1: Completion of the EOSD Land Cover Project". *Canadian Journal of Remote Sensing* 34(6): 549–562.

Wycisk, P., T. Hubert, W. Gossel, and Ch Neumann. 2009. "High-Resolution 3D Spatial Modelling of Complex Geological Structures for an Environmental Risk Assessment of Abundant Mining and Industrial Megasites". *Computers and Geosciences* 35(1): 165–182.

Xu, Zhao, and Volker Coors. 2012. "Combining System Dynamics Model, GIS and 3D Visualization in Sustainability Assessment of Urban Residential Development". *Building and Environment* 47(1): 272–287.

Yemshanov, Denys, Frank H. Koch, Kurt H. Riitters, Brian McConkey, Ted Huffman, and Stephen Smith. 2015. "Assessing Land Clearing Potential in the Canadian Agriculture-Forestry Interface with a Multi-Attribute Frontier Approach". *Ecological Indicators* 54: 71–81.

Yuan, Yumin, Mark Cave, and Chaosheng Zhang. 2018. "Using Local Moran's I to Identify Contamination Hotspots of Rare Earth Elements in Urban Soils of London". *Applied Geochemistry* 88: 167–178.

Zhang, Chaosheng, Lin Luo, Weilin Xu, and Valerie Ledwith. 2008. "Use of Local Moran's I and GIS to Identify Pollution Hotspots of Pb in Urban Soils of Galway, Ireland". *Science of the Total Environment* 398 (1–3): 212–221.

Zhang, Y.J., A.J. Li, and T. Fung. 2012. "Using GIS and Multi-Criteria Decision Analysis for Conflict Resolution in Land Use Planning". *Procedia Environmental Sciences* 13 (2011): 2264–2273.

Zhou, Xiaobing, Ni-bin Chang, and Shusun Li. 2009. "Applications of SAR Interferometry in Earth and Environmental Science Research". *Sensors* 9(3): 1876–1912.

Zhu, Zhe, Michael A. Wulder, David P. Roy, Curtis E. Woodcock, Matthew C. Hansen, Volker C. Radeloff, Sean P. Healey, et al. 2019. "Benefits of the Free and Open Landsat Data Policy". *Remote Sensing of Environment* 224: 382–385.

13

INDIGENOUS IMPACT ASSESSMENT

A quiet revolution in EIA?

Ciaran O'Faircheallaigh and Alistair MacDonald

Introduction

Indigenous impact assessment (IIA) is a recent development, with the first IIA we are aware of being conducted in 1991 by the Aboriginal community of Hopevale in far north Queensland (Holden and O'Faircheallaigh 1995). IIA now occurs regularly in Australia, Canada, and New Zealand, though its occurrence remains far from routine. It is also beginning to emerge in other parts of the globe (Lawrence and Larsen 2017). IIA displays considerable diversity, being used to reappraise existing projects as well as new ones that are applying for regulatory approval; it ranges from extensive, multi-disciplinary studies considering a wide range of impacts and generating reports numbering hundreds of pages (KLC 2010a), to much more limited studies that focus on a single issue or area, for example, cultural heritage (Jolley 2007). This chapter focuses mainly on Australia and Canada, where the majority of IIAs have been conducted to date and where our experience largely lies.

Drawing on Gibson et al. (2018, 10), we define IIA as follows:

> A process that assesses, and sets out to manage, the impacts of a project, existing or planned, where the assessment is designed and conducted with meaningful input and a significant degree of control by affected Indigenous parties. The Indigenous parties exercise such control over scoping, data collection, assessment, management planning, and/or decision-making about a project.

As implied in this definition, we are not focusing on Indigenous inputs, often in the form of contributions to baseline studies or provisions of written comments on draft reports studies, into environmental impact assessment that is conducted and controlled by project proponents or the regulatory authorities (referred to, as elsewhere in this volume, as 'EIA'). We are examining impact assessment (IA – by which we mean any form of impact assessment study) over which affected First Peoples have some capacity to determine what is assessed and how it is assessed.

Another definitional issue involves 'Indigenous'.

> Indigenous communities, peoples and nations are those which, having a historical continuity with pre-invasion and pre-colonial societies that developed on their territories,

DOI: 10.4324/9780429282492-14

consider themselves distinct from other sectors of the societies now prevailing on those territories, or parts of them.

(United Nations 2013, 6)

We use 'Indigenous' interchangeably with 'First Peoples', the term preferred by Indigenous people themselves in some jurisdictions.

As indicated by the use of the terms 'significant' and 'meaningful' in our definition of IIA, 'control' is not a black and white concept in this context. There are degrees of control, depending on which aspects of IA, and how many of these aspects, are subject to Indigenous influence and the extent of that influence. In the section on mechanisms of control, we identify four broad aspects of IA that are potentially subject to Indigenous control – scoping; who conducts IA; how it is conducted; and IA findings and recommendations. We discuss why these are important and the ways in which Indigenous control can be meaningfully exercised. In the process, we provide the reader with a picture of what IIA entails and the ways in which it can be conducted.

In the following section, we treat separately the very important issue of Indigenous control over decisions about whether, and under what conditions, a proposed project should proceed. This is because IA systems generally maintain a distinction between IA, which is one component that feeds into major project decision-making, which is conducted by an agency of the state based on multiple factors. This has implications for the aspirations and outcomes that First Peoples seek and associate with IIA.

Why has IIA emerged in recent decades? One negative force involves the fact that Indigenous peoples have, for reasons explored in detail in the next section, been excluded or marginalized from EIA. First Peoples have challenged this exclusion (Procter 2020), and in some cases responded by developing their own IA systems. A second more positive reason is the growing recognition of Indigenous legal and political rights in domestic jurisdictions and in international law (Gibson et al. 2016; O'Faircheallaigh 2016). This growing recognition has increased the ability of First Peoples to insist that their interests should be recognized in the assessment of projects that affect them, and that they themselves should have the ability to define and articulate those interests. It has also placed pressure on governments and corporations to ensure that there is effective consultation with Indigenous peoples about such projects. IIA provides one such mechanism.

There has also been a growing recognition among corporate managers in recent decades that they must demonstrate their 'corporate social responsibility' in order to sustain their long-term profitability (O'Faircheallaigh and Ali 2008; Moody's Investor Service 2020). One means of doing so involves providing financial support for IIA, a critical matter given resource limitations experienced by almost all Indigenous communities. Another involves negotiation of agreements with affected Indigenous landowners and communities; IIAs may be conducted as a precursor to such negotiations, or may result from them with agreements requiring the conduct of IIAs for any future project expansions.

We approach our discussion of IIA as follows. In the next section, we identify critical weaknesses in EIA from an Indigenous perspective. We then examine what renders IIA distinctive, before examining the way in which IIA is linked to decision-making by First Peoples, state regulators, and corporations about whether projects should proceed and, if so, under what conditions.

Next, we consider the broader benefits arising from IIA, which is followed by a consideration of a range of models that can be utilized when conducting IIA. We consider their potential and limitations, and discuss practical constraints and choices that First Peoples face in utilizing IIA. In concluding the chapter, we examine potential future trajectories of IIA.

Problems with EIA

By EIA we mean IA where the project proponent prepares an assessment of a project's expected impacts and makes commitments and proposes measures to mitigate these. The 'technical' work of compiling an EIA is exclusively or primarily conducted by specialist consulting firms employed, directed, and paid for by the proponent. A government statutory board and/or a minister with relevant portfolio responsibilities will, on the basis of the proponent's EIA filings and in some cases comments on them by interested parties, determine whether or not a project should proceed; and if it is to proceed what conditions will be attached to it.

This conventional approach creates serious problems for First Peoples. EIA has often entirely ignored their interests and even their existence, with the results that there is no competing story of likely change from affected First Peoples and that proponents almost entirely control the narrative. For instance, Weitzner (2008, 8) notes that the environmental and social impact assessment for the Bakhuis bauxite project in Suriname entirely ignored impacts on Indigenous Guyanese communities, an outcome the proponent justified by denying that there were Indigenous peoples anywhere near the project (for other examples see World Bank 2011, 21). Exclusion of Indigenous peoples from EIA is not just a historical phenomenon or confined to developing countries. In a current EIA for a new mine in Australia adjacent to an Aboriginal community, the consulting firm retained by the proponent planned to undertake the scoping and baseline phases of the IA with no input at all from the affected community.[1] Even where Indigenous peoples are not entirely excluded, there are still issues about whether their participation is meaningful (Gibson et al. 2018). In the Canadian context, Indigenous groups have raised a strong and consistent message that engagement by proponents and consultation by government has lacked meaning, and the Courts have criticized government for consultation processes that only allow First Peoples to 'blow off steam', meeting minimum consultation requirements rather than contributing substantially to project planning and conditioning (see Craik 2016).

There are a variety of reasons why the opportunities for First Peoples' participation in EIA may be severely restricted. Their access to EIA documents may be limited because they are only available in capital cities, are not translated into local languages, and are written in highly technical language. The definition of who has 'standing' in relation to a project and so has a right to comment on it may exclude affected Indigenous peoples, and time frames for public input may be unrealistic (Gibson 2012; Weitzner 2008). Proponents may conduct much or all of their baseline data collection prior to engaging First Peoples and may be reluctant to reactivate such studies thereafter. Methodologies used in IA field studies can also reduce participation by failing to identify affected groups (Tsuji et al. 2011).

Even where participation does result in Indigenous or community knowledge and perspectives being made available, there is no guarantee that proponents, consultants, or regulators will pay attention to it, or assign it appropriate weight in relation to their own 'scientific' studies (Weitzner 2008). Even if the existence of Indigenous knowledge is recognized, significant problems can arise in ensuring that it is appropriately incorporated into EIA and environmental management (O'Faircheallaigh 2007; Gibson et al. 2016). These include the risk that knowledge will be taken out of the hands of Indigenous knowledge holders and used out of context, being re-interpreted by non-Indigenous users in ways contrary to Indigenous interests (Nadasdy 2003). One of the biggest problems under EIA is that First Peoples' substantive inputs most often stop after baseline data collection. They produce studies that are then subject to proponent or state reinterpretation, and Indigenous peoples are disengaged from the assessment process.

One important underlying factor in EIA is the imbalance of resources between proponents and the state, on the one hand, and Indigenous communities, on the other. The former can mobilize extensive political, financial, and technical resources. First Peoples can almost never match the resources available to the proponents and the state, placing them at a serious disadvantage. Government funding for Indigenous participation is in most cases completely inadequate, forcing First Peoples to negotiate with proponents to provide funding. This allows proponents to substantially control the extent and form of public participation in EIA (Rodriguez-Garavito 2011, 298), and may also impose significant negotiation costs on First Peoples and/or result in proponents demanding that Indigenous communities moderate any criticism of proposed projects in return for funding (Gibson and O'Faircheallaigh 2015, 85).

Another issue involves the fundamental values that underpin IA. The dominant narratives around large-scale industrial development emphasize its importance as a source of employment and economic growth, devalue nature, and equate development with the 'public interest', making it difficult to oppose project approval and resulting in negative impacts being ignored or underestimated. For example, Devlin and Yap (2008, 22) show how proponents of the Pilar Dam in Brazil equated the project with 'modernization and progress', arguing that it would attract industries and employment and result in the provision of technical support to local farmers, and on this basis suggesting the project was a *fait accompli*. One expression of this tendency to privilege industrial development is that government regulators have a high propensity to accept proponents' assessments of the significance of expected impacts from projects. For example, Singh et al. (2019, 133) compared the significance determinations in proponent EIA reports to final regulator decisions in Canada and found that they are 'overwhelmingly identical (93–95%)'. They conclude that while regulators are financially independent of proponents, 'their decisions on significant are heavily dependent on the information and analysis provided by the proponent reports' (Singh et al. 2019, 133).

Linked to the question of values is the evidence and theories of knowledge that underpin the impact assessment and management of large industrial projects. The Western biophysical and human 'sciences' which constitute the basis for EIA are deficient in fundamental ways when considered from an Indigenous perspective. Deficiencies can include shallow time depth in observations of, and limited understanding of, the biophysical and human environments; a 'silo' approach that fails to appreciate the links between different components of those environments, between human and non-human elements of creation, and between the spiritual and material aspects of life; and short time horizons in considering future impacts (Candler et al. 2015; Hoogeveen 2016). Gaps in the consideration of Indigenous worldviews are seen throughout EIA but are perhaps most noticeable in the weak consideration of cumulative effects. From an Indigenous worldview, it is the total sum of effects on Indigenous values over time that are important to consider when determining whether a new project should be allowed to proceed. Cumulative effects alter the vulnerability and resilience of Indigenous peoples and the resources they rely upon in the pre-project situation. This suggests that a deep understanding of cumulative effects should be integral to the assessment of project-specific impacts. This is only rarely the case in most EIA, where a cumulative effects assessment is typically only conducted if substantial impacts on a valued component are first estimated for a project on a stand-alone basis. This means that existing impacts on Indigenous values and rights, often heavily damaged already, are hidden from view during impact assessment.

In the following sections, we consider how IIA is being used to overcome EIA shortcomings.

IIA: Mechanisms of control

'Control' under IIA should be understood as a spectrum. At one end, a First People might have minimal influence over just a few aspects of IA, a situation we would clearly not define as IIA. Conversely, extensive Indigenous influence across all aspects of IA would merit such a description. It is not helpful to nominate a specific point on the spectrum at which an IA should be labeled 'Indigenous'. Rather we highlight key areas where First Peoples should try to maximize their influence if IA is to work in their favor.

Control over scoping

The first area involves control over scoping, which includes defining what constitutes an impact, which impacts will attract the most attention and resources in an IA, and the space and time over which impacts will be assessed.

A key starting point involves defining the source of the impact to be assessed, which can be done more narrowly or more broadly. A narrow definition might include only the mining operation itself; a broader definition would also include, for instance, mine infrastructure and the impact of outsiders attracted to a region in the hope of gaining employment. A broader definition may be strongly preferred by Indigenous communities who may be affected as much, for instance, by roads built to access a mine site as by mining itself. Another issue involves whether a mine or oil field is considered as a discrete impact or whether the impact is defined as the new mine *in addition to* whatever mining or oil extraction is already occurring or is about to occur. These cumulative effects are a major issue in the Alberta oil sands industry, for example, where some First Nations, surrounded by large-scale industrial activity and have already lost much of the land they rely on for physical, cultural, and spiritual sustenance (Candler et al. 2015).

A second issue involves the space over which impacts are assessed and, related to this, defining the character of the feature that is impacted. In some cases, an area of impact may be narrowly and arbitrarily defined, for example, by way of a set distance (X kilometers) from the mine site. This may ignore the way in which impacts can be spread much more widely by, for instance, water flows or movements of people.

Space and the character of the feature which receives impacts interact in the case of cultural heritage in Australia, as illustrated by the following example, involving two EIAs of Liquefied Nature Gas (LNG) projects in adjacent regions of Western Australia. In the EIA conducted by a project proponent, possible damage to Aboriginal cultural heritage sites was recognized, but limited to heritage sites within the project area. The proposed management response was to create exclusion zones around the sites. This focus on sites within the project area ignores the reality that site damage could occur much more widely as a result of a predicted increase in the non-Indigenous population living in the region, and that in Aboriginal culture, sites in one area are almost inevitably linked to sites elsewhere and to cultural practices associated with such clusters of sites. In contrast, an IIA for a second LNG project followed Aboriginal understandings in adopting a *regional* approach to the identification and management of the project's impacts on cultural heritage (O'Faircheallaigh 2017).

A third aspect of scoping involves the relative importance assigned to different impacts and the resources allocated to assessing them. In one EIA, considerable attention was focused on microscopic worms believed to occur in the project area because of their rarity. At the same time, funds allocated to social impact assessment were limited. This led an Aboriginal community member to complain: 'The company and the government care more about worms you can't see than they do about people'.[2] An important aspect of IIA is that affected peoples and communities determine what should be the primary focus of assessment work. This is indicated, for example, by the fact that IIA typically has a strong focus on the cultural impacts of projects, an issue that often receives scant attention in EIA (Gibson et al. 2011; O'Faircheallaigh 2017).

Another key issue is the time period over which predicted impacts are assessed. In EIA, this is often driven by commercial factors that shape project design. Because future income is discounted and because proving ore reserves is expensive, mining companies often initially design a project for an operational life of about 15 years. However, once a project is operational, additional ore reserves are usually established, and mine life extended, sometimes for decades. In addition, it is now recognized that the environmental and social impacts of mining can last well beyond the period when minerals are being extracted, and a mining company has surrendered its leases, and that it is adjacent Indigenous communities that bear the brunt of that impact (Keeling and Sandlos 2015). Time depth must also be considered in the context of cumulative impacts arising from earlier industrial development and government policies. EIA tends to focus on changes that are occurring from the present day, while Indigenous groups focus more on total cumulative effects on resources from further back in time. For example, when the container ship MV Rena ran aground on Otāiti (Astrolabe reef) in the Bay of Plenty, New Zealand, in 2011, it resulted in what is considered to be New Zealand's worst maritime environmental disaster. The resultant environmental impacts due to the fuel oil spilled and flotsam cost more than NZ\$660 million to clean up. Faaui et al. (2017) show how an impact assessment using Maori methodologies that considered the disaster in the context of its cumulative impact in addition to earlier effects on Maori of colonization and industrial development resulted in an assessment considerably more negative than analysis of the maritime disaster in isolation. One result of this approach was that efforts at environmental and cultural remediation that appeared positive when the disaster was considered in isolation were assessed as inadequate from a 'cumulative effects' perspective (Faaui et al. 2017, 239–240).

A final point involves the heavy emphasis on avoiding or mitigating negative impacts that characterize EIA. This reflects an underlying and generally unstated assumption that the existing 'pre-project' situation is satisfactory and that the task of EIA is to prevent any deterioration in this situation. However, for many First Peoples, the existing situation is *not* satisfactory. Rather it is characterized by inadequate social services like health and education and limited employment opportunities; overcrowded housing; high rates of social trauma and incarceration; and limited control over their land and cultural resources. They want to know if a proposed project can *improve* the existing situation. Thus, IIA also tends to focus on whether and how positive impacts can be maximized and may be regarded as providing a foundation for the subsequent negotiation of benefit-sharing agreements with project developers (O'Faircheallaigh 2017).

In formal terms, these various aspects of scoping come together in the Terms of Reference or Scoping Document for an IA. An indication of Indigenous control of this area as a whole would be a requirement for affected First Peoples to approve this document before an IA could proceed.

Control over who conducts the IA

A key issue here is who conducts IA work and to whom will they be accountable. In EIA, assessment work is usually carried out by large environmental and engineering services companies, which often have a limited understanding of what is needed to facilitative Indigenous participation, or of the values and worldviews which shape Indigenous understandings of project impacts. A critical aspect of Indigenous control involves the right to choose, or to approve the choice of, the team that will undertake IA work, so as to ensure that the people involved understand Indigenous values and the requirements to achieve effective Indigenous participation. Ideally, a team will include:

- Individuals, Indigenous or non-Indigenous, with credentials in relevant professional areas and substantial experience in Indigenous communities;

- Community members with relevant knowledge and expertise, for example, elders with a deep knowledge of land use and culture, and individuals with extensive social networks and awareness of community dynamics; and
- Younger community members who provide energy and a capacity to engage in particular with youth, and who can use the IA as a learning experience.

An IIA team would normally report to a community-based governing entity, either an existing body such as a Community Council or one created specifically for the purpose. An example of the latter would be the Steering Committees set up to oversee IIAs in Cape York in far north Queensland. These usually include a substantial number of elders; and representatives of formal governing bodies and of significant interests in the community, for example, educators, health workers, women, and youth (O'Faircheallaigh 2000). A similar 'representational cross-section' approach was used by the Stk'emlúpsemc te Secwepemc Nation (SSN) for the Ajax Mine Project IIA in British Columbia, Canada (SSN 2017).

Control over how the IA is conducted

The first issue here involves the sort of information that is used in establishing baseline data and in documenting community concerns and aspirations in relation to a project. In EIA, heavy reliance is placed on documentary sources and on collection of data through methods such as surveys of households; flora and fauna surveys; chemical analysis of water; physical monitoring of water flows and tidal movements; and computer-based modeling of expected biophysical impacts. IIA may replace such methods, or supplement them, by documenting the experiential learning of elders and other community members with a deep knowledge, accumulated over generations, of environmental, cultural, and social dynamics. Indigenous observational or sensory indicators, sometimes criticized for being qualitative and imprecise in nature by people whose only frame of reference is Western science, are in fact detailed, rigorous, and replicable,[3] and much more likely to resonate with affected First Peoples.

The second issue involves how information about a proposed project and its potential impact is communicated to people, and how their concerns and aspirations are documented. EIA relies heavily on the provision of information in written form and on public meetings to solicit community input. Both have serious limitations in an Indigenous context, given that the dominant mode of communication is oral, that literacy in English or other official national languages is often limited, and that there may be serious inhibitions about expressing views in public, especially ones that contradict or challenge the perspective being shared by a proponent.

In IIA, a stronger emphasis is placed on communication of information orally and visually, and through small group interaction. Consultation may be conducted on the land or in people's homes, or through forums that lend themselves to the discussion of specific issues, for instance, meetings of elders, hunters and trappers, youth, or women. Strong emphasis is often placed on achieving consensus, not necessarily in terms of unanimity on matters of detail, but in terms of a shared understanding of issues that a proposed project raises and broad outcomes that are desirable, acceptable, and not acceptable.

A final issue involves the time available for IA. Communication and engagement practices used in IIA require considerably more time than those typical of EIA. This does not mean that IIA cannot occur within tight time frames if adequate resources are available to, for instance, pursue multiple strands of research and engagement simultaneously. However, the very limited time frames applied in many EIA processes, for instance, requiring responses to draft scoping documents or draft EIA reports within a matter of weeks, are unlikely to be compatible with IIA.

Control of IA findings

The culmination of an IA process is the preparation of a report that makes findings of facts and interpretations. Findings of significance are especially important, as they directly affect judgments as to whether or not impacts (and by extension, the project) are acceptable. They are also particularly open to multiple perspectives. For example, an expected decline of 10% in the regional population of a particular food species may not be regarded as 'significant' in an EIA. In contrast, if a specific Indigenous community relies heavily on this species, or if it has high cultural or spiritual importance, any impact on its population may be deemed highly significant. In addition, IIAs are much more likely to focus on total cumulative effects on a wildlife species from all causes of impacts, rather than to focus only on the effect of one proposed project.[4] Referring to the earlier discussion of cultural sites, the loss of a single site of a form that is widely represented across a region may not be rated as significant in an EIA. If that site is part of a complex group of sites that is in turn linked to important ceremonial activity or spiritual connections, an IIA may regard its loss as so significant that it renders the project unacceptable (KLC 2010b).

Recommendations regarding management strategies can also be of great importance. If accepted and imposed as conditions on project approval, they will play a key role in determining the ultimate residual impacts of a project. Here also there is considerable scope for divergence, especially as EIA, no matter how rigorous, is rarely informed by the understanding of ecological and social dynamics that IIA can apply in designing mitigation and management methods.

Finally, recommendations as to whether a project may proceed, and under what conditions, are of great importance. Though they may not formally bind decision-makers (see next section), in most cases they create the parameters within which regulatory decisions are taken. A key indicator of Indigenous control in this area would be a requirement that an EIA report and recommendations would have to be approved by the affected First Peoples before submission to the relevant regulatory authorities.

The relationship between IA and decision-making

IIA seeks to give Indigenous peoples greater control over decisions that affect their lives, including whether a major project should proceed and, if so, under what conditions. However, this aspiration needs to be reconciled against the reality that in almost all impact assessment regimes, it remains the responsibility of other parties to make the ultimate decision on these issues. State authorities, typically a minister, will approve, reject, or require changes to the project as proposed. This state control over the fate of the project means that there are circumstances where Indigenous decisions to withhold consent are not adopted by the state. It is also the case that final investment decisions are undertaken by corporate proponents. In a capitalist system, only they can decide if a project approved by the state will proceed, and they have considerable discretion in shaping project design within parameters acceptable to the state.

It can be difficult to reconcile state and corporate decision-making with the aspirations of Indigenous groups who have ancestral stewardship and governance responsibilities for their territory, but in our experience IIA increases the influence that First Peoples can have on these other layers of decision-making. IIAs can create judicial review pressures on statutory decision-makers to fully consider Indigenous evidence. Governments (and proponents) seeking to reduce legal risks have sought to more meaningfully engage Indigenous groups as a result, and IIA can allow them to do this. IIA has been used in Australia for over two decades to help influence proponent decisions on project design and on systems for environmental and cultural heritage protection (O'Faircheallaigh 2000, 2015). In addition, IIA can be of fundamental value in the assertion and pursuit of First Peoples' governance and stewardship responsibilities which cannot be shirked even if they are not recognized by others (O'Faircheallaigh 2016, 68–69).

Not every IIA has control over decision-making on proposed projects as its primary outcome, especially as the state may be more willing to concede influence or control in other areas. MacDonald et al. (2020) developed a 'control and responsibility' spectrum for IIA and, looking at case studies over the past decade, found that most commonly, IAs have seen Indigenous groups taking control over baseline studies. Less numerous are examples where Indigenous groups establish control over EIA processes (timelines, steps involved, information requirements), and least often are they involved in ultimate decision-making on a project (assessment findings, condition setting, project approvals/rejections).

Each of these foci for IIA can be valuable; Indigenous control over ultimate decision-making can remain elusive without necessarily negating the value of IIA (see next section). However, it is important for First Peoples considering an IA to understand these limitations on control over ultimate decisions. If they are not recognized, disappointment in the outcomes of the assessment among leaders and the community could lead to a reluctance to participate in future IIAs.

A final and important point in relation to decision-making is that IIA should be part of an overall strategy rather than the single tool used by Indigenous groups. Successful IIAs can be accompanied by activities on other fronts in order to help inform and influence final decisions on projects, including the development of a political and media strategy, finding allies with environmental and citizen's groups, and mobilizing pan-Indigenous action. All of these strategies were employed by the Kimberley Land Council in Western Australia, for example, to try and ensure that the findings and recommendations of the IIA of the proposed Browse LNG project were acted on by State and Federal government decision-makers (for details, see O'Faircheallaigh 2015).

Wider benefits of IIA

For Indigenous communities, a critical benefit of IIA is that if a proposed project proceeds, regardless of whether Indigenous consent is provided or withheld, they will be better prepared to deal with its impacts if they understand them as a result of IIAs they have conducted. Indigenous peoples may be more likely to engage in an IIA than in one run by a proponent or the state, as community members are more comfortable to engage within their own cultural group; processes are designed to share information in ways that make sense to First Peoples; and participants feel more confident that their information will be valued, acted on, and protected from misuse.

IIAs can also be used to develop internal unity, and to build impact assessment capacity that has benefits beyond the project concerned. They can be used to gather information critical to the Indigenous group both within and outside the project-specific context, and this information can be employed, for example, to lever additional funds from government or to develop new policies and plans to protect Indigenous territories. A well-structured and credible IIA can increase the leverage of the Indigenous group and allow it to negotiate an ongoing role in state IA processes. For example, the Carrier Sekani First Nations' multi-faceted IIA of the Coastal GasLink Project (Toth and Tung 2014; MacDonald 2014) was a critical precursor to negotiating a Collaboration Agreement with British Columbia for all future EIAs in their territories (BC, CSFNs and CSTC 2015).

The Indigenous worldview is often better aligned with principles of sustainable development and multi-generational planning than the shorter-term and more primarily economic focused approach of project proponents. When coupled with increased state recognition of the importance of sustainability principles (see, for example, Canada's 2019 federal *Impact Assessment Act*), these Indigenous perspectives gain greater weight and have increased informational value for state decision-makers.

Both project-specific and cumulative issues raised by IIAs can lead to better project decisions, more extensive conditions on approvals to protect and benefit Indigenous groups and the environment, and wider benefits in the form of tools such as cumulative effects management systems and co-management agreements. In the longer term, IIAs can blaze a path for entirely new ways of conducting impact assessment at the state level. A steady stream of both calls for, and conduct of, IIAs were two of the factors contributing to changes to impact assessment legislation in Canada at the provincial (British Columbia's new BC *Environmental Assessment Act 2018*) and federal levels (*Impact Assessment Act 2019*).

IIA can assist proponents to ensure that they meet their legal obligations, and provide a basis on which to build long-term, positive relationships with affected communities, reducing the risk of project delay or disruption. The ability to finance a project is increasingly being subject to consideration of the degree to which First Peoples' rights and interests are being recognized and respected (Moody's Investor Service 2020).

In drawing together the benefits of IIA and the wider discussion to this point, it is useful to summarize differences between EIA and IIA across a number of key variables (Table 13.1). Each feature should be thought of as describing a tendency rather than as an absolute or a situation that is manifested in every EIA or can be achieved in every IIA. For example, it is not the case that all EIA calculates impacts solely over a time frame driven by commercial discount rates, or that all IIA uses time frames spanning multiple generations. It is rather that EIA has a *tendency toward* the former approach and IIA has a tendency toward the latter. The priority for First Peoples is to shift IA as far as possible toward the right-hand column of Table 13.1 in order that decisions – those made by First Peoples, proponents, and state actors – are all informed by their worldview and knowledge.

Opportunities and obstacles for First Peoples undertaking IIA

There is no 'best practice' in IIA; it is about what works best and is possible at the time for each Indigenous group. This section examines some of the opportunities, obstacles, and choices First Peoples face in developing and undertaking IIAs. We break this down into a discussion of a series of 'whether, who, what, and how' choices. These choices are presented in a linear fashion here; in reality, they may occur in a different order and overlap and change during the course of an IIA.

Whether to conduct an IIA

There are a variety of specific enabling factors that will improve the chances of success of an IIA (Gibson et al. 2018). Indigenous groups need to gauge their situational context and the 'art of the possible' when determining whether to make an effort to take greater control and responsibility in an IA. Enabling factors include the following:

- Supportive legislation including self-government and co-management mechanisms and opportunities built into statutory mechanisms.
- High degree of Indigenous leverage, which can be associated with the degree of connection to place and the centrality of the project's location within a Nation's territory; rec-

Table 13.1 EIA and IIA – tendencies compared

Variable	EIA	IIA
Indigenous participation in IA process	Marginal to secondary	Central rationale and focus
Time frame for conducting IA	Driven by project and regulatory deadlines, often short	Driven by requirements for meaningful Indigenous participation
Time frame over which impacts assessed	Economic life of the project and driven by discount rates	Expected duration of impacts based on Indigenous knowledge; multi-generational; strong emphasis on capturing cumulative effects over the entire project life cycle
Sources and nature of knowledge	Short term, primarily quantitative data collection undertaken for EIA, written (often secondary) sources	Heavily reliant on knowledge of Indigenous peoples, substantial time depth; experiential and sensory; oral
Legal structures/orders	Written legislation and regulations; little latitude to expand the scope of assessment beyond written norms	Group-specific laws and stewardship rights/responsibilities; may be encoded in stories
Organization of knowledge and understanding of impact pathways	Disciplinary and siloed (examines separately impacts on water; air; vegetation; flora and fauna; people); use of biophysical proxies instead of sociocultural perspectives impacting harvesting	Holistic, recognizing interdependency of elements of the environment and of environment and people
Assessing for …	Avoidance of significant adverse effects from the project; preventative	Best future uses of Indigenous territory ('net gains'); aspirational
Assessment of significance	Project-specific, based on scientific or subjective 'professional opinion' definitions of, e.g., acceptable levels of contaminant releases; species 'rarity'	Cumulative, and based on the assessment of the impact on well-being and sustainability of environments, animals and people; more likely to be highly precautionary
Relative weight attached to economic, environmental, and social values	Economic values (local, regional and national) are heavily prioritized	Focus on protecting land-based subsistence economic livelihoods and social and cultural connection to land over the long term
Role for cumulative effects	Only considered (tangentially) if the project causes a residual adverse effect on a Valued Component	Central to the whole process; cumulative change to date helps understand sensitivity to future change, and cumulative effects from all sources drive decisions
Who conducts IA	Consultants selected by and reporting to the proponent	Community members supported by technical experts chosen by and accountable to the community
Indigenous control over project decisions	Very limited, key decisions lie with the regulator, proponent	Control over community-level decisions; increased to substantial control over process/project decisions

ognition of Indigenous rights through government agreements or legal precedents; and a history of community efforts to protect the territory.

- Having the whole of a project within one Indigenous group's territory, rather than across the territory of multiple Nations, may increase leverage. Where there are multiple Indigenous groups involved, a diversity of values and opinions may emerge, a single IIA process may be more difficult to establish, and having different Indigenous voices in an assessment may reduce the clarity and consistency of messaging. That said, where regional Indigenous unity and cohesion can be attained, having multiple Indigenous groups involved is not necessarily a disabling factor, and indeed pooling of resources and assessment capacity among Indigenous groups can prove beneficial. The IIA for the proposed Browse LNG in Western Australia is a case in point. Here the regional representative organization, the Kimberley Land Council, provided a coordinating role, facilitating mobilization of human and financial resources and mutual support across a large number of First Nations (O'Faircheallaigh 2015).
- High Indigenous group human resources capacity. This may include a stable cadre of experienced staff with substantial experience in EIA.
- Degree of funding available. The greater the engagement in an IA, the greater First Peoples' internal, legal, and consulting costs will be. Covering these costs may be difficult. A recent study has shown that in Canada, Indigenous groups have received guaranteed funding from state assessment bodies for only a very small portion of their IA costs (First Nations Major Projects Coalition 2018). Unless adequate and timely funding can be levered from the state or proponents, it may be difficult or impossible to conduct an IIA.

Not having one or more of these enabling factors does not mean that an IIA is not possible or advisable. Having as many in place as possible does, however, increase the likelihood that desirable outcomes are achieved, and to determine how far along the control spectrum the Indigenous group can venture. Indigenous groups considering conducting an IIA should first conduct an assessment of which enabling factors they have in place, what the implications are of their presence or absence, and identify means by which they can be augmented by, for example, building higher internal capacity or creating regional Indigenous alliances.

Who to partner with

There are three general models that can be adopted by Indigenous groups undertaking IIA (Gibson et al. 2018):

1. An *independent IIA* where the Indigenous group 'goes it alone' and makes its own final, independent decision on whether a project should proceed and under what conditions;
2. A *collaborative EIA* conducted with the state impact assessment agency; or
3. A *co-developed IA* where the Indigenous group teams up with the proponent to assess some or all of a project's impacts.

There are potential benefits and limitations to each approach. For example, an independent IIA may work when an Indigenous group has a substantial capacity, or funds to expand capacity, but it has drawbacks where the group lacks the leverage to enforce its decision or conditions at the end of the process. Collaborative IIAs to date have required some degree of acceptance of (often flawed) EIA systems, and while an Indigenous group may gain a seat at the table with the state, increased process involvement without decision-making control over final outcomes may not see the fundamental change many groups are seeking. Engaging primarily with the proponent requires strong relationship building from the outset of a proposed project and a willing and incentivized partner, but may put restrictions on Indigenous groups' ability to raise remaining concerns in public. Table 13.2 identifies some of the attributes we have encountered in examples from the three models.

It is important to remember that the choice of model is not 'once and for all'. For instance, early engagement with a proponent in the development of an EIA does not preclude an Indigenous group from later conducting its own IIA or engaging heavily in a partnership with the state.

What to focus the IIA on

First Peoples need to determine which aspects of an IA they are capable of undertaking, including conducting Indigenous baseline studies, taking more control over elements of the EIA process, or producing assessment outputs and decisions. Conducting Indigenous baseline studies is the least daunting of these tasks. Such studies have at least three decades of track record (Tobias 2000) and many First Peoples are familiar with engaging in Indigenous knowledge and land use studies.

In comparison, running a full IA is beyond the capacity of many Indigenous groups, and will inevitably duplicate elements of the EIA. A better choice may be to 'shadow' the EIA (Bruce and Hume 2015), and separate out key topics for Indigenous groups to control. These may include:

- Cultural impact assessment;
- Indigenous-specific socioeconomic impact assessment
- Indigenous knowledge and use studies;
- Indigenous rights and title impact assessments; and
- Cumulative effects assessments across multiple Valued Components.

Table 13.2 Comparing three 'partnership' models for conducting IIA

Factor	Independent IA ('go it alone')	Co-managed EIA with state	Co-developed IA with proponent
Degree of Indigenous control	High	Variable	Variable
Internal capacity/level of effort required	High to very high	Variable	Variable
Control over decision-making	Internal decisions – high; state and proponent – variable	Variable but potentially higher for state decisions	Variable but potentially higher for proponent decisions
Indigenous decision at the end of the process	Mandatory, highly structured	Optional	Optional
Funding sources	Greater requirement for self-funding; possible access to state and proponent funds	Greater access to state funding	Greater access to proponent funding
Level of community involvement	High to very high	Minimal to high	Moderate to high

Focusing IIA in this way has the advantage of prioritizing topics that are central to First Peoples' concerns, that Indigenous community members are strongly incentivized to engage with, and which EIA in general conducts poorly.

How to undertake the IIA – lenses and voices

Above we discussed some of the methods and approaches that are used in conducting IIA. Here we expand on the discussion of a key issue, that of determining significance. Two matters are involved here: how significance is assessed and how decisions are taken about whether or not anticipated impacts are acceptable to a community.

There are a variety of ways to assess significance. They include typical EIA tools, which focus on the imposition of professional judgment and/or quantitative thresholds of acceptable or manageable change. Both are problematic from an Indigenous perspective, given that First Peoples tend to use qualitative observations over a much deeper time depth to make their judgments, and that the professionals making significance estimations rarely share the Indigenous worldview or knowledge base. IIA, in contrast, may involve the development of very different community-specific metrics or lenses, often using the type of decision-making tools that the Indigenous group would use to make other decisions. Examples include:

- Consent: what level and type of impacts will result in the community consenting to, or rejecting, a proposed project.
- Whether the project will provide a net gain or net loss to the First Peoples or the resources they rely upon, for example, whether the project increases the risk that community ecological and sociocultural restoration goals may not be accomplished (TWN 2015).
- Whether the project will make Indigenous laws and norms difficult or impossible to adhere to (Okanagan Indian Band 2018).
- Whether the project will cause problems for future generations (intergenerational equity), or continue or exacerbate the existing imbalance of benefits and risks between Indigenous and non-Indigenous peoples (impact equity).
- Whether the project will contribute to or take away from reconciliation between Indigenous and non-Indigenous peoples (Wabun Tribal Council 2016).

When it comes to the question of how decisions regarding IIAs and the projects they assess will be made, answers will vary depending on what is deemed appropriate by each Indigenous group based on their own governance norms. Decision-making options include the following, elements of which can be combined:

- **Collaborative consensus** or similar joint decision-making approaches have been sought in state–Indigenous community engagement in some IAs, with the two parties seeking to find agreeable measures to fuel informed consent and protect the environment. However, power imbalances with the state retaining control over the ultimate decision may hamstring this process.
- **Community referendums** or other community voting or consensus processes, using the type of governance mechanisms that are appropriate to the specific community. For example, IIA reports prepared in preparation for negotiation of agreements governing Rio Tinto's bauxite mine in Western Cape York were endorsed by widely attended meetings in all the affected communities (O'Faircheallaigh 2016).

- **Customized review panels** may be struck from a broad cross-section of Indigenous community members, and a panel set up to hear evidence in a quasi-judicial or less formal setting. A panel of 26 community members was used by the Stk'emlúpsemc te Secwépemc Nation in their assessment of the proposed Ajax Mine in Canada; their recommendations were provided to and endorsed by community leaders (SSN 2017).
- **Leadership decision-making**. In some cases, elected or customary leaders may be empowered to make decisions on behalf of the community once the results of the IIA are available.

Conclusion

Indigenous impact assessment is an important emerging form of IA. The current growth of IIA is likely to continue because it helps address power imbalances related to proponent control over information provision and gives expression to growing recognition of Indigenous rights; because First Peoples have growing access to financial and human resources, they can apply to IIA, and because integrating Indigenous peoples into IA can be critical to long-term project security and viability.

IIA brings a variety of ingredients to the IA table that have been long neglected or under-used. They include the ability to ensure that Indigenous culture, language, and way of life are central to IA; and access to Indigenous perspectives regarding current environmental, cultural, and social conditions, how these might change as a result of the project, and the significance and acceptability of predicted change. IIA can also allow much higher levels of collaboration and better relationships with the state and with proponents, which in turn can result in tangible project changes and unique mitigation and benefit opportunities.

A continuing and serious challenge is to ensure that IIA has an impact on state and proponent decisions about projects on Indigenous land. Given this challenge, IIA should be designed to create benefits for communities even where it fails to shape project decisions. These benefits can involve increased capacity and knowledge of how to engage in IA, a more engaged community, and data collection that informs strategic initiatives to improve community well-being beyond the confines of the individual project (e.g., a community workforce capacity profile or an enhanced traditional land use database).

What can First Peoples do to maximize positive outcomes from IIA? First, decisions about whether to engage in IIA and what degree of control to seek must be informed by knowledge of the leverage available to the Indigenous group. Second, IIA should be part of a multi-pronged strategy designed to influence state and proponent decision-makers that is political as well as technical. Third, and related to both these points, the decision to conduct an IIA has to happen as soon as possible in the planning for a proposed project. Wherever possible, First Peoples should develop their visions and structures for IIA even in advance of a specific project being proposed. Increasingly, Indigenous groups that have engaged in prior IIAs are moving in this direction (see, for example, the 'Squamish Nation Assessment Model' for IIA (Bruce and Hume 2015)). First Peoples that have not conducted an IIA before can also use available tools and resources (e.g., Gibson et al. 2018) to identify their protocols, methods, and information and resourcing requirements for IIA in advance of new projects being proposed in their territories.

What does the future hold for IIA? As we have shown, there is no one 'IIA', but many possible IIA pathways, in terms of rationale, scope, focus, and methods. This diversity is likely to continue because of the diverse circumstances faced by First Peoples, and the fact that most IIA does not have a statutory basis and so First Peoples are free to experiment and innovate. IIA role as a testing ground for innovative methods and lenses makes it of interest to the whole IA community.

If current trends persist, we envision a growing legislative requirement for IIA and a greater willingness by state agencies to embrace collaborative EIA with First Nations. We also envisage greater collaboration with proponents, not just in EIA but also in the overall management of projects and their impacts. As collaboration with both state agencies and corporate actors grows, we expect an increased focus in EIA on Indigenous values and knowledge, with greater attention paid to oral histories and other forms of Indigenous knowledge, and new assessment frames such as sustainability and intergenerational equity, impact equity (who wins and who loses from a project), and the need for projects to offer net gains and contributions to the recovery of the ecosystem. As the participation of First Peoples in EIA grows, we would eventually expect to see them achieve a growing influence on the determination of impact significance, and on final decisions about whether, and on what terms, projects should be allowed to proceed.

Notes

1 We do not identify the project concerned because doing so might undermine negotiations currently under way between the community and proponent to ensure a much more substantive role for the community in the IA.
2 O'Faircheallaigh Field Notes, Browse LNG Project Aboriginal Social Impact Assessment, 2010.
3 For example, Nesbitt et al. (2018, 76) found that 'Inuit knowledge of their water sources and their ability to describe the relevant characteristics in terms of preferred taste, smell, mouth feel, temperature, and appearance set a foundation for improved quantification using IQ and western science together'.
4 For example, the First Nations Major Projects Coalition's (2019, 32) *Major Project Assessment Standard*'s criteria 8.10 rejects the 'project contribution approach', and requires a focus on total cumulative effects loading: 'The appropriate measure is … the total sum of all cumulative effects on each value from all sources'.

References

BC, CSFNs, and CSTC (British Columbia, the Carrier Sekani First Nations, and the Carrier Sekani Tribal Council). 2015. *Collaboration Agreement (the 'Agreement')*. April, 2015. http://www.carriersekani .ca/images/docs/Collaboration/The%20Agreements.pdf

Bruce, A. and Hume, E. 2015. *The Squamish Nation Assessment Process: Getting to Consent*. Ratcliffe & Company LLP, November, 2015. http://www.ratcliff.com/sites/default/files/publications/The %20Squamish%20Nation%20Process.%20Getting%20to%20Consent%20A%20Bruce%20and%20E %20Hume%20November%202015%20%2801150307%29.PDF

Candler, C., Gibson, G., Malone, M., The Firelight Group Research Cooperative with Mikisew Cree First Nation. 2015. *Wîyôw'tan'kitaskino (Our Land is Rich): A Mikisew Cree Culture and Rights Assessment for the Proposed Teck Frontier Project Update*. https://www.acee-ceaa.gc.ca/050/documents/p65505/102730E .pdf

Craik, A.N. 2016. 'Process and reconciliation: Integrating the duty to consult with environmental assessment', *Osgoode Legal Studies Research Paper Series*, 122, October, 2016. http://digitalcommons.osgoode .yorku.ca/olsrps/122

Devlin, J.F. and Yap, N.T. 2008. 'Contentious politics in environmental assessment: blocked projects and winning coalitions', *Impact Assessment and Project Appraisal*, 26, no. 1, 17–27. https://doi.org/10.3152 /146155108x279939

Faaui, T.N., Morgana, B., and Hikuroab, D. 2017. 'Ensuring objectivity by applying the Mauri Model to assess the post-disaster affected environments of the 2011 MV Rena disaster in the Bay of Plenty, New Zealand', *Ecological Indicators*, 79, 228–246.

First Nations Major Projects Coalition. 2019. *The Major Project Assessment Standard*. https://static1.squarespace.com/static/5849b10dbe659445e02e6e55/t/5cdc93e2fa0d6007b00b5a2d/1557959669570/ FNMPC+MPAS+FINAL.pdf

First Nations Major Projects Coalition. 2018. *EAO Revitalization Indigenous Engagement Costing Study*. Vancouver: First Nations Major Projects Coalition.

Gibson, G., Galbraith, L., and MacDonald, A. 2016. 'Towards meaningful Aboriginal engagement and co-management: The evolution of environmental assessment in Canada'. In K.S. Hanna (ed), *Environmental Impact Assessment: Process, Practice, and Critique* 3rd ed., Toronto: Oxford University Press, pp. 159–180.

Gibson, G., Hoogeveen, D., and Macdonald, A. 2018. *Impact Assessment in the Arctic: Emerging Practices of Indigenous-Led Review*. Gwich'in Council International. https://gwichincouncil.com/sites/default/files/Firelight%20Gwich%27in%20Indigenous%20led%20review_FINAL_web_0.pdf.

Gibson, G., Macdonald, A., and O'Faircheallaigh, C. 2011. 'Cultural considerations for mining and indigenous communities'. In P. Darling (ed), S*ME Mining Engineering Handbook* 3rd ed.,. Denver: Society for Mining, Metallugy and Exploration, pp. 1797–1816.

Gibson, G. and O'Faircheallaigh, C. 2015. *IBA Community Toolkit: Negotiation and Implementation of Impact and Benefit Agreements*. Ottawa: Walter & Duncan Gordon Foundation.

Gibson, R. 2012. 'In full retreat: The Canadian government's new environmental assessment law undoes decades of progress'. *Impact Assessment and Project Appraisal*, 30, no. 3, 179–188.

Government of Canada 2019. *Impact Assessment Act*. https://laws-lois.justice.gc.ca/PDF/I-2.75.pdf

Holden, A. and O'Faircheallaigh, C. 1995. *Economic and Social Impact of Mining at Cape Flattery*. Brisbane: Centre for Australian Public Sector Management, Griffith University.

Hoogeveen, D. 2016. 'Fish-hood: Environmental assessment, critical Indigenous studies, and posthumanism at Fish Lake (Teztan Biny), Tsilhqot'in territory'. *Environment and Planning D: Society and Space*, 34, no. 2, 355–370.

Jolley, D. 2007. *Cultural Impact Assessment for a Proposed Plan Change and Coastal Subdivision at Claverley*. Dyanna Jolley Consulting. https://www.qualityplanning.org.nz/sites/default/files/Cultural%20Impact%20Assessment%20for%20a%20Proposed%20Plan%20Change%20and%20Coastal%20Subdivision%20at%20Claverley.pdf

Keeling, A. and Sandlos, J. (eds) 2015. *Mining and Communities in Northern Canada: History, Politics and Memory*. Calgary: University of Calgary Press.

KLC (Kimberley Land Council) 2010a. *Indigenous Impacts Report: Kimberley LNG Precinct Strategic Assessment Six Volumes*. http://www.dsd.wa.gov.au/state-development-projects/lng-precincts/browse-kimberley/browse-lng---environment/appendices-to-strategic-assessment-report/strategic-assessment-report-appendix-e

KLC (Kimberley Land Council) 2010b. *Browse Liquefied Natural Gas Precinct Strategic Assessment Report: Indigenous Impacts Report Volume 4, Report on Heritage Impact Assessment Report*. https://www.jtsi.wa.gov.au/docs/default-source/default-document-library/browse_sar_appendix_e-4_1210.pdf?sfvrsn=686b1c_10.

Lawrence, R. and Larsen, R.K. 2017. 'The politics of planning: Assessing the impacts of mining on Sami lands'. *Third World Quarterly*, 38, no. 5, 1164–1180.

MacDonald, A., Tam, J. and The Firelight Group Research Inc. 2020. *Environmental Scan of Indigenous-led Impact Assessments in Canada*. Research for the Impact Assessment Agency of Canada. Ottawa: Impact Assessment Agency of Canada.

MacDonald, A. 2014. *Cumulative Effects on the Aboriginal Rights and Interests of Carrier Sekani Tribal Council First Nations: A Preliminary Re-assessment of the Coastal GasLink Project*. October, 2014, Carrier Sekani Tribal Council. http://www.carriersekani.ca/images/docs/cstc/Appendix%20B%20%20FL-CSTC-309_CEA_extended_summary_Oct3_2014_final.pdf

Moody's Investor Service. 2020. 'ESG – Canada: Focus on Indigenous rights increasingly vital for project execution, corporate activities'. Sector In-depth Report, June 22, 2020.

Nadasdy, P. 2003. *Hunters and Bureaucrats: Power, Knowledge and Aboriginal-state Relations in the Southwest Yukon*. Vancouver: UBC Press.

Nesbitt, R.A., Hutchinson, N.J., Klein, H.E., Parlee, B.L., Hart, J., Tulugak, J., and Manzo, L. 2018. 'The One Voice method: Connecting Inuit Qaujimajatuqangit with western science to monitor Northern Canada's freshwater aquatic environment'. *Polar Knowledge: Aqhaliat* 2018, Polar Knowledge Canada, p. 70–77.

O'Faircheallaigh, C. 2000. *Negotiating Major Project Agreements: The 'Cape York Model'*. Australian Institute for Aboriginal and Torres Strait Islander Studies, Research Discussion Paper No 11, Canberra.

O'Faircheallaigh, C. 2007. 'Environmental agreements, EIA follow-up and aboriginal participation in environmental management: The Canadian experience'. *Environmental Impact Assessment Review*, 27, no. 4, 319–342.

O'Faircheallaigh, C. 2015. 'ESD and community participation: The strategic assessment of the proposed Kimberley LNG Precinct, 2007–2013'. *Australasian Journal of Environmental Management*, 22, no. 1, 46–61.

O'Faircheallaigh, C. 2016. *Negotiations in the Indigenous World: Aboriginal Peoples and Extractive Industry in Australia and Canada*. New York: Routledge.

O'Faircheallaigh, C. 2017. 'Shaping projects, shaping impacts: Community controlled impact assessments and negotiated agreements', *Third World Quarterly*, 38, no. 5, 1181–1197.

O'Faircheallaigh, C. and Ali, S. (eds) 2008. *Earth Matters: Indigenous Peoples, Extractive Industries and Corporate Social Responsibility*. Sheffield: Greenleaf Publishing.

Okanagan Indian Band. 2017. *Okanagan Nation Rights and Interests: Submission to Part C of BC Hydro's Revelstoke 6 Environmental Assessment Application*. February 2017; prepared by the Okanagan Indian Band with support from Westbank First Nation, Penticton Indian Band, the Okanagan Nation Alliance and The Firelight Group. https://firelight.ca/wp-content/uploads/2016/04/EAC_Application-Revelstoke_Generating_Station_Unit_6-Volume_4.pdf

Procter, A., 2020. 'Elsewhere and otherwise: Indigeneity and the politics of exclusion in Labrador's extractive resource governance'. *Extractive Industries and Society*. doi.org/10.1016/j.exis.2020.05.018

Rodriguez-Garavito, C. 2011. 'Ethnicity.gov: Global governance, indigenous peoples, and the right to prior consultation in social minefields'. *Indiana Journal of Global Legal Studies*, 18, no. 1, 263–305.

Singh, G.G., Lerner, J., Murray, C., Wong, J., Mach, M., Ranieri, M., Peterson St-Laurent, J., Guimaraes, A., and Chan, K. 2019. 'Response to critique of "The insignificance of thresholds in environmental impact assessment: An illustrative case study in Canada"'. *Environmental Management*, 64, 133–37.

SSN (Stk'emlúpsemc te Secwépemc Nation) 2017. *Honouring Our Sacred Connection to Pípsell. SSN Ajax Decision Summary*. https://stkemlups.ca/files/2013/11/2017-03-ssnajaxdecisionsummary_0.pdf

Tobias, Terry N. 2000. *Chief Kerry's moose: a Guidebook to Land Use and Occupancy Mapping, Research Design, and Data Collection*. Vancouver: Union of BC Indian Chiefs.

Toth, Brian and Michelle Tung. 2014. *Assessment of the Proposed Coastal GasLink Pipeline Project's Effects on Select Fish and Wildlife Interests of the Carrier Sekani Tribal Council (CSTC) First Nations*. October, 2014. Prepared for the Carrier Sekani Tribal Council by the Upper Fraser Fisheries Conservation Alliance. http://www.carriersekani.ca/images/docs/cstc/Appendix%20A%20-Final%20Version%20UFFCA%20Report.pdf

TWN (Tsleil-Waututh Nation Sacred Trust Initiative, Treaty, Lands and Resources Department). 2015. *Assessment of the Trans Mountain Pipeline and Tanker Expansion Proposal*. North Vancouver, BC. https://twnsacredtrust.ca/wp-content/uploads/TWN_assessment_final_med-res_v2.pdf

Tsuji, L., McCarthy, D., Whitelaw G. and McEachren, J. 2011. 'Getting back to basics: The Victor Diamond Mine environmental assessment scoping process and the issue of family-based traditional land versus registered traplines'. *Impact Assessment and Project Appraisal*, 29, no. 1, 37–47.

United Nations. 2013. *The United Nations Declaration on the Rights of Indigenous Peoples: A Manual for National Human Rights Institutions*. https://www.ohchr.org/documents/issues/ipeoples/undripmanualfornhris.pdf

Wabun Tribal Council. 2016. *Wabun Tribal Council Submission to Ms Johanne Gelinas, Chair of the Expert Panel for the Review of the Environmental Assessment Process [Canada]*. December 23, 2016.

Weitzner, V. 2008. *Missing Pieces: An Analysis of the Draft Environmental and Social Impacts Reports for the Bakhuis Bauxite Project, West Suriname*. Ottawa: North-South Institute.

World Bank. 2011. *Implementation of the World Bank's Indigenous Peoples Policy A Learning Review (FY 2006–2008)*. OPCS Working Paper, August, 2011.

14

INNOVATIVE APPROACHES TO ACHIEVING MEANINGFUL PUBLIC PARTICIPATION IN NEXT-GENERATION IMPACT ASSESSMENT

A. John Sinclair, Alan P. Diduck, and John R. Parkins.

Introduction

Public participation is an essential element of impact assessment (IA) and has been explored and discussed since the advent of assessment, yet it still proves to be a vexing issue in terms of establishing meaningful processes for engaging the public. Despite years of assessment experience, considerable academic research, numerous guidebooks, and the existence of at least two relevant professional associations (International Association for Public Participation [IAP2] and the International Association for Impact Assessment [IAIA]), normal participation processes in impact assessment are often found wanting. For example, they tend to over-rely on passive opportunities for engagement, such as letter writing and open houses, rarely present opportunities for early and ongoing involvement, and typically afford few genuine opportunities to influence decisions (e.g., O'Faircheallaigh 2010; Morgan 2012; Sinclair and Diduck 2016). This situation manifests itself despite the fact that public participation is viewed as a cornerstone of assessment and that the basic legitimacy, effectiveness, and fairness of assessment are often in question when it does not provide for meaningful participation (e.g., Petts 1999, 2003; Morrison-Saunders and Early 2008; O'Faircheallaigh 2010; Morgan 2012; Lawrence 2013).

In this chapter, we contextualize *meaningful* public participation in relation to next-generation IA (see Sinclair et al. 2018), describe its key features, principles and benefits, and canvas innovative and promising approaches and tools for achieving such participation. Some of these approaches and tools are drawn from IA experiences, while others come from the broader environmental governance literature. In all cases, we have emphasized the practical aspects and implications of the approaches and tools. Our approach to the work included undertaking an integrative literature review (Torraco 2005; Wang 2019), inquiring into the essential elements of meaningful participation and case examples where such participatory processes had been

DOI: 10.4324/9780429282492-15

undertaken. We also drew on our years of experience of participating in governance processes, including IA, and processes of assessment reform.

Next-generation impact assessment and meaningful public participation

Public participation is thought to have benefits that increase the effectiveness and fairness of IA processes, key hallmarks of good IA. Sinclair and Diduck (2009) canvas many of the benefits of meaningful participation in IA, and note that from a theoretical perspective, it helps actualize the fundamental principles of democracy and strengthen the democratic fabric of society when people can participate in decisions that affect them (Sinclair and Diduck 1995; Shepard and Bowler 1997; Wiklund 2005; Forester 2006; Morgan 2012; Lawrence 2013). They note further that public participation in IA can have a wide range of practical benefits:

- access to local and traditional knowledge from diverse sources;
- enhance the legitimacy of proposed projects;
- better problem definition and identification of a wider array of possible solutions;
- a more comprehensive consideration of factors upon which decisions can be based;
- better alignment of the purpose and design of a project with the needs of the public;
- access to various ethical perspectives that can be brought into the decision-making process;
- a broader range of potential solutions considered;
- access to new financial, human, and in-kind resources;
- prevention of "regulatory capture" of IA agencies by project proponents;
- more balanced decision making;
- increased accountability for decisions made;
- facilitation of challenges to illegal or invalid decisions before they are implemented;
- illumination of goals and objectives, which is necessary for working through value or normative conflict;
- creation of venues for clarifying different understandings of a resource problem or situation, which is key to resolving cognitive conflict;
- avoidance of costly and time-consuming litigation;
- reduced controversy associated with a problem or issue.

(Sinclair and Diduck 2016)

Despite this long list, there are a number of enduring challenges facing participation in IA, and due to these challenges there has been an over-reliance in normal IA participation processes on simplistic and largely ineffective techniques, such as letters, surveys, and open houses. An overview of key concerns is presented in Table 14.1. One of the main reasons why more meaningful participation has not often been achieved in IA is because legislation and public policies have rarely extended beyond simply identifying meaningful public participation as a purpose or a goal. For example, in Canada's new federal Impact Assessment Act, passed in 2019, meaningful participation was established as a purpose but was not defined in the Act or accompanying regulations, although it was fleshed out in non-binding practice guidance documents. A helpful tool for understanding the root of many of the challenges of meaningful participation is Arnstein's (1969) classic ladder of citizen participation. Arnstein identified eight levels of public participation and associated degrees of power-sharing. Activists and ENGOs are often highly critical of processes that only consist of lower levels of involvement and limited power-sharing (see Sinclair and Diduck 2017, for a discussion of the application of Arnstein's ladder in IA).

Table 14.1 Challenges to achieving meaningful public participation in IA (adapted from Sinclair, Diduck, and Vespa 2015)

Challenge	Reasons for challenge	Overcoming the challenge
Perceived inefficiencies	Time required to undertake meaningful participation Added cost Poorly designed participation programs	Learn from examples of well planned and executed IA participation programs.
Accelerated decision processes	Need of decision makers to make decisions as quickly and efficiently as possible limiting opportunities for participation	Developing case-specific timelines that reflect the needs and desires of participants while ensuring efficiency.
Lack of broad and early participation	Lengthens timelines Participants wishing for short term results No concrete project to discuss	The use of extensive methods in tandem with intensive methods to determine the views of those not actively involved in the IA.
Lack of shared decision-making	Legal tools incorporated, e.g., placing decisions in the hands of Cabinet	Give hearing and other formal bodies more influence over the decision as some jurisdictions already practice.
Information and communication deficiencies	Reliance on traditional tools Perceived cost of some techniques Security concerns related to some new tools, e.g., Twitter	Learning from examples where information and communication have occurred in effective and fair ways.
Limited participant assistance	Many jurisdictions do not offer participant assistance Clear definition of scope required to receive funding	Changes to funding availability and application processes.
Weak participation in follow-up and monitoring	Sense that formal assessment ends with project approval and conditions Perceived costs Role of regulator and public not clearly defined	Clearly defined roles and responsibilities for all, including the public.
Power imbalances	Differential access to political and administrative authorities Unequal financial resources Privileged knowledge claims Variable skills and capacities	Take special steps to ensure an inclusive approach toward those who may not be otherwise included.
Limited capacity to engage in the IA process	Little mention of practical applications in theoretical work Use of non-deliberative methods that rely on proponents and not regulators	Expand opportunities for practical experiences by ramping up typical participatory processes used in traditional IA and testing innovative approaches.

As a result of these concerns and other shortcomings in IA worldwide, increasingly there have been calls for new approaches to, and a next generation of, IA processes and tools. The term *next-generation impact assessment* was coined by Gibson, Doelle, and Sinclair (2016) (see also Fonseca and Gibson 2020; Sinclair, Doelle, and Gibson 2019) as a way of thinking about the future of assessment processes and bringing together what we have learned over 50 plus years of IA practice into one package of 13 essential elements. The argument is that, while the literature documents years of experience with different elements of IA (e.g., public participation, tiered assessment, monitoring, and follow-up), these have not been brought together into a full package of core elements that every assessment process should strive for. As one of these elements, meaningful participation has been defined by Sinclair and Doelle (2018) as:

> Meaningful public participation establishes the needs, values, and concerns of the public, provides a genuine opportunity to influence decisions, and uses multiple and customized methods of engagement that promote and sustain fair and open two-way dialogue.

This definition is informed by what we have termed the essential elements of meaningful public participation in next-generation IA, presented in Table 14.2. These elements are drawn from the extensive literature on IA as well as our own IA practice. The essential elements reflect years of experience, can be tailored to the size and complexity of a project or undertaking, and are specific in terms of what is needed for meaningful participatory processes. Together they represent the overriding elements of an effective and fair participation process and need to be implemented as a package in each situation to achieve meaningful public participation.

Innovative and promising approaches

As mentioned above, the research literature and practice guidebooks establish numerous techniques and tools for engaging the public in IA. Table 14.3 provides a broad overview of the various techniques that are available and often promoted for use during assessments. Readers can probably think of examples of the use of these and other techniques. In Canada, for example, it was at one time common to set up field offices in jurisdictions where federal assessments occurred for large projects, such as Confederation Bridge, linking Prince Edward Island and Canada's mainland. One could also find fairly extensive literature on best practices or new models for implementing such techniques and their applicability in IA. Sinclair and Diduck (2017), for example, promote a "civics" approach involving broad, deep, and active participation throughout IA, including planning, communicating, implementing, monitoring, assessing, and adapting.

There is also a rich (and growing) literature on the effectiveness of emerging methods. For example, Roque de Oliveira and Partidário (2020) did a systematic review of the use of visual tools and their potential for promoting inclusive public participation. They found numerous studies in the broad environmental governance/management literature but fewer in the IA literature. They screened 170 articles and ended up analyzing 22, which included papers on environmental planning and management (including IA), land use, urban and community planning, and development projects. The articles were organized into four main groups: (1) maps, schematics, drawings, aerial photos, satellite photos, three-dimensional models; (2) photographs used in photo preference surveys, photo visioning, photo-elicitation, and photovoice; (3) GIS, including PPGIS (which includes the use of maps, interactive maps, photographs, videos, 3D models, and animation); and (4) visual narratives, such as filmed narratives and oral narratives combined with satellite images, photographs, and drawings. The methods were then appraised based on their technical, cognitive, social, and emotional dimensions to uncover their potential capacity to promote inclusive participation, whether they were fit for purpose, and to what extent the public was involved in their design. The authors found that the cultural, social, structural, and

Table 14.2 Essential elements for meaningful public participation in IA (adapted from Sinclair and Diduck 2016 – see also Sinclair and Doelle 2018; Environmental Assessment and Planning Caucus 2017)

Essential element of meaningful participation	Specific requirement for meaningful public participation
Adequate notice	Direct notice
	Use of phone, email, or social media
	Notice about assessment, where further information is located and where comments can be directed
Access to information	Ongoing and timely exchange of information among all parties
	Access to a public registry
Participant assistance	Need for assistance because of complex issues
Opportunities for public comment	Open to all interested parties and individuals
	Cover "need for" and "alternatives to"
	Interactive modes of participation beyond open houses and website submissions
Access to public hearings/ADR	Frequent and creative use of the hearing process
	Transparency and timely written decision
	Inclusive, informal venues for deliberation
	Negotiation and mediation
Early and ongoing participation	Public engaged in the design of the participatory program to be followed
	Extension to the follow-up stage
	Value judgments when choosing trade-offs
	Stakeholder involvement in many assessment choices
Deliberative forums	Emphasis on knowledge integration
	Face-to-face decision-making
	Open dialogue in a non-judgmental environment
	Establishing sustainability as a concept and a goal
	Include forms of alternative dispute resolution
	Incorporate future methods such as visioning and scenario development
Decision impact	Input is not treated as advisory only but can impact and change the course of the decision at hand
Learning oriented	Promotes learning "about" and "through" IA
	Fosters mutual learning among all participants
	Feedback to participants about how their input has, or has not, been used
	Ensures lessons from past assessments as well as process experiences are considered in future assessments and assessment reform
Fair and open	Engagement processes follow principles of natural justice and procedural fairness
	Transparent and open to all

political contexts in which visual tools are used may be the key determinants of their effectiveness. Further, visual tools are often pre-set and not sufficiently tailored to the case in which they are being applied, and the public is seldom involved in the design and choice of the tools.

In the following sections, we offer further details on how some of these techniques have been implemented in IA and other environmental governance processes. In choosing examples, we kept in mind the purpose of this volume, namely to provide feasible and practical guidance. We also aimed to cover a wide range of techniques and approaches, while emphasizing examples about which we have first-hand knowledge. It was a challenge to keep within the word limits for the chapter because meaningful public participation could easily be the subject of its own

Table 14.3 Public participation techniques available for use in IA (adapted from Sinclair and Diduck 2016 – sources include Rowe and Frewer 2005; Diduck et al. 2015; International Association for Public Participation 2015)

Passive public information techniques

Advertisements	Feature stories	Information repositories
News conferences	Newspaper inserts	Press releases
Print materials	Technical reports	Television
Websites		

Active public information techniques

Briefings	Central contact person	Community fairs
Expert panels	Field offices	Field trips
Information hotline	Open houses	Technical assistance
Simulation games		

Small-group public input techniques

Informal meetings	In-person surveys	Interviews
Small-format meetings		

Large-group public input techniques

Public hearings	Response sheets	Mail, telephone, and internet surveys

Small-group problem-solving techniques

Advisory committees	Citizen juries	Community facilitation
Consensus-building	Mediation and negotiation	Panels
Role-playing	Task forces	

Large-group problem-solving techniques

Workshops	Interactive polling	Sharing circles
Websites and chat rooms	Future search conference	

practical guidebook. Nevertheless, what follows is an array of methods and approaches we trust will serve the reader well in creating opportunities for meaningful participation.

Expert panels

The Expert Panel on environmental assessment, formed by Canada's Minister of Environment and Climate Change in 2016, provides an excellent example of how traditional participation tools, when used in a complementary manner, can be very effective. The Panel was struck to engage broadly with Canadians, Indigenous Peoples, provinces and territories, and key stakeholders to develop recommendations to the minister on how to improve federal EA processes. The Panel met in person in 21 cities across Canada and provided opportunities for the public and stakeholders to make presentations, submit written comments, and participate in workshop and dialogue sessions. A user-friendly web portal was launched, a "Choicebook" survey was developed, which could be completed online or in hard copy, Eventbrite (an event management and ticketing website) was used to manage events and record participation, and a toll-free phone number was set up. As well, the Panel was supported in their work by a secretariat that managed the events and input received. The Canadian Environmental Assessment Agency, on behalf of the Panel, also provided support to help participants get to meetings. Figure 14.1 provides an overview of the participants who were involved.

The Panel was clear on their mandate in all of their communications and developed background materials such as a "Suggested Themes for Discussion"[1] document to help inform peo-

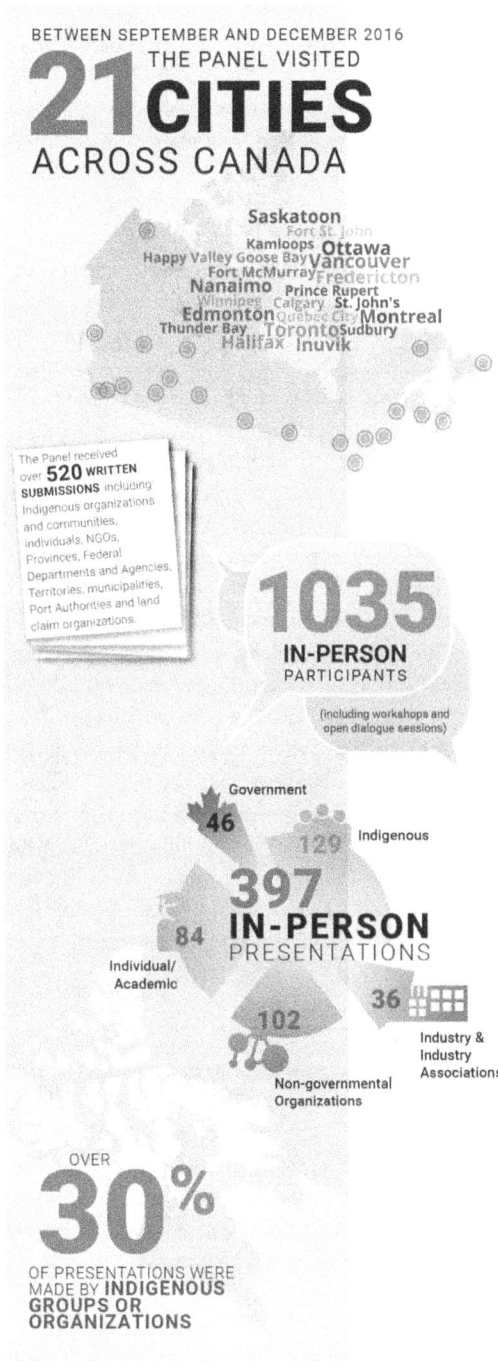

BETWEEN SEPTEMBER AND DECEMBER 2016
THE PANEL VISITED
21 CITIES
ACROSS CANADA

Saskatoon
Fort St. John
Kamloops Ottawa
Happy Valley Goose Bay Vancouver
Fort McMurray Fredericton
Nanaimo Prince Rupert
Winnipeg Calgary St. John's
Edmonton Quebec City **Montreal**
Thunder Bay **Toronto** Sudbury
Halifax Inuvik

The Panel received over **520 WRITTEN SUBMISSIONS** including Indigenous organizations and communities, individuals, NGOs, Provinces, Federal Departments and Agencies, Territories, municipalities, Port Authorities and land claim organizations.

1035
IN-PERSON
PARTICIPANTS

(including workshops and open dialogue sessions)

Government
46
129 Indigenous

397
IN-PERSON
PRESENTATIONS

84
Individual/
Academic

36 Industry & Industry Associations

102
Non-governmental
Organizations

OVER
30%
OF PRESENTATIONS WERE
MADE BY **INDIGENOUS**
GROUPS OR
ORGANIZATIONS

Figure 14.1 "Participation in the Expert Panel's Engagement Process"

ple of potential issues to discuss with the Panel (e.g., when assessment should be required; what the role of the proponent should be in an assessment; what sorts of information should inform assessment decision-making). Importantly, the Panel also published a "Public Engagement Plan" outlining the different ways that the public and stakeholders could participate. This plan established the principles of public engagement that they planned to follow during the engagement process. These included mutual respect, accessibility and inclusiveness, openness and transparency, good faith, and timeliness and efficiency, each of which they explained in plain language.

In regards to the presentations, workshops, and dialogue sessions held across the country, the usual approach was for Panel presentations and dialogue sessions to be during the day and workshops to be in the evening. These events allowed for active participation, and the public was welcome to observe if they wanted to just listen. Pre-registration was required for the presentations, workshops, and dialogues, but in relation to the latter walk-ins were also welcome. The Panel sought to hear from everyone who wanted to present in any given location, but sometimes this was not possible due to high demand, so some participants were asked to present in another city. Extra sessions were also added. People were asked to present for no longer than 10 minutes, leaving 5 minutes for questions from the panelists. Presenters were allowed extra time, and discussion often ensued when time permitted. The Panel often challenged people to be as specific as possible about their suggested proposals for reform.

The public workshop and dialogue sessions started with a short presentation from the Panel about the purpose of the session, key questions that the Panel wanted input on, and how the workshop was organized. This was followed by a presentation meant to set the context of the federal assessment process at the time, including its scope. The audience was then broken into groups facilitated by the secretariat. Each group worked on issues of their choice and/ or ones identified by the Panel, such as "what do you think federal EA should achieve?" and "what about the EA process needs to change or stay the same"? There was a recorder in each breakout group, the thoughts shared were presented to the Panel members in an open forum, and questions were answered as time allowed. Panel members circulated among the breakout groups in some instances. Flip chart paper was posted around the room for people to leave individual comments for the Panel. Response from workshop participants was generally positive in terms of how the sessions were run, the active engagement of the Panel, and the topics covered.

The dialogue sessions were less formal than the workshops and were designed to hear the views of Indigenous participants. These sessions were held since large resource projects are often located on or near the territories of Indigenous people, so people from these communities need to be engaged in IA processes or co-govern the decision process. The federal government also made a commitment to reconciliation with the Indigenous people of Canada. The Panel and secretariat made presentations, but the sessions that followed were less structured. Individuals could make formal statements or presentations if they wished, but the format emphasized circle discussions involving Panel members and participants. Comment posters were also made available so that people could provide more individual feedback to the Panel if they wished. Such an informal format that included culturally appropriate components would have also made it easier for Indigenous people to share their views.

In addition to the opportunity to provide written comments, of which the Panel received 520, a Choicebook survey was developed that provided a framework that individuals could use to structure their written input. As noted in Figure 14.1, 2,763 people provided input through the Choicebook option. The Choicebook started with a statement from the Chair and Panel and proceeded to request that people "think big" in terms of what federal EA should achieve and how it could be made to work better. A very short and general overview of federal EA was provided as well as the context for federal assessment. People were then led through a series of open and closed-ended questions. These started broadly, asking people in an open-ended question, for example, "what goals federal EA should achieve". After this, participants were led through six issue areas, each containing open and closed-ended questions. These covered issues like "when should federal EA apply", "what information should a federal EA consider", and "who should prepare the environmental impact statement". The Choicebook ended by posing the question, "Imagine you are sharing an elevator ride with a member of the Panel. What is the one message you would like to give to him or her"?

The Panel also sought other advice to help develop their recommendations. A Multi-Interest Advisory Committee had been struck by the Minister of Environment and Climate Change, and it provided a report covering a wide breadth of assessment issues. Panel members and the Chair met with this committee, updating them on the public engagement sessions and seeking their input on selected issues. Additionally, the Panel commissioned research from academics and EA specialists on topics such as monitoring and follow-up, enforcement, impact benefit agreements, and multi-jurisdictional assessment.

Lastly, and importantly, all materials received, including emails, written submissions, and transcripts from the in-person sessions, were available through the web portal. As well, at the end of the consultation the Panel reported back on what they heard and indicated how, in their final report, they had tried to respond to the input they received. This reporting back is captured in an "Annotated Compendium of Expert Panel's Responses to Participant Recommendations". This document, which includes 813 individual entries, provides each participant's name, the title of the submission, a hyperlink to the submission, a summary of the participant's main recommendations, and where in their final report the Panel dealt with the issue raised, including why they did not deal with it if that were the case. This aspect of the engagement process was lauded by many participants as this critical feedback aspect of engagement is often missed.

As a small postscript on the Expert Panel's work, while most participants felt the Panel and secretariat did a very good job with the engagement process and in capturing the voices of participants in their final report, there are few meaningful links between the Panel recommendations and the final Impact Assessment Act adopted. This is despite the fact that most of the recommendations to the minister and federal government had strong support from the public, and the final report was often referred to in consultations and debates during the passage of the new Impact Assessment Act. This raises a number of important questions from a participation/ governance perspective about the original intent of the Expert Panel in the minds of those who implemented this option. The Panel clearly did their best to fulfill their mandate of consulting with Canadians, which is no small task, yet their recommendations did not get the attention of participants in the process as expected. This disconnect could mean a number of things, but it underscores the need to be crystal clear and open about the purpose of a public consultation process and its potential impacts on subsequent management or policy decisions.

Managing large participant numbers

One of the many challenging aspects of implementing public participation programs is managing large numbers of participants. In the assessment context, this is often an issue because in many jurisdictions assessment only applies to large projects or undertakings of great significance, which in turn are often of interest to a large number of people. A fairly recent example of this in Canada was the Energy East pipeline proposal, which would have seen an existing pipeline refurbished and extended with new sections to move tar sands bitumen from Alberta to the east coast of Canada. The project proposal garnered significant public interest, with over 2,600 applications being made to participate in the review being done by the National Energy Board. Only 337 applicants were granted intervenor status (meaning they could present before the Board), with another 271 being provided commenter status. A success rate of only about 23% meant that many interested parties were left to find other means for sharing their views with decision makers.

There are now, though, more and more examples of governments and organizations finding ways to engage large numbers of people. In Hong Kong, for example, organizers can typically count on over 5,000 people wanting to participate in major development proposals (Peirson-Smith – personal communication). The Task Force on Land Supply, which was reported in 2018, provides an excellent example of a wide-ranging consultation that managed to involve a tremendous number of participants. The Task Force carried out a five-month (April 26–September 26, 2018) public engagement exercise entitled "Land for Hong Kong: Our Home, Our Say" (https://www.landforhongkong.hk/pdf/Report%20(Eng).pdf). The Task Force put forward 18 land supply options and invited an impressive array of stakeholders, communities, and the public to express views on these options and other land supply-related issues.

The Task Force had a secretariat that worked to initiate engagement activities with the help of private consultants. As outlined in their report, the Task Force used the following methods and materials to raise public awareness of their work and the issues they were considering.

- A public engagement booklet, pamphlets, and website (www.landforhongkong.hk);
- Twenty-five short videos, including public interest television announcements and animated infographics about land supply, land shortages, and land supply options;
- A Facebook page and YouTube channel;
- An 11-episode radio program; and
- Twelve online blogs prepared by or on behalf of the Task Force Chairman.

To collect input from participants, the Task Force used 185 public engagement activities; web-based and paper questionnaires; randomized telephone surveys; and opinions submitted by the public through the mail, facsimile, email, telephone, or in person. The 185 public engagement activities included:

- Four open public forums;
- Forty roving exhibitions each spanning three days in 18 districts of Hong Kong;
- Seventy-two meetings, workshops, seminars, and exchange sessions with different stakeholders, including:
  - Legislative Council, advisory and statutory bodies (15);
  - District Councils (4);
  - Professional groups (24), such as the Hong Kong Institute of Architects, Hong Kong Institute of Engineers, and the Hong Kong Institute of Landscape Architects;
  - Concerned groups/stakeholder organizations (29), such as The Conservancy Association; Green Power; Hong Kong Bird Watching Society, and Green Sense.
- Multiple and diverse outreach activities, such as public forums, roving exhibitions, youth sharing and exchange workshops, which focused on youth (20 activities), schools (23), community organizations (12), and the corporate sector (14). (https://www.devb.gov.hk/en/home/index.html)

Over the course of the six-month consultation, the extent of the input was astonishing. For example: 29,065 people responded to the questionnaires; over 3,000 people participated in telephone interviews; 68,300 people provided comments through email, mail, and fax; the videos were watched over 1.6 million times; and 100,000 responses were obtained through the public engagement channels noted above, which included presentations directly to the Task Force. This level of input can only be described as extraordinary in terms of participant numbers and quantity of data. The Task Force sought independent help to analyze the data, and this was provided by the Social Sciences Research Center of the University of Hong Kong. As well, the Hong Kong Institute of Asia-Pacific Studies of The Chinese University of Hong Kong (CUHK) conducted and analyzed the data from the randomized telephone survey. The Task Force reported the results of these studies in an appendix to their report and also used the data throughout their report. The Task Force concluded that although such a large-scale consultation cannot be perfect, they had been motivated by the large early response they received to provide as many opportunities to participate as possible. One of the critiques of the process was that the outcome was predetermined by the government, and the Task Force perhaps left themselves open to this by consulting on 18 short- and long-term options that were predetermined. Having said that, the critiques seem to mainly be about how the consultation was cast at the outset as opposed to the techniques of engagement.

Community-based impact assessment

One approach to involving people in assessment processes is to give people the tools to complete the assessment on their own and thereby build local capacity. As Spaling et al. (2011) explain, community-based impact assessment (also called community-based environmental assessment) follows the same sort of steps as conventional IA but, unlike an expert-based process, it is highly interactive, and it is the community that conducts scoping, considers alternatives, identifies impacts, assesses significance, selects mitigation measures, and decides on the environmental management plan all guided by a facilitator. This approach is being applied in many jurisdictions, particularity those in Africa, and continues to evolve and even be tested in new contexts such as strategic assessment (e.g., Ozoike-Dennis et al. 2019; Walker, Spaling, and Sinclair 2016). In the project context, however, this approach typically applies to small community-based undertakings, such as boreholes and sand dams, irrigation projects, and other local development activities particularly related to agriculture (Kilemo et al. 2014).

Biswal (2021) undertook two recent community-based assessments in Kenya – the GAKAKI small-scale irrigation project and the MIUKA irrigation project. In both cases, over 50 local people participated directly in the assessment and development of the project impact statements, led by Biswal and local assessment practitioners. The approach guided participants through the steps of assessment, allowing them to design the process and outputs. The community-based assessment started with an introductory workshop, where the proposed project was described, and the assessment approach was introduced. Also, at this meeting, people discussed their interests and concerns as well as the possible activities in which they might engage. This was followed some days later by a second workshop that set the groundwork for the assessment, which included a participatory mapping exercise. After the second workshop, transect walks were initiated to reveal and discuss the proposed location of water storage tanks and associated piping.

In a third workshop, participants identified and discussed local needs and issues affecting the sustainability of their community as they related to the proposed project. After these sessions, participants worked directly on aspects of the assessment and again used workshops and informal discussions with individual community members. Considerations included defining the project scope, consideration of project alternatives and cost-benefit analysis of each option, identifying potential impacts on valued components, which resulted in the development of summary impact tables and mitigation approaches, and finally the development of an environmental management plan as required under Kenyan law. Workshops involved individual postings of ideas on walls for discussion, small break out group dialogue and presentations, and utilizing flipcharts. Some voluntary one-on-one follow-up meetings in between the formal workshops were also undertaken to clarify participants' issues and concerns, which contributed greatly to the success of the assessment. As is common in these cases, outside expertise was also required for input on technical issues related to aquifers, in some cases wildlife and other related considerations, with the resulting data being discussed and integrated into the assessment by local peoples. All of this took place in the community over a period of six weeks.

Community response to the assessments was very positive, and there was active participation throughout the assessments in both cases. Community members indicated that the entire process was very useful, interactive, and informative. Community members, as well as the members of the GAKAKI management committee, suggested that the entire IA was a great learning experience. The MIUKA project committee even mentioned that the IA "was a one of its kind. We had never experienced anything like this. It would not be wrong to say that the entire process was unique in the entire Kirinyaga County we have ever attended".

Social learning processes and methods

Another promising approach for achieving meaningful public participation is to employ social learning processes and methods. Sharing features with community-based IA, social learning involves diverse actors coming together in an iterative process of knowledge sharing and co-production, sometimes resulting in mutual understanding and shared values pertaining to environmental challenges and solutions (e.g., Cundill and Rodela 2013; Ensor and Harvey 2015). Further, it involves learning outcomes that go beyond the individual, extending to organizations, communities, networks, and institutions (Reed et al. 2010; Vinke-de Kruijf and Pahl-Wostl 2016; Wolfram et al. 2019). Social learning is often motivated by environmental pressures and can be emergent through self-organized collaboration or it can result from a designed process planned by a facilitator (e.g., Rist et al. 2007; Cundill and Rodela 2013; Johannessen and Hahn 2013).

Processes supportive of social learning outcomes include bringing together differing perspectives, addressing identity differences, balancing power asymmetries, and embracing the complexity of social-ecological systems (e.g., Rist et al. 2007; Leys and Vanclay 2011; Pahl-Wostl et al. 2013; Suškevičs et al. 2019). Helpful methods include facilitative leadership, practice-based dialogues, intentional experimentation, and boundary objects (e.g., Armitage et al. 2011; Plummer and Baird 2013; Baird et al. 2014; Suškevičs et al. 2019). Much of the evidence about social learning comes from environmental governance contexts other than IA, such as climate change adaptation, watershed management, and community-based forestry (see reviews by Suškevičs et al. 2018, 2019), although an early leading study was based on IA experiences. Focusing on cognitive and moral development, Webler et al. (1995) examined learning by participants in an IA of a waste disposal facility in the Swiss Canon of Aargau and revealed the degree to which carefully facilitated public participation programs can enable the development of shared understandings, interests, and norms among stakeholders.

For the purpose of facilitating meaningful public participation in conventional forms of IA (i.e., those that are largely proponent and state-driven rather than community-based), deliberately planned social learning processes are obviously more pertinent than self-organized processes. Opportunities for incorporating planned social learning processes into IA could perhaps be found at the project planning and impact appraisal stages, although such opportunities would depend on having suitably structured IA systems and project proponents and regulators who are committed to early, ongoing, and deliberative involvement (Diduck and Sinclair 2021). Since opportunities like this are rare, the best chance for incorporating social learning processes in IA is probably at the follow-up stage, especially considering there is evidence showing that it can take up to five years for social learning outcomes to become manifest in practical management actions (e.g., Measham 2013).

During follow-up, opportunities exist for social learning about the effectiveness of mitigation measures, the accuracy of impact predictions, and ways to improve future project design, predictions, and decision-making. Additionally, the establishment of independent oversight bodies can help entrench and legitimize follow-up and its social learning processes and outcomes (Diduck et al. 2012; Sinclair and Diduck 2016). An early prominent example in Canada is the Institute for Environmental Monitoring and Research, a co-management organization involving local and Aboriginal peoples. The institute was established following an IA of a program of low-level military training flights over the Quebec–Labrador Peninsula. Another early notable example is the Independent Environmental Monitoring Agency, which includes representation from local and Aboriginal communities. The agency was created following an IA of diamond mining projects in the Northwest Territories, Canada. The agency reviews monitoring and management plans and results, encourages the use of traditional knowledge, shares concerns of Aboriginal peoples and the general public with the proponent and regulator, and keeps Aboriginal peoples and the wider public informed about agency activities (Sinclair and Diduck 2016; Independent Environmental Monitoring Agency 2020).

Online Tools

In a general sense, online engagement "refers to the use of Information and Communications Technology (ICTs) to support citizens' engagement in the definition of policymaking processes and contents" (Fedotova et al. 2012). In her review of the literature, Potamianos (2019) identifies a variety of online engagement tools that have great potential in the context of next-generation impact assessment. These include:

- Digital IA: Developed, and so far mostly used in Europe, this is a new approach to creating completely digital and interactive impact statements.
- Email: This tool can be used to answer stakeholder questions, allow stakeholders to subscribe to listservs, facilitate discussion groups and disseminate information like maps, agendas, and meeting minutes (Evens-Cowley and Conroy 2006).
- GIS maps: GIS software can be used to enable participants to collaborate in designing and creating maps and/or adding comments to maps directly (Gordon et al. 2011).
- Mobile participation: This involves the use of mobile devices to facilitate stakeholder participation. It includes SMS text messaging, blogs, apps, and social media platforms like Facebook and Twitter. Specific applications can be developed for a wide range of purposes like crowdsourcing (i.e., online polling and surveying), as well as information sharing (Fathejalali and Jain 2019).
- Online discussion forums: These can be used to facilitate dialogue both between and among participants and between participants and government. Charettes, often described as "multiday collaborative design workshop[s]" where consultants, planners, government staff, citizens, and other stakeholders come together to create a plan for a specific area or to solve a specific issue (Lennertz 2011) have also used webpages to enhance in-person

charettes. With COVID-19, more work is going into the design of online discussion tools, such as hackathons.

- Online portals: Stakeholders can submit comments to online portals about upcoming projects. Commentary submitted through portals is different from forums because the latter have the potential for two-way dialogue.

- RSS (Rich Site Summary or Really Simple Syndication): RSS can be used by governments to present messages and updates to subscribed stakeholders in feed format (Sinclair et al. 2017).

- Websites: Webpages can be launched by governments and/or government entities to provide information to stakeholders about upcoming projects and participation opportunities, and provide links to online commentary portals, related social media sites, and links to stream and/or participate in in-person public meetings through Skype, Zoom, or other platforms (Sinclair et al. 2017).

Readers will have experience with many of these techniques, especially the use of webpages and online portals. In the following, we highlight promising approaches in the IA context. It is worth noting at the outset that there are still issues with connectivity in some jurisdictions, especially in relation to complex files, portals, and websites. While some parts of the world are implementing 5G technology, many jurisdictions are still in transition to 3G service. In Canada, for example, there are still many homes without internet services, and one does not have to go to remote northern regions, the location of much development activity, to have little or no service. Many governments, including here in Canada, have pledged to improve on this, but progress is likely to be slow.

Cloud-based platforms

Cloud-based platforms can be an effective way to share information and permit a certain level of interaction for those people who have access. Examples are still not prevalent in IA, and this seems in part the result of the desire to be able to control access and thereby know who is posting comments, outcomes that are difficult to achieve once a cloud is opened up to the public, but they still seem to hold good potential (e.g., Sinclair, Peirson-Smith, and Boerchers 2017; Evans-Cowley, and Hollander 2010). Two brief examples are offered here.

In Australia, "Consultation Hubs" are being established in a number of sectors by the federal and state governments on issues ranging from health to the environment. Sometimes these hubs simply link to online surveys, but in other cases also incorporate cloud environments. The Environmental Protection Agency (EPA) in Western Australia, for example, launched a hub to coordinate all consultation activities for IAs (https://consultation.epa.wa.gov.au) (Sutton and Weston 2015). Potamianos (2019) noted that the hub is based on cloud technology, which can allow the EPA to create surveys, promote IA consultations, analyze and record public responses and provide responses to citizen inquiries. Our review of the hub reveals, however, that it is mainly being used for providing information about projects and gaining input from citizens through surveys, as opposed to creating opportunities for interactive exchanges. Citizens can also subscribe to certain webpages on the EPA website through RSS.

In the context of an IA case, Sinclair, Peirson-Smith and Boerchers (2017) note that Manitoba Hydro, a provincial crown corporation in Canada, developed a cloud-based approach for working with their project partners (four First Nations) and internal and external teams during the development of an impact statement for the 695-megawatt Keeyask hydroelectric project. Hydro established a virtual environment outside of its mainframe system that could be accessed by partners, company employees, and external team members (e.g., IA consultants). This cloud-based environment was used as a collaborative space for their team (but not the public) to work together on IA documents, coordinate activities, respond to and refine documents, and many other tasks. Hydro notes, for example, that over 150 authors were involved when the team was responding to the over 1,000 interrogatory questions on the environmental impact statement. Using the cloud environment, responses could be developed collaboratively among core authors. Draft and final documents were available when they needed to be, everyone had access to current versions, tasks were automated, and notifications were sent to review teams. This put an end to several rounds of often large email attachments (and the difficulty of tracking versions), and allowed for much more robust and collaborative responses, according to Hydro officials. In this case, an external company managed the cloud system for the Keeyask partnership team, and MB Hydro had internal employees who managed the site.

Visual and interactive impact statements

Much work has been done in Europe on what has been termed "digital IA". The approach, pioneered by Royal HaskoningDHV and led by Paul Eijssen, involves visualizing and sharing all data related to an assessment and final IA report in ways that are interactive, quick, transparent, and accessible. Using off the shelf software, Royal HaskoningDHV and now others are creating virtual landscapes and using visual animation to share proposals and gain feedback from the public. This is not simply a PDF version of an impact statement; it is a much more interactive digital platform. As Eijssen (2017) notes, it provides information using videos, photos, interactive maps, tables, infographics, 3D interactive renderings, and audio, moving far beyond traditional text-based impact statements and yet retaining their fundamental integrity. One can, for example, create different configurations of a project, such as the location of windmills in a wind farm and see the results in 3D. A new "eParticipate" platform allows communication among project participants and stakeholders in several ways. Options to provide input or ask questions are available throughout impact statements. Results of consultation events can also be directly shared. Designers are also working on a new way to engage people and allow discussion through the eParticipate platform. As well, multiple digital interactive publications can be added to the site and from every publication it is possible to easily create a downloadable PDF from the online content (with smart ways to deal with interactive elements like movies and other viewers). An introduction to the platform can be viewed at https://www.royal-haskoningdhv.com/en-gb/specials/digital-eis?utm_source=newsletter&utm_medium=email&utm_campaign=digitaleis

There are now over 100 projects around the world that have used a version of the platform, and interest continues to build. Royal HaskoningDHV is currently developing ways to make the platform they developed available to others. In 2019 the company won the Institute of Environmental Assessment and Management's award for "Innovation in Impact Assessment", with a judge noting that

> the pioneering technology helps stakeholders understand the impact of engineering projects by presenting complex information in a visual, dynamic way. As such, it minimises complexity and replaces a range of long, detailed written project reports, from feasibility studies and environmental impact statements (EIS) to masterplans and forecasting.

Instagram: georeferenced images and textual analysis

As a mode of bringing public information into impact assessment, researchers have also undertaken studies that draw on the capabilities of social media data. Unlike other online tools that are intended to enhance participation in IA, drawing from big data sources like Instagram or Twitter represents a type of passive data collection that has specific advantages over conventional methods of social research. Passive data collection often takes place without the knowledge of the participant, and has advantages in terms of limiting social desirability and reactivity biases that are common in conventional research methods. In particular, engaging younger people in traditional IA procedures can be challenging, but some of these challenges can be addressed through the information that is gathered on social media platforms.

In a recent study by Chen et al. (2018, 2019), researchers sought to integrate data from visual and textual data from Instagram to examine impacts from changes to current and proposed hydroelectric dams and surrounding landscapes. The work involved using geo-tagged Instagram posts that were collected in two regions of Canada (Mactaquac, New Brunswick, and Site C, British Columbia) to map out the landscape values in proximity to these areas and the associated landscape values that would be affected by changes to the dams. Putting social, cultural, and historic values on maps is not new, but this mode of passive data collection is novel, affording the opportunity to draw on existing information to gain understandings of potential impacts. After filtering results from the geo-tagged images, the analysis included 80 Instagram posts from the Site C region and 273 posts from the Mactaquac region. The variation in the number of posts is due largely to differences in population between these regions.

After coding the images and textual information, researchers ended up with landscape value maps that included seven distinct layers: aesthetics, sense of home, community attachment, cultural identity, lifestyle, memory, and hot-spot overlays of all values. Results further show that the two study areas are at different stages of hydroelectric development and therefore have different value patterns associated with related landscapes. In addition to identifying specific values at risk and specific locations that are more sensitive to impacts from hydroelectric development, researchers also learned about the strengths and weaknesses of Instagram as a strategy for engagement in IA.

The strengths of Instagram for impact assessment include utilizing large amounts of (almost) free data, including perspectives that are often hard to capture (e.g., youth), avoiding biases of conventional data collection, and linking values to specific landscape features and regions that can be directly impacted by proposed projects. Drawbacks are also important to consider, including substantial challenges in managing large amounts of data and dealing with biases that are directly linked to social media participation based on the self-censoring of images or the lack of engagement due to socioeconomic or communication limitations (Chen et al. 2018, 290).

Social media platforms such as Twitter and Instagram have much potential for enhancing public participation, especially if they are part of a multi-pronged approach to public engagement. The passive use of social media data can give insight into broad-based understandings, experiences, uses, and values within a region and how a proposed project or policy change might impact these social and cultural considerations.

Conclusions/implications

While there are some excellent examples of meaningful participatory tools and processes, these are, on the whole, far too few. In impact assessment, there is a dire need to implement the essential elements of meaningful public participation to bring assessment processes closer toward next-generation models. Such action would not be a huge leap because, as noted above, the literature is replete with guidebooks, examples, and other best practice tools and techniques for implementing meaningful participation, and in IA many practitioners have deep experience with participatory processes. What is needed is the willingness to reach into the existing toolbox to start to use and gain more practice with the tested techniques. This is especially needed for projects that attract significant public interest or have serious potential negative implications for sustainability.

It is also clear from the literature, and our own experiences, that a critical barrier to overcome in achieving these ends is created by the political and administrative pressures to accelerate timelines for assessment approvals. In fact, what is needed are flexible timelines designed to fit particular types of projects or undertakings and the level of public interest in participating in the assessment process. This does not mean endless timelines, but rather reasonable ones, negotiated at the outset to be sure that a plan for meaningful participation in a particular case can actually be implemented. Recognizing that people will want to participate in different ways, what is also needed are opportunities for both passive and active participation and the use of innovative means to connect with often hard-to-reach groups, such as youth.

Lastly, there will be the tendency to want to choose certain of the essential elements of meaningful participation, whether that be in designing IA legislation/regulations or in facilitating participation for an individual project or undertaking. While a selective approach may be necessary for certain contexts, it is certainly not desirable as the elements need to be implemented as a package to help ensure participation is meaningful. Collectively, IA practitioners, regulators, and scholars have the experience and know-how to implement each of these elements successfully, but if we need to draw on the experience of others we can reach outside of assessment to environmental governance and planning more broadly and even completely outside these areas to, for example, the health sector. There is a considerable experience out there to draw upon – we need the willingness, largely at political levels, to take action.

Note

1 These themes as well as the panel's final report and link to the "Annotated Compendium of Expert Panel's Responses to Participant Recommendations" can be found at https://www.canada.ca/en/services/environment/conservation/assessments/environmental-reviews/environmental-assessment-processes/building-common-ground.html

References

Armitage, D., Berkes, F., Dale, A., Kocho-Schellenberg, E. and Patton, E. (2011). Co-management and the co-production of knowledge: Learning to adapt in Canada's Arctic. *Global Environmental Change*, 21: 995–1004.

Arnstein, S.R. (1969). A ladder of citizen participation. *Journal of the American Institute of planners*, 35(4): 216–224.

Baird, J., Plummer, R., Haug, C. and Huitema, D. (2014). Learning effects of interactive decision making processes for climate change adaptation. *Global Environmental Change*, 27: 51–63.

Biswal, R. (2021). Actioning sustainability through next generation community-based environmental assessment. PhD Thesis. http://hdl.handle.net/1993/36030.

Chen, Y., Sherren, K. and Parkins, J.R. (2019). Leveraging social media to understand younger people's perceptions and use of hydroelectric energy landscapes. *Society & Natural Resources*, 32(10): 1114–1122.

Chen, Y., Parkins, J.R. and Sherren, K. (2018). Using geo-tagged Instagram posts to reveal landscape values around current and proposed hydroelectric dams and their reservoirs. *Landscape and Urban Planning*, 170: 283–292.

Cundill, G. and Rodela, R. (2013). A review of assertions about the processes and outcomes of social learning in natural resource management. *Journal of Environmental Management*, 113: 7–14.

Diduck, A.P., Fitzpatrick, P. and Robson, J. (2012). Guidance From Adaptive Environmental Management, Monitoring, and Independent Oversight For Manitoba Hydro's Upcoming Development Proposals: A Report Prepared for the Public Interest Law Centre of Legal Aid Manitoba. Winnipeg: Public Interest Law Centre of Legal Aid Manitoba, 46 pages.

Diduck, A.P., Reed, M.G. and George, C. (2015). "Participatory Approaches to Resource and Environmental Management." In B. Mitchell, ed., *Resource and Environmental Management in Canada: Addressing Conflict and Uncertainty*, 142–70. Toronto: Oxford University Press.

Diduck, A.P. and Sinclair, A.J. (2021). A learning-focused analysis of Canada's new Impact Assessment Act. In M. Doelle and A.J. Sinclair (Eds.), *The New Canadian Impact Assessment Act (IAA)*. Toronto: Irwin Law.

Evans-Cowley, J. and Hollander, J. (2010). The new generation of public participation: internet-based participation tools. *Planning Practice & Research*, 25(3): 397–408. doi: 10.1080/02697459.2010.503432.

Evans-Cowley, J. and Conroy, M. (2006). E-participation in planning: An analysis of cities adopting on-line citizen participation, *Environment and Planning C*, 24(3): 371.

Eijssen, P. (2017). Going digital. *The Environmentalist*, http://www.environmentalistonline.com.

Ensor, J. and Harvey, B. (2015). Social learning and climate change adaptation: evidence for international development practice. *WIREs Climate Change*, 6: 509–522.

Fathejalali, A. and Jain, A. (2019). Mobile participation (mParticipation) in urban development: The experience of Flashpoll app in Berlin. *Information Polity*, 24: 199.

Fedotova, O., Teixeira, L., and Alvelos, H. (2012). E-participation in Portugal: evaluation of government electronic platforms. *Procedia Technology*, 5, 152–161.

Fonseca, A. and Gibson, R.B. (2020). Testing an ex-ante framework for the evaluation of impact assessment laws: Lessons from Canada and Brazil. *Environmental Impact Assessment Review*, 81. doi:10.1016/j.eiar.2019.106355

Forester, J. (2006). Participatory governance as deliberative empowerment: The cultural politics of discursive space. *American Review of Public Administration*, 36(1): 19–40.

Gibson, R.B., Doelle, M., and Sinclair A.J. 2016. Fulfilling the promise: Basic components of next generation environmental assessment. *Journal of Environmental Law and Practice*, 29(1): 257–283.

Gordon, E., Schirra, S. and Hollander, J. (2011). Immersive planning: A conceptual model for designing public participation with new technologies. *Environment and Planning B: Urban Analytics and City Science*, 38(3): 505–511.

International Association for Public Participation (IAP2), (2016), *The IAP2 public participation spectrum*, https://iap2.org.au/wp-content/uploads/2019/07/IAP2_Public_Participation_Spectrum.pdf.

Johannessen, Å. and Hahn, T. (2013). Social learning towards a more adaptive paradigm? Reducing flood risk in Kristianstad municipality, Sweden. *Global Environmental Change*, 23: 372–381.

Kilemo, D.B., Parkins, J.R., Kerario, I.I. and Nindi, S.J. (2014). Making community based environmental impact assessment work: Case study of a dairy goat and root crop project in Tanzania. *International Journal of Development and Sustainability*, 4(4): 767–783.

Lawrence, D.P. (2013). *Impact Assessment: Practical Solutions to Recurrent Problems and Contemporary Challenges*. 2nd Edition. New Jersey: John Wiley and Son.

Lennertz, B. (2011). High-Touch/High-Tech Charettes. American Planning Association. https://www.canr.msu.edu/nci/uploads/files/High-touch_High-tech_Charrettes.pdf

Leys, A.J. and Vanclay, J.K. (2011). Social learning: A knowledge and capacity building approach for adaptive co-management of contested landscapes. *Land Use Policy*, 28: 574–584.

Measham, T.G. (2013). How long does social learning take? Insights from a longitudinal case study. *Society & Natural Resources: An International Journal*, 26: 1468–1477.

Morgan, R.K. (2012). Environmental impact assessment: the state of the art. *Impact Assessment and Project Appraisal*, 30(1): 5–14.

Morrison-Saunders, A. and Early, G. (2008). What is necessary to ensure natural justice in environmental impact assessment decision-making?, *Impact Assessment and Project Appraisal*, 26(1): 29–42.

O'Faircheallaigh, C. (2010). Public participation and environmental impact assessment: purposes, implications, and lessons for public policy making. *Environmental Impact Assessment Review*, 30(1): 19–27.

Ozoike-Dennis, P., Spaling, H., Sinclair, A.J., and Walker, H.M. (2019). SEA, urban plans and solid waste management in Kenya: Participation and learning for sustainable cities. *Environmental Assessment Management and Policy*, 34(3): 186–198.

Pahl-Wostl, C., Becker, G., Knieper, C. and Sendzimir, J. (2013). How multilevel societal learning processes facilitate transformative change: A comparative case study analysis on flood management. *Ecology and Society*, 18: 58.

Petts, J. (1999). Public participation and environmental impact assessment. In J. Petts (Ed.), *Handbook of Environmental Impact Assessment: Environmental Impact Assessment: Process, Methods and Practice*, Volume 1. Oxford: Blackwell Science, pp. 145–77.

Petts, J. (2003). Barriers to deliberative participation in EIA: Learning from waste policies, plans and projects. *Journal of Environmental Assessment Policy and Management*, 5(3): 269–293.

Potamianos, A. (2019). *Using online tools to engage the public*. Unpublished report, West Coast Environmental Law.

Plummer, R. and Baird, J. (2013). Adaptive co-management for climate change adaptation: considerations for the Barents Region. *Sustainability*, 5: 629–642.

Reed, M.S., Evely, A.C., Cundill, G., Fazey, I., Glass, J., Laing, A., Newig, J., Parrish, B., Prell, C., Raymond, C. and Stringer, L.C. (2010). What is social learning?. *Ecology and Society*, 15, [online]. http://www.ecologyandsociety.org/vol15/iss4/resp1/.

Rist, S., Chidambaranathan, M., Escobar, C., Wiesmann, U. and Zimmermann, A. (2007). Moving from sustainable management to sustainable governance of natural resources: The role of social learning processes in rural India, Bolivia and Mali. *Journal of Rural Studies*, 23: 23–37.

Roque de Oliveira, A. and Partidário, M. (2020). You see what I mean? – A review of visual tools for inclusive public participation in EIA decision-making processes. *Environmental Impact Assessment Review*, 83: 106413. doi:10.1016/j.eiar.2020.106413

Rowe, G., and Frewer, L.J. (2005). "A Typology of Public Engagement Mechanisms." *Science Technology and Values* 30(2): 251–290.

Shepard, A. and Bowler, C. (1997). Beyond the requirements: Improving public participation. *Journal of Environmental Planning and Management*, 40(6): 725–738.

Sinclair, A.J. and Diduck, A.P. (2009). Public participation in Canadian environmental assessment: enduring challenges and future directions. In *Environmental Impact Assessment Process and Practices in Canada*. Second edition, K.S. Hanna (ed.). Toronto: Oxford University Press, pp. 56–82.

Sinclair, J. and Diduck, A.P. (1995). Public education: an undervalued component of the environmental assessment public involvement process. *Environmental Impact Assessment Review*, 15(3): 219–240.

Sinclair, A.J. and Diduck A.P. (2016). Public participation in Canadian environmental assessment: enduring challenges and future directions. In K. S. Hanna (Ed.), *Environmental Impact Assessment: Practice and Participation*. Toronto: Oxford University Press. pp. 65–95.

Sinclair, A.J. and Diduck, A.P. (2017). Reconceptualizing public participation in environmental assessment as civics action. *Environmental Impact Assessment Review*, 62: 174–182. doi:10.1016/j.eiar.2016.03.009

Sinclair, A.J. and Doelle, M. (2018). *Meaningful public participation in the proposed Federal Impact Assessment Act*. https://blogs.dal.ca/melaw/2018/02/23/meaningful-public-participation-in-the-proposed-canadian-impact-assessment-act-ciaa/

Sinclair, A.J., Diduck, A.P. and Vespa, M. (2015). Public participation in sustainability assessment: Essential elements, practical challenges and emerging directions. In A. Morrison-Saunders, J. Pope and A. Bond (Eds.), *Handbook of Sustainability Assessment*. Camberley, UK: Edward Elgar, pp. 349–375.

Sinclair, A.J., Doelle, M. and Gibson, R.B. (2018). Implementing next generation assessment: A case example of a global challenge. *Environmental Impact Assessment Review*, 72(1): 166–176.

Sinclair A.J., Peirson-Smith T.J. and Boerchers, M. (2017). Environmental Assessment in the Internet age: the role of E-Governance and social media in creating platforms for meaningful participation. *Impact Assessment and Project Appraisal*, 35(2): 148–157.

Spaling, H., Montes, J. and Sinclair, A.J. (2011). Best practices for promoting participation and learning for sustainability: Lessons from community-based environmental assessment in Kenya and Tanzania. *Journal of Environmental Assessment, Policy and Management*, 13(3): 343–366.

Suškevičs, M., Hahn, T., Rodela, R., Macura, B. and Pahl-Wostl, C. (2018). Learning for social-ecological change: A qualitative review of outcomes across empirical literature in natural resource management. *Journal of Environmental Planning and Management*, 61: 1085–1112.

Suškevičs, M., Hahn, T. and Rodela, R. (2019). Process and contextual factors supporting action-oriented learning: A thematic synthesis of empirical literature in natural resource management. *Society & Natural Resources*, 32: 731–750.

Sutton, A. and Weston, D. (2015). Influence of Social Media in Australian EIA. (Paper delivered at the 35th Annual Conference of the International Association for Impact Assessment, Florence, Italy, 20–23 April

2015) at 3, online (pdf): https://conferences.iaia.org/2015/Final-Papers/Sutton,%20Anthony%20-%20Influence%20of%20Social%20Media%20in%20Australian%20EIA.pdf

Torraco, R.J. (2005). Writing integrative literature reviews: guidelines and examples. *Human Resource Development Review*, 4(3): 356–367.

Vinke-De Kruijf, J. and Pahl-Wostl, C. (2016). A multi-level perspective on learning about climate change adaptation through international cooperation. *Environmental Science & Policy*, 66: 242–249.

Walker, H., Spaling, H. and Sinclair, A.J. (2016). Towards a home-grown approach to strategic environmental assessment: Adapting practice and participation in Kenya. *Impact Assessment and Project Appraisal*, 34(3): 186–198.

Wang, J. (2019). Demystifying literature reviews: What I have learned from an expert. *Human Resource Development Review*, 18(1), 3–15. doi:10.1177/1534484319828857

Webler, T., Kastenholz, H. and Renn, O. (1995). Public participation in impact assessment: A social learning perspective. *Environmental Impact Assessment Review*, 15(5): 443–463.

Wiklund, H. (2005). In search of arenas for democratic deliberation: a Habermasian review of environmental assessment. *Impact Assessment and Project Appraisal*, 23(4): 281–292. doi:10.3152/147154605781765391

Wolfram, M., Van Der Heijden, J., Juhola, S. and Patterson, J. (2019). Learning in urban climate governance: Concepts, key issues and challenges. *Journal of Environmental Policy & Planning*, 21: 1–15.

PART II

Jurisdictional profiles

15

A PRACTITIONER'S GUIDE TO EIA IN DEVELOPING COUNTRIES

Reece C. Alberts, Francois P. Retief, Claudine Roos, and Dirk P. Cilliers

The need to contextualize EIA practice for developing countries

Over three-quarters of the world's population live in so-called developing countries. These are nations with weak and unstable economies, insecure energy supply, and low levels of technological development, and whose populations typically lack access to employment, food, water, education, housing, and healthcare (West and Desai 2002). Given that most, if not all, countries have now adopted some form of environmental impact assessment (EIA) (Morgan 2012; Yang 2019; Bond et al. 2020), it is the aim of this chapter to provide guidance to practitioners involved in EIA within these developing countries.

Despite the proliferation of classifications, we rely on the World Economic Situation and Prospects Report (WESP) and its classification of countries into three broad categories, namely developed economies, economies in transition, and developing economies (Fialho and Van Bergeijk 2017; UN 2020). In accordance with the WESP classification, the countries listed in Table 15.1 are considered to be developing countries, which are also the focus of this chapter (UN 2020).

It is acknowledged that even within the classification of developing countries significant differences do exist based on historical development trajectories, demographics, and political context. However, amidst these differences, there are certain shared common characteristics such as low per capita real income, high population growth rate, high rates of unemployment, high reliance on primary sectors, together with primary commodity export dependence (West and Desai 2002). It is these characteristics that influence EIA practice in these countries and which EIA practitioners must consider.

DOI: 10.4324/9780429282492-17

Table 15.1 UN world economic situation prospects: Developing country classification by region

Africa		Asia	Latin America and the Caribbean
North Africa	Southern Africa	East Asia[b]	Caribbean
Algeria	Angola	Brunei Darussalam	Bahamas
Egypt	Botswana	Cambodia	Barbados
Libya	Eswatini	China	Belize
Mauritania	Lesotho	Democratic People's Republic of Korea	Guyana
Morocco	Malawi	Fiji	Jamaica
Sudan	Mauritius	Hong Kong SAR[c]	Suriname
Tunisia	Mozambique	Indonesia	Trinidad and Tobago
Central Africa	Namibia	Kiribati	Mexico and Central America
Cameroon	South Africa	Lao People's Democratic Republic	Costa Rica
Central African Republic	Zambia	Malaysia	Cuba
Chad	Zimbabwe	Mongolia	Dominican Republic
Congo	West Africa	Myanmar	El Salvador
Equatorial Guinea	Benin	Papua New Guinea	Guatemala
Gabon	Burkina Faso	Philippines	Haiti
Sao Tome and Prinicipe	Cabo Verde	Republic of Korea	Honduras
East Africa	Côte d'lvoire	Samoa	Mexico
Burundi	Gambia	Singapore	Nicaragua
Comoros	Ghana	Solomon Islands	Panama
Democratic Republic of the Congo	Guinea	Taiwan Province of China	South America
Djibouti	Guinea-Bissau	Thailand	Argentina
Eritrea	Liberia	Timor-Leste	Bolivia(Plurinational State of)
Ethiopia	Mali	Vanuatu	Brazil
Kenya	Niger	Viet Nam	Chile
Madagascar	Nigeria	South Asia	Colombia
Rwanda	Senegal	Afghanistan	Ecuador
Somalia	Sierra Leone	Bangladesh	Paraguay
South Sudan	Togo	Bhutan	Peru
Uganda		India	Uruguay
United Republic of Tanzania		Iran (Islamic Republic of)	Venezuela (Bolivarian Republic of)
		Maldives	
		Nepal	
		Pakistan	
		Sri Lanka	
		Western Asia	
		Bahrain	
		Iraq	
		Israel	

(Continued)

Table 15.1 (Continued)

Africa	Asia	Latin America and the Caribbean
	Jordan	
	Kuwait	
	Lebanon	
	Oman	
	Qatar	
	Saudi Arabia	
	State of Palestine	
	Syrian Arab Republic	
	Turkey	
	United Arab Emirates	
	Yemen	

Box 15.1 Examples of developing country issues and challenges which may influence EIA (UNEP 2018; UN 2019)

Increased urbanization;

Growing middle class and changing consumption habits;

Prioritized economic development;

Lack of or weak legislation and enforcement;

Low public awareness and negative attitudes;

Political instability/interference and conflicts;

Inadequate provision of resources;

Climate change vulnerability;

Inefficient resource use and exploitation; and

Weak governance structures.

Many developing countries have a long history of EIA practice, with EIA introduced from the 1970s – mainly through donor countries or organizations, which include the Organisation for Economic Co-operation and Development (OECD) and the World Bank (Kamijo and Huang 2017, 8). There have subsequently been numerous studies evaluating EIA practice in these countries over the past 30 years (Lee and George 2000; Lloyd 2008; Kamijo and Hunag 2017).

Kennedy (1985) highlighted six constraints to effective EIA implementation in developing countries already in the early 1980s. These included low public awareness, lack of a framework of EIA law, lack of strong and well-organized institutions, lack of EIA trained personnel, poor availability of data, and lack of finance. Further constraints to effective EIA implementation such as lack of enforcement, late implementation, insufficient consideration of alternatives, weak consultation, and lack of information disclosure in developing countries have subsequently been identified and explored (World Bank 2006; Kamijo and Huang 2017). McCullough (2017) makes the point that despite wide adoption in developing countries, evidence suggest that EIAs are not having their intended impact and that implementation of procedures is weak. The author further contends that despite constraints related to legal and administrative frameworks, procedural correctness, and low level of resources and capacity of staff, the political context within which EIA is conducted in developing countries is also important in understanding its ineffectiveness (McCullough 2017). For the purpose of this chapter we apply a new and unique approach toward developing guidance for EIA practice in developing countries, namely theory of change (ToC). Applying ToC in this context allows for the identification of key assumptions underpinning EIA in developing countries, which must be understood and acknowledged by EIA practitioners to ensure that EIAs conducted within these contexts deliver on the intended outcomes. It is our expectation that by flagging these assumptions we could assist EIA practitioners to better navigate the unique challenges posed to EIA practice in these countries.

Applying ToC to guide EIA practice in developing countries

ToC has gained prominence in recent years as a method to evaluate the efficacy of policy implementation instruments (Alberts et al. 2020). Also, recently, it has been applied to EIA system evaluation within a specific developing country, namely South Africa (Alberts 2020). The benefit of using ToC is its ability to highlight challenges to EIA practice in particular contexts, in this case EIA in developing countries. In particular, the ToC approach identifies the causal linkages between implementation requirements and outcomes achieved while articulating the assumptions underlying the causal understanding (Biggs et al. 2017). That is to say that it allows us to better understand what causes what, or which inputs and actions result in which outcomes. Ultimately, the ToC method produces a conceptual framework in the form of a ToC map (Figure 15.2) and a related causal narrative (Box 15.2) to guide implementation or evaluation. The causal narrative is explained and structured around a sequence of different so-called system components, namely inputs, activities, outputs, outcomes, and impacts (Weiss 1995; Connel and Kubisch 1998; Thornton et al. 2017; Romero and Putz 2018). In some instances, a design component is also included, which deals with the contextual design of the intervention (DPME 2011). These components are typically illustrated and explained in the form of a pyramid, namely the "results-based pyramid" shown in Figure 15.1.

Applying the ToC method to EIA systems in developing countries requires adapting and contextualizing the following generic evaluation questions as shown in Figure 15.1, namely what we use to do the work (inputs)? What we do (activities)? What we produce or deliver (output)? What we wish to achieve (outcome)? What we aim to change (impact)?

Using the ToC approach and its focus on causal inference serves to reveal the underlying causal relationships within EIA systems in developing countries. We discuss these causal relationships next.

The ToC map and narrative

In applying the ToC to EIA in developing countries, a ToC map (as outlined in Figure 15.2) is developed. The content of the map is an exploded view of EIA practice in developing countries across the different components explained in Figure 15.1. It addresses the causal relationship systematically from left to right between the design, inputs, activities, output, outcome, and impact evaluation components. Importantly it also identifies the various key assumptions that underpin

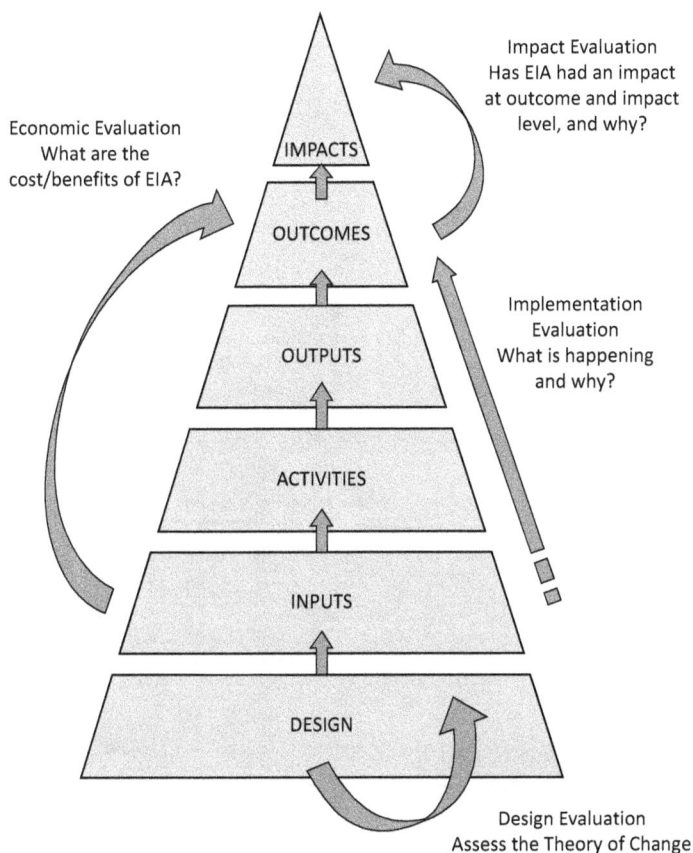

Figure 15.1 Results–based pyramid (adapted from DPME 2011).

the causal relationship. These assumptions (there are 18 assumptions across the different EIA system components – see Box 15.2) are described in the ToC narrative. The numbers indicated in brackets (...) after each assumption described in the ToC narrative (Box 15.2) relate to the numbered key assumptions on the ToC map (Figure 15.2).

The ToC narrative relates directly to the ToC map (Figure 15.2). The narrative is framed against the different system components, i.e., design and inputs, activities and outputs, as well as outcomes and impacts. The general design and functioning of EIA systems in developing countries have to a large extent been adapted from general international best practice and are therefore similar to the typical international understanding of how EIA systems function (IAIA 1998; Wood 1999, 2003; Morgan 2012; Kidd et al. 2018). It is the context within which the system operates that differs internationally. Accordingly, the ToC narrative presented here should be familiar and generally understood by regulators, academics, and professionals working in the EIA field. This causal statement is, after all, based on almost half a century of international EIA practice, and is therefore relatively straightforward and uncontroversial as presented in Box 15.3.

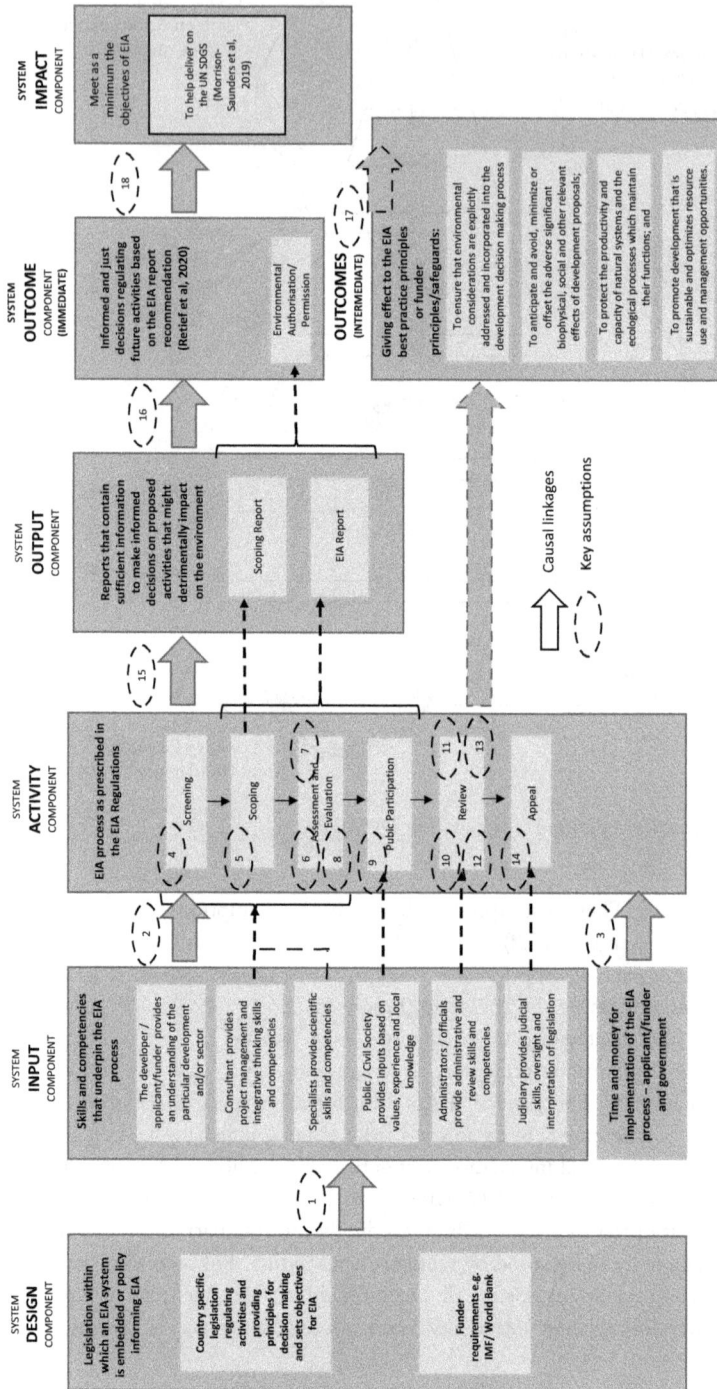

Figure 15.2 ToC map for EIA in developing countries.

Box 15.2 Summary of key assumptions underpinning the inner logic of EIA systems in developing countries

- Clear legislative or other requirements exist to effectively regulate and implement EIA practice (1);
- Sufficient skills and competencies exist to implement the EIA system (2);
- The benefits of doing EIA outweigh the costs (3);
- Effective screening mechanisms exist to identify projects or activities requiring EIA (4);
- It is possible to identify key issues during scoping (5);
- It is possible to determine significance during assessment and evaluation (6);
- EIA is a scientific exercise (7);
- EIA assumes the accuracy of prediction (8);
- The public is willing and sufficiently capacitated to participate (9);
- Reviewers read reports (10);
- Reviewers understand the content of reports (11);
- Reviewers are rational, impartial, and objective (12);
- Reviewers share the same value system (13);
- Appeal authorities are objective and impartial (14);
- Following the generic EIA process will produce good quality reports (15);
- Good quality reports will lead to informed decisions (16);
- Decisions will give effect to the EIA best practice principles or funder principles/safeguards (17); and
- Informed decisions regulating future activities that are just will assist in achieving the objectives of EIA (18).

Box 15.3 The ToC causal statement for EIA in developing countries

The EIA systems in developing countries are embedded in either legislation or donors/funder requirements or both (design component). These systems rely on a certain level of skill and competence (input component) to administer and implement a process (activity component) that produces sufficient information captured in an EIA report (output component), to inform decision-making (outcome component), on the authorization or refusal of future activities that might have a detrimental effect on the environment, toward achieving the UN sustainable development goals (impact component).

Based on the above statement, we now turn our attention to a more detailed narrative of the different components and the underlying context-specific assumptions and their implications for EIA practice in developing countries.

Design and input components

EIA systems have been around for 50 years, with the fundamental design being traceable back to the adoption of NEPA in the United States. Within the developing world context, EIA was introduced around 40 years ago, and although initial uptake was slow, a surge of EIA adoption was prevalent after 1992. As stated above, the design of the EIA system within the developing world was largely influenced by donor countries or donor institutions. This is mainly attributed to the fact that EIA appealed to donor governments or institutions as a well-defined, internally integrated procedure and planning tool (Horberry 1985; Kamijo and Huang 2017). For example, the OECD and the World Bank both adopted EIA as a prerequisite for funded developments in the 1980s (World Bank 2016), with other donor agencies adopting EIA guidelines on recommendation by the OECD. By 1996, OECD member countries had developed EIA guidelines on development cooperation, and had ensured coherence in EIA processes (OECD 1996). More recently, in August 2016, the World Bank reviewed the safeguard policies and approved a new Environmental and Social Framework (World Bank 2016). Often these funder requirements are applied as an extra layer over the existing country-specific EIA legislation. When considering country-specific legislation in so far as it relates to EIA, the ToC approach highlights the first assumption which is made regarding the existence of clear regulatory requirements for EIA practice.

Clear legislative or other requirements exist to effectively regulate and implement EIA practice (1)

Despite the proliferation of EIA related legislation in developing countries (McCullough 2017), EIA legislation was in most cases inadequate or overly complex and fragmented, and EIA performance has remained weak (Kolhoff et al. 2013; McCullough 2017). The main reason attributed to this fact is that often EIA legislation in developing countries is overly ambitious and cannot achieve its objectives in the light of constraining contexts (Kolhoff et al. 2013). To this end, EIA is often considered as being a tool for developed countries, which is expected to be applied within the developing world and is thus not contextualized to deal with developing world constraints. Practitioners in developing countries should, thus, not expect clear and concise EIA legislation, which is easily understood and implemented. Furthermore, as another layer of potential complexity, donor requirements relating to EIA may be overlaid on top of country-specific requirements, as mentioned earlier. This may result in practitioners having to ensure compliance to both country-specific law and donor requirements when conducting EIA processes.

The embedding of EIA in law and policy, thus making it a legal requirement, presupposes a command and control (CaC) approach. CaC based instruments are one of the oldest and most common approaches used by governments to achieve policy objectives (Schneider & Ingram 1990) and are still widely relied on by governments and regulators in developing countries (Blackman 2018). It is supported by the basic assumption that public policy almost always attempts to get people to do things that they might not otherwise do, or it enables people to do things that they might not have done otherwise. The use of CaC instruments, such as EIA, further assumes that citizens will do what is expected of them – even without tangible pay-offs (Kelman 1981). Furthermore, loyalty to duty is assumed to be an inherent virtue of citizens in civic life and of officials in government structures. In short, and in the context of EIA, we assume we can regulate and enforce our way to the achievement of policy aims such as sustainability.

Finally, and possibly most importantly, CaC instruments assume agreement on the outcomes to be achieved. The latter is possibly the most significant challenge and at the same time a flawed EIA assumption. No clearly defined and tangible outcome has been agreed for EIA, so how can it be effective as a regulatory instrument? We, therefore, assume agreement on what EIA should deliver, but actually in reality no common agreement exists. It is well acknowledged that the expectation of what EIA should deliver differs for different role players (Petts 1999; Cape et al. 2018). Acknowledging the limitations of CaC, opens the discussion to other policy approaches and instruments with which EIA could align and that are less regulatory and more incentive-based, such as fiscal- and civil-based approaches, a topic which has been explored in the developing world context (Nel and Alberts 2018). Practitioners should therefore be aware that within the developing world, specifically, CaC instruments, such as EIA, are ineffectively implemented as regulators often lack the capacity to implement, monitor, and enforce (Blackman 2018) – a constraint which brings us to assumption 2 discussed below.

Sufficient skills and competencies exist to implement the EIA system (2)

The main input to the functioning of any EIA system is skills and competencies. The causal logic that underpins this is that in order to make informed decisions, information is required which is derived from different role players with different skills and competencies, such as:

- Consultants/EIA practitioners provide project management and integrative thinking skills and competencies;
- Specialists provide scientific skills and competencies. These inputs are typically considered as the scientific basis for decision-making;
- Public/Civil society provides inputs based on values, experience, and local knowledge. These inputs are typically considered as evidence for decision-making;
- Administrators/officials provide administrative and review skills and competencies. This typically requires integrated thinking and understanding of issues;
- Judiciary provides judicial skills, oversight, and interpretation of legislation; and
- The developer/funder/applicant provides an understanding of the particular development and/or sector.

Skills and competencies to implement the EIA system in any given developing country is usually recognized as a potential weakness (Kamijo and Huang 2019). The skills and competencies required to operate within the EIA profession are integrated in nature, and studies have shown that there is a specific need for especially postgraduate EIA training for practitioners and officials dealing with EIAs in the developing country context (Duthie 2001; Machaka et al. 2013; Alberts 2020). The multidisciplinary nature of EIA makes it a contested field, and various disciplines have attempted to lay claim to this area of expertise, most notably the biological/ecological community, especially during the early years of EIA. However, the growing realization of the importance of the human dimension of EIA has supported the growing contribution of a body of practitioners from non-ecological backgrounds and the ongoing demand for postgraduate education in EIA for students from a wide range of disciplinary backgrounds. As an example, within the South African context, the integrated nature of EIA has been recognized and subsequently six core competencies (Box 15.4) required by EIA consultants and reviewers alike have been identified by the South African Qualifications Authority (SAQA).

Box 15.4 Core skills and competencies for EIA practitioners

1. Demonstrate a conceptual understanding of the environment, sustainable development, environmental assessment, and integrated environmental management.
2. Demonstrate the ability to think holistically, systemically, systematically, spatially, and in an integrative manner and to discern what is relevant to decision-making.
3. Identify and apply environmental assessment procedures and methods.
4. Review and monitor environmental assessment procedures and methods.
5. Conduct applied research in a specific context.
6. Meet specific communication requirements at all levels through environmental reporting processes and stakeholder engagement.

Notwithstanding the qualifications and training within a particular EIA system, it is important to consider the skills and competencies in so far as it relates to the number of EIA professionals/consultants/specialists/reviewers working in the private and public sectors as well as the level of skills/qualifications they have. EIA practitioners in developing countries must be aware that they might possibly be functioning within a low skill and capacity environment.

The benefits of doing EIA outweigh the costs (3)

An EIA system also relies on inputs related to time and money. In this regard, the main direct financial burden lies with the applicant or funder who pays for the appointment of the EIA consultants and specialists to conduct the EIA. Within many developing country contexts, the lack of a strong private sector results in projects which are subject to EIA being funded by either government, donors, or financial institutions, such as the World Bank, the European Investment Bank, or the IMF. Consultants operating within these contexts must be aware of the requirements placed on the EIA process by differing funding models. *He who pays the piper calls the tune*, and it is no different with EIA. Funding and the perception of bias has been identified as a potential constraint for EIA quality and effectiveness within developing country context (Badr et al. 2011; Kabir and Momtaz 2012; Chanty and Grünbühel 2015).

Notwithstanding who funds EIAs, the successful adoption thereof may be traced to the implicit or explicit assumptions that the benefits of doing EIA outweigh the associated costs and resultant economic impacts (Retief and Chabalala 2009). Although it has been stated internationally that the aim of any EIA system should be to maximize environmental benefits, to minimize environmental costs, and to minimize the costs to the proponent (i.e., economic impact), it has been argued that it would be impossible to establish precisely either the benefit or the cost of EIA (Wood 2003). The reason being that comparing the benefits and costs of EIA is ultimately a matter of judgment, dependent on how the factors are weighted, which was and still remains extraordinarily difficult to measure and largely unquantifiable (Arrow et al. 1996; Sadler 1996; Retief et al. 2007). Debates centered around EIA cost invariably raise basic fundamental and often contentious questions around the need for EIA, whether it has or adds value, and if EIA is ultimately worthwhile pursuing – especially so within the developing world context where there is an increased focus on developmental needs. Accordingly, recent years have seen increasing and renewed pressure building around EIA systems internationally and specifically within developing countries to become more efficient and to demonstrate its added-value (Retief et al.

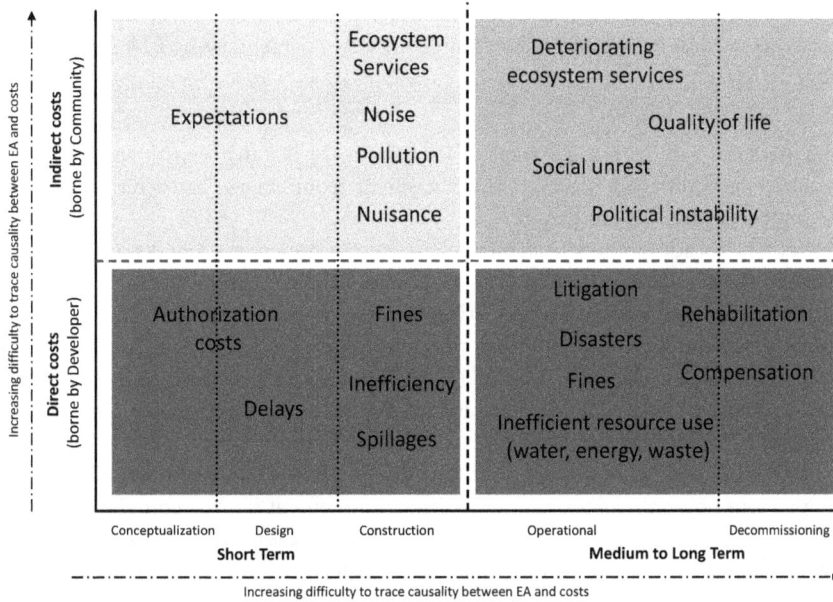

Figure 15.3 Conceptualizing the business case for EIA. Source: Retief et al. 2007.

2007; Bond et al. 2014; Morrison-Saunders et al. 2015; Loomis and Dziedzic 2018; Alberts et al. 2020). When considering EIA costs, the conceptualization regarding the business case for EIA (as shown in Figure 15.3) is important.

It suggests that when thinking about cost, one needs to consider not only the direct and indirect nature thereof but also the time element, since many of the effects of EIA only occurs long after the EIA has been completed. The conceptualization also suggests that direct costs are mostly borne by the developer/proponents and the indirect costs by communities and/or society in general and even future generations. Based on the conceptualization in Figure 15.3, Retief et al. (2007) suggest the following four categories of EIA costs.

- **Category 1:** Short-term direct costs borne by the developer in relation to aspects such as Authorization costs during the design phase as well as potential delays, fines, inefficiencies, spillages, etc. during the construction phase;
- **Category 2:** Medium- to long-term direct costs borne by the developer related to possible litigation, disasters, fines, inefficient use of resources (i.e., water and energy), etc., during operations, as well as direct rehabilitation and compensation costs during decommissioning;
- **Category 3:** Short-term indirect costs borne by the community, such as unrealized expectations, loss of ecosystem services, nuisance, and pollution, during design and construction phases; and
- **Category 4:** Medium- to long-term indirect costs during operational and decommissioning phases borne by the community such as deterioration ecosystem services, quality of life, political instability, and social unrest.

When considering the objectives of EIA in developing countries specifically, it becomes evident that categories 3 and 4 have particular relevance when considering the impacts on communities and society. Given the socioeconomic constraints faced by many developing countries, further factors need to be considered by consultants when contemplating EIA costs (Retief and Chabalala 2009), such as:

- **EIA process**: The level and extent of the assessment is a main cost consideration. Most countries have different processes and assessment requirements based on the outcome of the screening phase of the EIA.
- **Travel**: The location of a consultancy relative to the project is a key determinant of competitive pricing. In this regard, local consultancies have a distinct advantage.
- **Controversy and risk of application not succeeding**: From a financial feasibility perspective, consultants are sometimes reluctant to get involved with EIAs for controversial projects or projects that have a limited chance of success. The main reason for this is the need to avoid the potential quagmire created by public participation and ultimately appeal processes.
- **Anticipated public response and public participation requirements**: Linked to the previous point, the potential participation requirement is a key cost consideration. In a developing country context, this is especially important with complex social dynamics in terms of power relations, language, and cultural differences.
- **Sensitivity of the environment**: The sensitivity of the receiving environment is a key cost consideration, mainly because it affects the complexity of the EIA and the cost of potential specialist inputs.

As stated above, funding is a recognized constraint for EIA quality and effectiveness in the developing world context. It has been argued that the quality of reports on large scale projects within developing countries is usually of better quality than those of smaller-scale projects due to the funding available for the EIA (Kamijo and Huang 2019). This may also be attributed to the fact that large funded projects are also often subject to funder requirements in terms of EIA process and content.

Activity and outputs components

It has long since been argued internationally that environmental assessment is more about the process than about the product (Glasson et al. 2001; Owens et al. 2004). In other words, there is an agreed causal link between the activities related to the EIA process and the eventual quality of the EIA report (Bond et al. 2018a). The international so-called best practice operational principles (IAIA 1998) set out the different EIA activities, which are understood to underpin good quality reports. It is generally accepted that the activities within developing country systems should reflect the international best practice operational principles and should be prescribed as procedural phases in the EIA policy and legislation. Broadly speaking, they include screening, scoping, assessment and evaluation, public participation, review, and appeal.

The outputs of any EIA process are generally expected to be good quality reports that contain sufficient information to make informed decisions on proposed activities that might detrimentally impact the environment. Therefore, the causal relationship suggests that an EIA system primarily produces good quality reports as the main output or "product". EIA report quality in developing countries is widely and well researched, and is considered a major constraint to EIA systems within such countries (Kamijo and Huang 2020). In South Africa, for example, the quality of EIA reports is probably one of the most researched aspects of EIA (see for example Hallatt et al. 2015; Sandham and Pretorius 2007; Sandham et al. 2008a, 2008b, 2010, 2013a, 2013b). It is, furthermore, recognized that there is a clear relationship between the quality of the EIA reports and the effectiveness of the EIA system (Wende 2002; Alberts et al. 2020).

Some of the factors influencing report quality in developing countries include, among others: the experience of EIA practitioners, consultants, authorities and stakeholders, the size of projects, availability of guidance, political will, and EIA practitioner independence. Further factors include the complexity of the EIA, quality of baseline data, access to data and EIA funding (Kamijo and Huang 2019). It is globally conceded that improving the quality of EIA reports is a major concern (Kamijo and Huang 2020). An exception seems to be EIAs conducted for projects in protected areas specifically (Sandham et al. 2020). The key assumptions that underpin the activity and output components are therefore particularly important to guide EIA practice in developing countries.

Effective screening mechanisms exist to identify projects or activities requiring EIA (4)

Internationally, it is recognized that an effective screening mechanism is one of the most important components of a well-functioning EIA system (IAIA 1998; Wood 2003; Pinho et al. 2010). The main aim of screening is to match the need for and extent of the assessment to the environmental significance of impacts related to a particular project. Therefore, EIA systems with weak screening mechanisms are usually characterized by a large number of EIA applications for projects with low impact significance. The consequent pressure placed on public and private sector capacity and resources is typically blamed for poor quality EIA reports, inefficient and lengthy decision-making processes, as well as increased costs (Retief et al. 2011). It is important to ensure that large high impact projects are effectively screened and assessed, while avoiding an over-assessment of smaller, less intensive projects. An assumption facing EIA in developing countries is the one that effective screening mechanisms exist to identify those projects or activities which require an EIA. The assumption also alludes to the fact that where screening mechanisms exist, these are effective, unambiguous, and are consistently applied. The context within which developments take place often compound issues rented to screening. The developing country context is often characterized by big infrastructure projects within sectors such as energy, transport, and infrastructure. As stated earlier, these projects often take place within a complex and ambiguous legal framework for EIA with intense political interest. However, where clear screening processes and criteria exist, they are of great benefit to EIA practitioners in avoiding misunderstandings and controversies.

It is possible to identify key issues during scoping (5) and to determine significance during assessment and evaluation (6)

The failure of effective scoping is a well-researched topic and is not only an assumption applicable to EIA in developing countries. Studies on EIA quality have highlighted the fact that EIA reports score lower grades in the more analytical areas (i.e., scoping and significance) compared to the higher grades in the more descriptive and presentational areas of evaluation (Sandham and Pretorius 2007; Sandham et al. 2008a, 2008b, 2010; Barker and Wood 1999; EC 1996; Lee 2000; Pőlőnen et al. 2010). This also correlates with the assumption that we make regarding the accuracy of prediction. As stated above, the types of developments that are subject to EIA in the developing world are often large and complex, and often take place within areas of high biodiversity and are often credited with high impacts on neighboring communities.

Dealing with significance is recognized as one of the major shortcomings of EIAs in general (Erlich and Ross 2015). The subjective nature of significance determinations lies at the heart of the difficulties in dealing with significance. Within the developing country context, any discussion around significance will be influenced by a myriad of contextual factors, including but not limited to societal values, political climate, and professional judgments (Weston 2000; Erlich and Ross 2015). EIA consultants practicing within the developing world must be cognizant of the fact that these factors are magnified within a developing country context.

EIA is a scientific exercise (7)

EIA recognizes the validity of quantitative and qualitative data, thereby accommodating more subjective elements of impact predictions, values, and views as well as objective evidence. EIA is understood to be both an "art" and a "science" (Morgan 2012). This alludes to the assumption that EIA is a scientific exercise. Arguably, the aim of developing an EIA system within a particular context was to make decision-making affecting the environment more accountable through the use of objective scientific evidence (Bond et al. 2020). The challenge for EIA in developing countries is that science is in many countries poorly developed amidst a very complex biodiversity and socioeconomic environments. In cases where science is weak and information limited, it leads to the application of principles such as precaution and the "best practicable environmental option". The expectation is in many cases to rely on evidence-based information rather than science-based information to reach a decision (Bond et al. 2016; Bond et al. 2018b).

EIA assumes accuracy of prediction (8)

EIA is fundamentally a predictive policy implementation instrument. Therefore, based on this understanding, and for EIA to be effective, we have to assume a high level of accuracy in prediction. Notwithstanding, it has been generally understood and acknowledged since the early days of EIA that the accuracy of prediction is at the best of times low (Holling 1978). Given rising uncertainty in the future, the accuracy of prediction is expected to become even more difficult to achieve, especially in developing countries (Retief et al. 2016). To this end, it has been recommended that EIA rely more on disciplines such as scenario planning and that EIA professionals typically consider multiple possible futures to deal with the uncertainties encountered when making future predictions. Moreover, an even greater focus on adaptive management is recommended to deal with this acknowledged lack of accuracy in prediction. Challenging the core assumption around the accuracy of prediction potentially requires fundamental changes (or paradigm shifts) in thinking about the nature and function of EIA as a predictive decision support instrument (Alberts et al. 2020).

The public is willing and sufficiently capacitated to participate (9)

Most EIA systems in developing countries prescribe extensive public participation requirements, and where these requirements are lacking, funding agency requirements supplement them. However, public participation poses various challenges such as language barriers, cultural barriers, dysfunctional nation-states and illiteracy (Kanu et al. 2019; Aucamp and Lombard 2018). Many developing countries demonstrate democratic values with dictatorial type systems that militate against the principles of public involvement and transparency and accountability. Cilliers et al. (2020) and Roos et al. (2020) have reported that government officials perceive political interference in the EIA process as a barrier to EIA achieving its potential benefits. Moreover, according to Zuhair and Kurian (2016), a politically influenced decision-making process reduces the willingness of the public to participate in the EIA process. In most cases and because of its inherent complexities EIA practitioners in developing countries typically rely on local public participation and/or social impact assessment specialist to ensure the effectiveness of public participation processes.

Reviewers read reports (10) and understand the content of reports (11)

Very little empirical data exists to address the question as to whether reviewers read and understand reports. Research by Alberts (2020) has shown that out of 42 EIA cases in South Africa, the majority of the decisions that have been made by the authorities has a meaningful correlation to the information contained within the report. The research, furthermore, showed that despite poor quality substantive report content, the regulators still used these reports as the basis for decision-making. This begs the question as to why decisions are made based on poor quality EIA reports. A potential answer may be that EIA is still perceived as a tick box exercise within developing countries, with more focus being placed on report completeness rather than report substance (Alberts 2020). Understanding the content of reports relates back to assumption 2, dealing with skills and competencies. It is reasonable to conclude that within developing countries where there are low levels of skills and competencies, the ability by regulators to question weak substantive EIA quality reports will be limited. This places a high level of moral responsibility on EIA practitioners to ensure that good quality EIA reports are presented to decision-makers since the capacity of the consultants in many cases exceeds that of regulators.

Reviewers are rational, impartial, and objective (12) and share the same value system (13); appeal Authorities are objective and impartial (14); following the generic EIA process, will produce good quality reports (15); and good quality reports will lead to informed decisions (16)

Effective EIA depends on the assumption that decision-makers, including reviewers and appeal authorities, will act rationally when provided with objective information – the so-called technical rational paradigm (Owens et al. 2004). Furthermore, and maybe even more fundamental, it is assumed that decisions are impartial, unbiased and objective, and, ultimately, we assume that good quality EIA reports will lead to good, rational decisions (i.e., a positive correlation between report quality and good decision-making). However, it has been generally recognized in the decision-making sciences, since the 1950s, that people are not rational beings and that we are all limited in, for example, our ability to understand complex information (Etzioni 1964). Admittedly, various EIA scholars have also pointed this out over the years, notwithstanding, practice and policy design seem to doggedly continue along the technical rational paradigm based on the assumption of rational decision-making (Kornov and Thissen 2000; Nitz and Brown 2001).

A pervasive phenomenon that is highly relevant within developing country contexts, and which is not well researched in the field of EIA, is corruption and malfeasance within EIA systems. However, corruption in developing countries in general is well researched and is recognized to be a well-known problem in public–private sector interaction, and therefore prudent to be raised here (Olken and Pande 2012). The large scale and budgets of many mega infrastructure projects suggest that EIA systems are particularly vulnerable to corrupt practices (Williams and Dupuy 2017). Indeed, emerging empirical evidence suggests that EIAs are being influenced by corrupt practices, including bribery, collusion, and conflicts of interest (Desta 2019; Williams and Dupuy 2017). Should corruption enter and manifest within a particular EIA system, it would fundamentally destroy the integrity and causal relationships discussed in this chapter rendering the system mute or even worse complicit. As mentioned earlier, research by Cilliers et al. (2020) and Roos et al. (2020) highlighted the concern of government officials responsible for EIA review, that often EIA processes are subjected to political interference and undue influence in support of unsustainable developments. EIA practitioners in developing countries should be aware of the potential for corrupt practices to proactively avoid any involvement.

Outcome and impact components

The main immediate outcome of EIA is a decision informed by the content of the EIA report. Although decision-making happens incrementally throughout the EIA process, the final outcome is a single decision authorizing future activities. It could therefore be argued that the immediate outcome of the EIA system is numerous authorizations or refusal decisions on listed activities related to individual projects. The decision is communicated through an environmental authorization or similar document that provides the rationale for the decision.

Decisions will give effect to the EIA best practice principles or funder principles or safeguards (17), and informed decisions regulating future activities that are just will assist in achieving the objectives of EIA in delivering on the SDGs (18)

As an intermediate outcome, it is assumed that decisions made as a result of EIA practice in developing countries will give effect to the best practice principles of EIA or funder principles/ safeguards. Intermediate outcomes are generally accepted to be short to medium-term outcomes. The argument is that the implementation of EIA practice within developing countries should result in certain short- to medium-term benefits. These benefits align with the best practice principles of EIA and include benefits such as the protection of biodiversity, pollution prevention, reduced resource consumption, increased education and awareness, improved governance, and mitigation of environmental impacts. More broadly, most EIA systems in developing countries do aim at achieving sustainable development (Morrison-Saunders and Retief 2012). However, measuring the contribution of EIA to sustainability in developing countries is difficult to conceptualize and evaluate empirically (see, for example, Retief 2011, 2013). Attempts to do so have, however, resulted in confirming the international understanding (discourse) that EIA systems produce incremental gains contributing toward sustainability (Pope et al. 2018). The difficulty in determining the contribution of EIA to sustainability is compounded by the fact that the impact of the EIA cannot be measured and/or evaluated until such time as well-defined and measurable objectives are developed. This is due to the lack of quantifiable and well-defined targets and objectives for the EIA systems as a whole.

In pursuing the question as to what the perceived benefits of EIA are within the developing country context, recent research indeed suggests that, from a regulators' perspective, EIA is seen as contributing toward the achievement of aspects such as increased public participation and awareness, protection of biodiversity, access to information, and certain immediate economic benefits (Roos et al. 2020; Cilliers et al. 2020). However, the research indicates that the benefits of EIA are perceived to be short term project-specific, with regulators battling to identify and articulate well-defined long term contributions by EIA to sustainable development *per se* (Roos et al. 2020; Cilliers et al. 2020). This is unfortunate since it would be reasonable to assume that the eventual impact of EIA should be to contribute as a decision aiding tool of choice for developing countries toward achieving the UN SDGs. Morrison-Saunders et al. (502019) recommended that for this to happen EIA needs to be more strategic and more integrated. In a developing country context, this would mean that EIA systems would need to provide a strong strategic and planning context to inform project-level decision-making as well as ensure integration of policy and institutional arrangements. The planning and policy development in many developing countries is notoriously weak or even absent, which means that EIAs are in many instances conducted in a policy and planning vacuum. Moreover, developing countries also tend to have very silo-based governmental structures with limited coordination and integration of decision-making. Therefore, aiming EIA at the UN SDGs as the ultimate impact seems correct and logical but might be overly ambitious.

Conclusions

This chapter highlights the underlying assumptions we make when undertaking EIAs in developing countries, with a view to provide guidance to EIA practitioners. The assumptions are discussed in relation to different EIA system components. In conclusion, the chapter highlights the following considerations for EIA practitioners when operating in developing country contexts:

- In terms of the design and input components, EIA in developing countries is often hampered by complex and overly ambitious legal frameworks, while the resources, required skills and competencies needed to implement these systems are often lacking and considered to be a major constraint.
- The activity and output components of the EIA systems together with the underlying assumptions highlight that: EIA report quality remains poor, especially regarding the substantive components such as dealing with significance and scoping, while the EIA process appears to be satisfactory. It is assumed that poor quality of EIA reports may lead to poor, uninformed decisions. Furthermore, political interference and corrupt practices may place EIA processes at risk.
- Regarding the immediate and long term outcome component, the typical outcomes are the assessment and mitigation of environmental impacts, public participation, contribution to informed decision-making, access to information, legal compliance and enforcement, as well as its contribution to the protection of the environment. The policy emphasis in developing countries is typically on socioeconomic development and not environmental protection *per se*, which suggest a very broad and challenging mandate for practitioners.
- The impact component expectation is that EIA should contribute to sustainable development and more specifically the achievement of the UN SDGs. However, we caution against overly optimistic expectations of what EIA is expected to achieve in these contexts. It might be more effective in developing country contexts for practitioners to set well-defined and realistic immediate- and medium-term expectations, rather than pursue vague and poorly defined long-term impacts.

Although the ToC approach paints a very challenging and almost grim picture of EIA practice in these countries, it must be pointed out that EIA in many cases provides the only mechanism which provides a voice to marginalized and vulnerable communities. Moreover, even amidst the typical governance challenges, EIA does promote a level of transparency and accountability, albeit imperfectly and with much room for improvement. It is easy to criticize and question the role of EIA in some of these countries, but the general conclusion has to be that these countries and affected communities are better off with EIA than without.

The identification of underlying assumptions for effective EIA in developing countries is in essence the identification of underlying risks in those systems. Regulators in these countries will be well advised to understand these assumptions and their causes, and in dealing with these they may serve to address in such a manner that the risks to EIA are mitigated. Further research is thus required in order to effectively determine how the assumptions may be dealt with within each specific developing country context.

References

Alberts, R.C. 2020. *An Application of Theory of Change to EIA System Evaluation, Doctor of Philosophy (PhD).* Research Unit for Environmental Science and Management, North West University, Potchefstroom, South Africa.

Alberts, R.C., Retief, F.P., Roos, C. & Cilliers, D.P. 2020. Re-thinking the fundamentals of EIA through the identification of key assumptions for evaluation. *Impact Assessment and Project Appraisal.* DOI: 10.1080/14615517.2019.1676069.

Arrow, K.J., Cropper, M.L., Eads, G.C., Hahn, R.W., Lave, L.B., Noll, R.G., Portney, P.R., Russel, M., Schmalensee, R., Smith, V.K. & Stavins, R.N. 1996. Is there a role for benefit-cost analysis in environment, health and safety regulations? *Journal of Science*, 272: 221–222.

Aucamp, I. & Lombard, A. 2018. Can social impact assessment contribute to social development outcomes in an emerging economy? *Journal Impact Assessment and Project Appraisal*, 36(2): 173–185.

Badr, E.A., Zahran, A.A. & Cashmore, M. 2011. Benchmarking performance: Environmental impact statements in Egypt. *Environmental Impact Assessment Review*, 31: 279–285. DOI: 10.1016/j.eiar.2010.10.004.

Barker, A. & Wood, C. 1999. An evaluation of EIA system performance in eight EU countries. *Environmental Impact Assessment Review*, 19: 387–404.

Biggs, D., Cooney, R., Roe, D., Dublin, H.T., Allan, J.R., Challender, D.W.S. & Skinner, D. 2017. Developing a theory of change for a community-based response to illegal wildlife trade. *Conservation Practice and Policy*, 31(1): 5–12.

Blackman, A., Zhengyan, L. & Liu, A.A. 2018. Efficacy of command and control and market based environmental regulation in developing countries. *Annual Review of Resource Economics*, 10: 381–404.

Bond, A., Pope, J., Morrison-Saunders, A., Retief, F. & Gunn, J. 2014. Impact assessment: Eroding benefits through streamlining? *Environmental Impact Assessment Review*, 45: 46–53.

Bond, A., Pope, J., Morrison-Saunders, A. & Retief, F. 2016. A game theory perspective on Environmental Assessment: What games are played and what does this tell us about decision making rationality and legitimacy? *Environmental Impact Assessment Review*, 57: 187–194.

Bond, A., Retief, F., Cave, B., Fundingsland Tetlow, M., Duinker, P.N., Verheem, R. & Brown, A.L. 2018a. A contribution to the conceptualisation of quality in impact assessment. *Environmental Impact Assessment Review*, 68: 49–58.

Bond, A., Pope, J., Retief, F. & Morrison-Saunders, A. 2018b. On legitimacy in Impact Assessment: An epistemologically-based conceptualisation. *Environmental Impact Assessment Review*, 69: 16–23.

Bond, A., Pope, J., Retief, F., Morrison-Saunders, A., Fundingsland, M. & Hauptfleisch, M. 2020. Explaining the political nature of environmental impact assessment (EIA): A neo-Gramscian perspective. *Journal of Cleaner Production*, 244: 1–10.

Cape, L., Retief, F., Lochner, P., Bond, A. & Fischer, T. 2018. Exploring pluralism – Different stakeholder views of the expected and realised value of strategic environmental assessment (SEA). *Environmental Assessment Review*, 69: 32–41.

Chanthy, S. & Grünbühel, C.M. 2015. Critical challenges to consultants in pursuing quality of Environmental and Social Impact Assessments (ESIA) in Cambodia. *Impact Assessment and Project Appraisal*, 33: 226–232. DOI: 10.1080/14615517.2015.1049488.

Cilliers, D.P., Van Staden, I., Roos, C., Alberts, R.C. & Retief, F.P. 2020. The perceived benefits of EIA for government: A regulator perspective. *Impact Assessment and Project Appraisal* 38 (5): 358–367. DOI: 10.1080/14615517.2020.1734403.

Connell, J. & Kubisch, A. 1998. Applying a theory of change approach to the evaluation of comprehensive community initiatives: Progress, prospects and problems. In: Fulbright-Anderson, K., Kubisch, A., and Connell, J. eds. *New Approaches to Evaluating Community Initiatives, Vol. 2, Theory, Measurement, and Analysis*. Aspen Institute, Washington, DC.

Desta, Y. 2019. Manifestations and causes of civil service corruption in developing countries. *Journal of Public Administration and Governance*. Macrothink Institute, 9(3): 23–35.

DPME (Department of Planning, Monitoring and Evaluation) 2011. *National Evaluation Policy Framework*. National Department of Performance Monitoring and Evaluation, Pretoria.

Duthie, A.G. 2001. A review of provincial environmental impact assessment administrative capacity in South Africa. *Impact Assessment and Project Appraisal*, 19(3): 215–222.

Erlich, A. & Ross, W. 2015. The significance spectrum and EIA significance determinations. *Impact Assessment and Project Appraisal*, 33(2): 87–97.

Etzioni, A. 1964. *Modern Organizations*. Prentice Hall, Englewood Cliffs.

EC. European Commission 1996. Environmental Impact Assessment in Europe. A Study on Costs and Benefits, Brussels.

Fialho, D. & Van Bergeijk, P.A.G. 2017. The proliferation of developing country classifications. *The Journal of Development Studies*, 53(1): 99–115. DOI: 10.1080/00220388.2016.1178383.

Glasson, J., Therivel, R. & Chadwick, A. 2001. *Introduction to Environmental Impact Assessment*, 2nd edition. UCL Press, Philadelphia.

Hallatt, T., Retief, F. & Sandham, L. 2015. A critical evaluation of the quality of biodiversity inputs to EIA in areas with high biodiversity value – Experience from the Cape Floristic Region, South Africa. *Journal of Environmental Assessment Policy and Management*, 17(3): 1–26.

Holling, C.S. 1978. *Adaptive Environmental Assessment and Management*. Blackburn Press, New Jersey.

Horberry, J. 1985. International organization and EIA in developing countries. *Environmental Impact Assessment Review*, 5(3): 207–222.

IAIA (International Association for Impact Assessment) 1998. *International. EIA Best Practice Principles*, International Association for Impact Assessment, IAIA, Fargo.

Kabir, S.M.Z. & Momtaz, S. 2012. The quality of environmental impact statements and environmental impact assessment practice in Bangladesh. *Impact Assessment and Project Appraisal*, 30(2): 94–99. DOI: 10.1080/14615517.2012.672671.

Kamijo, T. & Huang, G. 2017. Focusing on the quality of EIA to solve the constraints on EIA systems in developing countries: A literature review. In: *Improving the Planning Stage of JICA Environmental and Social Considerations*. JICA Research Institute, Tokyo Japan.

Kamijo, T. & Huang, G. 2019. Determinants of the EIA report quality for development cooperation projects: Effects of alternatives and public involvement. In: *Improving the Planning Stage of JICA Environmental and Social Considerations*. JICA Research Institute, Tokyo Japan.

Kamijo, T. & Huang, G. 2020. Decision factors and benchmarks of EIA report quality for Japans cooperation projects. *Environment, Development and Sustainability* 23(2): 2552–2569. DOI: 10.1007/s10668-020-00686-1.

Kanu, E.J., Tyonum, E.T. & Uchegbu, S.N. 2018. Public Participation in environmental impact assessment (EIA): A critical review. *Architecture and Engineering*, 3(1): 7–12.

Kelman, S. 1981. *What Price Incentives?* Auburn House, Boston.

Kennedy, W.V. 1985. Environmental impact assessment: The work of the OECD. *Environmental Impact Assessment Review*, 5(3): 285–290.

Kidd, M., Retief, F. & Alberts, R. 2018. Integrated environmental impact assessment and management. In: Strydom, H., King, N., and Retief, F. eds. *Environmental Management in South Africa*. Juta Publishing, Cape Town.

Kolhoff, A.J., Driessen, P.P.J. & Runhaar, H.A.C. 2013. An analysis framework for characterisizing and explaining development of EIA legislation in developing countries – Illustrated for Georgia, Ghana and Yemen. *Environmental Impact Assessment Review*, 38: 1–15.

Kornov, L. & Thissen, W. 2000. Rationality in decision- and policy-making: Implications for strategic environmental assessment. *Impact Assessment and Project Appraisal*, 18(3): 191–200.

Lee, N. 2000. Reviewing the quality of environmental assessments. In: Lee, N., and George, C. eds. *Environmental Assessment in Developing and Transitional Countries*. Wiley & Sons, Chichester, pp 137–148.

Lee, N. & George, C. 2000. *Environmental Assessment in Developing and Transitional Countries*. John Wiley and Sons, Chichester.

Li, J.C. 2008. *Environmental Impact Assessment in Developing Countries. An Opportunity for Greater Environmental Security*. Foundation for Environmental Security and Sustainability.

Loomis, J.J. & Dziedzic, M. 2018. Evaluating EIA systems' effectiveness: A state of the art. *Environmental Impact Assessment Review*, 68: 29–37.

Lloyd, A. 2008. *Exploring the Evolution of Debates on Environmental Assessment in Developing Countries*. Masters in Environmental Management thesis. North West University, South Africa.

Machaka, R.K., Ganesh, L. & Mapfumo, J. 2013. Managing the expectations between the regulatory authority and consultants in the environmental impact assessment system in Zimbabwe. *International Journal of Scientific and Engineering and Research*, 2(10): 107–112.

McCullough, A. 2017. Environmental Impact Assessments in developing countries: We need to talk about politics. *The Extractive Industries and Society*, 4(3): 448–452.

Morgan, Richard K. 2012. Environmental impact assessment: The state of the art. *Impact Assess Proj Apprais*, 30(1): 5–14.

Morrison-Saunders, A. & Retief, F. 2012. Walking the sustainability assessment talk – Progressing the practice of environmental impact assessment (EIA). *Environmental Impact Assessment Review*, 36: 34–41.

Morrison-Saunders, J., Bond, A., Pope, J. & Retief, F. 2015. Demonstrating the benefits of impact assessment for proponents. *Impact Assessment and Project Appraisal*. DOI: 10.1080/14615517.2014.981049.

Morrison-Saunders, A., Sanchez, L.E., Retief, F., Sinclair, J., Doelle, M., Jones, M., Wessels, J. & Pope, J. 2019. Gearing up impact assessment as a vehicle for achieving the unsustainable development goals. *Impact Assessment and Project Appraisal* 38(2): 113–117. DOI: 10.1080/14615517.2019.1677089.

Nel, J. & Alberts, R. 2018. An introduction to environmental management and law. In: Strydom, H., King, N., & Retief, F. eds. *Environmental Management in South Africa*. Juta Publishing, Cape Town.

Nitz, T. & Brown, A. 2001. SEA must learn how policy making works. *Journal of Environmental Assessment Policy and Management*, 3(3): 329–342.

Olken, B.A. & Pande, R. 2012. Corruption in developing countries. *Annual Review of Economics*, 4(1): 479–504.

Owens, S., Rayner, T. & Bina, O. 2004. New agendas for appraisal: Reflections on theory, practice, and research. *Environment and Planning. Part A*, 36(11): 1943–1959.

Perdicoúlis, A. & Glasson, J. 2006. Causal networks in EIA. *Environmental Impact Assessment Review*, 26(6): 553–569.

Petts, J. 1999. Public participation and environmental impact assessment. In: Petts, J. ed. *Handbook of Environmental Impact Assessment, Vol. 1 Environmental Impact Assessment: Process, Methods and Potential.* Blackwell Science, Oxford, 145–177.

Pinho, P., McCallum, S. & Cruz, S.S. 2010. A critical appraisal of EIA screening practice in EU Member States. *Impact Assessment and Project Appraisal*, 28(2): 91–107.

Pőlőnen, I., Hokkanen, P. & Jalava, K. 2010. The effectiveness van the Finnish EIA system – What works, what doesn't and what could be improved? *Environmental Impact Assessment Review*, 31(2): 120–128.

Pope, J., Bond, A., Cameron, C., Retief, F. & Morrison-Saunders, A. 2018. Are current effectiveness criteria fit for purpose? Using a controversial strategic assessment as a test case. *Environmental Impact Assessment Review*, 70: 34–44.

Retief, F. 2007. A performance evaluation of strategic environmental assessment (SEA) processes within the South African context. *Environmental Impact Assessment Review*, 27: 84–100.

Retief, F., Marshall, R. & Morrison-Saunders, A. 2007. "Avoiding extinction – Proving the business case for environmental assessment (EA)", IAIA International conference, "Growth, Conservation and Responsibility", 3–9 June. Seoul, South Korea.

Retief, F. & Chabalala, B. 2009. The cost of environmental impact assessment (EIA) in South Africa. *Journal of Environmental Assessment Policy and Management*, 11: 51–68.

Retief, F., Welman, C. & Sandham, L.A. 2011. Performance of environmental impact assessment (EIA) screening in South Africa: A comparative analysis between the 1997 and 2006 EIA regimes. *South African Geographical Journal*, 93(2): 1–18.

Retief, F. 2013. Sustainability assessment in South Africa. In: Bond, A., Morrison-Saunders, A., and Howitt, R. eds. *Sustainability Assessment: Pluralism, Practice and Progress*. Routledge, London.

Retief, F., Bond, A., Pope, J., Morrison-Saunders, A. & King, N. 2016. Global megatrends and their implications for environmental assessment practice. *Environmental Impact Assessment Review*, 61: 52–60.

Retief, F., Fischer, T.B., Alberts, R.C., Roos, C. & Cilliers, D.P. 2020. An administrative justice perspective on improving EIA effectiveness. *Impact Assessment and Project Appraisal* 38(2): 151–155. DOI: 10.1080/14615517.2020.1680042.

Romero, C., Putz, F.E. 2018. Theory-of-Change development for the evaluation forest stewardship council certification of sustained timber yields from natural forests in Indonesia. *Forests*, 9(547): 1–15.

Roos, C., Cilliers, D.P., Retief, F.P., Alberts, R.C. & Bond, A.J. 2020. Regulators' perceptions of environmental impact assessment (EIA) benefits in a sustainable development context. *Environmental Impact Assessment Review*, 81. DOI: 10.1016/j.eiar.2019.106360.

Sadler, B. 1996. *International Study of the Effectiveness of Environmental Assessment - Final Report: Environmental Assessment in a Changing World: Evaluating Practice to Improve Performance*. International Association for Impact Assessment and the Canadian Environmental Assessment Agency, Ottawa.

Sandham, L. & Pretorius, H.M. 2007. A review of environmental impact report quality in the North West Province of South Africa. *Environmental Impact Assessment Review*, 28(4–5): 229–240.

Sandham, L., Moloto, M. & Retief, F. 2008a. The quality of environmental impact assessment reports for projects with the potential of affecting wetlands. *Water SA*, 34(2): 155–162.

Sandham, L., Hoffman, A. & Retief, F. 2008b. Reflections on the quality of mining EIA reports in South-Africa. *The Journal of the Southern African Institute of Mining and Metallurgy*, 108: 701–706.

Sandham, L., Carrol, T. & Retief, F. 2010. The contribution of environmental impact assessment (EIA) to decision making for biological pest control in South Africa – The case of Lantana camara. *Biological Control*, 55: 141–149.

Sandham, L.A., Van Heerden, A.J., Jones, C.E., Retief, F.P. & Morrison-Saunders, A.N. 2013a. Does enhanced regulation improve EIA report quality? Lessons from South Africa. *Environmental Impact Assessment Review*, 38: 155–162.

Sandham, L.A., Van der Vyver, F. & Retief, F.P. 2013b. The performance of environmental impact assessment in the explosives manufacturing industry in South Africa. *Journal of Environmental Assessment Policy and Management*, 15(03): 1350013.

Sandham, L.A., Huysamen, C., Retief, F., Morrison-Saunders, A., Bond, A., Pope, J. & Alberts, R.C. 2020. Evaluating environmental impact assessment (EIA) report quality in South African national parks. *Koedoe: African Protected Area Conservation and Science*, 62(1): a1631. DOI: 10.4102/koedoe.v62i1.1631

Schneider, A. & Ingram, H. 1990. Behavioral assumptions of policy tools. *Journal of Politics*, 52(2): 510–529.

Thornton, P.K., Schuetz, T., Forch, W., Cramer, L., Abreu, D., Vermeulen, S. & Campbell, B.M. 2017. Responding to global change: A theory of change approach to making agricultural research for development outcome-based. *Agricultural Systems*, 152: 145–153.

UN Environment 2019. Global environment outlook-GEO-6: Health planet, healthy people. Nairobi. Kenya.

UNEP 2018. *Africa Waste Management Outlook*. United Nations Environment Programme, Nairobi, Kenya.

United Nations 2020. *World Economic Situation Prospects*. New York.

Weiss, C.H. 1995. Applying a theory of change approach to the evaluation of comprehensive community initiatives: Progress, prospects, and problems. In: Fullbright-Anderson, K., Kubisch, A.C., & Connell, J.P. eds. *New Approaches to Evaluating Community Initiatives: Concepts, Methods and Contexts*, Vol. 1. The Aspen Institute, Washington DC, pp 65–92.

Wende, W. 2002. Evaluation of the effectiveness and quality of environmental impact assessment in the Federal Republic of Germany. *Impact Assessment and Project Appraisal*, 20(2): 93–99.

West, J. & Desai, P. 2002. *Diverse Structures and Common Characteristics of Developing Nations*. Inggris, Oxford University, 39.

Weston, J. 2000. EIA, decision-making theory and screening and scoping in UK practice. *Journal of Environmental Planning and Management*, 43(2): 185–203. DOI: 10.1080/09640560010667.

Williams, A. & Dupuy, K. 2017. Deciding over nature: Corruption and Environmental Impact Assessments. *Environmental Impact Assessment Review*, 65: 118–124.

World Bank 2006. *Environmental Impact Assessment Regulations and Strategic Environmental Assessment Requirements: Practices and Lessons Learned in East and Southeast Asia*. World Bank, Washington, DC.

World Bank 2016. Environmental and social framework. http://consultations.worldbank.org/Data/ hub/ files/consultation-template/review-and-update-world-bank-safeguard-policies/en/material s/the_esf_clean_final_for_public_disclosure_post_board_august_4.pdf (accessed September 16, 2016).

Wood, C. 1999. Comparative evaluation of environmental impact assessment systems. In: Petts J, *Handbook of Environmental Impact Assessment*, 2, Oxford: Blackwell. pp. 10–34.

Wood, C. 2003. *Environmental Impact Assessment: A Comparative Review*. Prentice Hall, Harlow.

Yang, T. 2019. The emergence of the environmental impact assessment duty as a global legal norm and general principles of law. *Hastings Law Journal*, 70(2): 525–572.

Zuhair, M.H. and Kurian, P.A. 2016. Socio-economic and political barriers to public participation in EIA: implications for sustainable development in the Maldives. *Impact Assessment and Project Appraisal 34*(2), 129–142.

16

THE EUROPEAN UNION ENVIRONMENTAL IMPACT ASSESSMENT DIRECTIVE

Strengths and weaknesses of current practice

Gesa Geißler, Johann Köppel, and Marie Grimm

Introduction

The European Union (EU) Environmental Impact Assessment (EIA) Directive has been in place for more than 35 years. When compared to earlier EIA legislation, such as the US National Environmental Policy Act (NEPA), it has been ground-breaking. While the EIA Directive focuses on public and private development projects, attention to the wider scope, including plans, programs, policies, and strategies, has been helped by the passage of the EU Strategic Environmental Assessment (SEA) Directive, in 2001. The EU EIA Directive has been the subject of much research and evaluation and extensive consideration in academic literature since its inception. In this chapter, we focus on the most recent major changes to the Directive in 2014, discussing its adoption and implementation by EU member states.

After a brief introduction to the Directive's main objectives and development over the past 35 years, we discuss the latest legislative changes to the Directive as of 2014. A case study of a recent project-based EIA in Germany, the Tesla Gigafactory in Berlin-Brandenburg, helps illustrate the Directive's implementation and practice challenges that have evolved over the 30 years of German and European EIA experience.

Development of the EU EIA Directive

After several years of discussions and development, the European Union EIA Directive (85/337/EWG) was adopted in 1985. By this time, several EIA-like instruments were already applied to some extent in the EU member states at the time (e.g., the Netherlands, UK, France, and Germany). The EIA experiences of the EU member states have helped shape the final EIA Directive (Fischer 2016; Wood and Lee 1988).

Directives in the European Union provide goals to be met by all EU countries but require member states to implement them through their own legislation. While the EIA Directive provides harmonized standards and rules for EIA in the EU, in practice, member states have discretion

DOI: 10.4324/9780429282492-18

in defining and implementing the details. This practice follows the *principle of subsidiarity*. The *principle of subsidiarity* holds that actions are taken at or by the national, regional, or local level, and not by the EU; except for areas that fall within its exclusive competence (remit) or where it is more effective for the EU to act (cf. Boyes and Elliott 2014). The goal of *harmonization of EIA practice* is supported by guidelines issued by the EU (e.g., European Commission 2017a, 2017b, 2017c) and by the enforcement of the Directive's regulations through the European Court of Justice (ECJ). However, the ambitions or goals of the Directive are not always apparent, or necessarily effective, when it comes to the interpretations or implementing legislation of all member states.

The EU EIA Directive applies a polluter pays framework and the precautionary principle as stated in Recital 2 of the Directive's preamble. As is outlined in Article 1, the Directive applies to public and private projects likely to have significant effects on the environment. A project is

---

Box 16.1 EU Directive 2011/92/EG on the assessment of the effects of certain public and private projects on the environment

The overall goal of conducting environmental impact assessments under the EU EIA Directive is to take into account "effects on the environment […] at the earliest possible stage in all the technical planning and decision-making processes" (Recital 2). EIAs shall contribute "to a high level of protection of the environment and human health" (Recital 1). The standards and procedures established by the EIA Directive shall provide for a minimum of harmonization among the member states but at the same time allow for freedom to adopt the rules to the individual needs and conditions of the national states. As national planning and project permitting regimes vary among the EU member states, integration of EIA into these national procedures can take different shapes. Overall, EIA must be an integrative part of permitting and consenting procedures for projects.

To achieve these goals, the EIA Directive contains first a preamble providing for the background and reasoning for the subsequent provisions, and subsequently 16 articles and 5 annexes provide the detailed provisions.

Article 1 – providing definitions
Article 2 – detailing the integration of the EIA into permitting regimes and coordination with other assessments
Article 3 – outlining the scope of the EIA
Article 4 – defining the screening process (in conjunction with Annexes I and II)
Article 5 – outlining the content of the EIA report (in conjunction with Annex IV)
Article 6 – outlining agency and public consultation
Article 7 – outlining transboundary agency and public consultation
Articles 8, 8a, and 9 – regulating the consideration of the EIA in the decision-making and information requirements; as well as monitoring obligations
Article 9a – regulating the quality of staff and competent authorities
Articles 10, 10a, and 11 – requiring rules on penalties in case of infringements, regulating access to administrative review procedures for the public
Article 12 and 13 – setting reporting obligations to the EU for member states on EIA practice
Articles 14 to 16 – terms and time frames for transposing the Directive

---

defined as "the execution of construction works or of other installations or schemes" or "other interventions in the natural surroundings and landscape including those involving the extraction of mineral resources" (EIA Directive[1] Art. 1). As described by Glasson and Therivel (2019), the process of conducting EIAs under the EU EIA Directive follows the generally accepted steps of screening, scoping, environmental assessment, development of the EIA report, public and agency consultations, and decision-making.

The introduction of the EIA Directive in Europe has generally been regarded as a beneficial policy, leading to better considerations of environmental concerns in project-related decision-making, which has facilitated greater levels of transparency (COWI 2009). The requirement for public information and consultation in environmental decision-making established in the Directive is a key component. Various evaluations of EIA practice in Europe pointed out that the EIA process leads to modifications of projects (e.g., Wood et al. 1996; Barker and Wood 1999; Wende 2002; Christensen et al. 2007). At the same time, work examining EIA effectiveness and quality identified several weaknesses and needs for improvement. For example, a recurring issue is the weak consideration of alternatives in EU EIA practice (e.g., Barker and Wood 1999; Köppel, Geißler, Helfrich, and Reisert 2012; Jiricka-Pürrer et al. 2018). Regardless of the challenges of putting the Directive into practice, by 2008 over a three-year period (2005–2008) estimates showed that 16,000 EIAs were being completed (GHK 2010). During the same period, it was estimated that some 34,000 EIA screening decisions were made (GHK 2010).[2]

Since the initial adoption of the EU EIA Directive, the EU has changed substantially, and so has the Directive. In 1985 the EU consisted of ten member states (Belgium, France, Germany, Italy, Luxembourg, the Netherlands, Denmark, Ireland, UK, and Greece). By 2022 there are 27 member states, with the United Kingdom having left the EU at the end of January 2020 (EU 2022). Five countries (Albania, Montenegro, North Macedonia, Serbia, and Turkey) are in the process of joining the EU. Although the five countries are in different stages of the EU joining process, to some degree, they have already implemented the EIA Directive (Fagan and Sircar 2015; Drenovak-Ivanović 2016; Savaşan 2020; Güneş 2020).

Since its adoption, the EIA Directive has been shaped by court rulings (cf. COM 2017). It has been amended four times: In 1997, the amendment implemented the UNECE Espoo Convention on environmental impact assessment in a transboundary context (Directive 97/11/EC). In 2003 amendments integrated the requirements of the UNECE Aarhus Convention on Access to Information, Public Participation in Decision-Making and Access to Justice in Environmental Matters (Directive 2003/35/EC) and in 2009 among others the lists of projects being subject to an EIA have been amended (Directive 2009/31/EC).[3] The last changes through the amendment of the Directive occurred in 2014.

This latest review of the EIA Directive started with the development of a review report on the application and effectiveness of the EIA Directive conducted between 2007 and 2008, which was published in 2009 (COWI 2009). The 2009 study focused on the "organizational and legal arrangements and their effectiveness" and included information regarding the experiences of conducting EIAs by EU member states (COWI 2009). The study was based on information provided by the member states, but also used interviews, literature review, and document analysis (COWI 2009).

In the following year (2010), an online consultation process was held for three months. It was open to the public and interested stakeholders in all EU member states,. The goal of this consultation process was to collect feedback on the weaknesses and strengths of the Directive. The online consultation received about 1,365 responses, with about 50% from individual citizens, 35% from organizations/companies, and 15% from public authorities and administrations (COM n.d.).

During both the review and the consultation process, certain points of critique were highlighted, which the amended EIA Directive was asked to improve. For example:[4]

- Implementation of the screening procedure: Many authors and stakeholders criticized the practice of "salami-slicing" of projects to avoid a full EIA process (e.g., COWI 2009; Justice & Environment 2012; later also Enríquez-de-Salamanca 2016). This practice of splitting up projects into smaller sub-applications to stay below the EIA thresholds was reported from several EU member states. For example, this was commonly seen in the case of road project applications, which were split into sections (Justice & Environment 2012) and permitted individually without an EIA.

- Assessment of alternatives: It was evident from many studies that alternatives were often not assessed in EIAs, ignoring alternatives proposed by the public and stakeholders (e.g., Justice & Environment 2012). In addition, the justification of the chosen alternative was regarded as a weak point that needed legislative change (e.g., CEEweb for Biodiversity 2013).

- Quality control of the EIA: A lack of objectivity from decision-making bodies was reported, as well as in some cases, the limited expertise of consultants and agency staff undertaking the EIA. For example, participants reported quality issues of the EIA report and process, which has resulted in an uneven picture among the member states, which is linked to the lack of a formal *quality control mechanism* in the EIA Directive for reports and processes (e.g., CEEweb for Biodiversity 2013; COWI 2009).

- Post-decision monitoring: The lack of monitoring requirements was seen as a weakness, as the verification of the predicted environmental impacts would be able to improve future assessments (e.g., COWI 2009). This criticism was already voiced before, e.g., by Barker and Wood (2001).

The review report, including results from the online consultations, and other information, resulted in the first proposal to update the EIA Directive by the Commission in 2012 (EU Commission 2012). After discussions in the European Parliament and negotiations about changes to the Commission's draft, the final amended EIA Directive (2014/52/EU) was approved in April 2014 and came into force on May 15, 2014. Following this, EU member states were given three years to implement the changes (cf. Table 16.1) into their national legislation (Arabadjieva 2016).

Changes to the 2014 EIA Directive and its implementation at the national level

This chapter presents and discusses the main changes to the EIA Directive by the 2014 amendment. In doing so, we feature the ways in which these changes have made their way into EU member states national legislation, highlighting the experiences of select member states. Our overview is based on a review of academic literature and reports and reflects discussions at the 2018 EIALAW Conference in Helsinki[5] and the 2018 Nordic-Baltic Impact Assessment Conference in Tallinn.[6]

Application of the EIA and screening

Overall, the EIA Directive applies to certain public and private projects likely to have significant effects on the environment. The 2014 amendments did not change the basic concept of applicability and screening (see Bond and Walthern 1999). This is still regulated by Article 4, in conjunction with Annexes I and II of the Directive. Annex I lists projects for which EIA is

Table 16.1 Major changes to the EIA process by Directive 2014/52/EU

Selected, substantive changes to the EIA requirements by Directive 2014/52/EU

General aspects
- requiring a joint and/or coordinated procedure with assessments under the EU Habitats Directive and EU Wild Birds Directive (Article 2(3))
- requiring member states to set penalties in case of infringement of national regulations implementing the EIA Directive (Article 10a)

Screening
- introduction of maximum timeframe for screening of 90 days (Article 4(6)), allowing for extension under certain circumstances
- amendment of Annex III – criteria to be considered in making a screening determination
- requiring the developer to submit the information detailed in Annex II.A (Article 4(4))
- requiring the competent authority to justify their screening decision (Article 4(5))

EIA report
- adding "biodiversity" and "land" to the list of factors that need to be considered in the EIA (Article 3(1))
- adding the requirement to assess "expected effects deriving from the vulnerability of the project to risks of major accidents and/or disasters" (Article 3(2))
- rewriting Annex IV, including more details and examples of information to be included in the EIA report
- requiring "a description of the reasonable alternatives studied by the developer" Article 5(1)

Public participation
- introduction of minimum time frame for public consultation of 30 days (Article 6(7))
- requirement to make EIAs electronically available via central EIA portals or access points (Article 6(5))

Quality control/management
- developer must ensure that EIA report is written by competent experts and competent authority must ensure sufficient expertise to be able to review the report (Article 5(3))
- the competent authority shall request additional information from the developer if needed (Article 5(3))
- requirement to ensure competent authorities conduct their duties in an objective manner avoiding conflict of interest (Article 9a)

Decision-making
- rewording the requirement to consider the results of the EIA and consultations in decision-making to "shall be duly taken into account" (Article 8)
- requesting that the decision includes a reasoned conclusion on the significant effects as well as mitigation measures and if appropriate monitoring (Article 8a)

Monitoring
- introducing the requirement to develop monitoring of significant environmental effects (Article 8a)

mandatory; thus, no screening process is necessary. Annex II is different in that it lists projects for which member states would have to determine whether an EIA is needed or not – using case-by-case determinations, setting thresholds/criteria indicators, or a combination of these approaches (EIA Directive Art. 4, European Union 2017a). To address the issue of "salami-slicing" projects, the new Directive requires that when determining the need for an EIA, there must also be a consideration of the cumulative effects in combination with existing and/or approved projects (Annex III).

Member states have implemented different approaches for screening, but most involve some form of thresholds for the inclusion or exclusion of projects from the EIA requirement. While the 2014 amendments of the EIA Directive regarding screening are aimed at clarifying and easing this step, this was not necessarily achieved in every country. For example, the German EIA Act anticipates 24 different possible situations in the screening process (Hartlik 2020). While the decision for new projects reaching the defined thresholds for an EIA are rather straightforward; the decision about the need of an EIA for projects which individually do not reach the EIA requirement thresholds but could if considered in combination with existing projects or projects in the permitting process. This determination is left with the EIA practitioners in member states. Thus, the determination if an EIA is required is rather a complex process in practice and can lead to uncertainty among those involved in the EIA process.

Determining the scope of the assessment

Scoping is a crucial element of determining the necessary content and level of detail of the EIA process and EIA report (Morrison-Saunders et al. 2014; Lyhne and Kørnøv 2013; Enríquez-de-Salamanca 2016; Fonseca and Fernandez 2020). The treatment of scoping in the Directive as non-mandatory has been criticized for some time. Barker and Wood, when referring to scoping, stated that "'key best practices' stages have been omitted from legislative provisions" (2001, 246). The amended EIA Directive still does not make scoping mandatory. In the review of the EIA Directive, mandatory scoping was part of the proposal by the European Commission (EC 2012) but was not included in the final version adopted by the European Parliament and the Council of Ministers. The proposed mandatory scoping phase and its content requirements were regarded by most member states as too strict and inflexible (Sangenstedt 2014). The Directive now states that a developer (also called a *proponent* in other chapters) may request a scoping opinion from the responsible agency. So, it is up to the member states to decide if scoping is mandatory, and if it should be required in a member state's national legislation.[7] This means that in practice, unlike in other parts of the world, in the EU scoping is not necessarily a step completed in every EIA process (Köppel, Geißler, Helfrich, and Reisert 2012). The EU, however, does promote scoping as an important early step in the EIA process. The EU's 2017 scoping guidance provides a rationale for conducting scoping (European Union 2017b), but it is uncertain the extent to which the EU scoping guidance is implemented by EIA practitioners and authorities.

In the Netherlands, scoping in EIA processes has been mandatory since 1987 and has always included public participation (Wood 1995; Arts et al. 2012). Likewise, in Slovakia, the scoping process is mandatory, and its content and process are further detailed through law. The determination of the scope of the assessment is published online by the competent authority[8] and the public, including opportunities for stakeholders to submit comments on the scoping determination (Justice & Environment 2016). While in Germany, scoping is conducted only if requested by the developer or regarded as necessary by the competent authority (EIA Act Art. 15, para. 1). If a scoping process is carried out in Germany, the competent authority may provide stakeholders with the opportunity of commenting. While the law lists experts, other agencies, and environmental NGOs, public consultation is neither required nor promoted by the law. Moreover, the competent authority needs to document the outcomes of scoping, but it is not indicated how the documentation is to be made, and there is no requirement to make scoping documents publicly available (EIA Act Art. 15), meaning that scoping information is not regularly available to the public (Köppel et al. 2018).

Quality control and management of the EIA process and report

In response to the call for a *quality control mechanism*, the 2014 Directive implemented new provisions in Article 5 of the Directive, binding both the competent authority, and the developer, including their consultants to quality control mechanisms. For example, the EIA Directive now requires that the developer makes sure that competent experts are working on the EIA to ensure a sufficient quality of the EIA report (Art. 5(3)). The provisions, however, do not specify ways in which this expertise or competence must be proven, and there is no EU guidance on this so far. Once again, it depends on the actions of member states to implement the Directive through its own legislation. In practice, quality control varies from certification schemes for EIA consultants (for example, in the UK) (Lonesdale et al. 2017; Fischer and Fothergill 2014; Bond, Fischer, and Fothergill 2017), to no specific requirements (for example, Germany or Denmark) (Köppel 2016; Günther, Geißler, and Köppel 2017; Kørnøv and Kjellerup 2016). It should be noted that in Denmark, the development of national EIA guidance documents and professional training has been initiated (Kørnøv and Kjellerup 2016). While in Austria, for example, the environmental report and other expert statements are reviewed by the Environment Agency Austria, which issues a statement regarding quality control (Jiricka, Bösch, and Völler 2016).

The competent authority

A member state determines in its legislation who the competent authority is and puts provisions/regulations in place to ensure that the competent authority is independent from a developer. The goal is to safeguard that the competent authority, so it can act in an objective manner, avoiding conflicts of interest. The amended EIA Directive of 2014 states that the competent authority "shall ensure that it has, or has access as necessary to, sufficient expertise to examine the EIA report" (Art. 5(3) Directive 2014/52/EU). This is important as the competent authority is not necessarily an environmental agency but often a sectoral, e.g., transportation, building, planning, or economic development agency without expertise in environmental questions and perhaps a specific interest in certain projects. Here again, the Directive leaves the details of implementing this requirement to the member states. In putting the EIA Directive into German law, the government argued that no changes to the EIA Act are needed with respect to competent authorities and their expertise even though there were no obligations in place. This was strongly criticized by the German EIA Association, and doubts were raised in regard to if the German interpretation meets the EIA Directive's mandate stated in Recital 33 to safeguard sufficient expertise in competent authorities as no specific measures for ensuring expertise in authorities exist (UVP Gesellschaft e.V. 2016; UVP Gesellschaft e.V. 2015).

New requirements for the EIA report (climate change, land, disasters)

In terms of the environmental issues to be included in the EIA report, the required *coverage* has changed to some degree with the 2014 amendment of the Directive. Population and human health, biodiversity, land, soil, water, air and climate, material assets, cultural heritage, landscape, and the interaction between these factors have been considered for some time. Building upon these, new requirements regarding climate change, consideration of land use and change (Recital 9 & Art. 3(1c)), and the risk of major accidents and/or disasters have also been added.

Climate has always had a place in the EIA Directive, and 2013 EU guidance also dealt with the integration of climate change into EIA (EC 2013). For example, the practice was often focused on micro-climate issues, and not so much on climate change (e.g., Geißler, Köppel, and Odparlik 2011; Enríquez-de-Salamanca, Martín-Aranda, and Díaz-Sierra 2016). In Germany, case law reinforced this and upheld that there is no need to consider greenhouse gas emissions in an EIA under the pre-2014 version of the Directive (Verheyen and Schayani 2020). This situation will improve with the 2014 amendments, as there is an explicit call for considerations of the possible contributions of projects to climate change (e.g., from greenhouse gas emissions) and possible climate change-induced effects on the projects (e.g., the vulnerability of the project to extreme weather events, water scarcity, or other impacts).

Although the regulations have been in place since 2014, and in 2017 were transferred into national EIA legislation, current EIA practice in many EU member states still performs rather poorly in considering climate change effects and adaptation (Rasmussen et al. 2020; Jiricka-Pürrer, Wachter, and Driscoll 2019; Enríquez-de-Salamanca, Martín-Aranda, and Díaz-Sierra 2016). Some member states, e.g., Germany (Rasmussen et al. 2020, Köppel 2016) and Austria (Jiricka-Pürrer et al. 2018) have been reluctant to meaningfully implement this requirement and use EIA to implement climate mitigation and adaptation goals for consideration in decision-making.

Assessment of alternatives

According to the 2014 EIA Directive, the EIA report must include documentation of the reasonable alternatives that have been studied by the developer and reasons for the selection of the chosen alternatives (Art. 5(1d) EIA Directive). In preamble 31, the authors of the Directive explain that this requirement is seen as "a means of improving the quality of the environmental impact assessment process and of allowing environmental considerations to be integrated at an early stage in the project's design". Thus, the assessment and consideration of alternatives is regarded as an important part of the EIA process and has been a longstanding element of the EIA Directive. Prior to the 2014 amendment, many scholars have called for requirements for developers to identify different alternatives, as developers only had to document alternatives if they had considered some (e.g., Baker and Wood 1999; Baker and Wood 2001; Köppel, Geißler, Helfrich, and Reisert 2012). During the latest amendment process of the EIA Directive in 2012, an obligatory assessment of alternatives initially had been proposed by the EU Commission (COM 2012) but did not make it into the final version of the Directive because it was watered down in the legislative process (Fischer 2016). However, the Directive still does not clearly require the developer to identify alternatives, but only to document "the reasonable alternatives [...] studied by the developer" (Annex IV, 2), limiting the requirement to identify alternatives. This provides a leeway for developers, allowing them to just state that no reasonable alternatives were studied (cf. Lonsdale et al. 2017). Thus, whether to consider alternatives or not and which alternatives are included is still up to the developer. Despite this, some still argue that the 2014 amendment has somewhat strengthened the consideration of alternatives in EIA; for example, Annex IV now provides examples of reasonable alternatives and requires a comparison of the environmental effects of alternatives (e.g., Fischer et al. 2016).

In implementing the EIA Directive, member states in their legislation have mainly abstained from making the identification and assessment of alternatives mandatory for developers, and do not go beyond the EU Directive (e.g., for Germany: Köppel 2016; for Austria: Jiricka, Bösch, and Völler 2016; Jiricka, Bösch, and Pröbstl-Haider 2018). So, in practice, the consideration of alternatives has a *voluntary quality* and is limited to the ones the developer has considered.

Public participation

The general goals and procedural requirements for public consultation and participation in the EIA Directive have not been changed substantially by the 2014 amendment. Other researchers have discussed participation requirements (e.g., Glasson and Therivel 2019; Köppel, Geißler, Helfrich, and Reisert 2012; Geißler, Köppel, and Günther 2013; Hartley and Wood 2005), and these are briefly outlined in Table 16.2.

Table 16.2 Public information and consultation mandates as regulated by the EU EIA Directive

Screening
- the screening determination by the competent authority must be made available to the public (Article 4(5))
- however, timeframes are not regulated, and in the case of a positive screening determination, it is allowed to make the screening determination available to the public at the time when the EIA report is made available for consultation

Scoping
- no public participation in the scoping phase, or public information about the authority's scoping opinion, is required (Article 5)

Public and agency consultation
- concerned authorities with environmental responsibility or local or regional competence have to be granted the opportunity to express their opinion on the information provided by the developer (Article 6(1))
- to ensure effective participation of the public, the public must be informed electronically and by public notices early in the decision-making process, at the latest when information can be provided on:
  - request for development consent
  - result of the screening process
  - details on the competent EIA authority
  - nature of the possible decision or draft decision
  - the EIA report
  - times and places where relevant information will be made available
  - details on arrangements for public participation (Article 6(2))
- the concerned public "shall be given early and effective opportunities to participate" and for this shall be "entitled to express comments and opinions when all options are open to the competent authority" and thus before any decision about the developer's request is taken (Article 6(4))
- there must be reasonable time given for informing authorities and the public and for authorities and the public concerned to "prepare and participate effectively in the environmental decision-making" (Article 6(6))
- the timeframe for consultation of the concerned public on the EIA report has to be at least 30 days (Article 6(7))
- the "relevant information" (incl. EIA report and "main reports and advice issued to the competent authority" – Article 6(3)) must be made electronically available via central EIA portals or easily accessible points of access (Article 6 (5))

Decision-making
- the decision must be "promptly" made available to the public (Article 9)
- this must include the decision and conditions attached to it as well as the main reasons for the decision and how also results from the public consultations have been incorporated or addressed (Article 9(1))

One innovation in the 2014 amendment, however, was the requirement to make EIAs electronically available via central EIA portals and thus improves public access to them (Art.6(5)). Recital 18 of the EIA Directives Preamble states:

> (18) With a view to strengthening public access to information and transparency, timely environmental information with regard to the implementation of this Directive should also be accessible in electronic format. Member States should therefore establish at least a central portal or points of access, at the appropriate administrative level, that allow the public to access that information easily and effectively.

The Directive does not go into detail about which information is required to be made available in such portals, but only refers to "the relevant information" (Art. 6(5)). This requirement to establish EIA portals meant actual changes to previous practice in many EU member states, although examples of best practices already existed (Köppel et al. 2018; Odparlik and Köppel 2012; Odparlik, Köppel, and Geißler 2013). In Austria, for example, the federal environment agency has been hosting an EIA portal website[9] since 1997,[10] which provides comprehensive data on EIA processes (cf. Köppel et al. 2018). However, the Austrian website does not provide the actual EIA documents directly (cf. Köppel et al. 2018). In Italy, a comprehensive EIA/SEA portal[11] is hosted by the Ministry of the Environment (Ministerio dell'ambiente e della tutela del territorio e del mare), providing information on both current and completed EIAs (and SEAs) and providing all relevant EIA documents and information (cf. Ceoloni and Pucci 2015). For example, EIA reports are available, but also screening determinations, scoping information, additional reports, and submissions by the public and other stakeholders. Comments on the screening determination and the EIA report can be submitted online via the EIA portal (Köppel et al. 2018). In the UK, for nationally significant infrastructure projects, a central web page[12] has existed since 2009 – where *relevant* information is provided to the public (cf. Köppel et al. 2018; Glasson and Therivel 2019).

In other EU member states, this new requirement resulted in the development of central electronic EIA portals.[13] For example, in late 2017, Germany developed a national EIA portal,[14] as did several German states.[15] The German EIA portals differ from the Italian example in that only information on the formal consultation of the EIA report, project proposal, negative screening determinations, and final decisions are made available – and they can be accessed only for a certain time.[16] For Germany, the not-for-profit organization "Independent Institute for Environmental Issues" (Unabhängiges Institut für Umweltfragen – UfU) presented an approximate figure of around 2,000 approval and planning procedures for infrastructure projects with EIA and public participation in 2018; however, only 190 of those were documented in the EIA portal (UfU 2020). Moreover, the quality of the information included in the German EIA portals differs substantively (UfU 2020; Cyperski 2020). In practice, the goal of requiring central electronic EIA portals is to "allow the public to access […] information easily and effectively" and "strengthening public access to information and transparency" (Recital 18 EIA Directive), but despite the development of online access, this might not be fulfilled in every member state. For Germany at least, the implementation was rather half-hearted, and followed a strict "one-to-one" (1:1) approach,[17] also often referred to as a "no gold-plating" policy (Jans et al. 2009). What this means is that German legislators adopted the minimum level required by the European standard, rather than using the EU Directive as an opportunity to implement more rigorous national standards.

Monitoring

A major change in the 2014 EIA Directive is the introduction of post-decision monitoring. Before this, it was only required by the SEA Directive and not for project-based EIAs. Despite being a tenet of good EIA practice (e.g., cf. Jones and Fischer 2016; Tinker, Cobb, Bond, and Cashmore 2005; Arts 1998), and having been part of a 1985 draft of the first EIA Directive (COM 1980), the inclusion of a monitoring requirement in practice has only recently become part of the Directive. Article 8 of the Directive now states:

> 4. [...] Member States shall ensure that the features of the project and/or measures envisaged to avoid, prevent or reduce and, if possible, offset significant adverse effects on the environment are implemented by the developer, and shall determine the procedures regarding the monitoring of significant adverse effects on the environment. The type of parameters to be monitored and the duration of the monitoring shall be proportionate to the nature, location and size of the project and the significance of its effects on the environment. Existing monitoring arrangements resulting from Union legislation other than this Directive and from national legislation may be used if appropriate, with a view to avoiding duplication of monitoring.

While member states added this new requirement to their national legislation (e.g., for Greece, see Pediaditi, Banias, Sartzetakis, and Lampridi 2018, and for Germany, see Albrecht, Wende, and Grahn 2021), it is still difficult to judge the successes of implementation in practice, because evaluations or audits are needed. With time, the monitoring requirement may provide evidence and information about successful mitigation measures and will support learning and the provision of better baseline information for future EIAs (Glasson and Therivel 2019; Fischer et al. 2016). Others expect that monitoring approaches will remain very limited and vague and will only refer to existing monitoring activities, without mandating any further actions or consequences of unfavorable monitoring results (e.g., Albrecht, Wende, and Grahn 2021).

Before the 2014 amendment came into force, some member states, or at least certain competent agencies, already had monitoring requirements in place. For example, in the Netherlands, monitoring measures have been mandated by the competent authority for over 20 years; these are called post-decision evaluation (Verheem 1992). There have also been sector-specific monitoring requirements. For example, in Belgium, a joint monitoring program for several offshore wind projects in the North Sea was established by the competent authority (with funds from several proponents) (Glasson and Therivel 2019). Similarly, in Germany, the competent authority responsible for issuing permits for offshore wind farms mandates monitoring of the construction, operation, and demolition phase where developers are obliged to fulfill, collect, and publish[18] the monitoring data (Köppel et al. 2019).

Challenges of implementing the 2014 EIA Directive in practice: a case study of the Tesla EIA in Brandenburg, Germany

In mid-November 2019, Tesla CEO Elon Musk announced the intent to construct the Tesla Gigafactory 4 in Grünheide, Brandenburg – just outside the city of Berlin, Germany (BBC 2019). With a reputation as an innovative and *green* company, the production of Tesla's electric vehicles for the European market has been hailed as a spark for the e-mobility transition in Germany and Europe (Clean Energy Wire 2019).

However, the construction of a facility to produce eclectic cars (about 500,000 per year) will have impacts on the environment. The Tesla Berlin-Brandenburg factory would entail the loss of about 2 km² of pine forest,[19] which is the habitat of certain protected species (in particular sand lizards [*Lacerta Agilis*] and bat species), and the facility will use large quantities of water during operation (about 4,627 m³/day[20] in the first stage[21]). Because the plant was classified as "melting non-ferrous metals", and because of its significant forest conversion, it was determined that there is an obligation to conduct an environmental impact assessment according to § 7 of the Germany EIA Act (Umweltverträglichkeitsprüfungsgesetz – UVPG) (GfBU Consult 2020). With an ambitious timeline plan for the permitting and construction of the facility (about one year), a scoping meeting between the applicant (Tesla Manufacturing Brandenburg SE) and the competent permitting authority, the Brandenburg Environment Agency (Landesamt für Umwelt – LFU), was held in mid-November 2019. The developer requested this scoping meeting, but it was not open to the public and the supporting documentation was not publicly available. The EIA report was developed within a month after the scoping meeting by a consultancy and was submitted to the agencies in mid-December 2019. In January 2020, the EIA report was made publicly available through the German EIA portal website[22] and the competent authority announced the period of public and agency consultation (LfU 2020a). The posting was scheduled for one month, and comments could be submitted for two months, which fulfills the minimum requirements of the EIA Directive and the German EIA Law. This is just the public display of documents and includes options to provide comments. But there is no requirement to respond to comments, although a hearing could come later.

Soon after the announcement of Tesla's permit application, public opposition to the plans emerged and was covered in the local and national media. In response to local opposition, Tesla opened a citizen information office to answer questions from the public (Tagesspiegel 2020). More than 370 comments were submitted during the consultation phase (LFU 2020b). While the public hearing had to be postponed due to emergent COVID-19 health protocols, the planning process continued, and an amended EIA report was submitted to the competent agency in June 2020. The EIA report was published for a second public and agency review phase, lasting from July to August 2020 (the period of summer holidays in the region). Over eight days, in September and October 2020, the public hearing took place, with considerations of the 414 submitted comments (for both consultation periods) from agencies and 110 individuals. Due to restrictions during the COVID-19 pandemic, the number of participants in the hearing was limited, although an online meeting would have been possible.[23] But despite the emergence of internet-based consultation tools and the growing use of virtual meetings during the pandemic, the hearing was not made available to the public through a video livestream, or even recordings. Instead, only representatives of local and national media were allowed to follow the meeting via livestream.

Analysis and discussion of the Tesla EIA process

With respect to the 2014 EIA Directive amendments, the Tesla EIA case shows some of the major limitations in the implementation of the new regulations, e.g., meaningful quality control, substantive scoping phases, and effective consultation. Several issues and lessons are apparent.

First, the Tesla case shows the difficulties in providing for *meaningful quality control* of the EIA process and for the report from the competent authority. The EIA report submitted to the permitting agency in December 2019 displayed several major mistakes and failed to address relevant environmental impacts. While the project will use considerable amounts of groundwater, an assessment of the impacts of the proposed factory on groundwater resources and its compatibility with EU water-related goals and regulations (e.g., EU Water Framework Directive), was missing completely in the first EIA report (December 2019). Given the pre-existing stresses on limited water resources and projections for even greater challenges in the future coming from reduced precipitation and higher temperatures due to climate change (Grünwald 2010), this omission was difficult to comprehend or accept. Although the permitting agency has the responsibility to ensure the quality of the information in the EIA documents, the agency made no requests for additional information or studies before the documents were published for consultation. In response to these apparent inadequacies, the public and environmental NGOs quickly raised the issue of water use and impacts on groundwater resources, and the EIA report had to be amended substantively. Similarly, only an environmental NGO raised concerns about the measures developed for the case of hazardous incidents in the Tesla factory. The competent agency thus ordered Tesla to provide an independent expert opinion evaluating the plausibility of the Tesla documents. This expert statement was published in May 2021 and revealed major deficits.[24] The permitting documents and the EIA will need to be amended again and possibly another phase of public display will be required. This also shows that ensuring the quality and independence of the environmental consultants is important and challenging.

The case also emphasizes the relevance of a *substantive and participative scoping phase*. For example, in the Tesla EIA, if a comprehensive scoping phase would have involved the public and environmental NGOs from the onset of the assessment, there is little doubt the issues of water resources and climate change-related effects clearly would have been raised earlier and been made more prominent. Some of the deficits of the EIA report could have been anticipated, and uncertainties reduced if more weight would have been given to this early scoping phase. These and other shortfalls led to substantive changes of the EIA report and a second and also third round of public and agency consultations (Landkreis Oder-Spree 2020). The lack of initial public engagement in scoping and for determining the necessary content of the EIA documents and identifying issues of possible concern did not lead to a shorter and streamlined EIA process, but instead led to the opposite. In the absence of mandatory scoping, public involvement requirements, documentation standards, and transparency during the scoping process, uncertainty, less trust in the process, and possibly more time spent by developers during the assessment permitting stages of their project were the result. EU scoping guidance is not familiar to German EIA practitioners at large, and no German scoping guidance or Germany-wide general EIA guidance exists. As there is no clear standard established and the federal administrative rule on EIA dates back to the 1990s, it is only an interpretation of the law and not a substantial guidance document, nor is it up to date. However, the Directive also does not explicitly state that scoping opinions need to be provided to the public through EIA portals or other venues (Art. 6(5) EIA Directive), so it is up to EU member states to decide if scoping documents are included, and as we see in Germany this has not been the case.[25]

Although Tesla's stated mission is "to accelerate the world's transition to sustainable energy" (Tesla 2020), *climate change and climate adaptation* were not really dealt with in the EIA report. Implications of the project for climate, and *vice versa*, have not been assessed in detail. Emissions from the construction and operation of the plant, including potential effects on the climate have not yet been quantified. These analyses were also requested during the public consultation period, but so far they have not been provided in the amended EIA documents nor the hearing. There are potential climate change adaptation issues too. The proposed project is located in a region that has been affected by summer droughts and frequent wildfires in recent years, and as we noted, it will require large quantities of fresh water for its operations. The implementation of the new climate change and climate adaptation requirements of the 2014 EIA Directive has been far from effective. While research on climate change in EIA and SEA has been conducted by the German Federal Environmental Agency (Schönthaler, Balla, Wachter, and Peters 2018), clear guidelines, methods, or toolboxes for practitioners to help advance climate change consideration in German EIA do not exist (Rasmussen et al. 2020). The implementation of this requirement has not really taken place in Germany, but studies have shown that this is a challenge for other EU member states too (e.g., Rasmussen et al. 2020; Jiricka-Pürrer, Wachter, and Driscoll 2019; Enriquez-de Salamanca, Martín-Aranda, and Díaz-Sierra 2016).

Finally, the Tesla case study illustrates weaknesses regarding *information provisions and public participation* in the German EIA process. We noted that making information available to the public and other non-government stakeholders started only with the announcement of public consultation, while no early information and participation before or during the scoping phase took place. Moreover, the information uploaded to the central EIA portal website was incomplete and omitted several documents. These were only available in paper format in the relevant agencies. The EIA Directive does not specify what is meant by "relevant information" (Art. 6(5) EIA Directive) to be provided. Practice has been inconsistent, and in Germany the actual project application is not always included in the information provided to the public. The effectiveness of the EIA Directive in having member states create EIA portals to "allow the public to access [...] information easily and effectively" (Recital (18) EIA Directive) is at best variable, and for Germany it has been weak (Köppel 2019). This is shown by the limited number of EIA examples where EIA reports have been uploaded to the German EIA portals (e.g., only 190 out of approximately 2,000 in 2018; with more recent data still to come) (UfU 2020).

The Tesla EIA case study shows prominently the disconnect between certain provisions of the Directive, and German EIA regulations, including the requirements for quality control, easy and effective access to EIA information, and consideration of climate change and climate adaptation in EIA – which have not yet been fully implemented in practice. Moreover, the case is illustrative of the longstanding issue of misjudging the relevance of comprehensive, inclusive, and deliberative scoping, which has had an impact on EIA effectiveness, and on public trust in the EIA process. With strong political support for the Tesla project from national and state governments, and an ambitious timeline for the permitting and EIA process, it might have been wiser to invest more time at the start of the process instead of rushing quickly to a first version of the documents resulting in several iterations in the process now. Nevertheless, as of May 2021, the Tesla Gigafactory in Grünheide is almost complete, having used options for pre-authorizations, with a final permit not likely before the end of 2021 (RBB24 2021).

Conclusions and outlook

With the adoption of the EIA Directive more than 35 years ago, clear benefits have been achieved. A harmonized system of EIA has been established across EU member states, albeit with some differences. EIA has led to modifications of projects, has improved transparency of decision-making, and provided increased opportunities for public input regarding project approvals that affect the environment. However, there is room for improvement.

The vagueness of the EU EIA Directive in some areas, and the discretion of the member states in implementing it, poses a challenge. Efforts in achieving the goals of the Directive depend on the national will, and therefore outcomes have been inconsistent across the member states. Many have directly implemented the Directive "one-to-one" and have not used the opportunity to go beyond the minimal requirements to develop more robust EIA processes. Three areas of need stand out:

1. In many countries, the EIA scoping phase remains voluntary, even though scoping is expected to increase effectiveness and timeliness and reduce uncertainties.

2. Providing easy and timely access to all relevant EIA information (e.g., through internet portals) remains a challenge. This is something which in Germany, for example, has not yet been achieved.

3. Ensuring adequate expertise in competent authorities and consultancies working on EIA reports, with independence from developers, is a quality where practice and national legislation in several countries remain weak.

So, while the procedural obligations of the EIA Directive and the 2014 amendment can be said to have been adopted by the member states, what is less certain is the impact of implementation on environmental protection and decision-making. This depends on the preferences of member states and their dedication to EIA best practices and capacity building.

Notes

1 EIA Directive noted in this chapter refers to DIRECTIVE 2011/92/EU of the European Parliament and of the Council of December 13, 2011 on the assessment of the effects of certain public and private projects on the environment As amended by: Directive 2014/52/EU of the European Parliament and of the Council of April 16, 2014.

2 A list of EIA activities in member states between 2005 and 2008 is, for example, provided in European Union (2012).

3 For details on the development of the EIA Directive, refer to Glasson and Therivel (2019).

4 For further points of critique, refer to, e.g., COWI 2009.

5 EIALAW 2018, How to combine streamlining with environmental effectiveness? – Sharing First Experiences with the New EIA Directive, March 23, 2018, Helsinki.

6 Nordic–Baltic Impact Assessment Conference 2018, September 30–October 2, 2018, Tallinn, https://www.tlu.ee/sites/default/files/Konverentsikeskus/Nordicbaltic18/Nordic%20Baltic%20Impact%20Assessment%20Conference%20Programme%20Sep%202018.pdf.

7 Scoping was, for example mandatory even before the 2014 amendments in the following countries: Czech Republic, Estonia, Hungary, Latvia, Lithuania, Malta, Poland, Slovakia, Bulgaria, and Romania, while, for example, in Germany, Croatia, Slovenia, and Cyprus it is voluntary (COWI 2009).

8 A **competent authority** (may be referred to as an authority, agency, regulatory agency, permitting agency) is one or more designated by a member state as having the powers and responsibilities for transposing or implementing the Directive and its provisions/requirements.

9 https://www.umweltbundesamt.at/umweltthemen/uvpsup/uvpoesterreich1/uvp-dokumentation

10 earliest project with EIA documented in the database is from 1997, https://secure.umweltbundesamt.at/uvpdb/maps/index.html

11 https://va.minambiente.it/en-GB

12 https://infrastructure.planninginspectorate.gov.uk/

13 For example, in France with the platform "projets-environnement.gouv.fr", https://www.projets-environnement.gouv.fr/pages/home/

14 https://www.uvp-portal.de/

15 https://www.uvp-verbund.de

16 A federal ordinance for the EIA portals was adopted only in November 2020; three and a half years after transposition of the 2014 EIA Directive. It details that the documents must be available via the EIA portal for at least the duration of the consultation period (min. 60 days). Permanent storage and availability of the documents and information was not mandated although requested by several actors. However, the position promoted by industry associations "that industrial espionage could take place if information about projects would be available online" prevailed.

17 The approach of a "one to one" (1:1) transposition of EU Directives into national legislation was adopted by the German governing coalition of social democrats and Christian democrats in 2018 in relation to environmental and climate policies of the EU (cf. Krohn 2018). Literally speaking this refers to taking the language of the EU directives and implementing it without deviation into the national legislation to avoid "gold-plating", thus adopting stricter standards as mandated by the EU. This policy, however, has drawn much criticism as the EU Directives in general provide much leeway for adopting

the EU standards to member states' situations and arguing that the "one to one" policy in fact means a "de minimis" approach and adopting the lowest standard compatible with EU provisions (Krohn 2018).

18 The data is made available via the Marine Life Investigator (MARLIN), https://lindevmarlin61.bsh.de /MARLINDMZ/publicSites/MainAppPublic.jsf

19 This amount is required for the first expansion stage; for a second stage further forest will be cut.

20 In comparison, the daily water consumption per person in Germany is about 125 l/day (Statista 2021).

21 In the first proposal submitted for consultation a battery factory was proposed as part of the facility, which was removed from the proposal later reducing in particular the water footprint of the project. However, it is expected that the battery factory will still be proposed at a later stage.

22 https://www.uvp-verbund.de/portal/

23 With the passage of the Act to Ensure Proper Planning and Licensing Procedures During the COVID 19 Pandemic (Planning Assurance Act – PlanSiG) of May 2020, planning authorities have the option to hold digital meetings instead of in-person hearings. So far, the act is only valid until end of December 2022; however, it remains to be seen if a permanent regulation allowing for online hearings and online participation options will follow.

24 One example that was pointed out by the experts concerns possible accidents by which the refrigerant tetrafluoropropene could be released. Other than stated in the application documents, this chemical would not be released in gaseous form but in liquid which makes a complete reassessment of the possible consequences and necessary safety measures necessary.

25 Some EU member states regularly provide scoping reports through their portals, e.g., in Italy these are uploaded on the EIA/SEA Portal web page (https://va.minambiente.it/en-GB).

References

Albrecht, Juliane, Wende, Wolfgang, and Grahn, Doris. 2021. "Die neue Überwachungspflicht für Projekte nach § 28 UVPG: Rechtsgrundlagen, fachliche Anforderungen, Durchsetzbarkeit und erste Einblicke in die praktische Umsetzung (The new monitoring obligation for projects according to Art. 28 UVPG: Legal basis, technical requirements, enforceability and first insights into practical implementation)". *UVP-report* 35(1): 3–17, DOI: 10.17442/uvp-report.035.02.

Arabadjieva, Kalina. 2016. "'Better Regulation' in Environmental Impact Assessment: The amended EIA directive". *Journal of Environmental Law* 28(1): 159–168, DOI: 10.1093/jel/eqw001.

Arts, Jos, Runhaar, Hens A.C., Fischer, Thomas B., Jha-Thakur, Urmila, Van Laerhoven, Frank, Driessen, Peter P., and Onyango, Vincent. 2012. "The effectiveness of EIA as an instrument for environmental governance: Reflecting on 25 years of EIA practice in the Netherlands and the UK". *Journal of Environmental Assessment Policy and Management* 14(04): 1250025.

Arts, Jos. 1998. *EIA Follow-Up. On the Role of Ex Post Evaluation in Environmental Impact Assessment.* Groningen.

Barker, Adam, and Wood, Christopher. 1999. "An evaluation of EIA system performance in eight EU countries". *Environmental Impact Assessment Review* 19(4): 387–404.

Barker, Adam, and Wood, Christopher. 2001. "Environmental assessment in the European Union: Perspectives, past, present and strategic". *European Planning Studies* 9(2): 243–254, DOI: 10.1080/713666468.

BBC. 2019. "'Berlin rocks', says Elon Musk as he chooses European factory". *BBC*, November 13, 2019. https://www.bbc.com/news/business-50400068.

Bond, Alan, and Wathern, Peter. 1999. "EIA in the European Union". In: *Handbook of Environmental Impact Assessment*, edited by Judith Petts, 223–248. Oxford: Blackwell Science.

Bond, Alan, Fischer, Thomas, and Fothergill, Josh. 2017. "Progressing quality control in environmental impact assessment beyond legislative compliance: An evaluation of the IEMA EIA Quality Mark certification scheme". *Environmental Impact Assessment Review* 63: 160–171.

Boyes, Suzanne J. and Elliott, Michael. 2014. "Marine legislation--the ultimate 'horrendogram': international law, European directives & national implementation". *Marine Pollution Bulletin* 86(1–2): 39–47. DOI: 10.1016/j.marpolbul.2014.06.055.

CEEweb for Biodiversity. 2013. "Implementation of Environmental Impact Assessments in Central and Eastern Europe. Lessons learnt from Estonia, Bulgaria and Hungary". http://www.ceeweb.org/wp -content/uploads/2012/01/Implementation-of-Environmental-Impact-Assessments-in-Central-and -Eastern-Europe.pdf.

Ceoloni, Paola, and Pucci, Valentina. 2015. "La nuova direttiva via 2014/52/UE e la valutazione degli impatti sulla salute umana". In: *La VIS in Italia: Valutazione e partecipazione nelle decisioni su ambiente e salute*, edited by Liliana Cori, Adele Ballarini, Nunzia Linzalone, Marinella Natali, and Fabrizio Bianchi, 26–31. Emilia Romagna: Arpae.

Clean Energy Wire. 2019. "Tesla gigafactory in Germany hailed as gamechanger for lagging e-Mobility transition". November 18, 2019. https://www.cleanenergywire.org/news/tesla-gigafactory-germany -hailed-gamechanger-lagging-e-mobility-transition.

COM – Commission of the European Communities (ed.). 1980. "Proposal for a Council Directive concerning the assessment of the environmental effects of certain public and private projects". COM(80) 313 final, June 11, 1980, Brussels.

COM – European Commission (ed.). 2012. "Proposal for a Directive of the European Parliament and of the Council amending Directive 2011/92/EU on the assessment of the effects of certain public and private projects on the environment". COM/2012/0628 final – 2012/0297 (cod). 52012PC0628. Explanatory memorandum. http://eur-lex.europa.eu/legal-content/EN/TXT/HTML/?uri=CELEX:52012PC0628

COM – European Commission. 2013. "Guidance on integrating climate change and biodiversity into environmental impact assessment". https://ec.europa.eu/environment/eia/pdf/EIA%20Guidance.pdf (30/12/2020).

COM – European Commission. 2017. "Environmental Assessment of projects and plans and programmes. Rulings of the Court of Justice of the European Union". https://ec.europa.eu/environment/eia/pdf/ EIA_rulings_web.pdf.

COM – European Commission. n.d. "Results of the consultation on the review of the EIA Directive". https://ec.europa.eu/environment/eia/pdf/results_consultation.pdf.

COWI. 2009. "Study concerning the report on the application and effectiveness of the EIA Directive". Final Report.

Cyperski, Anna. 2020. "Informationsbereitstellung in Deutschlands UVP-portalen – Good practice beispiele und Verbesserungsvorschläge". Bachelor thesis.

Drenovak-Ivanović, Mirjana. 2016. "Environmental Impact Assessment in Serbian Legal System: Current issues and prospects for revision". *Anali Pravnog Fakulteta u Beogradu* 64(3): 126–139, DOI: 10.5937/ AnaliPFB1603126D.

Enríquez-de-Salamanca, Álvaro, Martín-Aranda, Rosa M., and Díaz-Sierra, Rubén. 2016. "Consideration of climate change on environmental impact assessment in Spain". *Environmental Impact Assessment Review* 57: 31–39.

Enríquez-de-Salamanca, Álvaro. 2016. "Project splitting in environmental impact assessment". *Impact Assessment and Project Appraisal* 34(2): 152–159, DOI: 10.1080/14615517.2016.1159425.

European Union. 2012. "Commission staff working paper: Impact assessment". COM/ 2012/ 628 Final. Brussels: EU.

European Union. 2017a. "Environmental impact assessment of projects. Guidance on screening". https://ec .europa.eu/environment/eia/pdf/EIA_guidance_Screening_final.pdf [accessed June 27, 2020].

European Union. 2017b. "Environmental impact assessment of projects. Guidance on scoping". https://ec .europa.eu/environment/eia/pdf/EIA_guidance_Scoping_final.pdf [accessed June 27, 2020].

European Union. 2017c. "Environmental impact assessment of projects. Guidance on the preparation of the Environmental Impact Assessment Report". https://ec.europa.eu/environment/eia/pdf/EIA_guid-ance_EIA_report_final.pdf [accessed June 27, 2020].

European Union. 2022. "Country profiles". https://european-union.europa.eu/principles-countries-his-tory/country-profiles_en [accessed January 2, 2022].

Fagan, Adam, and Sircar, Indraneel. 2015. "Environmental impact assessment (EIA) processes in Serbia". In: *Europeanization of the Western Balkans*, edited by Fagan Adam, and Indraneel Sircar, 126–149. UK: Palgrave Macmillan.

Fischer, Thomas, and Fothergill, Josh. 2014. "Das IEMA-UVP-Gütezeichen im Vereinigten Königreich: Ein Beispiel freiwilliger Akkreditierung. (The IEMA EIA Quality Mark in the United Kingdom. An Example of Voluntary Accreditation)". *UVP-report* 28(3+4): 113–118.

Fischer, Thomas, Therivel, Riki, Bond, Alan, Fothergill, Josh, and Ross, Marshall. 2016. "The revised EIA Directive – possible implications for practice in England". *UVP-report* 30(2): 106–112.

Fischer, Thomas. 2016. "Implications of the revised EIA Directive – Editorial". *UVP-report* 30(2): 59–60.

Fonseca, Alberto, and Fernández, German Marino Rivera. 2020. "Reviewers' perceptions of the volume of information provided in environmental impact statements: The case for refocusing attention on what is relevant". *Journal of Cleaner Production* 251: 119757.

Geißler, Gesa, Köppel, Johann, and Odparlik, Lisa. 2011. "Addressing green-house gas emissions in Environmental Impact Assessments – The discursive making of guidance in the United States". *UVP-report* 25(4): 215–221.

Geißler, Gesa, Köppel, Johann, and Gunther, Pamela. 2013. "Wind energy and environmental assessments – A hard look at two forerunners' approaches: Germany and the United States". *Renewable Energy* 51: 71–78, DOI: 10.1016/j.renene.2012.08.083.

GfBU consult – Gesellschaft für Umwelt- und Managementberatung MbH. 2020. Kurzbeschreibung für das Vorhaben „Gigafactory Berlin-Brandenburg". https://www.uvp-verbund.de/documents/ingrid -group_ige-iplug-bb/94AFADF0-92F1-44EA-AA54-E1CD7C0FF6AD/Tesla%20Manufacturing %20Brandenburg%20SE_V2_24-06-2020_Kurzbeschreibung.pdf.

GHK. 2010. "Collection of information and data to support the Impact Assessment study of the review of the EIA Directive". https://ec.europa.eu/environment/eia/pdf/collection_data.pdf.

Glasson, John, and Therivel, Ricki. 2019. *Introduction to Environmental Impact Assessment*, 5th edition. Oxon, New York: Routledge.

Grünwald, Uwe. 2010. „Wasserbilanzen der Region Berlin-Brandenburg". Diskussionspapier 7, Berlin-Brandenburgische Akademie der. *Wissenschaften.* https://edoc.bbaw.de/files/278/diskussionspapier_7 _gruenewald_online.pdf.

Güneş, Şule. 2020. "Environmental impact assessment in Turkey: A principal environmental management tool". In: *Environmental Law and Policies in Turkey* vol 31, edited by Zerrin Savaşan, and Vakur Sümer, 83–97, The Anthropocene: Politik – Economics – Society – Science. Cham: Springer, DOI: 10.1007/978-3-030-36483-0_5.

Günther, Markus, Geißler, Gesa, and Köppel, Johann. 2017. "Many roads lead to Rome: Selected features of quality control within environmental assessment systems in the US, NL, CA, and UK". *Environmental Impact Assessment Review* 62: 250–258, DOI: 10.1016/j.eiar.2016.08.002.

Hartlik, Joachim. 2020. "Inhalte und Methoden bei der Bearbeitung von Verwaltungsverfahren nach § 5 bis 25 UVPG". In: *Umweltverträglichkeitsprüfung, Strategische Umweltprüfung. Bearbeitung umweltrechtlicher Praxisfälle. Erläuterungswerk. Abschnitt III. Inhalte und Methoden der Umweltprüfungen*, edited by Wolfgang Sinner, Ulrich M. Gassner, Joachim Hartlik, and Albrecht Juliane. Wiesbaden: Kommunal- und Schul-Verlag.

Hartley, Nicola, and Wood, Christopher. 2005. "Public participation in environmental impact assessment – Implementing the Aarhus Convention". *Environmental Impact Assessment Review* 25(4): 319–340.

Jiricka, Alexandra, Bösch, Martin, and Völler, Sonja. 2016. "Learning from the past and upcoming challenges – the implementation of the amendment of the EIA Directive in Austria". *UVP-report* 30(3): 143–151.

Jiricka-Pürrer, Alexandra, Czachs, Christina, Formayer, Herbert, Wachter, Thomas F., Margelik, Eva, Leitner, Markus, and Fischer, Thomas B. 2018. "Climate change adaptation and EIA in Austria and Germany – Current consideration and potential future entry points." *Environmental Impact Assessment Review* 71: 26–40.

Jiricka-Pürrer, Alexandra, Bösch, Martin, and Pröbstl-Haider, Ulrike. 2018. "Desired but neglected: Investigating the consideration of alternatives in Austrian EIA and SEA practice". *Sustainability* 10(10): 3680.

Jiricka-Pürrer, Alexandra, Wachter, Thomas, and Driscoll, Patrick. 2019. "Perspectives from 2037 – Can environmental impact assessment be the solution for an early consideration of climate change-related impacts?". *Sustainability* 11(15): 4002, DOI: 10.3390/su11154002.

Jones, Robert, and Fischer, Thomas B. 2016. "EIA Follow-Up in the UK – A 2015 update". *Journal of Environmental Assessment Policy and Management* 18(1): 1650006.

Justice & Environment. 2012. "The EIA in selected member states". http://www.justiceandenvironment .org/_files/file/2012/EIA%20comprehensive%20report%202012_1.pdf.

Justice & Environment. 2016. "Collection of examples of good practice in EIA procedures". http://www.jus ticeandenvironment.org/fileadmin/user_upload/Publications/2016/EIA_good_practices_collection.pdf.

Köppel, Johann, Geißler, Gesa, Helfrich, Jennifer, and Reisert, Jessica. 2012. "A snapshot of Germany's EIA approach in light of the United States archetype". *Journal of Environmental Assessment Policy and Management* 14(4): 1250022, DOI: 10.1142/S1464333212500226.

Köppel, Johann, Günther, Markus, Geißler, Gesa, Grimm, Marie, and Möller-Lindenhof, Theresa. 2018. ""The Right to Know" – zur Einführung von UVP-portalen in Deutschland ("The Right to Know" – on implementing EIA portals in Germany)". *UVP-report* 32(1): 24–33.

Köppel, Johann, Biehl, Juliane, Dahmen, Marie, Geißler, Gesa, and Michelle, E. Portman. 2019. "Perspectives on marine spatial planning". In *Wildlife and Wind Farms: Conflicts and Solutions, Volume 4*, edited by Martin R. Perrow, 281–317. Exeter: Pelagic Publishing.

Köppel, Johann. 2016. "Wishful thinking on the potential of the amended EU Directive 2014 for reviving EIA in Germany?". *UVP-report* 30(2): 61–62, DOI: 10.17442/uvp-report.030.10.

Köppel, Johann. 2019. "UVP-Portale in Deutschland – wer sucht, der findet?". *UVP-report* 33(1): 32–33.

Kørnøv, Lone, and Kjellerup, Ulf. 2016. "Observations and reflections upon the Danish transposition of the EIA Directive: Focus on quality and competence enhancement". *UVP-report* 30(3): 130–132.

Krohn, Susan. 2018. "Die Eins-zu-Eins-Umsetzung des europäischen Umweltrechts, oder: Wenn für Deutschland das Nötigste gut genug ist". *Zeitschrift für Umweltrecht* 29(7–8): 385–386.

Jans, Jan H., Squintani, Lorenzo, Aragão, Alexandra, Macrory, Richard, and Bernhard, W. Wegener. 2009. "'Gold Plating' of European Environmental Measures?" *Journal for European Environmental & Planning Law* 6(4): 417–435.

Oder-Spree, Landkreis. 2020. "Auslegung der geänderten Unterlagen für tesla-Fahrzeugwerk beginnt". https://www.landkreis-oder-spree.de/Service-Aktuelles/Aktuelles/Mitteilungen/Weitere -Beteiligung-der-%C3%96ffentlichkeit-f%C3%BCr-Tesla-Fahrzeugwerk-beginnt-Neu-Auslegung -zus%C3%A4tzlich-im-Internet.php?object=tx,2426.5.1&ModID=7&FID=2689.2932.1&NavID =2689.203&La=1&call=suche.

LfU – Brandenburgisches Landesamt für Umwelt. 2020a. "Errichtung und Betrieb einer Anlage für den Bau und die Montage von Elektrofahrzeugen mit einer Kapazität von jeweils 100.000 Stück oder mehr je Jahr am Standort 15537 Grünheide (Mark)".

LfU – Brandenburgisches Landesamt für Umwelt. 2020b. "Insgesamt 414 Einwendungen zur Errichtung der tesla-Fahrzeugfabrik in Grünheide (Mark) – Erörterungstermin ab 23. September". https://mluk .brandenburg.de/mluk/de/aktuelles/presseinformationen/detail/~10-09-2020-eroerterungstermin -ab-23-september [accessed January 27, 2021].

Lonsdale, Jemma, Weston, Keith, Blake, Sylvia, Edwards, Ruth, and Elliott, Michael. 2017. "The Amended European Environmental Impact Assessment Directive: UK marine experience and recommendations". *Ocean & Coastal Management* 148: 131–142.

Lyhne, Ivar and Kørnøv, Lone. 2013. "How do we make sense of significance? Indications and reflections on an experiment". *Impact Assessment and Project Appraisal* 31(3): 180–189. DOI: 10.1080/14615517.2013.795694.

Morrison-Saunders, Angus, Pope, Jenny, Gunn, Jill A.E., Bond, Alan, and Retief, Francois. 2014. "Strengthening impact assessment: A call for integration and focus". *Impact Assessment and Project Appraisal* 32(1): 2–8, DOI: 10.1080/14615517.2013.872841.

Odparlik, Lisa F., and Köppel, Johann. 2013. "Access to information and the role of environmental assessment registries for public participation". *Impact Assessment and Project Appraisal* 31(4): 324–331, DOI: 10.1080/14615517.2013.841028.

Odparlik, Lisa F., Köppel, Johann, and Geißler, Gesa. 2012. "The grass is always greener on the other side: Der Zugang zu Umweltprüfungs-Dokumenten in Deutschland im internationalen Vergleich". *UVP-report* 26(5): 236–243.

Pediaditi, Kalliope, Banias, Georgios, Sartzetakis, Eftychios, and Lampridi, Maria. 2018. Greece's reformed EIA system: Evaluating its implementation and potential. *Environmental Impact Assessment Review* 73: 90–103.

Rasmussen, Andrew, Langkau, Alina, Lehmler, Sebastian, Lindemann, Philippa, Wessel-Bothe, Eva, Varol, Nermin, and Köppel, Johann. 2020. "Handreichung zur Berücksichtigung des Klimawandels in der Umweltverträglichkeitsprüfung". *UVP-report* 34(2): 92–99.

RBB24. 2021. "Tesla-Serienproduktion Soll erst 2022 starten." https://www.rbb24.de/studiofrankfurt/ wirtschaft/tesla/2021/05/tesla-auto-gruenheide-verzoegerung-baustelle-produktion.html [accessed May 18, 2021].

"Revised EIA Directive – Directive 2014/52/EU of the European Parliament and of the Council of 16 April 2014 amending Directive 2011/92/EU on the assessment of the effects of certain public and private projects on the environment". *Official Journal of the European Union* L 124: 1–18.

Sangenstedt, Christoph. 2014. "Die Reform der UVP-Richtlinie 2014: Herausforderungen für das Deutsche Recht". *Zeitschrift für Umweltrecht* 10: 526–535.

Savaşan, Zerrin. 2020. "The development process of environmental law in Turkey: The EU impact". In: *Environmental Law and Policies in Turkey* vol 31, edited by Zerrin Savaşan, and Vakur Sümer, 7–31, The Anthropocene: Politik – Economics – Society – Science. Cham: Springer. DOI: 10.1007/978-3-030-36483-0_2.

Schönthaler, Konstanze, Balla, Stefan, Wachter, Thomas F., and Peters, Heinz-Joachim. 2018. "Grundlagen der Berücksichtigung des Klimawandels in UVP und SUP". *Umweltbundesamt*. https://www

.umweltbundesamt.de/sites/default/files/medien/1410/publikationen/2018-02-12_climate-change _04-2018_politikempfehlungen-anhang-4.pdf.

Statista. 2021. "Entwicklung des Wasserverbrauchs pro Einwohner und Tag in Deutschland in den Jahren 1990 bis 2019". https://de.statista.com/statistik/daten/studie/12353/umfrage/wasserverbrauch-pro -einwohner-und-tag-seit-1990/ [accessed May 18, 2021].

Tagesspiegel. 2020. "Tesla eröffnet Bürgerbüro in Grünheide". https://www.tagesspiegel.de/berlin/nach -protesten-gegen-waldrodung-tesla-eroeffnet-buergerbuero-in-gruenheide/25437060.html [accessed January 31, 2021].

Tesla. 2020. "About tesla". https://www.tesla.com/about [accessed June 27, 2020].

Tinker, L., Cobb, D., Bond, A., and Cashmore, M. 2005. "Impact mitigation in environmental impact assessment: Paper promises or the basis of consent conditions?". *Impact Assessment and Project Appraisal* 23(4): 265–280.

UfU – Independent Institute for Environmental Issues. 2020. "Monitoring Report 2018 – Öffentlichkeitsbeteiligung bei Infrastrukturprojekten in Deutschland". https://www.ufu.de/wp-content/uploads/2020/04/2018_Monitoringreport.pdf [accessed January 31, 2021].

UVP Gesellschaft e.V. 2015. "Paderborner Erklärung – Forderungen zur Novellierung des UVP-Gesetzes." *UVP-report* 29(2): 104–107.

UVP Gesellschaft e.V. 2016. "Stellungnahme der UVP-Gesellschaft e.V. zum Entwurf des Bundesministeriums für Umwelt, Naturschutz, Bau und Reaktorsicherheit für ein Gesetz zur Modernisierung des Rechts der Umweltverträglichkeitsprüfung". *UVP-report* 30(4): 222–233.

Verheem, Rob. 1992. "Environmental assessment at the strategic level in the Netherlands". *Project Appraisal* 7(3): 150–156.

Verheyen, Roda and Schayani, Kilian. 2020. „Der globale Klimawandel als Hindernis bei der Vorhabengenehmigung: Entscheidungsbesprechung zum Urteil „Heathrow Airport" und Einordnung in die internationale Rechtsprechung". *Zeitschrift für Umweltrecht* 31(7–8): 412–418.

Wende, Wolfgang. 2002. "Evaluation of the effectiveness and quality of environmental impact assessment in the Federal Republic of Germany". *Impact Assessment and Project Appraisal* 20(2): 93–99, DOI: 10.3152/147154602781766735.

Wood, Christopher, Barker, Adam, Jones, Carys and Hughes, Joanna. 1996. *Evaluation of the Performance of the EIA Process*. Final Report. https://ec.europa.eu/environment/archives/eia/eia-studies-and-reports/pdf/eiaperform.pdf [accessed January 2, 2022].

Wood, Christopher, and Lee, Norman. 1988. "The European Directive on environmental impact assessment: Implementation at last?". *Environmentalist* 8(3): 177–186, DOI: 10.1007/BF02240251.

Wood, Christopher. 1995. *Environmental Impact Assessment: A Comparative Review*. Harlow: Longman.

17

THE UNITED STATES NATIONAL ENVIRONMENTAL POLICY ACT

History, Process, and Politics

Matt Lindstrom and Ben West

Prior to 1970, there was no requirement for the US federal government to account for the environmental impacts of their decisions and projects. If negative environmental impacts existed, there was also no requirement for the federal government to inform the public or seek their input and comments on the federal project. All of this changed on January 1, 1970, when Republican President Richard Nixon signed the National Environmental Policy Act (NEPA).

This chapter reviews the purpose of NEPA and provides an overview of the basic environmental assessment process in the United States. In order to provide a context for NEPA's impacts and meaning, the chapter discusses the legislative, executive, and judicial history of NEPA, as well as debates over how NEPA and the environmental assessment processes should most effectively provide and facilitate ecological and economic goals and values. Also addressed in this chapter is an analysis of recent US judicial case law guiding NEPA's judicial enforcement as well as legislative and executive branch attempts at NEPA reform during US President Trump's administration (2016–2020). Finally, this chapter discusses NEPA's future under an administration led by President Joe Biden and Vice-President Kamala Harris.

After Congress passed NEPA by overwhelming bi-partisan majorities and President Nixon signed the bill on January 1, 1970, a new era of environmental policy began. For the first time in US history, NEPA required every office and entity of the US national government agency to collect, assess, and share information about possible environmental impacts of large-scale federal agency decisions *before* they made those decisions.

The passage of NEPA did three important things. First, it declared a comprehensive national policy for the environment modeled around an understanding of the balance between humans and ecosystems (Title I of NEPA). Second, NEPA introduced the environmental impact assessment process (EIA) as a new procedural assessment tool for federal agencies to use in order to comply with the environmental policy. NEPA specifically applies only to federal agencies and their decisions, including permit approvals and funding of projects. When an individual or corporate actor needs a federal permit or gets federal funding, and if this project meets the NEPA threshold of significant environmental impacts, then non-governmental actors are pulled into the NEPA EIS process. The EIA process dramatically opened the federal decision-making process to public participation and increased the transparency of federal decision-making. Third, the Act established the Council on Environmental Quality (CEQ)

DOI: 10.4324/9780429282492-19

to supervise NEPA's implementation in the executive branch and to monitor environmental quality (Title II of NEPA).

The National Environmental Policy Act of 1970 was created during a flurry of US environmental activism and national government initiatives around environmental issues. Besides NEPA, President Nixon established the Environmental Protection Agency (EPA), bringing together over a dozen federal agencies under one organization to address water and air pollution and together with other federal agencies address industrial chemicals, climate change, urban planning, and much more. While the EPA is not the primary agency enforcing NEPA, the agency is routinely involved with the most controversial national projects requiring NEPA's EIA process. This involvement is primarily an advisory role but an important one, nonetheless.

Prior to the 1970s, there were multiple ethical and political questions for the US Congress to consider in the late 1960s when Congress was considering environmental impact assessment. United States citizens were demanding action by Congress related to a national policy for the environment and environmental assessment. Early questions and Congressional debates revolved around numerous questions. Once the information about potential impacts is completed, to what extent should that information influence the final decision of the federal agency? If environmental protection is not a direct part of the agency mission, is that federal organization obligated to even gather or consider environmental impacts? Should federal agencies be required to "look before they leap" into a significant project? Until NEPA was passed by Congress in 1969 and signed by President Nixon on January 1, 1970, there was no comprehensive requirement for federal agencies to at least *consider* the environmental impacts of those decisions. Congress fundamentally responded to the public demand for an ethical and informed national environmental policy for the United States.

Not only that, but there was also no requirement for federal agencies to inform and consult with public stakeholders during the federal agency decision process. NEPA's requirement to inform the public is a critical aspect of NEPA's procedural policy goals. This public disclosure requirement is especially important for groups that do not have lobbyists and others paying close attention to government actions and decisions. NEPA's founders intentionally wanted to include information sharing not only as an accountability or watchdog mechanism but also as a means of gathering more information and ultimately making better decisions. For less politically powerful groups such as indigenous peoples and low-income communities in the United States, NEPA's public disclosure requirement is a crucial step toward comprehensive democratic decision-making. While NEPA does not guarantee this occurs, without NEPA (or a similar law) it is guaranteed that the vast majority of the public would be left entirely out of the decision-making process for major federal projects that significantly impact the environment. Having the public informed and invited to participate creates government incentives and a level of transparency that is better able to mitigate excessive ecological impacts while also balancing other legal requirements and economic benefits. As we will see later in this chapter, NEPA does not guarantee ecologically harmonious outcomes or environmentally sound decisions, but it is certain that without NEPA there would be much less ecological accountability as well as democratic participation and transparency, especially for the least powerful members of the public.

NEPA requirements

NEPA directly applies only to actions of federal agencies. It does not directly regulate private firms or individuals, only indirectly when their proposals require a federal action like a permit or funding and only when that federal action is considered major and significant. The determination of major and significant is answered by the primary agency, or the lead agency with assistance from other federal agencies such as the EPA. The NEPA process can take several routes, often all resulting in the same final result, such as the road or pipeline is built, but the length of review time and associated review costs will vary greatly depending on the unique characteristics of each route or option in the NEPA review process. As the NEPA process illustration indicates, there are three primary avenues in the NEPA process (EPA n.d.):

1. Categorial Exclusion: in this level of analysis, it is determined that NEPA review is not required per the statutory language of a "major federal action that significantly impacts the environment". If the case in point does not elevate to this standard, NEPA does not apply.
2. Environmental Assessment: The primary purpose of an EA is to determine whether a full environmental impact statement is necessary. If the lead agency determines there is no significant impact, they will issue what is called a Finding of No Significant Impact or FONSI, thus ending the NEPA review process. The lead agency would then issue the permit or allow whatever the proposed action is, pending other regulatory requirements, of course. If the EA finds that the project meets the requirements for a full EIS – a major federal project significantly impacting the environment – then an EIS will be prepared by the lead agency in consultation with private sector partners (assuming they exist) and other federal agencies to the proposed federal project.
3. Environmental Impact Statement: If the proposed action clearly requires an EIS, an agency may go straight to the full EIS process. If not, they go through the EA process and then determine that a full review is required. As the diagram illustrates, the EIS process has several stages and includes the requirement for public review and commentary after the draft EIS is published (see Figure 17.1).

As this illustration indicates, there are many parts to the NEPA requirement. At its core, NEPA forces agencies to gather and consider information in the form of empirical facts as well as public opinions. In the end, the agency decides how to use this information. However, in many cases, without NEPA, the agency's proposed action would undergo little to no review from other federal agencies or the public. NEPA's public participation requirements offer a crucial tool for not only enhanced transparency but also smarter final results. No result is guaranteed, however. The EIS provides information for the decision-making authority in the federal agency as well as other stakeholders, but the US judicial system has consistently ruled that NEPA is a procedural statute and does not mandate a specific outcome or decision. Therefore agencies who want to avoid having their decisions changed by the courts put together thorough EIS statements sometimes reaching over 1,000 pages.

One example of this is a planned highway construction around the Hoover Dam in the state of Arizona. After soliciting and considering the data and opinions collected after a public comment process, the Federal Highway Administration changed the route to minimize negative ecological impacts and incorporated more highway design measures such as sidewalks, bike paths, and parking. The project manager stated, "the federal highway administration grossly underestimated some of the alternatives and too quickly dismissed them. Oftentimes the public is a huge influence on the project. NEPA is certainly the foundation for public participation" (Pepper 2015).

A full EIS process can be a long, litigious, and expensive process for all parties involved – the federal government, proposed action proponents from the private sector, and proposed action opponents often from environmental law groups. Some EIS statements can be over 1,000 pages long and over ten years due to extensive legal delays and the desire of lead agencies to follow

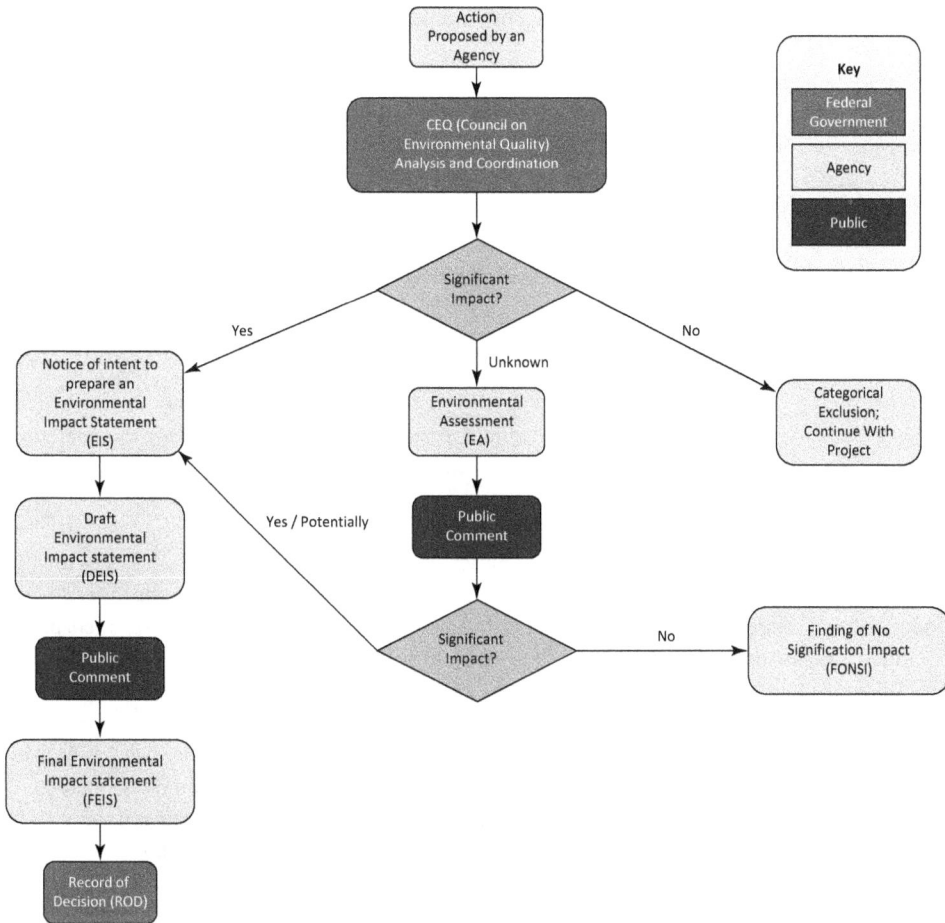

Figure 17.1 The NEPA process (by Ben West).

every procedure to an exacting depth so as to protect their discretionary power and (hopefully) make the final recommended decision using the best science available while also incorporating public input and meeting NEPA's stated goal of "creating productive harmony between man and nature" (CEQ n.d.).

NEPA's legislative history, origins, and purpose

NEPA provides an environmental policy framework and procedural guidance for enhancing the importance of environmental values in federal agency decision-making. Prior to NEPA, ecological values and goals were only incorporated into federal agency decision-making if, and only if, they were a core part of the agency's mission. Developed by Senator Henry Jackson (D-WA), with the invaluable and crucial assistance of Dr. Lynton K. Caldwell, a former political science professor at Indiana University, whose contribution is analyzed later in this chapter.

NEPA legislation passed both chambers of Congress by huge majorities (something almost unheard of today) and was signed by President Nixon on national television on January 1, 1970, marking the beginning of a new era of environmental governance. When President Nixon signed the NEPA bill, he exclaimed:

> It is particularly fitting that my first official act in this new decade is to approve the National Environmental Policy Act [...] By my participation in these efforts I have become further convinced that the 1970s absolutely must be the years when America pays its debt to the past by reclaiming the purity of its air, its water, and out living environment. It is literally now or never [...] We are determined that the decade of the seventies will be known as the time when this country regained a productive harmony between man and nature.
>
> *(The American Presidency Project n.d.)*

Fifty years later, NEPA's relatively brief text (only a few pages) remains largely unchanged by any significant Congressional amendment, and approximately 25 states, 80 other national governments and numerous economic and political institutions like the World Bank and European Union have emulated the National Environmental Policy Act (Andrews 1999).

While NEPA endures plenty of criticism, the law's accolades reach the highest levels of praise. While arguably hyperbolic, some commentators compare the 1970s US law with the 1215 English Magna Carta agreement, which dispersed powers from the King of England and fundamentally shifted the English political landscape (Mandelker 2002). Other exalted praise includes comparisons to the Bible and the Ten Commandments (Peter Borrelli 1989). Despite the laudatory metaphors, many environmental advocates suggest that the implementation and enforcement of the law has failed to meet the intentions of the law's framers. As will be reviewed later in this chapter, the US federal courts, particularly the Supreme Court, have essentially rendered impotent NEPA's substantive purposes for the supremacy of environmental quality.

The passage of NEPA was part of a legal movement away from a dependence on common law to public law as a means of pursuing and enforcing more consistent environmental quality initiatives across all federal agencies. NEPA, like many of the other environmental statutes passed by Congress in the 1970s, has proved to be a mixed success. A former chair of the Council on Environmental Quality summarized NEPA's evolution as an "obliteration of substance in NEPA that has occurred in the courts and agency implementation" (Dinah Bear 1993).

NEPA certainly is no panacea for the constellation of environmental problems that exist today. Arguably the most cogent yet holistic environmental law ever written, NEPA provides a progressive vision for this nation's future and a practical action-forcing environmental assessment mechanism for shaping US federal agency choices, but it rests on a very shallow jurisprudence. However, if the courts and presidents enforced NEPA's substantive principles, which form NEPA's foundation, agency decisions would reflect a much greater desire to achieve environmental quality.

The following chapter section discusses NEPA's substantive purpose and declaration of environmental policy from the Act's formation to implementation and enforcement. The chapter summarizes the content of NEPA's policy goals and examines the intentions of the framers and authors of NEPA. The core goals of NEPA and the goals of the Act's founders are assessed within the context of judicial rulings and decisions within various presidential administrations.

Congressional intent and aims of NEPA

It is common for legal analysts to interpret and debate the legislative intent that shapes our laws. This is especially true in NEPA's case because of its multi-layered statutory framework. While interpreting Congressional intent is rarely *absolutely* clear, one can accurately surmise from the legislative record and from a simple reading of the text's objectives that Congress intended to make NEPA more than a procedural paper trail. Instead, as its Senate sponsor, US Senator Henry M. Jackson, a Democrat from the State of Washington, wrote in 1971, "Adoption of the Act constituted Congressional recognition of the need for a comprehensive policy and a new organizing concept by which governmental functions can be weighed and evaluated" based on a systematic analysis of environmental impact (Jackson 1971).

As political scientists have concluded, US national public policy is most often developed incrementally (Lindblom 1959; Smith 2018; O'Leary 1993). However, NEPA, according to Senator Jackson, was written in order to "break the shackles of incremental policymaking in the management of the environment" (Jackson 1971). The timing was perfect for NEPA's passage. Congress was ready to respond to a growing public concern for improvements in environmental quality as air pollution filled cities, rivers caught fire due to hazardous flammable materials, and oil spills drenched beaches. This call for environmental action was particularly strong among economic and political elites. The following section examines how and why NEPA's ethical and administrative framework was developed in order to clarify how the Act's authors responded to the public's call to "do something" in the face of environmental calamity in the late 1960s.

Congress started to react to the concern for environmental quality expressed by the public, especially white economic elites, (Taylor 2002) in the late 1950s and early 1960s. In 1959, Montana Democrat US Senator James Murray introduced an early version of NEPA titled the Resources and Conservation Act. Although never signed into law, this bill was an important precursor to NEPA. Senator Murray's bill included two components that eventually ended up in NEPA: a statement of national policy and an office for executive oversight of environmental policy. Wisconsin Senator Gaylord Nelson introduced another version of the Murray bill in 1965 titled the Ecological Research and Survey's Act. Although this bill never received a vote, its ideas for an executive environmental oversight agency are incorporated into NEPA's Title II, specifically the creation of the White House level Council on Environmental Quality.

NEPA's declaration of a national environment policy can be traced back to Dr. Lynton Caldwell's 1963 article titled "Environment: A New Focus for Public Policy?" (Caldwell 1963). Augmenting Caldwell's call for a coherent national environmental policy were US writers such as Stewart Udall, Barry Commoner, Rachel Carson, among others, who argued that the environment must be assessed and regulated in its entirety as an ecosystem rather than the sum total of all its parts.

The centerpiece of this recognition is found in NEPA's Congressional statement of purpose. The framers' intent is profound, clear, and ecologically progressive. NEPA establishes a national environmental policy to promote "productive harmony" that balances social, economic, and environmental goals. This policy is then broken down into six different objectives imposed on federal "plans, functions, programs, and resources" and requires an impact statement to ensure NEPA's goals when there is a "major federal action that significantly affects the environment" (FedCenter 2020).

According to NEPA, federal officials in every agency and action implementing NEPA are responsible to

1. "fulfill the responsibilities of each generation as trustee of the environment for succeeding generations";
2. "assure for all Americans safe, healthful, productive and esthetically and culturally pleasing surroundings";

3. "attain the widest range of beneficial uses of the environment without degradation, risk to health or safety, or other undesirable and unintended consequences";
4. "preserve important historic, cultural, and natural aspects of our national heritage, and maintain, wherever possible, an environment which supports diversity and variety of individual choice";
5. "achieve a balance between population and resource use which will permit high standards of living and a wide sharing of life's amenities";
6. "enhance the quality of renewable resources and approach the maximum attainable recycling of depletable resources".

(FedCenter 2020)

These six objectives are the heart of NEPA's directives to federal agencies. The goals of NEPA are not to create more paperwork or bureaucratic hurdles within the US government. NEPA's goals are specifically to advance and elevate environmental values and sustainability within federal decision-making systems regardless of the agency mission. These objectives are not flowery pointless sentiments created for short-term public relations points; they are ecological goals and values built into US law, binding on all parts of the federal government.

The framers of NEPA intended to substantively redirect the goals and policy decisions generated within federal agencies. The framers' intention, however, was also to challenge the incremental gridlock of environmental policymaking. Prior to NEPA, federal agencies lacked effective tools or mechanisms for incorporating environmental values. In fact, the importance of ecological assets rested in the hands of agency officials and, by default, the mission of the agency (most of which have nothing to do with environmental quality).

By the end of 1967, it seemed certain that some form of national legislation would be enacted to address the growing dissatisfaction with environmental decision-making (Caldwell 1982). Indeed, by the late 1960s, 120 members of Congress had bills referring to 19 separate committees of the House and Senate dealing with environmental issues. Congressional observers felt this constituted a jurisdictional nightmare for the formation of a comprehensive national environmental policy both within Congress (there was initially little coordination among the competing bills) and, within the numerous agencies and departments designated in the bills, should any of them become law. With a profound, albeit ambitious, national environmental policy statement, NEPA mandated the wholesale re-arranging of institutional decision-making within federal agencies – agencies were now held accountable for the environment.

The legislative movement for a broad national governmental policy increased in momentum when the Subcommittee on Science, Research and Development of the House Committee on Science and Astronautics published a report in 1968 titled "Managing the Environment" (Staff 1968). This report was instrumental in getting members of Congress to realize that something had to be done to respond to the growing concerns over environmental quality and management.

The most significant springboard for coordinating some type of national environmental policy occurred in a "Joint House-Senate Colloquium to Discuss the National Policy for the Environment" on July 17, 1968. This seminal meeting between the Senate's Committee on Interior and Insular Affairs and the House of Representative's Committee on Science and Astronautics was convened at the Capitol Building. Co-chaired by Jackson, chair of Senate Interior and Insular Affairs Committee, and Congressman George Miller (D-CA), chair of the House Committee on Science and Astronautics, this meeting was an "informal study session" called to address environmental management, committee jurisdictions, the need for sounder environmental information. Representatives from the 90th Congress, academia, private, and public groups were invited to participate in this event. Of central concern for the conference participants was the special report issued by the Senate's Committee on Interior and Insular Affairs, titled "A National Policy for the Environment", written by Caldwell with assistance from William Van Ness and Senator Jackson (Senate 1968). This was a comprehensive essay stating the goals, visions, and agendas for a relatively unified federal environmental policy. In a condensed but substantial form, major concepts of this influential essay were later written into the first section of NEPA.

It was clearly the intention of NEPA's authors that a coordinated and comprehensive analysis of ecological and social impacts of federal decision-making replace the ad-hoc, fragmented policymaking status quo. In his influential essay to the Special Report to the Senate Committee on Interior and Insular Affairs, titled "A National Policy for the Environment", Caldwell explains his thinking behind NEPA:

> Our present governmental organization has not been designed to deal with environmental policy in any basic or coherent manner. The extent to which governmental reorganization may be necessary cannot be determined absolutely in advance of experience. But it does seem probable that some new facility will be needed to provide a point at which environmental policy issues cutting across jurisdictional lines of existing agencies can be identified and analyzed, and at which the complex problems involved in man's relationships with his environment can be reduced to questions and issues capable of being studied, debated, and acted upon by the President, the Congress, and the American people.
>
> *(Caldwell, A National Policy for the Environment*
> *1968)*

However, Caldwell notes, the EIS "may facilitate, but would not lessen, the political task of reconciling a great diversity of interests and values" (Caldwell, A National Policy for the Environment 1968).

According to NEPA's founders, to effectively reconcile a wide range of political and economic interests, a policy model for the environment

> must be compatible and consistent with many other needs to which the nation must respond. But it must also define the intent of the American people toward the management of their environment in terms that the Congress, the President, the administrative agencies and the electorate can consider and act upon. A national policy for the environment must be a principle that can be applied in action. The goals of effective environmental policy cannot be counsels of perfection; what the nation requires are guidelines to assist the government, private enterprise and the individual citizen to plan together and to work together toward meeting the challenge of a better environment.
>
> *(Caldwell, A National Policy for the Environment*
> *1968)*

In 1969, Senator Jackson reintroduced the Senate version of NEPA. On April 16, 1969, Jackson held a hearing on his bill S. 1075. It was here that the EIS was integrated into the bill after the strong recommendation and work of Dr. Caldwell. Caldwell and other Jackson staff like William Van Ness and Daniel A. Dreyfus argued that a declaration of policy needed some enforcement mechanism to ensure that NEPA's principles could not be ignored by federal agencies.

In his written report to the Senate Interior and Insular Affairs Committee, Caldwell explained his support for a national policy but insisted that it be enforced:

> We ought to think of a statement which is so written that it is capable of implementation: that it is not merely a statement of things hoped for; not merely a statement of desirable goals or objectives; but that it is a statement which will compel or reinforce or assist all of these things, the executive agencies in particular, but going beyond this, the Nation as a whole, to take the kind of action which will protect and reinforce what I have called the life support system of this country.
>
> *(Senate Staff 1969)*

Caldwell argued that NEPA's general goals would have no impact unless the bill had some sort of "action-forcing mechanism". Caldwell recommended to the committee "an action-forcing, operational aspect" to ensure agency compliance of NEPA goals. This specifically became the EIA process required by NEPA when there is a major federal action significantly impacting the environment. This EIA process is the method Congress requires to achieve the desired end of informed federal government decision-making. Doing the EIA is not an end in itself; it is the process for gathering information to make the best decision.

Besides forcing action or decisions that incorporate environmental factors, Section 103 of Senate Bill 1075 requires all federal agencies to examine their regulations, statutes, and policies to ensure "conformity with the intent, purposes, and procedures set forth in this Act". This measure is intended to circumvent agency and administrative indifference to environmental effects by installing institutional requirements for including environmental effects, values, and alternatives in federal decision-making. The "action-forcing" environmental assessment requirements and procedures in the second part of NEPA are specifically intended to be interpreted and administered considering the core ecological and common-good goals of NEPA's mission. They are a means to an end, not the end.

As Richard N.L. Andrews argues, Caldwell's recommendation was a "radical and unprecedented innovation" (Andrews 1976). The link between procedure and substance is of utmost importance. Caldwell called the connection between mandatory procedures and substantive criteria, "the genius of NEPA" (Caldwell, Science and the National Environmental Policy Act: Redirecting Policy Through Procedural Reform 1982). Senator Jackson's staff members, including William Van Ness and Daniel Dreyfus, discussed a need for such a mechanism but, as Liroff notes, "the Caldwell testimony lent new impetus to their considerations" (Liroff 1976). Caldwell envisioned the action-forcing requirement to be a mechanism that would force federal agencies to comply with the bill's goals – especially the consideration and evaluation of proposed projects consequential environmental ramifications.

On June 18, 1969, the Senate Interior and Insular Affairs Committee unanimously approved the amended version of S.1075 and issued their report to the full Senate. On July 10, 1969, S.1075 passed the Senate by a voice vote. With the tactful ability of Senator Jackson and William Van Ness, S.1075 was voted on during the "morning hour" – a period reserved for routine matters. Jackson's bill was passed without debate with no amendments. Meanwhile, the House was still debating similar legislation offered by Rep. John Dingell.

Action in the US House of Representatives on a NEPA bill originated in the 91st Congress with Representative John Dingell's (D-MI) HR 6750. Dingell's bill was written as an amendment to the Fish and Wildlife Coordination Act, a law under the jurisdiction of Dingell's Subcommittee on Fisheries and Wildlife Conservation of the Committee on Merchant Marine and Fisheries. Dingell's HR 6750, a less comprehensive version of NEPA, was co-sponsored by all but one member of the Fisheries and Wildlife Conservation Subcommittee of the Merchant Marine and Fisheries Committee. During this time, environmental policy was supported by conservation groups focused on fishing and hunting as well as hiking and preservation groups such as the Sierra Club. Congressman Dingell's bill authorized the establishment of an executive council for the environment and contained a statement detailing a national environmental policy, although it was much shorter and weaker than what was finally adopted. Dingell's HR 6750 contained no "action-forcing" mechanism requiring the preparation of environmental impact statements. However, between the third and fourth sessions of Congressional hearings on HR 6750, Congressman Lucien Nedzi introduced HR 12143, which contained investigative environmental findings requirements and was also referred to Dingell's subcommittee. Congressman Nedzi's HR 12143 was important because Dingell's bill had no "findings" requirement. In an executive session, the Subcommittee on Fisheries and Wildlife Conservation created the compromise bill, HR 12549. This was reported by the entire Merchant Marine and Fisheries Committee on July 11, 1969, and reached the floor on September 23, 1969. On October 8, 1969, the House of Representatives passed Dingell's HR 12549 by a vote of 372-15 with 43 abstentions. After signing NEPA into law on January 1, 1970, the focus shifted to implementing and enforcing this new law.

Because of the comprehensive nature of NEPA's environmental policy and its multiple mandates for executive administration, most agencies viewed NEPA with caution at best and with contempt at worst. Six years after NEPA was enacted, Richard A. Liroff concluded that:

> Several general patterns of agency response to NEPA are observable. First, there were those agencies like the AEC [Atomic Energy Commission] prior to *Calvert Cliffs* and the FPC [Federal Power Commission] who felt that compliance might interfere with the achievement of their traditional missions. Second, there was a lack of procedural response on the part of environmental agencies like the EPA [Environmental Protection Agency] that regarded NEPA as superfluous because their decisions were already infused with environmental considerations.
>
> Third, there were a few agencies, like the AEC after *Calvert Cliffs* and the [Army] Corps [of Engineers], in which some concerted efforts to implement NEPA was [sic] made …
>
> Fourth some agencies showed a lack of interest in NEPA because ecological considerations did not seem germane to their principle missions, and there was little reward to be gained by allocating scarce agency resources to environmental concerns.
>
> *(Liroff 1976)*

Although NEPA makes no direct provision for judicial review and its founders did not imagine the onslaught of NEPA litigation, the courts would forever play a role in shaping NEPA's meanings and mandate. Before this chapter covers NEPA's judicial record, let us review some of the basic NEPA requirements as set forth by the statute and the Council on Environmental Quality.

The NEPA process today

Despite the acronym and common public misunderstanding, NEPA does not stand for the National Environmental Policy Agency. Like the EPA, NEPA was officially born in 1970 and initially managed by President Nixon. Similar to other agencies and laws, NEPA's 50-year history has been significantly shaped by the federal courts. The courts create legal meanings and interpretations, and thus guide the implementation of NEPA. The US Supreme Court determined NEPA is predominantly a procedural statute, thus compelling the federal government to follow NEPA's informational gathering and public input processes. The most common legal challenge concerns the procedures and the allegation that agencies did not properly follow the procedures. Federal courts are reticent to challenge the technical interpretations and decision-making conclusions of federal agencies, but they have no problem requiring agencies to use due diligence and show that they consider the environmental (and some social and cultural) impacts. This is of course a nebulous area and subject to unique interpretations depending on the judge, lawyer, capitalist and/or activist. Almost every key word and requirement of NEPA has been subject to judicial review. The Council on Environmental Quality has compiled a thorough list of these cases on its website (Quality, Major Cases Interpreting the National Environmental Policy Act n.d.).

NEPA's implementation requirements are primarily determined through the guidance of the Council on Environmental Quality. In 2016, President Barack Obama's CEQ issued guidance to federal agencies requiring them to include cumulative, long-term climate change in their NEPA processes. On August 15, 2017, President Trump signed Executive Order 13807, directing the CEQ to reform and revise the NEPA implementation regulations to speed them up and reduce cost. A year later, when the Republicans controlled the majority in the House of Representatives (prior to November 2018 elections), there was a hearing on reforming NEPA procedures.

Over the 50 years since President Nixon signed the law, NEPA remains one of the most influential environmental laws in the United States. Despite the initial overwhelming support NEPA received from Democrats and Republicans in Congress, as well as Republican President Richard Nixon, over the last several decades NEPA and the environmental review process it created have been subject to criticism by the Congress. The initial bi-partisan Congressional support was in response to the growing public concern about air and water pollution in the United States. As NEPA was incorporated throughout the agencies and litigated in many judicial arenas, a common criticism of NEPA and the environmental impact assessment related to the often lengthy and expensive EIA process.

In a collaboration between the US Environmental Protection Agency and the US Army Corps of Engineers, the FedCenter.Gov website is a thorough resource for all things NEPA-related (FedCenter 2020). If one reviews Congressional bills over the last decade, even just the last year, it is evident that Congress continues, especially Republicans, to seek changes in the way NEPA is implemented and understood. While there are numerous and constant attempts to amend NEPA and exempt projects from NEPA requirements, there is not consistent bi-partisan opposition, and thus these legislative efforts to change NEPA are usually not successful. Regardless of who is president or which party controls Congress, the general importance of environmental assessment is baked into the essence of how the US federal government does business.

In 2018, a Republican-led Committee on Natural Resources in the US House of Representatives held a hearing with the title "Weaponization of the National Environmental Policy Act and the Implications of Environmental Lawfare" (Resources 2018). The memo announcing the hearing summarizes many of the anti-NEPA sentiments. Several prominent memo components include:

- Although originally intended to increase awareness regarding the effects of federal actions on the environment, NEPA's vague and ambiguous language has exposed the federal government to excessive litigation and resulted in perverse outcomes for agencies, the environment and taxpayers.

- Increasingly, NEPA has become a weapon of choice by litigation activists to stop, delay, restrict, or impose additional costs on all types of federal actions. This has resulted in the expansion of prolonged environmental reviews, mounting paperwork, detrimental project delays and a range of adverse fiscal and economic impacts.
- The associative costs of NEPA litigation prevent the federal government from undertaking actions or approving major projects in a timely fashion. For example, excessive litigation can negatively affect critical activities relating to national security, energy development, and infrastructure construction. (Resources 2018)

The US House Committee on Natural Resources Chair Rob Bishop (R-UT) said NEPA was a "weapon" used by environmental activists to stop good projects. He said the EIA process requires too much paperwork, takes too long and "NEPA was never intended to be a weapon for litigants to force delays and denials on all sorts of activities within the federal nexus, but NEPA as currently implemented provides just that" (U.S. House of Representatives, 2018).

A former associate director of NEPA oversight with the Council on Environmental Quality, Horst Greczmiel, testified at the House hearing that full NEPA reviews resulting in a comprehensive EIS only occur about 1% of the time. He told committee members that categorical exclusions cover 95% of environmental reviews (thus exempting projects from NEPA review), and preliminary environmental assessments occur only 4% of the time (U.S. House of Representatives, 2018 .

Mr. Greczmiel said only around 1% of NEPA cases are litigated each year, and the federal agencies usually always win in court. Others at the hearing noted that just the threat of a court battle and a long EIA process can discourage developers and other economic investments. With some EIA processes taking over four to five years, the US EIA process is often criticized as among the slowest in the world (U.S. House of Representatives, 2018.

Related to this Congressional hearing is law professor Richard Epstein's critique of NEPA in 2018 titled "The Many Sins of NEPA" (Epstein 2018). In this essay, Epstein states:

> NEPA was passed with the intention of avoiding the catastrophic consequences that could come from either Rachel Carson's Silent Spring [agricultural chemicals poisoning wildlife] or Santa Barbara's botched drilling operations [massive oil spill]. In both cases, the risks [industrial chemicals and oil spills] are real, but so are the perils of selecting the wrong institutional arrangement to deal with them.
>
> *(Epstein 2018)*

He goes on to say:

> early legal barricades put into place as a part of NEPA caused innovation to move too slowly […] NEPA's fetish for complete information often leads to searching every nook and cranny for potential risks and perils while ignoring risks from the status quo that were far greater.
>
> *(Epstein 2018)*

The criticism from Trump administration officials, industry, libertarians, and a variety of NEPA practitioners led to pressure to change the way NEPA is implemented. This culminated in significant proposed revisions to NEPA's guidelines by President Donald Trump's Council on Environmental Quality, with the final rule published in the Federal Registrar on July 16, 2020 (Quality 2020). This next section reviews these proposed changes.

On July 15, 2020, the Trump administration announced its final rule to "streamline the development of infrastructure projects and promote better decision making by the Federal government" (Council on Environmental Quality NEPA Modernization 2020). In his announcement about these changes Trump characteristically referred to himself as a victim by saying his building projects were delayed "because of mountains and mountains of bureaucratic red tape in Washington, DC", saying later "we are reclaiming America's proud heritage as a nation of builders and a nation that can get things done" (Brady 2020).

According to a law firm representing the interests of developers:

> The most significant aspects of the CEQ's final rule for project developers are that the CEQ: (1) clarified which undertakings should and should not be subject to NEPA environmental analysis; (2) created new time limits for environmental assessments (EAs) and environmental impact statements (EISs); (3) eliminated the requirement to consider whether a project is "highly controversial"; (4) revamped and streamlined the environmental "effects" analysis; and (5) revised the definition of a "reasonable alternative" to limit alternatives to those that are technically and economically feasible and consistent with the goals of the applicant.
>
> *(Firm 2020)*

These proposed changes will be sorted out in the federal courts over the next few years. For many, these changes are a big step in the right direction. In the White House press release, several supporter quotes were included, including a response from Mike Dunleavy, Republican Governor of the State of Alaska:

> I thank the Trump Administration for working to modernize and clarify the 40-year-old NEPA regulations. All Alaskans will benefit from an update to NEPA as it impacts many facets of our state, from the construction of roads and highways to energy projects, to land and forest management. We look forward to seeing this process unfold and the impact it will have on furthering Alaska's opportunity for business and resource development projects within our state.
>
> *(Department of Interior 2020)*

President Trump rescinded Obama's CEQ guidance, but the federal courts have rendered mixed results for President Trump. In a review of 12 climate-related NEPA cases issued prior to Trump's final CEQ guidance, Goldfuss, Hardin, and Rehmann determined "the courts have made it eminently clear that the Trump administration must consider greenhouse gas emissions when conducting the environmental review. Whether by intention or not, NEPA has become the strongest climate policy in the Trump era" (Christy Goldfuss 2019).

Challengers to NEPA's implementation in the agencies is most successful when it challenges an agency's decision that an EA or EIS was not required. Once the EIS is completed, the federal government almost always defeats any challenges to the EIS document and decision, if the environmental analysis is reasonably complete. An analysis of all United States Forest Service NEPA documents confirmed that most litigation is based on EISs and that the agency was more likely to lose a CE (threshold) case than any other type of case (Keele 2018).

Legal scholar Denise Keele studied 99 judicial cases between 1989 and February 2017, where NEPA was the primary centerpiece of the case. In this analysis published in the *Journal of Environmental Law*, Keele found:

Overall, the government defendants won the majority of the cases (54%). Plaintiffs won only 14% of their challenges; however, they settled with the government in another 8% of cases. Excluding the open cases from calculations, the government won almost two-thirds (68%) of all claims, and in the 67 cases decided by a judge, the government won the vast majority of the time (80%).

(Keele 2018)

The next chapter for NEPA remains to be determined. For the last 50 years, NEPA has been the centerpiece of US environmental policy despite the courts pulling out the sharpest green teeth of NEPA's statutory language. The 2020 presidential election between Democrat Joe Biden and President Trump drew sharp contrasts between a Trump administration that fast-tracked oil, gas, and mining proposals, and backed out of the United Nation's Paris climate agreement. Because Joe Biden beat Donald Trump, he will rejoin the Paris climate agreement and seek to change Trump's weakening of NEPA processes. Because the Democrats did not regain control of both the Senate and the House, there will be little chance Congress will rescind Trump's NEPA rules through the Congressional Review Act, which allows Congress to quickly review federal regulations during a new administration. Prior to the national elections on November 3, 2020, the US Senate Minority Leader from New York, Democrat Chuck Schumer, stated:

Donald Trump, his administration and congressional Republicans have done an unconscionable amount of damage to our country, especially when it comes to addressing the climate crisis, health care, voting rights, income inequality, immigration and other areas. Senate Democrats are committed, as we have been, to looking at every tool in our toolbox, which includes using the Congressional Review Act, to find ways to prevent the president's most egregious policies from becoming a reality.

(Davenport 2020)

The administration led by Joe Biden and Kamala Harris is committed to reinvigorating the Council on Environmental Quality, especially amplifying the importance of environmental justice and the voices of the most marginalized and vulnerable populations and communities. The Biden/Harris website describes their goals as follows:

Currently, the federal government has two key environmental justice groups. Biden will elevate and reestablish the groups as the White House Environmental Justice Advisory Council and the White House Environmental Justice Interagency Council, both reporting to the Chair of the White House Council on Environmental Quality, who reports directly to the President". Biden promised to add senior staff to the CEQ to focus on environmental justice.

(JoeBiden.Com 2020)

The same day that President Trump announced his NEPA rule changes in the summer of 2020, Joe Biden announced a series of environmental plans in direct contrast to Trump. In addition to discussing his plans for renewable energy investments and more aggressive climate goals, Biden also indicated he wanted to end Trump's NEPA changes. However, since Trump's NEPA CEQ regulatory changes went through the multi-year rule-making process, President Biden will need to prioritize which environmental regulations he can easily change with an executive order and which changes require the more formal and embroiled rule-making process governed by the Administrative Procedures Act.

NEPA will remain a bedrock of environmental law, and the EIA process will continue to be a critical planning and policy tool that ensures federal agencies gather and consider the information about short- and long-term impacts of their decisions. Individual agency and NEPA professionals will interpret and execute NEPA in nuanced ways, but the fundamental throughline continues to exist: look before you leap.

References

Andrews, Richard N.L. 1976. *Environmental Policy and Administrative Change: Implementation of the National Environmental Policy Act*. Lexington, MA: DC Health and Company.

Andrews, Richard N.L. 1999. *Managing the Environment, Managing Ourselves*. New Haven, CT: Yale University Press.

Borrelli, Peter. 1989. "Environmental Ethics – The Oxymoron of Our Time". *Amicus Journal* 39: 41.

Brady, Jeff. 2020. "National Public Radio". *National Public Radio*. July 2020. Accessed September 4, 2020. https://www.whitehouse.gov/ceq/nepa-modernization/.

Caldwell, Lynton K. 1968. *A National Policy for the Environment*. US Congress, Senate, Committee on Interior and Insular Affairs, 90th Cong., 2d session, Washington, DC: S Government Printing Office, 19.

—. 1982. *Science and the National Environmental Policy Act: Redirecting Policy Through Procedural Reform*. Alabama: University of Alabama Press.

Caldwell, Lynton K. 1982. The National Environmental Policy Act: Redirecting Policy Through Procedural Reform. Tuscaloosa: University of Alabama Press.

Caldwell, Lynton K. 1963. "Environment: A New Focus for Public Policy". *Public Administration Review* 23.

CEQ. n.d. *Council of Environmental Quality-National Environmental Policy Act*. Accessed August 15, 2020. https://ceq.doe.gov/laws-regulations/laws.html.

Goldfuss, Christy, Hardin, Sally and Rehmann, Marc. 2019. "12 Climate Wins from the National Environmental Policy Act". *Center for American Progress*. May 29. Accessed September 4, 2020. https://www.nrdc.org/resources/never-eliminate-public-advice-nepa-success-stories.

2020. *Council on Environmental Quality NEPA Modernization*. Accessed September 3, 2020. https://trump-whitehouse.archives.gov/ceq/nepa-modernization/

Davenport, Coral. 2020. "New York Times". *New York Times*. July 30. Accessed September 10, 2020. https://www.nytimes.com/2020/07/17/climate/trump-regulations-election.html.

2020. "Department of Interior". *Department of Interior "What They are Saying: CEQ Issues Proposed Rule to Modernize its NEPA Regulations"*. January 14. Accessed September 1, 2020. https://www.doi.gov/press-releases/what-they-are-saying-ceq-issues-proposed-rule-modernize-its-nepa-regulations.

Bear, Dinah. 1993. "NEPA: Substance or Merely Process". *Forum for Applied Research and Public Policy* 85.

EPA. n.d. *National Environmental Review Process*. Accessed August 28, 2020. https://www.epa.gov/nepa/national-environmental-policy-act-review-process.

Epstein, Richard. 2018. "The Many Sins of NEPA". *Texas A and M Law Review* 6(1): 1–27.

2020. "FedCenter". *FedCenter*. Accessed Sept 7, 2020. https://www.fedcenter.gov/programs/nepa/.

Firm, Gibson Dunn Law. 2020. "Gibson Dunn". August 3. Accessed September 10, 2020. https://www.gibsondunn.com/nepa-review-revamp-what-developers-should-expect-from-the-ceq-new-rule-and-incoming-litigation-storm-front/.

Jackson, Henry M. 1971. "Environmental Policy and the Congress". *Natural Resources Journal* 11: 407.

Keele, Denise. 2018. "Climate Change Litigation and NEPA". *Journal of Environmental Law* 30: 285–309.

Lindblom, Charles. 1959. "The Science of Mudding Through". *Public Administration Review* 19: 79–88.

Liroff, Richard A. 1976. *A National Policy for the Environment: NEPA and Its Aftermath*. Bloomington: Indiana University Press.

Mandelker, Daniel R. 2002. *NEPA Law and Litigation: The National Environmental Policy Act*. St. Paul: West Group.

Nixon, Richard 1970. "The American Presidency Project". *The American Presidency Project*. Accessed September 1, 2020. https://www.presidency.ucsb.edu/documents/statement-about-the-national-environmental-policy-act-1969.

O'Leary, Rosemary. 1993. "The Progressive Ratcheting of Environmental Laws: Impact on Public Management". *Review of Policy Research,* 12 (3/4): 118–136 (Autumn/Winter, 1993).

Pepper, Elly. 2015. *Never Eliminate Public Advice: NEPA Success Stories.* February 1. Accessed August 15, 2020. https://www.nrdc.org/resources/never-eliminate-public-advice-nepa-success-stories.

Quality, Council on Environmental. 2020. "Federal Register". *Federal Register.* July 16. Accessed September 3, 2020. https://www.federalregister.gov/documents/2020/07/16/2020-15179/update-to-the-regulations-implementing-the-procedural-provisions-of-the-national-environmental.

—. n.d. *Major Cases Interpreting the National Environmental Policy Act.* Accessed September 10, 2020. https://ceq.doe.gov/docs/laws-regulations/Major_NEPA_Cases.pdf.

US House of Representatives Committee on Natural Resources. 2018. "S House of Representatives Committee Repository". April. Accessed September 3, 2020. https://docs.house.gov/Committee/Calendar/ByEvent.aspx?EventID=108215.

Senate Staff, 91st Congress. 1969. *Hearings on S.1075, S.237 and S.1752, Senate Committee on Interior and Insular Affairs.* Washington, DC: S Government Printing.

Senate, US 1968. *A National Policy for the Environment, A Special Report to the Committee on Interior and Insular Affairs.* US Senate Committee, Washington, DC: S Government Printing Office.

Smith, Zachary. 2018. *The Environmental Policy Paradox.* London: Routledge.

Staff. 1968. *Managing the Environment.* Washington, DC: Subcommittee on Science, Research and Development, 90th Congress Committee Print.

Taylor, Dorceta E. 2002. "Race, Class, Gender and American Environmentalism". *United States Department of Agriculture U.S. Forest Service.* April. Accessed September 3, 2020. https://www.fs.fed.us/pnw/pubs/gtr534.pdf.

18

EIA IN ENGLAND

Josh Fothergill and Thomas B. Fischer

Introduction

Environmental impact assessment (EIA) in England can be seen to draw its origins from the development of the formal land-use ('town and country') planning system applied by local councils here since 1906. In terms of being present in a form we may recognize more as the tool we currently use, the origins stretch back to the 1970s where voluntary environmental assessments were conducted by developers in the exploration and development of offshore oil and gas reserves (Clark et al. 1976; Glasson 1999). It is not, however, until the mid-1980s and the adoption of the European Community's (today European Union) original EIA Directive (85/337/EEC) that England had the push needed to formalize EIA within legislation by 1988 (Jha-Thakur and Fischer 2016).

The original 1988 EIA legislation took the form of a number of sets of regulations requiring EIA for certain developments across a range of existing consenting regimes. Since then, the legislative basis of EIA in the United Kingdom (UK) has gone through several amendments, renewals, and iterations. Up until 2021, such changes have been driven both by amendments to the overarching European Union (EU) EIA Directive, or because of the outcomes of significant legal rulings that meant the way the English EIA legislation functioned had to be amended. As a result of the UK's decision to leave the EU, affected in late January 2020 and completed in 2021 (see Fischer 2016), the future direction of English EIA legislation lies, mainly, within the hands of the UK government. It is the authors understanding that – alongside broadscale changes proposed to the functioning of the English spatial planning system – the EIA (and strategic environmental assessment) system in England is 'under review', which could lead to significant changes as to its basis, form, and function within the next few years (Fischer et al. 2018).

Despite this context of 'being on the cusp of potential change', any such change would need to ensure that the English EIA system remained in line with the requirements of the Espoo Convention – in the context of EIAs with transboundary effects (ratified by the UK in 1997[1]). It should also be noted that while to the outside world the UK is a single nation-state, the country operates within the process of devolution with powers to make environmental law and the majority of development consent devolved to the administrations in each of the other three devolved nations (Scotland, Wales, and Northern Ireland). As such, while EIA legislation in England may be under scrutiny and on the cusp of reforms, this is not the case in the other

DOI: 10.4324/9780429282492-20

nations, and any changes to the assessment in these parts of the UK would need to be developed and passed by their own administrations. As such, any future changes to English EIA legislation would also need to consider how any new approach to the assessment's legislative requirements would interact with the existing (EU EIA Directive) oriented regimes that are likely to persist for some time across England's land borders with Wales and Scotland (see, e.g., Cowell, Fischer, and Jackson 2021).

A brief history of the EIA Directive – the current basis of English EIA legislation

The original EIA Directive was released in 1985. This was subsequently amended three times (in 1997, 2003, and 2009). A consolidated Directive was published by the European Commission in 2011 (2011/92/EU) to form a single text upon which further substantive amendments were made in 2014 2014 (2014/52/EU).

The original 1985 Directive defined an EIA process that included screening, an evaluation of the likely significant environmental effects, the preparation of an EIA report, consultation of statutory and other bodies, public participation, and the consideration of the EIA report in decision-making. Changes brought about by the subsequent amendments include:

- The inclusion of the requirements of the UNECE Espoo Convention on EIA in a Transboundary Context by Directive 97/11/EC. Furthermore, this 1997 revision increased the types of projects covered, and the number of projects requiring mandatory EIA (through Annex I; see above). It also introduced new screening arrangements in Annex III. Finally, minimum information requirements were established.
- An alignment with the provisions on public participation with the Aarhus Convention (on public participation in decision-making and access to justice in environmental matters) was sought through Directive 2003/35/EC.
- In 2009, Annexes I and II were amended through Directive 2009/31/EC. In this context, projects related to transport, as well as capture and storage of carbon dioxide, were added.
- Finally, 2014 made the most substantive amendments since 1997, with a formal definition of EIA introduced, new topics added, expectations on the competence of consultants and competent authorities and changes to the details of Annexes III and IV, plus the inclusion of a new Annex added (II-A) – setting out the minimum content in the information a developer must provide at screening, to help determine whether EIA is required.

A further discussion on how several of the 2014 EIA Directive amendments are playing out in English EIA practice is set out at the end of the next section of this chapter.

English EIA arising from the European Union EIA Directive: existing requirements and current and anticipated changes

The EIA Directive applies to certain public and private projects. These are defined in Annexes I and II of the Directive in terms of (Anx I) mandatory EIA and (Anx II) projects that require screening for arriving at a decision on whether or not EIA is required. Mandatory EIA applies to projects that are considered to potentially have significant effects on the environment. These include, for example,

> long-distance railway lines, motorways and express roads, airports with a basic runway length at or beyond 2100 m, installations for the disposal of hazardous waste, installa-

tions for the disposal of non-hazardous waste over 100 tons per day, as well as waste-water treatment plants of over 150.000 p.e.

(Annex I to the EIA Directive)

In England's EIA regimes, the content of Annex I is transcribed as Schedule 1 of the implementing regulations linked to the over-arching legislative act for the relevant consenting process, with the assessment most applied to determining whether planning permission is awarded under the Town and Country Planning Act 1991 (as amended).

Application of EIA to development consents in England (screening)

When a screening of projects is required, a screening procedure needs to be applied to ensure that the competent authority has made its own judgment on whether the development seeking consent is or is not likely to have significant environmental effects on the environment. This screening procedure is applied to a broad range of development sectors, with more specific development types listed under each, as presented in Annex II of the EIA Directive and transcribed as Schedule 2 within the English regulations. If a proposed development falls within the categories listed in Schedule 2 – or within a 'wide scope and broad purpose' interpretation of them, as defined by UK and EU case law – then it may be a Schedule 2 Development and needs to be screened. In England, as allowed by the flexibility of the EIA Directive in this area, the trigger factor of this need for screening is determined based on a series of thresholds or other criteria.

The *other criteria* within the English EIA regulations apply the screening process to its broadest reach, as they define a series of sensitive areas where all development that falls under a Schedule 2 category will need to be screened by the competent authority. These *sensitive areas* relate to specific locations that have been given an environmental designation and protection arising from either international/EU legislation (i.e., World Heritage Sites, Special Areas of Conservation and Special Protection Areas) or English legislation (i.e., The Broads, National Parks, Areas of Outstanding National Beauty, Sites of Special Scientific Interest, Scheduled Monuments). It should be noted that the list of sensitive areas in English EIA has remained unchanged since the introduction of new regulations in 1999, which transposed significant amendments made to the EU's EIA Directive in 1997. As such, the EIA regulations fail to recognize a range of persistent and growing environmental problems in England that have, sometimes legally, defined geographic designations. The most notable of such absences are Air Quality Management Areas that designate areas where air quality has regularly failed to meet legal standards over multiple years. The UK government has been taken to court and found to be failing in its legal duties to manage and improve air quality in such areas for well over a decade, and yet the EIA regimes have not been adjusted to ensure it is applied to help address the existing significant cumulative effect of adverse air quality on both, human health, and the natural environment.

Outside of sensitive areas, the requirement to screen for EIA is triggered by one or more thresholds related to each of the specific types of development listed under the sector categories in Schedule 2. These thresholds are set out in the second column of Schedule 2 and were originally devised for England's 1999 regulations, with a small number of additions and amendments implemented as the regulations have been updated over the years. The most notable being in relation to housing developments, where the threshold was raised from a proposed development site of 0.5 ha, or greater, to one of 5 ha, or above, in 2014. The government did not provide a scientific evidence base or findings to demonstrate that English housing developments between 0.5 and 4.99 ha had never been found to generate likely significant effects on the environment to justify this order of magnitude change. This was an early pre-Brexit referendum signal that the thresholds used to determine whether the English EIA regime should apply were potentially at risk of arbitrary change, driven by motivation outside of environmental/scientific evidence, and one that highlights a future risk now that Brexit has been completed.

Where development of a type listed is within a sensitive area or above the relevant threshold, it is officially deemed Schedule 2 Development, and the competent authority is prohibited from granting consent until it has been through the screening process. The screening process is effectively a publicly available judgment made by the competent authority as to whether the project is likely to lead to significant environmental effects – notably, these can be either adverse or beneficial effects, although the vast majority of English EIAs is triggered due to adverse effects. To assist the competent authority in making its judgment on the significance, they are required to consider a range of criteria set out in Schedule 3 of the English EIA regulations, where they transpose the content of Annex III of the EU EIA Directive. These 'significance criteria' are set out in three categories related to: details about the development and its different phases (construction, operation, decommissioning), details about the location where the development is proposed identifying sensitive environmental locations, and details about the impacts likely to occur in terms of scale, likelihood, frequency, persistence, and cumulation. The outcome of the screening of Schedule 2 Development is either that there is no likelihood of significant environmental effects occurring and thus EIA is not required, or that likely significant effects are anticipated, and the development is formally reclassified as EIA Development, thus requiring it to comply with the full English EIA regulatory procedure before the competent authority can legally determine its consent application.

Perceived failings in the application of this screening process are the most common area of legal challenge across both, the English EIA regime and within the EU. This is unsurprising, given the broad application of the screening procedure to Schedule 2 Developments, and the fact that evidence from Northern Ireland's regime indicates that less than 10% of projects that undergo screening are found to likely have significant environmental effects and thus require EIA. As such, despite the screening regime having existed for well over 20 years in England, failures to apply it at all, and to apply it correctly are still high across the 300+ competent authorities required to implement it. A series of legal challenges in the UK and EU – some of which making links to the Aarhus Convention[2] – led in the last decade to the need for the publicly available screening decision to include a clear explanation justifying its conclusions as to why EIA was or was not required. In theory, this means that all English screening decisions are now available for the public to review, and where concerned seek to challenge, via online databases maintained by the 300+ councils and other consenting authorities; however, the quality, accessibility and upkeep of such systems varies considerably across the country.

The developer's assessment and production of their Environmental Statement

As indicated in Figure 18.1, once a proposal is screened as requiring an EIA, it passes through a series of stages that are led by the developer, these being: scoping, iterative assessment, and writing of the Environmental Statement (ES). The first and last of these stages have some procedural requirements defined by the 2017 Regulations but leave a considerable range of decisions open to the developer's team of environmental professionals (most commonly a team of hired consultants). This flexibility is limited, however, by good practices established by EIA and pre-application engagement professionals.

The 2017 EIA regulations only define a voluntary scoping process and minimum requirements related to the production of the ES. Developers have the option to request a *scoping opinion* from the competent authority. Where a developer chooses to make such a request, formal

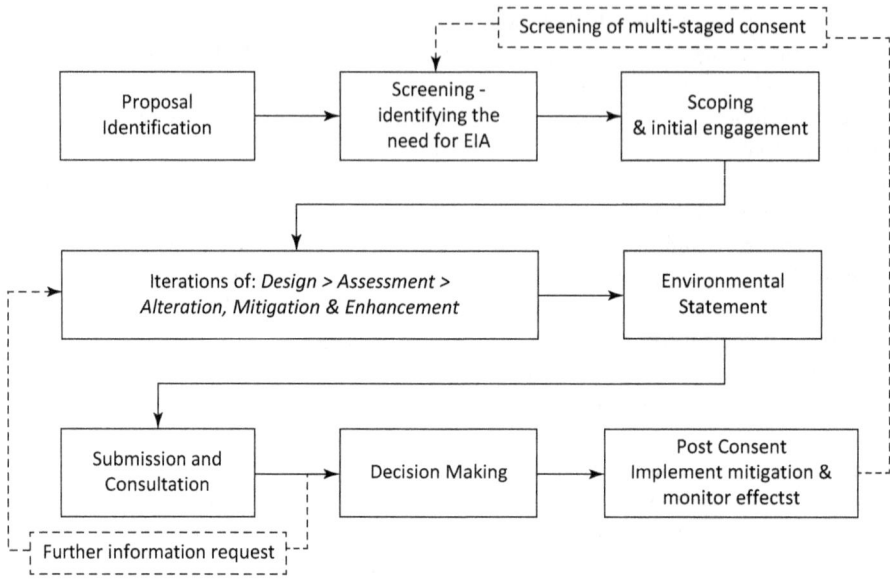

Figure 18.1 Diagram of EIA process applied to development consenting in England. IEMA special report – *The State of EIA Practice in the UK* (Fothergill 2011).

requirements kick-in that the competent authority must follow, including conducting a consulta-tion to gain the views of statutory bodies with environmental responsibilities, such as the English Environment Agency and Natural England. Informal research conducted by one of this chapter's authors during EIA related webinars in 2016/2017 found that approximately 85% of English EIAs do undertake this voluntary step. Interestingly, however, many scoping opinion requests in England come toward the end of the scoping activity undertaken by the developer's EIA team. As such, scoping requests are now regularly accompanied by hundreds of pages of scoping reports and appendices where the developer's ecologists, archaeologists, etc., have already undertaken baseline surveys, informal engagement and defined their view on the proposed assessment con-tent and methods. The formal scoping procedure in the EIA regulations is therefore more com-monly used as a confirmation of the EIA's terms of reference than a truly open scoping exercise.

While it is therefore fair to consider the majority of EIAs in England to be scoped on a 'developer-led' basis, the introduction of the 2017 regulations – implementing the 2014 EU EIA Directive amendments – requires developers to use competent experts to write their ES. As such, the scoping process is most often led by a team of expert environmental professionals, whose work is managed by an experienced EIA coordinator. A challenge arising from this approach is that the expert-led content presented in a formal scoping request is commonly detailed and written by environmental topic experts. The competent authority which is required to provide a formal scoping opinion on this content will generally have less expertise and tend to avoid removing content (scoping down/out) and often add wider topics that are of limited relevance due to a lack of confidence leading to a precautionary approach. This tends toward broadly scoped assessments, which contributes to England's problems with disproportionate EIA (Jha-Thakur and Fischer 2015).

The developer's assessment considers the construction, operational (and where relevant) decommissioning impacts of the development on the topics and issues scoped into the EIA, and seeks to influence the design and apply the mitigation hierarchy to reduce adverse effects. While the English EIA system expects the assessment to consider all likely significant environmental effects – positive as well as negative – the regulations currently provide little expectation, or even encouragement, to improve or enhance predicted positive environmental outcomes. As such, this is a role that the EIA team will try to progress with the developer, but lacks the legislative backing afforded to the expectation the developer will adopt mitigation measures.

The developer's assessment is formally reported in an Environmental Statement, with the English regulations retaining the original EIA Directive's terminology, rather than adopting the new EIA report wording of the 2014 amendments adopted by the EU – and notably applied within the Scottish EIA regulations, since 2017. The developer's ES is produced by their team of competent experts – although not defined in the regulations, it is generally linked to experience and senior membership of an appropriate professional body – i.e., IEMA, CIEEM, CIFA, LI, etc. The detailed potential content of an ES is established by Schedule 4 of the English regulations and covers considerable detail; however, the minimum legislative requirements remain unchanged from the 1999 regulatory updates, and must include:

(a) a description of the proposed development comprising information on the site, design, size, and other relevant features of the development;
(b) a description of the likely significant effects of the proposed development on the environment;
(c) a description of any features of the proposed development, or measures envisaged in order to avoid, prevent or reduce and, if possible, offset likely significant adverse effects on the environment;
(d) a description of the reasonable alternatives studied by the developer, which are relevant to the proposed development and its specific characteristics, and an indication of the main reasons for the option chosen, taking into account the effects of the development on the environment;
(e) a non-technical summary of the information referred to in sub-paragraphs (a) to (d); and
(f) any additional information specified in Schedule 4 relevant to the specific characteristics of the particular development or type of development and to the environmental features likely to be significantly affected.

(Regulation 18(3) of the Town and Country Planning (Environmental Impact Assessment) (England) Regulations 2017[3])

Where a developer requests a scoping opinion, the resultant ES must be based on this opinion, or indicate why it deviates from this, and the ES must also outline the expertise and qualifications of the competent experts who authored it. The developer is required to submit their complete ES alongside their application for development consent, at which point the formal EIA process defined by the regulations reverts back to being led by the competent authority.

The actions a competent authority must take in processing and determining consent for EIA development in England

The 2017 EIA regulations formalized the activities a competent authority had previously been expected to undertake upon receiving and determining an EIA development's application for consent. It did this by setting out the steps to be taken – including consultation, examination, reasoned significance conclusions, and EIA information provision about the decision – within the formal definition of EIA. As such, if the decision-making body fails to meet any of these requirements, the regulatory prohibition to granting development consent for EIA develop-

ments remains in place and – if consented – the development may be vulnerable to legal challenge on EIA grounds.

The first actions of the competent authority are to check that the document the developer
claims to be an Environmental Statement does in fact meet the minimum standards defined
by the regulations (see above) and is aligned to the latest scoping opinion, where one has been
issued. After this has been confirmed, the competent authority must organize a consultation on
the application and its ES, informing the public via public notices, including an advert in a local
paper, and providing copies of the documents to the statutory environmental consultees. Both
of these consultee groups must be given a minimum consultation period of 30 days, in which
time they can provide responses and views to the competent authority.

Once these consultations are complete, the competent authority is required to review the
environmental information – a formal term defined in the regulations covering: the ES, consultation responses and any other/further information. Where the competent authority considers
that it requires more information to come to its own conclusions on significant environmental
effects – often due to missing information or conflicting views between ES and consultees –
they can request the developer provide *further information* to the ES. This is a formal request
– backed by 2017 EIA regulations – that the developer must respond to, or their application
cannot be consented to. Where the developer provides further information, or of their own
volition provides *other* environmental information after consultation on their ES has begun –
then the competent authority is required to undertake a further round of public and statutory
body consultation for a minimum of 30 days. As such, a poorly scoped EIA can lead to a series
of further information requests that can considerably delay the 16-week time period within
which an EIA development application should normally be determined within. While it might
be assumed that such further information requests would be limited due to ES regularly running
to beyond 5,000 pages, the disproportionate nature of the content provided can often mean that
the key evidence needed by the competent authority or sought by stakeholders is difficult to
locate within the morass of data and contextual material.

Once the competent authority is satisfied that it has the information needed to make
its own reasoned conclusions on the application's significant environmental effects and any
anticipated mitigations/monitoring should it be consented to, the process moves on to
decision-making. In relation to EIA developments, decision-making in England is generally
undertaken by one or more publicly elected officials – either a council's planning committee
of elected councilors or the relevant Secretary of State for large infrastructure development.
This decision-making process must be informed by the environmental information and the
competent authorities – reasoned conclusions of the proposal's likely significant effects, proposed mitigation measures and any monitoring expectations related to significant adverse
effects. The English EIA system remains a decision-*support* tool; as such, if the EIA process has
correctly followed the required regulatory procedures, it cannot prevent a development with
significant adverse effects being consented to. It may, however, influence the decision based
on its findings, or identify where the proposal would breach other environmentally related
legislation, some of which have stronger powers to prevent consent from being awarded – for
example, the Habitats Regulations that protect habitats and species of European/international
significance.

Following the determination of an EIA development's consent, the 2017 EIA regulations
introduced minimum requirements for the provision of environment related information alongside the competent authority's decision notice. This now ensures that both the developer and
the public are aware of the decision-maker's reasoning – including environmental matters – for
choosing to consent or not consent to the application, the opportunity to challenge this deci-

sion, and where consented details of environmental mitigation and monitoring that are required to be undertaken.

Recent progress in English EIA based on 2017 regulatory changes

Changes brought to English practice in May 2017 – as a result of revisions to the EIA Directive in 2014 – are now reasonably well embedded in the country's consenting systems. It should, however, be noted that it is still too early to determine whether any major implications will result from case law arising from legal challenges to EIAs delivered since the 2017 changes.

The main changes to pre-2017 practice were identified by Fischer et al. (2016). These include the inclusion of 'land' as a natural resource next to air, water, soil, and biodiversity. Furthermore, construction waste must now be routinely considered. Another amendment was to the description of 'reasonable alternatives' (to include design iteration and technology – alongside location, size, and scale). An implication of this change is that a comprehensive assessment should now be provided of at least the (preferred) development, any major iterations of its design, and the zero alternative (i.e., the future in the absence of the development). Also, up until the 2017 changes, England's EIA system had only been expected to consider impacts on human beings, alongside flora and fauna, water, soil, etc. This has changed and impacts on both population and human health are now expected to be considered in English EIA. The latter has generated enhanced discussion, events, articles, and guidance from professional practice, seeking to encourage better consideration of this broad issue, which both the government and developers had previously sought – under the human being heading – to focus on socio-economic factors, including job creation and schooling provision (see e.g. Fischer et al. 2021). Furthermore, climate change mitigation and adaptation are explicitly mentioned in England's 2017 EIA regulations. Prior to this, climate change (adaptation/resilience) was considered in a limited manner at best across English practice. Climate change adaptation was rarely covered, and the significance of climate change mitigation effects was often only tested against national levels of emissions.

While the amendment requirements were formulated to consider impacts on fauna and flora, as a topic this has been refocused to consider biodiversity, providing the context for a potential wider scope to this area of the EIA process, such as genetic diversity. The 2017 changes also expect English EIA to consider whether the development's predicted significant environmental effects would be different if the proposal suffered a major accident or disaster. While this is unlikely to be a major issue for many developments that require EIA, it is something that had not been previously mentioned within England's EIA legislation.

Strengths and challenges within English EIA – a focus for practitioners

EIA in England has a long history, and practice is characterized by both significant achievements and ongoing challenges (see, e.g., Arts et al. 2012). The English EIA system is among the most extensively reported and commented on in the world. A considerable number of authors of papers on EIA in the English language international professional literature are based in England (Fischer and Onyango 2012; Olagunju and Gunn 2016). Furthermore, in 2012, over a quarter of all master-level degrees revolving around EIA and management in the EU were found to be offered by UK-based institutions, most of which were in England (Fischer and Jha-Thakur 2013). There is therefore good reason to believe that any more dramatic changes to EIA in England may have consequences that go beyond England with possible wider implications for EIA systems elsewhere.

The UK has a strong professional community across EIA, with many practitioners in England volunteering time to share knowledge. Much of this is through the Institute of Environmental Management and Assessment (IEMA), with frequent practice articles on assessment available for over 20 years and a regular webinar series dedicated to EIA running for more than a decade. English practice also provides the majority of ESs that are assessed with the UK's EIA Quality Mark – a voluntary code of practice scheme created by Josh Fothergill and run by IEMA since April 2011. The scheme enables organizations (consultancies and developers) that coordinate the EIA assessment and reporting process to gain independent recognition of their practices through regular review. The scheme has been demonstrated to help improve the quality of practice in English EIA practice (Bond et al. 2017) and won IEMA an International Association for Impact Assessment (IAIA) global award in 2012. The scheme has recently been updated – to celebrate its 10th anniversary on April 19, 2021 – at which time over 60 organizations were registered to it, whose combined outputs represent the majority of ES currently produced in England.

Despite this positive context, however, EIA in England faces a number of challenges, many of which are also common in other national and international settings. These challenges include the generation of unfocused baseline data, disproportional Environmental Statements, and occasional compliance problems leading to legal challenges. According to Arts et al. (2012) and Phylip-Jones and Fischer (2013), these include the (insufficient) consideration of reasonable alternatives, lack of clarity about impact significance, an underdeveloped monitoring regime, insufficient input by the public, and overly lengthy documentation. A particular challenge for EIA was the prolonged period of austerity, which started following the financial crisis of 2008. This has led to ongoing capacity limitations of local authorities to deal with voluminous EIAs and other documentation (e.g., of health impact assessment – HIA). Over time, there may be a threat not just in organizational and legal terms, but also in behavioral, lifestyle, and ethical terms.

In response to issues both driving and arising from disproportionate assessment and reporting that have mounted since the 2008 crisis, a national proportionate EIA strategy was published in 2017 and rolled out over the following 12 months (Fothergill 2017). The aim of the strategy was not just to tackle long ESs, but to provide a focus around a common problem – recognized by all parties – that would enable the coalescing of action across the EIA community to address many of these challenges from multiple angles. The strategy highlights that disproportionate EIA is not considered to be an inherent problem within the EIA tool/process itself, but rather as the 'the cumulative consequence of the way that it is used, and the cultural context embedded within the [...] consenting system'. While the issue can appear to be one of the long and inaccessible Environmental Statements, this is simply the most visible aspect of a consent system – and thus EIA process – that continually seeks to build on experience and thus accumulates a need for more evidence and information over time. The result of which can be seen in evidence from the EIA outcomes in relation to the UK's offshore energy projects, in Figure 18.2.

The UK's proportionate EIA strategy highlights in the first couple of years after EIA regulations were introduced into England – July 1988 – the assessment was often undertaken by a single environmental expert, with a third of Environmental Statements from the first 18 months being less than 20 pages long (Jones et al. 1991). A clear message went out to EIA practitioners – if the tool was to be taken seriously and influence decision-making, it must improve its professionalism. Initially, there was a need to enhance both the breadth of expert inputs and the quantity of the related assessment outputs to drive improvement in the quality of ES and their influence on engagement and decision-making. Unfortunately, the concept that quantity has a strong relationship with quality has become embedded in the EIA psyche. This concept has been challenged over the years both within the EIA profession and across the English planning

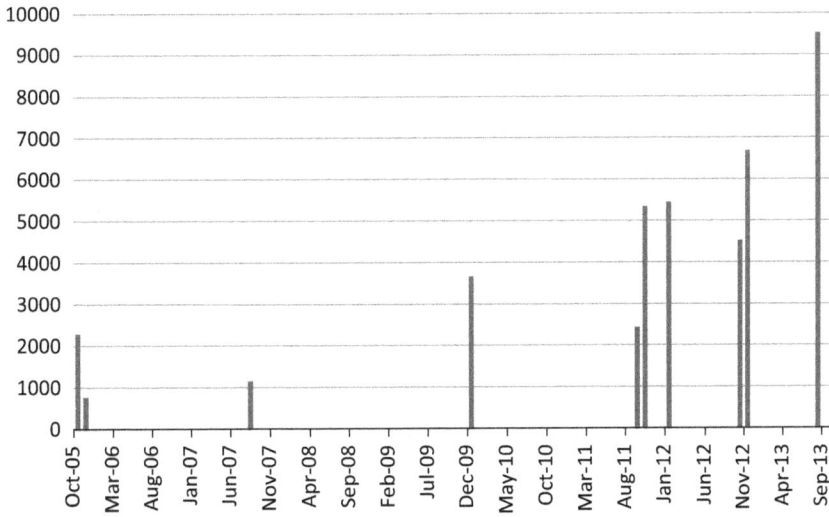

Figure 18.2 Growth of Environmental Statements (total page length) for offshore energy projects in the UK, 2005–2013. Source: Reproduced from Howard (2017).

system (Barker 2009), but such challenges have never effectively arrested the momentum toward longer and broader scoped consent, including the related EIA process.

IEMA sought to identify the drivers of disproportionate EIA and thus develop multiple solutions across the English EIA system to counter these. The 2017 strategy that resulted from this work set out four thematic action areas across which progress would be needed to arrest disproportionate assessment and reporting, thus delivering more proportionate and effective EIA. The themes being:

1. **Enhancing people** – So that those involved in EIA have the skills, knowledge, and confidence to avoid an overly precautionary approach.
2. **Improving scoping** – To generate a more consistently focused approach to this critical activity throughout the EIA process.
3. **Sharing responsibility** – Recognizing that disproportionate EIA is driven by many factors and that enabling proportionate assessment will require collaborative actions that work toward a shared goal.
4. **Embracing innovation and digital** – Modernizing EIA to deliver effective and efficient assessment and reporting that adds value to projects and their interaction with the environment.

Generating momentum behind the strategy's implementation in a coordinated manner has, however, proved more challenging in England in terms of practitioners coordinating with the government and the wider public sector (Fothergill 2018). One response linked to the strategy in Scotland saw the establishment of a free to attend annual national EIA conference as a partnership between the Scottish government, the EIA statutory consultees, and Fothergill Training & Consulting. One of the aims of *Scotland's EIA Conference* being to directly aid the first theme in the strategy and help enable discussion around progress across the wider themes. Unfortunately, despite a greater level of EIA activity in England, no similar event/activities have been able to be established with the government and statutory bodies in England at the time of writing. There have, however, been good examples of progress on individual EIAs of projects and by the activities undertaken across IEMA's impact assessment volunteer network. In terms of the latter, regular webinars and guidance are developed, including a 'Primer on Digital EIA' (IEMA 2020) and the re-establishment in 2018 of the EIA thought-piece series Impact Assessment Outlook, with the first document focused on perspectives around proportionate EIA (Fothergill 2018).

As indicated, multiple English EIAs have embraced one, or more, of the strategy's themes since its publication, with the uptake of digital presentation of the ES, or more commonly, the non-technical summary. While not exclusively limited to larger-scale developments, it is fair to say that more progress (experimentation) has been seen within the EIA of Nationally Significant Infrastructure Projects than of those that go through the normal local authority planning consent process.

A good example of this is the use of a digital EIA approach to the presentation of the ES for the reconfiguration of the A303, a major road currently directly adjacent to Stonehenge (AECOM 2018). The proposal that underwent EIA was to build a new road tunnel to enhance the capacity of the A303, a major route to the southwest of England, while at the same time attempting to reduce the impacts on the world heritage site and its surrounding contextual landscape. The ES was produced as a digital document to maximize the use of visual presentation and the inter-connectivity that can be achieved within a web-based approach.

A more comprehensive example of applying a proportionate approach with direct links to the strategy can be seen in another nationally significant infrastructure project's EIA. Related to a proposal for a major offshore wind energy development approximately 40km off the English coast as part of the UK's third round of offshore wind energy licensing. The developer and their consultants – RoyalHaskoningDHV – sought to embed the strategy's themes into the ethos of the whole approach to the EIA. This was achieved by agreeing to a *Proportionate Approach Document* at the start of the EIA process with the client and key stakeholders and then regularly reviewing progress, including the update of the document at the end of each major milestone in the EIA process – scoping opinion, preliminary environmental information report. An example of the issues considered in this EIA's proportionate approach is presented in Figure 18.3, with clear reference to the four themes of the 2017 strategy.

Enhancing People	**Improving Scoping - Reducing Uncertainty**
• Generate a positive proportionate culture	Scoping must evolve from a stage in the assessment to a core process running throughout an EIA.
• Upskilling the client, the supporting technical team and stakeholders.	
	• Use the Evidence plan process through EIA process
• Stakeholders: Proportionality Roadshow - following scoping response	• Use of Commitments Register and, Impacts and Effects Register
	• Assessments to use an agreed 'simpe' or 'detailed' approach
Sharing Responsibility	**Embedding Innovation and Digital**
Broad engagement across the EIA community to define both the individual and collaborative actions necessary to create a coordinated action plan for proportionate assessment.	Review of established practices and traditions with a focus on identifying areas that could be improved through the adoption of digital platforms and technology.
• Stakeholders: Post-Scoping Proportionality Roadshow	• Electronic submission of documentation to be allowed by PINS
• Assessment Team: Guidance Provided	• Embedded hyperlinks
• Client: Buy-in from senior management	• Digital platform for Section 42 consultation

Figure 18.3 EIA proportionate approach. Source: Information shared with the authors by Howard, personal communication 2019.

While the challenges outlined above indicate that EIA in England is by no means perfect, it is generally recognized to operate at a high level of quality, in line with other national systems that are often considered to represent good global practice (Jha-Thakur and Fischer 2016). EIA is well established and well understood. It leads to improvements – sometimes significant improvements – in projects and plans. It has also been said to lead to an increase in public understanding of and input to projects and plans. Their benefits are generally seen as outweighing their costs (IEMA 2011).

The future of EIA in England

The UK government is expected to use Brexit as an opportunity to change the English EIA system, possibly driven to cut perceived 'green tape' by weakening, for example, screening thresholds and habitat regulations assessment. As such, on a regulatory basis, English EIA could be in for several turbulent years of change – especially as it aligns with consent gained through England's land-use planning system, which is undergoing a major overhaul. Such change, however, is unlikely to change how EIA legislation is designed across the whole of the UK due to devolution placing control of the regulations within the control of the Governments in Northern Ireland, Scotland, and Wales.

Further to the above, on a practice basis, the UK has been playing a leading role internationally with regards to practice and guidance since well before the implementation of the EU Directives. Within the professional practice community, opportunities are already being taken to deliver consolidation of compliance requirements across multiple pieces of environmental legislation. This work includes exploration of more integrated assessment and actions to deliver proportionate and shorter Environmental Statements, utilizing the rapidly evolving digital technology. Furthermore, the Espoo Convention means that English EIA requirements will need to remain largely unchanged where transboundary impacts – with countries other than the UK – need to be included in the scope of the assessment.

There is still a possibility that the current government will aim to weaken environmental policy, as the government has been observed to be 'not a particularly keen advocate of EIA as a non-market based regulatory tool' (Fischer et al. 2016, 106). For EIA, this could mean, for example, raising thresholds for projects requiring assessments further than what has already happened or reducing timescales for preparing them. There is also the possibility that the UK government (but possibly not the other devolved governments) could abolish them.

Comments in this direction have been made by senior government officials on multiple occasions in recent years. For example, the Foreign Secretary, in a speech on February 14, 2018 (The Road to Brexit: A United Kingdom), stated that

> We can simplify planning, and speed up public procurement, and perhaps we would then be faster in building the homes young people need; and we might decide that it was indeed absolutely necessary for every environmental impact assessment to monitor two life cycles of the snail and build special swimming pools for newts – not all of which they use – but it would at least by our decision.

Conclusions

England has had a well-established and respected EIA regulatory regime for over 35 years. While the EIA system continues to have efficiency challenges – notable around proportionate assessment – it is generally effective and generates improvements to both project design and the environment. As part of the UK, English practice has a strong reputation on the international stage and both its academic and professional practitioners produce literature that has a positive influence both at home and around the world.

Despite this longevity and reputation, English EIA is on the cusp of change that has been enabled by the UK's decision to leave the EU, which previously provided it with the core structure of its legislative approach to EIA. While parts of the UK will continue to mirror their EIA regulations on the requirements of the EU's EIA Directive for years to come, due to the devolution of environmental law-making to Northern Ireland, Scotland, and Wales, this appears to be unlikely in England. At the time of writing, significant reforms to England's approach to land-use planning, related development consents, and the approach to both EIA and SEA are all proposed, which could result in the biggest change to England's EIA system since its legal adoption in the late 1980s. Whatever that future holds, it is also clear that England's EIA professional community will continue to work to improve the environmental performance of projects that require assessment and share their knowledge both locally and internationally to further enhance the development of practice.

Notes

1 https://treaties.un.org/Pages/ViewDetails.aspx?src=TREATY&mtdsg_no=XXVII-4&chapter=27 &clang=_en
2 UNECE Convention on Access to Information, Public Participation in Decision-Making, and Access to Justice in Environmental Matters (Aarhus Convention)
3 Available to access here: https://www.legislation.gov.uk/uksi/2017/571/regulation/18/made

References

AECOM 2018. https://eia.aecom-digital.com/A303/intro

Arts, J.; Runhaar, H.; Fischer, T. B.; Jha-Thakur, U.; van Laerhoven, F.; Driessen, P. & Onyango, V. 2012. The effectiveness of EIA as an instrument for Environmental Governance – A Comparison of the Netherlands and the UK, *Journal of Environmental Assessment Policy and Management*, 14(4): 1250025-1-40.

Barker, K. 2006. *Barker Review of Land Use Planning: Final Report – Recommendations*, HM Treasury.

Bond, A.; Fischer, T. B. & Fothergill, J. 2017. Progressing quality in environmental impact assessment: The IEMA Quality Mark as an example of voluntary accreditation, *Environmental Impact Assessment Review*, 63: 160–171.

Clark, B. D.; Chapman, K.; Bisset, R. & Wathern, P. (1976): The Assessment of Major Industrial Applications: A Manual, London (Department of Environment Research Report, 13).

Cowell, R.; Fischer, T. B. & Scott, A. 2021. Embedding the Environment in the Local Development Plan Process – A Think Piece, NatureScot.

Fischer, T. B.; Muthoora, T.; Chang, M. and Sharpe, C. 2021. Health Impact Assessment in Spatial Planning in England –Types of Application and Quality of Documentation, *Environmental Impact Assessment Review*, 90: 106631; https://doi.org/10.1016/j.eiar.2021.106631.

Fischer, T. B. 2016. Lessons for impact assessment from the UK referendum on BREXIT, *Impact Assess Project Appraisal*, 34: 183–185.

Fischer, T. B.; Glasson, J.; Jha-Thakur, U.; Therivel, R.; Howard, R. & Fothergill, J. 2018. Implications of Brexit for environmental assessment in the UK – results from a one-day workshop at the University of Liverpool, *Impact Assessment and Project Appraisal*, 36(4): 371–377.

Fischer, T. B.; Therivel, R.; Bond, A.; Fothergill, J. & Marshall, R. 2016. The revised EIA Directive – possible implications for practice in England, *UVP Report*, 30(2): 106–112.

Fischer, T. B.; Jha-Thakur, U. & Hayes, S. 2015. Environmental impact assessment and strategic environmental assessment research in the UK, *Journal of Environmental Assessment Policy and Management*, 17(1): 1550016.

Fischer, T. B. & Jha-Thakur, U. 2013. Environmental assessment and management related higher education master level degree programmes in the EU – an analysis, *Journal of Environmental Assessment Policy and Management*, 15(4): 1350020.

Fischer, T. B. & Onyango, V. 2012. SEA related research projects and journal articles: an overview of the past 20 years, *Impact Assessment and Project Appraisal*, 30(4): 253–263.

Fothergill, J. 2011. *The State of Environmental Impact Assessment Practice in the UK*, IEMA – Institute for Environmental Management and Assessment

Fothergill, J. 2017. *Delivery Proportionate EIA – A Collaborative Strategy for Enhancing UK EIA practice*, IEMA – Institute for Environmental Management and Assessment

Fothergill, J. 2018. Streamlining EIA A Proportionate National Example The UK's Proportionate EIA Strategy. Presented at EIALAW18 conference in Finland, March 23, 2018. Accessible here: https://d665ab80-22e6-4add-ad92-76b02481d18c.filesusr.com/ugd/4da99e_a3898f8a84534d0baad8e34d90a3689b.pdf

Fothergill, J. 2018. Impact Assessment Outlook Journal Volume 1: December 2018 Perspectives upon Proportionate EIA Thought Pieces from UK practice, IEMA – Institute for Environmental Management and Assessment

Glasson J. 1999. The first 10 years of the UK EIA system: strengths, weaknesses, opportunities and threats. *Planning Practice and Research*, 14: 363–375.

Jha-Thakur, U. & Fischer, T. B. 2016. 25 years of the UK EIA system: Strengths, weaknesses, opportunities and threats, *Environmental Impact Assessment Review*, 61: 19–26.

Howard, R. 2017. Environmental Impact Assessment – Proportionate Assessment. Graphic reproduced from presentation delivered at the launch of IEMA's Proportionate EIA Strategy on July 18, 2017, hosted at the University of Liverpool.

Jones, C., Lee, N. & Wood, C. 1991. UK Environmental Statements 1988–1990: An Analysis, EIA Centre, University of Manchester

Olagunju, A. & Gunn, J. A. E. 2016. Integration of environmental assessment with planning and policy-making on a regional scale. *Literature Review Environ Impact Assess Review*, 61: 68–77.

Phylip-Jones, J. & Fischer, T. B. 2013. EIA for wind farms in the United Kingdom and Germany, *Journal of Environmental Assessment Policy and Management*, 15(2): 1340008.

Revised EIA Directive – Directive 2014/52/EU of the European Parliament and of the Council of April 16, 2014 amending Directive 2011/92/EU on the assessment of the effects of certain public and private projects on the environment. *Official Journal of the European Union* L 124: 1–18.

19

ENVIRONMENTAL ASSESSMENT REFORM IN CANADA

Jeff Nishima-Miller

Introduction

Environmental impact assessment (EIA) in Canada has evolved over 50 years; from a guidelines-based process to one embedded in legislation and regulation. Today, there are 14 EIA processes, one in each of the ten provinces and the three territories, and a federal process. Federal assessment legislation applies across the country to projects and activities that fall under areas of federal responsibility, while the provinces and territories each have their own unique EIA processes. These sometimes intersect with areas of federal jurisdiction, meaning that projects may require both a federal and provincial EIA.

This chapter provides a very brief overview of the Canadian EIA setting, but the focus is on the federal and British Columbia (BC) processes. Within Canada, BC is where most federal assessments occur and the two jurisdictions must cooperate in conducting project reviews, but the main reason for the emphasis in this chapter is that both these jurisdictions have recently created new assessment laws that reflect a demand for changes to the ways projects are evaluated and approved. These updates are significant to assessment practice in general; they are arguably conflict-driven, reflect the importance of Indigenous participation, and provide illustrations of innovation and best practices that can be of interest and relevance to other jurisdictions. Both arguably represent the most advanced approaches to EIA in Canada.

The national setting

Canada's federal government first adopted EIA in the early 1970s with periodic project reviews. The approach had ad hoc qualities applied on a project-by-project basis, but by 1984 the *Environmental Assessment and Review Process Guidelines Order* (EARPGO) was enacted as a cabinet policy. The Guidelines Order was initially treated by federal agencies as a non-binding process that could be used to help decision-making for large-scale natural resource developments, but this discretionary interpretation was problematic. The courts came to view the Guidelines Order as being something that needed to be applied whenever the federal government had a regulatory duty for proposed developments or activities (Noble 2015). This injected uncertainty into the federal assessment setting and made it quite clear that conducting project-based assessments was not optional – it was required. So, the federal government introduced the *Canadian*

DOI: 10.4324/9780429282492-21

Environmental Assessment Act, which was eventually brought into force in 1992. This became the first legislated basis for applying and interpreting assessments for federally regulated projects and activities (Chamberlain and Hail 2016; Hanna 2016).

The provinces began implementing EIA laws and processes in the 1970s and 1980s. Ontario was the first; it introduced its *Environmental Assessment Act* (EAA) in 1973, well before the federal government and before the other provinces. Historically, in Ontario, EAs have been required for almost all public sector projects regardless of size. In contrast to BC and federal EIA (i.e., focusing on the assessment of large industrial developments), Ontario does not generally require assessments for many private sector projects (Northey 2020). In 2020, amid the COVID-19 pandemic, Ontario undertook a process of environmental assessment (EA) reform. With these updates, many smaller/low-risk projects will become exempt from EA, assessment timelines will be shortened, and decision-makers will have additional discretion – all with the aim of enhancing EA efficiencies, rather than facilitating a more comprehensive review process. Ontario will now also use a designated project list (like BC and Canada) for activities (including private sector projects) that require an assessment.

Across Canada, provincial processes evolved over several decades, and the territories followed most recently with a transition from using the federal process to unique regional approaches. All provinces and territories now have an EIA process in place (see Table 19.1). Provincial EIA requirements operate under either regulation emanating from an environmental/resource management law, such as New Brunswick's *Clean Environment Act*, or as a standalone EIA law, such as British Columbia's *Environmental Assessment Act* or Canada's *Impact Assessment Act* (Noble 2015).

In the territories, the EIA processes are part of federal–territorial agreements (i.e., the *Yukon Territory and Environmental and Socio-economic Assessment Act*) and regulations facilitated by Indigenous land claims and co-management boards (i.e., the *Nunavut Land Claims Agreement*) (Noble 2015). EIA systems in the territories have been developed and are administered through co-management boards and agencies such as the Nunavut Impact Review Board, the Mackenzie Valley Environmental Impact Review Board, and the Western Arctic and Yukon North Slope Environmental Impact Screening Committee and Environmental and Socio-economic Assessment Board. These hybrid federal/territorial approaches also mean that, unlike in the provinces, the federal Assessment Act does not apply to these regions.

Of particular interest to the evolution of EIA practice in Canada, the Northwest Territories (NWT) *Mackenzie Valley Resource Management Act* (MVRMA) accelerated innovative transitions in EIA practice, with considerable influence on the reach of federal EIA (Noble 2015). Following the proposal for the Mackenzie Valley Pipeline in the 1970s, an inquiry into the project was initiated. The inquiry was noteworthy in accelerating the settlement of land claim agreements with Indigenous Nations/groups in the NWT. As a result, proclaimed in 1998, the MVRMA was implemented by the federal government with the intent of giving Indigenous Nations/groups increased decision-making authority over natural resources and associated developments throughout the territory through co-management boards (Noble 2015). With this innovative co-management setting, federal EIA laws were no longer applied for the Mackenzie Valley region, except for trans-boundary issues.

As was noted above, there are instances where both federal and provincial levels have overlapping responsibilities and parallel legislative authority (Fitzpatrick and Sinclair 2016). Along with the authority to make laws, both federal and provincial governments have *ownership* over a natural resource, which in practice provides a strong level of control over environmental issues (Fitzpatrick and Sinclair 2016). As such, provincial, territorial, and federal governments in Canada play a dual role as the legislators and regulators, but also "owners" of natural resources

Table 19.1 Federal, provincial, and territorial EIA laws and assessment bodies

Jurisdiction	Associated EIA legislation or regulations	Assessment body
Canada	Impact Assessment Act	Impact Assessment Agency of Canada
British Columbia	Environmental Assessment Act, 2018	Environmental Assessment Office
Alberta	Environmental Protection and Enhancement Act- Environmental Assessment Regulation	Alberta Environment and Parks, Assessment and Continuations Unit or the Alberta Energy Regulator (for upstream oil, gas and coal projects) or Alberta Environment and Sustainable Development
Saskatchewan	Environmental Assessment Act	Environmental Assessment Branch
Manitoba	Environment Act- The Classes of Development Regulation	Environmental Approvals Branch
Ontario	Environmental Assessment Act	Environmental Assessment Branch
Quebec	Environmental Quality Act- Regulation Respecting the Environmental Impact Assessment and Review of Certain Projects	Environmental Assessment Directorate
New Brunswick	Clean Environment Act- Environmental Impact Assessment Regulation 87-83	Environmental Impact Assessment Branch
Nova Scotia	Environment Act- Environmental Assessment Regulations	Environmental Assessment Board
Prince Edward Island	Environmental Protection Act	Department of the Environment, Water, and Climate Change
Newfoundland and Labrador	Environmental Protection Act- Environmental Assessment Regulations	Environmental Assessment Division
Yukon	Environmental and Socio-Economic Assessment Act	Yukon Environmental and Socio-economic Assessment Board
Northwest Territories	Mackenzie Valley Resource Management Act	Mackenzie Valley Environmental Impact Review Board
Nunavut	Nunavut Planning and Project Assessment Act, and Nunavut Land Claims Agreement Act	Nunavut Impact Review Board
Inuvialuit Settlement Region	Inuvialuit Final Agreement/Claims Settlement Act	Environmental Impact Screening Committee and the Environmental Impact Review Board

and vast tracts of land (Fitzpatrick and Sinclair 2016; Noble 2015). This can also create situations where the government is a proponent or developer.

Within this multi-jurisdictional approach to environmental management, federal, provincial, and territorial EIA laws and processes can differ considerably in practice and scope. What are considered environmental effects or undertakings, the connection of EIA to planning and decision-making, what types of projects require assessment, considerations of alternatives, engagement and participation opportunities, and even how the environment is defined can vary greatly depending on the jurisdiction. Regardless, all processes follow some form of a rationalized comprehensive planning approach (see Chapter 1) (Table 19.1).

EIA reform in Canada

Globally, EIA practices have evolved and expanded over time, with an increasing number of diverse and distinct forms of practice (Pope et al. 2013). The same pattern is seen in Canada (Table 19.2). As social, political, and environmental landscapes in Canada have evolved, so have EIA processes, but change has not always been progressive. Some jurisdictions have sought to weaken the application of EIA, often by reducing the types of projects it covers, or emphasizing efficiency characteristics, such as time limits to process stages, which can limit or weaken the coverage, scope, and efficacy of EIA (Hanna 2016).

At the time, Gibson and Hanna (2016) held that the fourth stage does not exist as a consistent practice anywhere in the world. Today, the recent updates in Canada's federal and BC's provincial assessment process fall somewhere between the third and fourth stages, with some clear inclusive qualities emerging, and the move by governments to add sustainability as a quality that proponents should, or must, factor into their projects – but this is not present in all EIA processes in Canada.

Table 19.2 Stages of EIA reform (adapted from Gibson and Hanna 2016, 15–16)

Four stages of EIA policy and reform	*Thematic qualities*
Reactive	Reactive pollution control through measures responding to locally identified problems (most often air, water, or soil pollution), with technical solutions considered and issues addressed through closed negotiation of abatement requirements between government officials and the polluters.
Proactive	Proactive impact identification and mitigation through relatively formal EIA and project approval/licensing, but still focused on biophysical concerns (though now integrating consideration of various receptors) and still treated mostly as a technical issue with no serious public role (but perhaps expert review).
Integrative	Integration of broader environmental considerations in project selection and planning through environmental assessment processes that include:
	• consideration of social, cultural, historical, and economic as well as biophysical effects;
	• consideration of alternatives, aiming to identify the best options environmentally as well as socially and economically; and
	• public reviews, which would reveal conflicts and uncertainties among experts and recognize the significance of public choice.
Inclusive	noindent
	An emerging emphasis on planning and decision-making for sustainability, addressing policies and programs as well as projects and cumulative local, regional, and global effects, with review and decision-making processes that are more inclusive and comprehensive in defining the environment and:
	• Provide empowering assessment and decision-making options for Indigenous peoples;
	• Enable meaningful public participation;
	• Recognize uncertainties and favor precaution, innovation, and adaptability;
	• Yield transparent decisions; and
	• Integrate sustainability into planning and decision-making processes.

Restoring confidence

Understanding the motivation behind Canada's current federal *Impact Assessment Act* (2019) requires knowing a bit about the legislation and controversies that came before it. Before 2019, the federal process was based on the provisions facilitated by the *Canadian Environmental Assessment Act 2012* (CEAA 2012). CEAA 2012 was itself driven by shifting policy objectives and ideology. It was introduced by the Conservative government of the day "to support jobs and growth and to sustain Canada's economy" (HoC Standing Committee on Finance, Sub-committee on Bill C-38 2012).

The process that existed before 2012 had been criticized as inefficient and preventing or discouraging economic development. CEAA 2012 was intended to ensure "an expedited review process and removing barriers to resource development ventures" (Noble 2015, 13), such as the highly contested Trans Mountain Pipeline and Tanker Expansion project, were among the primary drivers of the 2012 Act (Noble 2015).

Noble, 2015, outlines seven major changes under CEAA 2012 from the 1992 *Act*:

1. Fewer federal authorities involved in EIA: Under the 1992 *Act*, all federal departments had some responsibility for the EIA process, but under CEAA 2012, there were three designated *responsible authorities* – the Canadian Environmental Assessment Agency (CEAA), the National Energy Board (NEB), or the Canadian Nuclear Safety Commission (CNSC).
2. Elimination of EIA for small projects: CEAA 2012 had more strictly defined thresholds for what would kind of project would require assessment. Only "designated projects" (those on the *Designated Projects Regulations*) would be assessed, and this eliminated more than 90% of federal assessments.
3. Delegation, substitution, and equivalency of EIA: CEAA 2012 allowed for increased opportunities to delegate part of the EIA process to other jurisdictions and the ability to substitute a provincial EIA process for federal assessment (or vice versa).
4. Established time limits: CEAA 2012 established time limits for government review responsibilities.
5. Changes to the definition of "environmental effect": Under the 1992 *Act*, environmental effects included any change that the project could cause to the environment, including the social and economic effects associated with the change. Under CEAA 2012, environmental effect was given a narrower definition focusing on components of the environment under federal jurisdiction (i.e., fish and migratory birds), changes to federal lands, and effects on Aboriginal peoples resulting from changes to the environment.
6. Focus on "interested parties": Under CEAA 2012, only those who would be directly impacted by the project or those with expertise can take part in the review panel hearing.
7. Enforcement: Under CEAA 2012, the decision statement by the minister regarding the project conditions, impact mitigations, or other commitments made by the proponent, became enforceable.

In practice, it cannot be said that CEAA 2012 moved projects toward approval any faster through the assessment process than its predecessor, nor is there evidence it necessarily encouraged new development or investment. Indeed, major contentious projects still encountered approval challenges and litigation, some of which linger today.

From its start, CEAA 2012 encountered significant criticism from many stakeholders, Indigenous organizations,[1] and environmental groups. Five main critiques emerged:

1. Thresholds and scoping: The regulatory elements of CEAA 2012 applied a more restricted assessment regime, drastically reducing the number of projects requiring EIA. Projects not

included in the regulations were left to the discretion of the Minister of the Environment to decide if an assessment would be required – even if the anticipated impacts were significant (Stacey 2015). Scoping was limited to narrowly defined "environmental effects", only including effects within federal jurisdiction, and primarily focusing on biophysical impacts (Stacey 2015; WCEL 2012).

2. Indigenous participation: Indigenous participation under the 2012 *Act* was driven by the "duty to consult" and was limited to meeting the minimum consultation requirements, with little Indigenous involvement in EIA decision-making (IAAC 2019; Stacey 2015; WCEL 2012). Furthermore, the *Act* was criticized for failing to align with the *United Nations Declaration of Rights of Indigenous Peoples* (UNDRIP), overlooking Indigenous jurisdiction and rights to self-determination. This resulted in several court challenges, most notably *Tsleil-Waututh Nation v. Canada (Attorney General)* and *Coastal First Nations v. British Columbia (Environment)* (see Box 19.1 for two examples of consultation failures). Such cases illustrate the inadequate integration of Indigenous perspectives into EIA reviews and how Indigenous participation in EIA is driven by meeting the Crowns obligations of the duty to consult.

3. Public participation: The issues associated with public participation under CEAA 2012 are twofold. First, the 2012 *Act* redefined a class of participant to the "interested party", drastically limiting opportunities for the public to influence projects when compared to the 1992 *Act*. Second, public participation under the 2012 *Act* was limited to tightly legislated timelines (Stacey 2015). For example, the public only had the opportunity to comment on whether a designated project should require an EIA under the *Act* for 20 days.

4. Discretionary power of ministers and Canadian Environmental Assessment Agency: CEAA 2012 changed the discretionary powers in two ways. First, decisions on projects where there where the anticipated environmental impacts were significant, and the project approval were often done so on the basis of "justified in the circumstances" (Stacey 2015). While this was true under CEAA 1992, under CEAA 2012, Minsters were only allowed to consider a set of narrowly defined environmental impacts but could use their discretion when considering a wide variety of social, economic, and cultural factors used for the justification of such environmental impacts. Critics note that the justification of impacts was left at the minister's discretion, without substantive explanation – therefore lacking transparency. Second, under CEAA 2012, designated projects could be exempt from a federal EIA if they are assessed (i.e., substituted) by an equivalent provincial process at the discretion of the minister (Stacey 2015). Some point out that this is problematic, as the assessment findings, conclusions, and recommendations under federal and provincial processes have sometimes been inconsistent.

5. The responsible authorities conducting federal EIA: Under CEAA 2012, the National Energy Board and the Canadian Nuclear Safety Commission were the responsible authority for assessments of projects which they were also in charge of regulating (i.e., NEB for energy projects and CNSC for nuclear projects), while CEAA was responsible for the assessment of all other designated projects. This was highly criticized for being limited in scope, restricting public involvement, and was considered as a means for cutting costs rather than improving the quality of assessments (Expert Panel Review of Environmental Assessment Processes 2015). Most notably, while all three responsible authorities were responsible for EIA under CEAA 2012, how the provisions of the *Act* were carried out differed and resulted in inconsistent assessment processes (Johnston 2018).

Box 19.1 Two examples of inadequate Indigenous consultation and associated court challenges in federal and BC EIA

In *Tsleil-Waututh Nation v. Canada*, the Federal Court of Appeal withdrew Kinder Morgan's Trans Mountain Pipeline expansion approval on two grounds: The Court decided that the NEB was unlawfully complacent in their decision to exclude increased marine traffic from the EIA report, and Canada did not adequately discharge the duty to consult and accommodate affected Indigenous peoples (Morales 2019). Having completed a supplemental assessment and consultation process, the federal government approved the project. That approval, while still controversial, survived subsequent court challenges, and the pipeline expansion is currently under construction.

In *Coastal First Nations v. British Columbia (Environment)*, the Supreme Court of BC found that the BC government abdicated its duty to make a decision under the provincial EA legislation and therefore failed also to consult with Indigenous peoples on the Enbridge Northern Gateway pipeline review (Morales 2018). In this case, the courts sided with the Indigenous jurisdictions ruling that the government had failed to adequately discharge the duty to consult. The federal government withdrew approval of the project.

Canada's New Federal Impact Assessment Act

The criticism of CEAA 2012 persisted, and in 2015 a newly elected government *promised to restore public trust* in the federal EIA process through review and reform of CEAA 2012. In 2016, the Minister of Environment and Climate Change announced the designation of an *Expert Panel Review of Environmental Assessment Processes* to undertake a review of the federal EIA process. Based on a detailed review process, the Expert Panel proposed several recommendations for an updated EIA process by producing a final report. There were a number of detailed recommendations made in the report, including:

- Transition from environmental assessment to impact assessment (IA[2]), which meant that federal reviews should place greater emphasis on social, cultural, economic, and health impacts in coordination with environmental impacts.
- Enhance opportunities for public participation by facilitating a collaborative process that involves the public in the design and selection of engagement plans. Also recommended was making participation available to all interested parties, therefore the updated regime should remove the "interested party" designation under CEAA 2012.
- Federal IA should move from three authorities to a single federal assessment authority.
- Implementing UNDRIP in IA and recognizing the rights and title of Indigenous peoples. A new IA process should implement the standards set out by UNDRIP and implement measures guaranteeing the protection of the rights and title of Indigenous people.
- The panel recommended that regional IA should be required on both federal and marine areas outside of federal lands where there is potential for cumulative impacts. The panel further recommended that strategic IAs should be required for federal programs, plans, and policies that would have "consequential implications" for project or regional IA.

Following the release of the Expert Panel Review, the federal government introduced *Bill C-69* in 2018, which was titled *An Act to Enact the Impact Assessment Act, to Amend the Navigation*

Protection Act and to Make Consequential Amendments to Other Acts. Bill C-69 was passed in 2019, and with this a new IA[3] process was put into place through the enactment of the *Impact Assessment Act* (IAA), *Canadian Energy Regulator Act*, and *Navigation Protection Act*. While the updated IAA aligns with many of the recommendations made by the Expert Panel Review, not all were adopted. See Table 19.2 for the main differences to the federal practice of EIA between CEAA 2012 and the IAA (Table 19.3).

Table 19.3 Primary differences between CEAA 2012 and the IAA (adapted from IAAC 2019b)

Canada Environmental Assessment Act, 2012	*Impact Assessment Act, 2019*
No early planning and engagement phase	A new mandatory early planning and engagement phase. This is intended to facilitate early dialogue with Indigenous peoples, provinces, the public, and stakeholders to identify and discuss issues early, leading to better project design.
Three responsible authorities conduct an environmental assessment	The introduction of the Impact Assessment Agency of Canada (IAAC), replacing the CEAA, who will act as the single responsible agency to lead assessments and coordinate Crown consultations with Indigenous peoples. Introduction of the Canadian Energy Regulator, Nuclear Safety Commission, and the Offshore Board as life-cycle regulators to work collaboratively with IAAC to provide expertise as needed.
Availability, accessibility and integration of science and knowledge varies Indigenous knowledge is not consistently considered	Decisions on projects are guided by science, evidence, and Indigenous knowledge. This includes mandatory consideration and protection of Indigenous knowledge along with other sources of evidence in impact assessments. Introduction of the Canadian Impact Assessment Registry, a platform for open science, data and plain-language summaries of the facts that support assessments.
Legislated timelines	While timelines remain legislated, there is increased flexibility on timelines for impact assessments and an extended planning phase.
Environmental assessments focus only on minimizing adverse environmental effects	A transition from environmental assessment to impact assessment based on the principle of sustainability. This includes a broader scope of assessments, including the positive and negative impacts on environmental, social, economic, and health factors. There are also new requirements for gender-based analysis and the assessment of impacts of projects on Indigenous peoples and their rights. Increased emphasis on the assessment of cumulative effects of existing or future activities in a specific region through increased provisions for regional assessments and of federal policies, plans and programs that are relevant to the IA of designated projects through strategic assessments.
Indigenous participation in reviews driven by the Duty to Consult	Early and inclusive engagement and participation at every stage of the assessment through specific consensus mechanisms and process-based recognition of Indigenous rights and interests from the start. Indigenous governments have greater opportunities to exercise powers and duties under the *Act*. This includes new opportunities for Indigenous-led impact assessment.

The stated purposes of the federal *Act* are to:

- Foster sustainability;
- Protect the components of the environment, and the health, social and economic conditions that are within federal authority;
- Establish a fair, predictable, and efficient process for conducting impact assessments;
- Take into account all effects – both positive and adverse;
- Address adverse direct or incidental project effects within areas of federal jurisdiction;
- Promote cooperation and coordinated action between federal and provincial governments;
- Promote communication and cooperation with Indigenous peoples of Canada with respect to impact assessments;
- Ensure respect for the rights of Indigenous peoples of Canada recognized and affirmed by section 35 of the *Constitution Act, 1982*;
- To ensure that opportunities are provided for meaningful public participation;
- Ensure that an impact assessment takes into account scientific information, Indigenous knowledge, and community knowledge;
- Ensure that an impact assessment takes into account alternatives;
- Avoid significant adverse environmental effects from *projects* carried out on federal lands, or those outside Canada carried out or financially supported by a federal authority;
- Encourage the assessment of the cumulative effects of physical activities in a region and the assessment of federal policies, plans, or programs; and
- Provide for follow-up programs.

The federal IA process, under the IAA, typically follows a five-phase process (IAAC 2019a). The federal IA process under the IAA has several similarities to the EIA process under CEAA 2012. For example, the process under the IAA follows a similar approach of designating projects as reviewable based on thresholds prescribed by the updated Physical Activities Regulation. The Regulations still focus on larger projects, and despite their apparent simplicity the criteria used by the Impact Assessment Agency to determine if a project meets a threshold is not publically accessible and be can be convoluted. As with the previous Act, the new one allows for ministerial discretion to designate projects not included in the regulations.

Table 19.4 provides a brief overview of the five phases of the federal IA process, as well as new features to the federal assessment process (i.e., key updates) included within each phase.

Major procedural differences between the IAA and CEAA 2012 can be seen in the introduction of a new "planning phase" (phase 1), timeline changes, and changes to decision-making, where reasons for decisions must be indicated in an "assessment report" explaining how the decision best reflects "public interest".

Overall, the process also provides increased opportunities for Indigenous participation throughout the IA process, but the Act does not explicitly mention aligning with UNDRIP. However, in 2021 Canada passed the *United Nations Declaration on the Rights of Indigenous Peoples Act*, which seeks to ensure the federal laws of Canada are consistent with the Declaration. This will also have an impact on EIA practice.

Table 19.4 Phases of the Canada Impact Assessment Process (IAAC 2019a; IAAC 2019b; IAAC 2019c; IAAC 2019e; IAAC 2020a; IAAC 2020b)

Phases	Assessment activities	Key updates of each phase
Phase 1: Planning	• Give participating levels of government, Indigenous Nations/groups, and the public opportunities to learn more about the project and provide feedback on their initial concerns and interests. • Communicate with provincial, territorial, and Indigenous jurisdictions that a project is being contemplated and may require collaboration and consultation for an assessment. • Determine whether an IA will be required or if the project should be exempt from IA or terminated altogether.	• This is a new phase, which focuses on reducing conflict, streamlining cooperation, and facilitating a more comprehensive review process. This includes seeking consensus with participating Indigenous Nations/groups on the work plans and collaborations protocols, which will be utilized throughout the assessment. • Where projects are contentious, decision-making (by either terminating or exempting the project from IA) can be done early on to save resources that would otherwise be utilized on the assessment of a project which has little chance for approval from the onset (or vice versa). • Indigenous Nations/groups may indicate that they would like to collaborate on or complete certain studies/assessment procedures through an Indigenous-led approach. In doing so, Indigenous Nations/groups will be recognized as independent jurisdictions.
Phase 2: Impact statement	• The proponent begins collecting information and conducting studies on the project for the preparation of the Impact Statement – a technical report detailing the assessment requirements, including the proposed activities anticipated impacts. • The proponent engages Indigenous Nations/groups for information sharing, participatory studies, and to gather information and knowledge. • Once the Impact Statement is completed, it is issued to the public, and all relevant studies are made publicly available.	• The proponent's Impact Statement must incorporate scientific information, evidence, community knowledge, and Indigenous knowledge. The most significant update here is the increased emphasis placed on integrating community and Indigenous knowledge into the proponents Impact Statement.
Phase 3: Impact assessment	• This phase can be led by either the "Agency" or a "Review Panel".	• If an Indigenous-led assessment is being completed in parallel or collaboratively with the Agency or Review Panel-led IA, IAAC will consider the results of the Indigenous-led assessment when drafting the Impact Assessment Report and draft conditions.

(Continued)

Table 19.4 (Continued)

Phases	Assessment activities	Key updates of each phase
	• For Agency-led assessments, IAAC is responsible for the review of the proponent's Impact Statement and finalizes an Impact Assessment Report– a technical report describing the potential positive/negative impacts and draft conditions for approval. The Agency is also responsible for working with participating Indigenous Nations/groups on a Consultation Summary – a document detailing the adequacy of consultations throughout the review process. • If the Minister refers the assessment to a Review Panel[4] (consisting of independent experts and sometimes members of relevant regulatory agencies), the Panel holds a public hearing, and then prepares the Impact Assessment Report. During Review Panel assessments, IAAC is still responsible for leading Crown consultations with participating Indigenous Nations/groups to discuss potential mitigation and accommodation measures and for preparing the Consultation Summary.	• New public engagement opportunities, where IAAC must post a copy of its draft report for a public commenting prior to finalizing the Impact Assessment Report. • Sustainability factors must be accounted for during the impact assessment and included in the Impact Assessment Report. These factors include: interactions between projects, need for the project, other means of implementing the project, alternatives to the project, the extent to which the project contributes to sustainability, extent to which the project affects the Government of Canada's ability to meet its environmental obligations and commitments to climate change, requirements to look at matters that are related to issues of gender and vulnerable communities, and any relevant regional or strategic assessments initiated by the federal government of another jurisdiction.
Phase 4: Decision-making	• This phase dictates the future of the proposed project in terms of either approval or rejection. Decision-making authority is held by the Minister of Environment and Climate Change, or if circumstances require, the Governor in Council (the federal Cabinet). • IAAC provides the Minister or Governor in Council with the Impact Assessment Report, Consultation Summary, and potential conditions for approval. • The final decision is noted through an Impact Assessment Decision Statement, which includes explanations of the Minster's or Governor in Council's determination and the conditions of approval (follow-up program and mitigation measures), which is then made publicly available.	• During this phase, Indigenous Nations/groups can inform the Minister or Governor in Council their views regarding the adequacy of consultation and whether consensus was achieved throughout the assessment process by collaborating with IAAC on the delivery of the Consultation Report. • Indigenous-led assessments or studies are submitted to the Minister or Governor in Council with the Impact Assessment Report. • Decision-making involves a determination of whether the adverse effects of an activity reflect public interest (see section "Public interest decision-making" for a description of the public interest determination factors).
Phase 5: Post decision	• This phase is only relevant if the project is approved. Here, the follow-up program and mitigation measures outlined in the "Decision Statement" are implemented.	• There are new opportunities for Indigenous Nations/groups and other communities to have an expanded role in monitoring impacts. If required, IAAC may establish an Environmental Monitoring Committee to do so.

British Columbia Environmental Assessment Act and key issues

British Columbia had one of the first EIA Acts in Canada. The 1994 legislation was innovative; it provided considerable procedural guidance, and had thresholds for reviewable projects. Its objectives included promoting sustainability by protecting the environment and fostering a sound economy and social well-being, and prevention and mitigation of adverse effects of reviewable projects, and a timely and integrated assessment of the environmental, economic, social, cultural, heritage and health effects of projects (BC Environmental Assessment Act, S.B.C. 1994, c.35, s.2).

The BC EIA legislation was changed in 2002. This was part of a government promise to overhaul the province's regulatory system to reduce bureaucracy and improve efficiency (Rutherford 2016). The user guide for the 2002 process emphasized themes such as flexibility, efficiency, and timely reviews of proposed major projects, and helping "revitalize the provincial economy" (EAO 2003, 1). Rutherford (2016, 299) comments, that as might be expected, given these origins, the 2002 EAA and its regulations are lean documents, providing a basic framework for EIA, but leaving much of the detail to policies and practices developed by the Environmental Assessment Office (EAO), the agency that administers the Act.

Haddock (2010) outlines the six major changes in the 2002 EAA from the 1994 *Act*:

1. The elimination of provisions requiring engagement of local governments and Indigenous Nations/groups on project committees and provisions allowing for inclusion of other stakeholders on public advisory committees.
2. Greater decision-making flexibility in the minister and executive director of the EAO in deciding whether a project requires an EIA and what terms of reference for those assessments would be.
3. A reduction from 93 sections of the 1994 *Act* to 51 within the 2002 *Act*, including the purpose clause, which guided decision-makers and courts in determining the reason/justification of the EIA.
4. Elimination of the requirement for assessment reports to consider and evaluate alternatives.
5. A requirement that the EAO executive director consider and reflect government policy identified by a government agency or organization responsible for the policy area when determining the scope, procedures, and methods of an EIA.
6. Increased thresholds in the *Reviewable Projects Regulation*, which had the effect of excluding many projects from an EIA requirement.

The BC EIA process under the 2002 EAA was generally proponent driven and was associated with high levels of proponent satisfaction. Contrasting the satisfaction of proponents, Indigenous Nations/groups and environmental groups were more critical. Criticism of the 2002 *Act* was generally based on seven main points (Haddock 2010; Rutherford 2016; First Nations Energy and Mining Council 2009; Booth and Skelton 2011):

1. Lack of consideration of sustainability: Under the 2002 *Act* and associated EIA regulations, ensuring sustainability/sustainable development was not listed as a goal. This was problematic, as it made it less likely that review panels, courts, responsible ministers, and others would use principles of sustainability when interpreting the *Act*, conducting/reviewing assessments, and making decisions.
2. Triggers and scoping: The 2002 EAA applied a project-threshold approach in determining what types and sizes of activities require an EIA set out by the *Reviewable Projects Regulation*. The thresholds for some types of projects were increased with the 2002 *Act*, with the result

being many projects were exempt from EIA. As many individual smaller projects no longer triggered EIA, the cumulative effects of multiple projects became a persistent issue.

3. Public participation: Public participation throughout the EIA process under the 2002 *Act* was limited. For example, the 2002 *Act* eliminated project committees and public advisory committees, therefore limiting opportunities for the public to provide analysis, advice, and recommendations to the responsible ministers regarding the potential effects of the project. Furthermore, public consultation under the 2002 *Act* was placed primarily on the proponent, and only required two formal commenting periods. This resulted in a lack of public trust and confidence in the BC EIA process.

4. Monitoring, compliance, and enforcement: Critiques note that under the 2002 *Act*, the BC EAO struggled to ensure effective monitoring, compliance, and enforcement as it was primarily the responsibility of the project proponent. Issues were identified in relation to the measurability and execution of EIA conditions, unclear monitoring requirements, and unproductive compliance and enforcement activities.

5. Oversight and decision-making: Both EIA oversight and decision-making under the 2002 *Act* lacked clear evaluation criteria. For example, the EAO is the responsible authority for the oversight and review of proponent applications, but how these responsibilities were carried out was not defined by legislation or clearly stated by rules. In the absence of clear evaluation criteria, the review of proponent applications was left to the interpretation of EAO staff members. Likewise, critics point out that there were no substantive criteria for decision-making regarding project approval.

6. Indigenous Participation: Under the 2002 *Act*, Indigenous participation in the assessment, decision-making, and monitoring processes was limited. This was reflected through a lack of mandated participation opportunities for Indigenous Nations/groups to engage throughout project reviews (including scoping and decision-making), inconsistent considerations of Indigenous Nations/groups interests, and inadequate funding mechanisms to support Indigenous Nations/groups participation in the EIA process. More recently, critiques have further pointed out that the BC EIA process under the 2002 *Act* failed to recognize Indigenous jurisdiction or meet UNDRIP standards.

7. Contradictory outcomes: Critics have pointed out that BC and federal EIA processes arrived with different findings in terms of project impacts, as BC's EIA process under the 2002 *Act* did not ensure effective coordination and collaboration with other jurisdictions. For example, the Taseko "Prosperity Mine" was reviewed separately by BC and federal EIA processes (CEAA 2010). In the BC review process, BC determined that the proposed project would not result in any adverse effects beyond the loss of fish and fish habitat. The federal review on the other hand determined that the project would result in significant adverse effects on fish and fish habitat, navigation, current and traditional uses of the land and resources by Indigenous peoples, and existing and potential Indigenous rights and title.

BC environmental impact assessment reform

Following the formation of a new minority government in 2017, the EAO was instructed to begin developing a new assessment framework, including supporting policies and regulations. This process was referred to as environmental assessment revitalization. The EA[5] revitalization was designed to achieve three objectives (EAO 2018):

1. Enhance public confidence, transparency, and meaningful participation;
2. Advance reconciliation[6] with First Nations; and
3. Protect the environment while offering clear pathways to sustainable project approvals.

The motivations for change were similar to the ones that spurred change at the federal level. There was one notable exception; BC arguably had a stronger focus on Indigenous representation in the new process. Following a similar process of the federal EIA reform, the EAO commissioned an Environmental Assessment Advisory Committee (EAAC) in 2018 to review and identify key areas of improvement for a new EAA (Government of British Columbia 2018). The EAAC produced a report, which included 33 recommendations for the purpose of EA revitalization.

The EAAC report included several recommendations relating to how the practice of EA in BC can be improved upon. For example, by emphasizing reconciliation with Indigenous peoples through specific Indigenous engagement and participation mechanisms, by fixing substantive issues within the EA process itself (i.e., the lack of opportunities for public participation, etc.), implementing best practices when transitioning to a new EA regime (i.e., through extensive Indigenous, stakeholder, public, and industry engagement), and provincial support programs which would help build EA capacity within government agencies (Government of British Columbia 2018).

Through a series of direct engagements with Indigenous Nations/groups, industry and business associations, non-governmental organizations, and EIA practitioners, a new EA process was created. The new process was designed to achieve the three goals listed above, while also mirroring the updates to the federal IA process under the IAA. With this, the BC government introduced Bill 51 – *2018 Environmental Assessment Act*, which was brought into force in late 2019. Unlike the federal IAA, the 2018 EAA does not have a section dedicated to the purposes of the *Act* itself. It does, however, have a section under "Part 2 – Administration", which outlines the two main purposes of the EAO. As outlined by the 2018 *Act*, the purposes of the EAO are to (EAA 2018; EAO 2020c):

1. Promote sustainability by protecting the environment and fostering a sound economy and the well-being of British Columbians and their communities by:
    *   Carrying out assessments in a thorough, timely, transparent, and impartial way, considering the environmental, economic, social, cultural and health effects of assessed projects;
    *   Facilitating meaningful public participation throughout assessments;
    *   Using the best available science, Indigenous knowledge, and local knowledge in decision-making under the *Act*; and
    *   Coordinating assessments with other governments, where appropriate, including Indigenous Nations, and with other provincial ministries and agencies.
2. Support reconciliation with Indigenous peoples in British Columbia by:
    *   Supporting the implementation of the *United Nations Declaration on the Rights of Indigenous Peoples* (BC has passed broad legislation to align BC laws with the Declaration);
    *   Recognizing the inherent jurisdiction of Indigenous Nations and their right to participate in decision-making in matters that would affect their rights, through representatives chosen by themselves;
    *   Collaborating with Indigenous Nations in relation to reviewable projects, consistent with the *United Nations Declaration on the Rights of Indigenous Peoples*; and
    *   Acknowledging Indigenous peoples' rights recognized and affirmed by section 35 of the *Constitution Act, 1982* in the course of assessments and decision-making under this *Act*.

The 2018 EAA has several key differences from the 2002 EAA. Many of the changes to the BC's EA process respond to the criticism associated with the 2002 *Act*. Explained in Table 19.3 are some of the new features of BC's revitalized EA process .

The new BC EA process under the *Environmental Assessment Act 2018* involves seven phases (EAO 2020a). Among the most prominent changes to the 2018 *Act* are the new opportunities

Table 19.5 A brief comparison of BC's 2002 Act and the new 2018 *Act*

EAA (2002)	EAA (2018)
Sustainability not noted as a goal	A defined purpose of the EAO is to promote sustainability by protecting the environment and fostering a sound economy and the well-being of British Columbians and their communities.
Weak strategic and regional qualities.	Regional and strategic assessments are now possible under the *Act*.
Thresholds do not trigger assessment for many projects of smaller scale, or those just below the regulatory threshold, possibly contributing to cumulative effects.	Notification to be provided by proponents to track projects that should potentially require an environmental assessment before proceeding despite being below the reviewability threshold. Criteria include land clearance, linear distance, workforce, design and effects thresholds, federal reviewability, and potential greenhouse gas contributions. However, the notification threshold may be too low in some instances to assess many smaller projects.
Limited public participation opportunities	A new early engagement phase to identify interests, issues and concerns of Indigenous Nations/groups, stakeholders and the public that can inform project design, siting and alternative approaches to developing the project. New public comment periods and engagement tools, including easier to understand information, earlier involvement of the EAO in local communities and funding to support public participation.
Criteria for decision-making not evident	An early decision following early engagement to determine whether a project is ready to proceed with an EA. Requirements on what needs to be assessed (including cumulative effects) and considered in decisions.
Inadequate opportunities for Indigenous participation throughout the EA process	A defined purpose of the EAO is to support reconciliation with Indigenous peoples in British Columbia and the implementation of UNDRIP. A clearly defined process for seeking consensus with participating Indigenous Nations, including a number of opportunities identified throughout the process that aims to secure consent on decisions. New opportunities for Indigenous-led assessments. Commitment ensuring adequate funding for participating Indigenous Nations/groups.
Alignment between the federal and BC EIA processes was weak	The 2018 EAA has alignment with the federal IAA .
Issues with compliance and enforcement	Modernized compliance and enforcement tools. Monitoring and effectiveness review of EA certificates to ensure mitigations are effective.

for Indigenous involvement throughout the EA process, as it strives to achieve consensus with and consent from affected Indigenous Nations/groups. Most notably, the updated consensus and consent seeking mechanisms align with UNDRIP, and as such, impacted Indigenous Nations/groups will be consulted as consistent with the Declaration. For example, mechanisms for consent are built into multiple stages of the seven-step process. See Table 19.6 for a brief description

Table 19.6 Phases of the BC EA process (EAO 2020a; EAO 2020b)

Phases	Assessment activities	Key updates of each phase
Phase 1: Early engagement	• Following the submission of the proponents Initial Project Description, all participants (proponents, public, Indigenous Nations/groups, etc.) are given opportunities to voice initial concerns and interests through a public commenting period – guiding the development of engagement approaches to be used throughout the assessment. • Following initial engagement, the EAO will produce a "Summary of Engagement", from which the Proponent has up to 1 year to submit a Detailed Project Description, which integrates the information collected through early engagement and details regarding project design.	• Like the federal process, this is a new phase, which focuses on fostering dialogue between the EAO, other jurisdictions (including Indigenous), the proponent, and the public. In doing so, this phase aims to reduce conflict, streamline cooperation, and facilitate a more comprehensive review process. • Following the submission of the Initial Project Description, it is posted for public commenting.
Phase 2: EA readiness decision	• A decision is made regarding whether the project should move forward through an assessment. Under some circumstances, projects may be terminated or exempt from an EA, according to a Minister's decision. • If sufficient information has been collected and the project has been neither terminated nor exempt from an EA, the project may proceed to the next phase.	• Where projects are contentious, decision-making – by either terminating or exempting the project from EA – can be done early on to save resources that would otherwise be utilized on the assessment of a project which has little chance for approval from the onset (or vice versa). • Participating Indigenous Nations/groups are given their first opportunity to provide a notice of consent or lack thereof regarding if the project should become terminated or exempt from EA.
Phase 3: Process planning	• Information requirements, roles and responsibilities of participants, and engagement plans are determined. • A Process Order is created, which outlines how the assessment will be conducted (including the scope and methods).	• Consensus with participating Indigenous Nations/groups on the details of the Process Order is a key requirement prior to the completion of this phase. • Indigenous Nations/groups may notify the EAO that they intend to carry out either part or all of the assessment (i.e., relevant studies, information gathering, etc.) through an Indigenous-led approach. • The EAO must engage in public consultation regarding the Process Order, seeking input on the proposed scope and methods of the assessment.

(Continued)

Table 19.6 (Continued)

Phase 4: Application development and review	• For the application development, the proponent engages with participants for the development of their EA Certificate Application. In doing so, participants are given the opportunity to provide feedback on data collection and analysis, identify any key issues, and suggest issue resolution. • The application is then reviewed by the EAO, participating Indigenous Nations/groups, and two advisor committees (Technical and Community). The involved parties then give feedback to proponents on revision requirements for their final EA Certificate Application. • The EAO will directly consult participating Indigenous Nations/groups to ensure consensus that the EA Certificate Application requirements have been satisfied. • The Community Advisory Committee is required by default (the EAO may establish more than one Community Advisory Committee if necessary). The role of the Community Advisory Committee is to advise the EAO on the potential effects of a project on communities, with the goal of fostering increased public participation.
Phase 5: Effects assessment and recommendation	• An effects assessment (positive and negative) is conducted, resulting in a Draft Assessment Report and Draft EA Certificate (which includes draft legally binding conditions for approval). • Following the effects assessment, recommendations are given to provincial decision-makers (this is a package that includes a Draft Assessment Report and Draft EA Certificate). • Indigenous Nations/groups, the public, and all other stakeholders are given the opportunity to comment on the Draft Assessment Report and Draft EA Certificate. • When recommendations are given to provincial decision-makers, participating Indigenous Nations/groups are given a second opportunity to provide notice of consent or lack thereof. • Indigenous-led studies or assessments are submitted to provincial decision-makers to help guide their decision.
Phase 6: Decision	• This phase dictates the future of the proposed project in terms of either approval or rejection. Decision-making authority is held by the Minister of Environment and a "Responsible Minister".[7] • The decision can be to approve the project by issuing an EA Certificate, or to refuse the project. As decision-making can be contentious, Ministers are required to provide the reasons for their decision, which are made publicly available. • Decision-making is to be based on considerations of sustainability and reconciliation (aligning with the purposes of the EAO described within the EAA), and any other matters which relate to decisions that reflect public interest. • If the Ministers' decision contradicts the notice of consent or lack thereof of the participating Indigenous Nations/groups, reasons must be given, including an opportunity to meet with the Nation/group that provided the notice for resolution.
Phase 7: Post-certificate, compliance, and enforcement	• This phase is only relevant if the project is approved. If an EA Certificate is issued, it will include a deadline (up to 10 years) in which the project must be substantially started. Post-certificate activities include mitigation effectiveness reports and may involve audits, certificate amendments, extensions, and transfers. • Enhanced compliance and enforcement tools such as the introduction of administrative monetary penalties, court-imposed fines, and options for prohibitions and directions to take specific actions if the conditions for approval are not being satisfied.

of each phase, including some of the key updates to BC's revitalized assessment process, as well as some of the new features (key updates) included within each phase.

The distinct qualities of EIA reform in Canada and BC

The EIA reform of the federal and BC processes has been accompanied by several new and unique qualities. These include Indigenous-led assessment, regional and strategic assessment, gender-based analysis, increased recognition of the substantial goal of EIA being sustainable development, greater emphasis on early planning and engagement prior to the onset of the EIA, new consent and consensus mechanisms with affected Indigenous Nations/groups throughout each EIA process, and updated decision-making criteria based upon public interest factors. This section will briefly explain these new qualities and what these features have the potential to achieve.

Indigenous-led assessment

In advancing reconciliation with Indigenous peoples, both sets of legislation have created opportunities for Indigenous-led assessments (ILAs), which will be completed in coordination with the BC and federal review processes. Through agreements with ministers and government agencies, Indigenous Nations/groups may conduct either part or all of an assessment on behalf of or with the IAAC or EAO. Here, Indigenous Nations/groups may design assessment processes that reflect their unique cultures, histories, natural resource issues, and perspectives.

While there has not yet been an ILA completed as part of a review under the IAA and EAA, there have been ILAs completed under the previous *Acts*, which illustrate how ILA could be integrated into future reviews. For example, IAAC explains how

> during the course of the environmental assessment for the Ajax Copper-Gold Mine in BC, the Stk'emlu'psemc te Secwepemc Nation (SSN) Joint Council carried out its own assessment process for the project. This assessment process was designed to assess the impacts of the project in a way that respected SSN knowledge and perspectives and facilitated informed decision-making by their communities consistent with their laws, governance, traditions, and costumes. The SSN's assessment was provided to the minister. The agency engaged deeply with SSN and made efforts to align federal environmental assessment with SSN's assessment process to the extent possible.
>
> *(IAAC 2019e)*

For more details on Indigenous-led assessment, see Chapter 13 in this book.

Regional and strategic assessment

To meet both the federal and BC goals of aligning with sustainability principles, particularly mitigating cumulative effects of human activities, both the IAA and 2018 EAA have integrated the use of regional and strategic assessment into EIA reviews. As explained by the EAO, regional assessments

> evaluate how different scenarios for development, protection and restoration in a region will cumulatively affect values and rights compared to historic and current conditions. They can identify management objectives and limits based on scientific

and Indigenous knowledge, which can be directly applied in project-level assessments and regulatory decision-making and serve as an input to land use or marine planning.

(EAO 2018, 11)

While strategic assessments "evaluate how higher-level policies, plans, and programs impact values and rights" (EAO 2018, 11). Under both the IAA and EAA, anyone may request a regional or strategic assessment, including members of the public, Indigenous Nations/groups, non-governmental organizations, industry associations, or other jurisdictions. For a more detailed description of regional assessment, see Chapter 10 in this book, and for strategic assessments, see Chapter 2 in this book.

Gender-based assessment

Included in the IAA are provisions for Gender-Based Analysis Plus (GBA+). GBA+ within the federal IA process seeks to assess considerations of the intersections between sex, gender, and other identity factors in relation to project proposals. As explained by IAAC, "GBA+ is an analytical tool – a way of thinking, as opposed to a specific set of prescribed methods. For example, in quantitative statistical analysis, specific methods help analysts to understand relationships between variables. In GBA+, the method helps analysts to understand issues related to employment or other opportunities in a community or industry and can include the use of descriptive statistics (e.g., percentage of women employed, disaggregated by age, ability, ethnic origin, or other factors), interviews (e.g., to contextualize statistics and understand why women or specific subgroups find a sector difficult to find work in), and community forums (e.g., to discuss findings and propose solutions)" (IAAC 2019a). GBA+ will be used within federal IA to determine if there are different impacts of projects on subsets of the population (IAAC 2019a). For more details on gender-based assessment, see Chapter 11 in this book.

Sustainability

Both the federal IAA and BC EAA explicitly mention sustainable development as an explicit goal. For example, this is communicated in the IAA in section 6 "Purposes", which states "to foster sustainability". Furthermore, under the IAA one of the factors that is required to be considered in IA is "the extent to which a designated project contributes to sustainability" (IAAC 2019d). This consideration can be measured through four sustainability principles (IAAC 2019d):

- Principle 1: Consider the interconnectedness and interdependence of human-ecological systems.
- Principle 2: Consider the well-being of present and future generations.
- Principle 3: Consider the positive effects and reduce adverse effects of the designated project.
- Principle 4: Apply the precautionary principle and consider uncertainty and risk of irreversible harm.

Under the IAA, practitioners are required to characterize the reviewed projects contribution to sustainability within the Impact Statement (IAAC 2019d). Proponents are also encouraged to describe how they have applied the sustainability principles (IAAC 2019d). For decisions related to whether a project should proceed, the minister or governor in council must consider the extent to which a designated project contributes to sustainability (IAAC 2019d).

Likewise, the EAA was designed with the explicit goal of "protecting the environment while offering clear pathways to sustainable project approvals" (EAO 2018), and again under the "purposes of the EAO", which states "promote sustainability by protecting the environment and fostering a sound economy and the well-being of British Columbians and their communities" (EAO 2018). As explained by the EAO, this sustainability "purpose" will be achieved by (EAO 2018):

- Carrying out assessments in a thorough, timely, transparent, and impartial way;
- Considering the environmental, economic, cultural, and health effects of assessed projects;
- Facilitating meaningful public participation throughout assessments;
- Using best available science, Indigenous knowledge, and local knowledge in decision-making; and
- Coordinating assessments with other governments.

Most significantly, under the EAA, before a project is referred to the ministers for a decision, the EAO is required to make a recommendation on whether the project is true to the purpose of sustainability. With this, the ministers must adequately consider this recommendation from the EAO when making project decisions.

Early planning and engagement

A key feature of the updated BC EAA and federal IAA is the newly added planning/early engagement phases. Following the submission of the proponent's application, the planning/early engagement phases aim to facilitate meaningful dialogue regarding a proposed project between the proponent, Indigenous Nations/groups, the public, municipalities, federal/provincial EIA agencies, and all other stakeholders. By doing so, upfront planning and engagement provides an opportunity for various parties to establish cooperative relationships, shaping the substantive concerns to be addressed throughout the assessment, which then may guide scoping and alternative approaches to developing the proposed project (Stacey 2020). The new emphasis placed on early planning is a valuable tool that can reduce conflict, streamline cooperation, and facilitate a more comprehensive review process. Following the early planning and engagement phases, Canada and BC may decide if an EIA is required or if the project should be exempt from EIA, or terminated from the onset due to significant contention, issues, or concerns (Stacey 2020).

New consensus and consent mechanisms

The updated EAA and IAA both reflect the emphasis placed on centering reconciliation within all aspects of British Columbia and Canada's activities. Most significantly, this is driven by BC and Canada's commitment to the implementation of the *United Nations Declaration on the Rights of Indigenous Peoples*, with the aim to secure free, prior, and informed consent for decisions that affect Indigenous peoples' rights and interests. This is reflected through ensuring that consensus with and consent from Indigenous Nations/groups is achieved throughout project-based assessments. It is important here to differentiate between consensus and consent. Consensus is an outcome or approach that is supported by Indigenous Nations/groups and the respective assessment agency, where Indigenous Nations/groups maintain the right to give their consent or lack thereof following the assessment (IAAC 2019e). Consent is specifically referring to Indigenous Nations/groups determination of if and how a project should proceed (IAAC 2019e).

The updated federal IA regime appears to more broadly apply a consensus-based approach when compared to the BC EA process, which offers distinct opportunities for both consensus and consent. Federally, process points for consensus are in the "planning" and "decision-making" phases of the assessment process. During the "planning phase", an Indigenous Engagement and Partnership Plan is collaboratively designed by IAAC and affected Indigenous peoples, outlining work protocols and partnership agreements between the two parties for the purpose of the assessment (IAAC 2019e). In addition, during the "decision-making phase", IAAC and affected Indigenous Nations/groups work to achieve consensus by collaboratively producing the Consultation Summary or Crown Consultation and Accommodation Report, which is used to provide views on the adequacy of consultation throughout the IA process (IAAC 2019e). As a final step to seeking consensus within the "decision-making" phase, to inform the ministers decision on whether to approve or reject the proposed project, Indigenous Nations/groups are consulted on the proposed conditions for the project being reviewed (IAAC 2019e).

Likewise, the BC process has dispute resolution points to facilitate consensus with affected Indigenous Nations/groups during the early engagement, EA readiness, process planning, and effects assessment and recommendation phases. The BC EA process, on the other hand, more explicitly mentions consent, with two separate process points for affected Indigenous Nations/groups to notify the EAO of their consent or lack thereof for how and if the proposed project should proceed. The first opportunity for Indigenous Nations/groups to provide consent is during the "EA readiness phase" where the EAO may recommend exempting or terminating the proposal from EA altogether (EAO 2018). The second is during the "decision-making phase", where recommendations are made to the responsible ministers regarding whether to issue an EA Certificate (i.e., determining project approval/conditions for approval, or in its absence, rejection) (EAO 2018).

Although the updated Canada and BC processes both place a greater emphasis on Indigenous decision-making, final decision-making authority is still maintained by federal/provincial decision-makers. For example, while the federal process requires IAAC to seek consensus with Indigenous Nations/groups at multiple points throughout an assessment, it is unclear if this will be effective in meaningfully implementing the minimum standards of UNDRIP. Likewise, this issue also arises when examining the BC process, which requires the EAO to seek consent through two formal stages as it remains unclear if BC will still allow for projects to proceed in situations where consent is withheld by Indigenous Nations/groups. As these updates are still new, and have not been tested through practical application, time will ultimately tell how successful these new consensus and consent mechanisms are when implemented.

Public interest decision-making

A key aspect of the new BC and federal EIA regimes is that final decision-making is to be done with considerations of whether or not the project reflects the public interest. This differentiates significantly from the previous BC and federal EIA regimes, where decision-making was done based on whether the project would result in justifiable environmental impacts – with the overarching goal of minimizing adverse environmental effects. For example, in moving beyond decision-making based on minimizing adverse environmental effects, decision-making under the updated federal process must consider the following public interest factors (IAAC 2020b):

- The extent to which the designated project contributes to sustainability.
- The extent to which adverse effects within federal jurisdiction and the adverse direct or incidental effects that are indicated in the impact assessment report in respect to the designated project are significant.
- The implementation of the mitigation measures that the minister or governor in council considers appropriate.
- The impact that the designated project may have on Indigenous Nations/groups and any adverse impact that the designated project may have on the rights of Indigenous peoples of Canada are recognized and affirmed by section 35 of the *Constitution Act, 1982*.
- The extent to which the effects of the designated project hinder or contribute to the Government of Canada's ability to meet its environmental obligations and its commitments in respect of climate change.

Final comments

The evolution of EIA in Canada is the product of 50 years of practice, participation, and reform. This has involved a process of progressive evolution, transforming from a set of guidelines to legislation. While the past federal and BC EIA regimes aimed to facilitate quick approvals, the reformed EIA regimes strive to build public trust through responsible resource development while also remaining efficient and predictable. The recent updates to both Canada's federal and British Columbia's EIA processes are at the forefront of EIA reform and have included several legislated changes that illustrate the progression of EIA moving toward best practices.

The success of and satisfaction of proponents, the public, and Indigenous Nations/groups with the updates associated with the IAA and EAA is yet to be determined. What we do know is both *Acts* include several updates which have the potential to increase opportunities for public participation throughout the review process and enhance reconciliation with Indigenous peoples while placing greater emphasis on sustainable development. It should be noted, however, that updated EIA processes still place a high degree of discretion in the hands of the associated assessment bodies and ministers. As such, the success of each new process will be determined by how this discretion is used by assessment agencies, ministers, and future governments.

Notes

1 The Indigenous-led social movement *Idle No More* movement played a significant role in galvanizing action against the Canadian government's dismantling of environmental protection laws, and specifically two omnibus Acts that brought in the changes to federal EA.
2 In staying consistent with the language used by the Expert Panel, IA will be used instead of EIA when referring to the recommendations of the report. EIA will still be used when referring to assessment practices federally and in BC.
3 IA will be used when discussing assessment practiced under the IAA.
4 This is most applied to the assessment of large projects and/or those with substantial public interest.
5 Environmental Assessment (EA) will be used when discussing the BC process to stay consistent with the language used by the EAO.
6 In Canada, reconciliation is a concept tied to improving provincial and federal governments relationship with Indigenous peoples while simultaneously establishing, maintaining, and enhancing a mutually respectful relationships between Indigenous and non-Indigenous peoples (Truth and Reconciliation Commission of Canada, 2015).
7 The Responsible Minister is the one with responsibilities for activities of the project being reviewed. For example, for a mine the Responsible Minister would be the Minister of Energy, Mines, and Petroleum Resources

References

Booth, A, and Skelton, N. 2011. "'We are fighting for ourselves' – First Nations' Evaluation of British Columbia and Canadian Environmental Assessment Processes". *Policy Management*, 13(3), 367–404.

CEAA. 2010. "Report of the Federal Review Panel: Prosperity Gold-Copper Mine Project". http://www.fonv.ca/media/pdfs/CEAA2010.pdf

CEAA. 2012. "Report of the Federal Review Panel: New Prosperity Gold-Copper Mine Project". https://ceaa-acee.gc.ca/050/documents/p63928/95790E.pdf

Chamberlain, A, and Haile, W. "Environmental Assessment and Environmental Impact Review in the Post-Devolution Northwest Territories". In *Environmental Impact Assessment: Practice and Participation*, edited by Hanna, K. Don Mills, Ontario: Oxford University Press.

Environmental Assessment Office (EAO). 2003. "Guide to the British Columbia Environmental Assessment Process". http://www.eao.gov.bc.ca

EAO. 2018. "British Columbia Environmental Assessment Revitalization Intentions Paper". https://www2.gov.bc.ca/assets/gov/environment/natural-resource-stewardship/environmental-assessments/environ-mental-assessment-revitalization/documents/ea_revitalization_intentions_paper.pdf

EAO. 2020a. "EAO User Guide: Introduction to Environmental Assessment Under the Provincial Environmental Assessment Act, 2018". https://www2.gov.bc.ca/assets/gov/environment/natural-resource-stewardship/environmental-assessments/guidance-documents/2018-act/eao_user_guide_v101.pdf

EAO. 2020b. "2018 Act- Environmental Assessment Process". https://www2.gov.bc.ca/gov/content/environment/natural-resource-stewardship/environmental-assessments/the-environmental-assessment-process/2018-act-environmental-assessment-process

EAO. 2020c. "EAO User Guide: Introduction to Environmental Assessment Under the Provincial Environmental Assessment Act (2018)". https://www2.gov.bc.ca/assets/gov/environment/natural-resource-stewardship/environmental-assessments/guidance-documents/2018-act/eao_user_guide_v101.pdf

Environmental Assessment Act, 2018 (S.B.C. 2018). https://www.leg.bc.ca/parliamentary-business/legislation-debates-proceedings/41st-parliament/3rd-session/bills/third-reading/gov51-3

Fitzpatrick, P, and Sinclair, J. "Multi-Jurisdictional Environmental Assessment in Canada". In *Environmental Impact Assessment: Practice and Participation*, edited by Hanna, K. Don Mills, Ontario: Oxford University Press.

First Nations Energy and Mining Council. 2009. *Environmental Assessment and First Nations in BC: Proposals and Reform*. https://www.ceaa-acee.gc.ca/050/documents/42766/42766E.pdf

Gibson, R, and Hanna, K. "Progress and Uncertainty: The Evolution of Federal Environmental Assessment in Canada". In *Environmental Impact Assessment: Practice and Participation*, edited by Hanna, K. Don Mills, Ontario: Oxford University Press

Government of British Columbia. 2018. "Final Report of the Environmental Assessment Advisory Committee". https://www2.gov.bc.ca/assets/gov/environment/natural-resource-stewardship/envi-ronmental-assessments/environmental-assessment-revitalization/documents/revitalization_eaac_report.pdf

Haddock, Mark. 2010. "Environmental Assessment in British Columbia". Environmental Law Centre: University of Victoria. http://www.elc.uvic.ca/wordpress/wp-content/uploads/2014/08/ELC_EA-IN-BC_Nov2010.pdf

Hanna, K (Ed). 2016. *Environmental Impact Assessment: Practice and Participation*. Don Mills, Ontario: Oxford University Press.

House of Commons Standing Committee on Finance, Sub-committee on Bill C-38. 2012. *Bill C-38, Part 3. Responsible Resource Development*. Ottawa: House of Commons.

Impact Assessment Act, 2019. S.C. 2019, c. 28, s. 1. https://laws-lois.justice.gc.ca/eng/acts/I-2.75/page-1.html

Impact Assessment Agency of Canada (IAAC). 2019a. "Gender-based Analysis Plus in Impact Assessment (Interim Guidance)". https://www.canada.ca/en/impact-assessment-agency/services/policy-guidance/practitioners-guide-impact-assessment-act/gender-based-analysis-plus.html

IAAC. 2019b. "Impact Assessment Act and CEAA 2012 Comparison". https://www.canada.ca/content/dam/iaac-acei/documents/policy-guidance/pg-gp/ceaa-vs-iaa-en.pdf

IAAC. 2019c. "Impact Assessment Process Overview". https://www.canada.ca/en/impact-assessment-agency/services/policy-guidance/impact-assessment-process-overview.html

IAAC. 2019d. "Interim Framework: Implementation of the Sustainability Guidance". https://www.canada
.ca/en/impact-assessment-agency/services/policy-guidance/practitioners-guide-impact-assessment
-act/interim-guidance.html

IAAC. 2019e. "Interim Guidance: Indigenous Participation in Impact Assessment". https://www.canada.ca
/en/impact-assessment-agency/services/policy-guidance/practitioners-guide-impact-assessment-act/
interim-guidance-indigenous-participation-ia.html

IAAC. 2020a. "Impact Assessment Process Overview". https://www.canada.ca/en/impact-assessment
-agency/services/policy-guidance/impact-assessment-process-overview/phase1.html

IAAC. 2020b. "Policy Context: Public Interest Determination under the *Impact Assessment Act*". https://
www.canada.ca/en/impact-assessment-agency/services/policy-guidance/public-interest-determina-
tion-under-impact-assessment-act.html

Johnston, Anna. 2018. "Canada's Proposed New Impact Assessment Act: Good from Afar but Far from
Good?" *West Coast Environmental Law*, 1–13. https://www.wcel.org/blog/canadas-proposed-new
-impact-assessment-act-good-afar-far-good.

Morales, Sarah. 2019. "Indigenous-Led Assessment Processes as a Way Forward". Centre for International
Governance Innovation. https://www.cigionline.org/articles/indigenous-led-assessment-processes
-way-forward

Noble, B. 2015. *Introduction to Environmental Impact Assessment: A Guide to Principles and Practice*. Don Mills,
Ontario: Oxford University Press.

Northey, R. 2020. "What is and is Not Modernized in Ontario's Re-Write if its *Environmental Assessment
Act*". https://gowlingwlg.com/en/insights-resources/articles/2020/modernized-ontario-environmen-
tal-assessment-act/

Pope, J, Bond, A, Morrison-Saunders, A, and Retief, F. 2013. "Advancing the Theory and Practice of Impact
Assessment: Setting the Research Agenda". *Environmental Impact Assessment Review*, 41, 1–9. https://doi
.org/10.1016/j.eiar.2013.01.008

Rutherford, M. 2019. "Impact Assessment in British Columbia". In *Environmental Impact Assessment: Practice
and Participation*, edited by Hanna, K. Don Mills, Ontario: Oxford University Press.

Stacey, J. 2020. "The Deliberative Dimensions of Modern Environmental Assessment Law". *Dalhousie Law
Journal*, 43(2), 865–900. https://digitalcommons.schulichlaw.dal.ca/cgi/viewcontent.cgi?article=2152
&context=dlj

Stacey, J. 2015. "The Environmental Emergency and the Legality of Discretion in Environmental Law".
Osgoode Hall Law Journal, 52(3), 985–1028; Osgoode Legal Studies Research Paper no. 20/2015
https://doi.org/10.2139/ssrn.2619688.

Truth and Reconciliation Commission of Canada. 2015. "Truth and Reconciliation Commission of
Canada: Calls to Action". http://trc.ca/assets/pdf/Calls_to_Action_English2.pdf

Tsleil-Waututh Nation v. Canada (Attorney General), [2018] FCA 153. 2018. https://decisions.fca-caf.gc
.ca/fca-caf/decisions/en/item/343511/index.do#_Remedy

WCEL. 2012. "Failing Grade: New Federal Approach to Environmental Assessment Leaves Canadians
at Risk Without a Voice". https://www.wcel.org/sites/default/files/publications/Report%20Card
%20June%2020%202012%20Legal%20Analysis%20Report.pdf

INDEX

For Product Safety Concerns and Information please contact our EU
representative GPSR@taylorandfrancis.com
Taylor & Francis Verlag GmbH, Kaufingerstraße 24, 80331 München, Germany